Hypersonic Aerothermodynamics

John J. Bertin
Visiting Professor at the
United States Air Force Academy,
and Consultant to the
United States Air Force

EDUCATION SERIES
J. S. Przemieniecki
Series Editor-in-Chief
Air Force Institute of Technology
Wright-Patterson Air Force Base, Ohio

Published by
American Institute of Aeronautics and Astronautics, Inc.,
370 L'Enfant Promenade, SW, Washington, DC 20024-2518

American Institute of Aeronautics and Astronautics, Inc., Washington, DC

Library of Congress Cataloging-in-Publication Data

Bertin, John J., 1938—
    Hypersonic aerothermodynamics / John Bertin.
      p. cm.—(AIAA education series)
    Includes bibliographical references and index.
    1. Hypersonic planes—Design and construction—Mathematical
models. 2. Aerothermodynamics—Mathematics. I. Title.
II. Series.
TL571.5B47 1994   629.132′306—dc20   93-26658
ISBN 1-56347-036-5

Third Printing

Texts Published in the AIAA Education Series

Re-Entry Vehicle Dynamics
Frank J. Regan, 1984

Aerothermodynamics of Gas Turbine and Rocket Propulsion
Gordon C. Oates, 1984

Aerothermodynamics of Aircraft Engine Components
Gordon C. Oates, Editor, 1985

Fundamentals of Aircraft Combat Survivability Analysis and Design
Robert E. Ball, 1985

Intake Aerodynamics
J. Seddon and E. L. Goldsmith, 1985

Composite Materials for Aircraft Structures
Brian C. Hoskins and Alan A. Baker, Editors, 1986

Gasdynamics: Theory and Applications
George Emanuel, 1986

Aircraft Engine Design
Jack D. Mattingly, William Heiser, and Daniel H. Daley, 1987

An Introduction to the Mathematics and Methods of Astrodynamics
Richard H. Battin, 1987

Radar Electronic Warfare
August Golden Jr., 1988

Advanced Classical Thermodynamics
George Emanuel, 1988

Aerothermodynamics of Gas Turbine and Rocket Propulsion,
Revised and Enlarged
Gordon C. Oates, 1988

Re-Entry Aerodynamics
Wilbur L. Hankey, 1988

Mechanical Reliability: Theory, Models and Applications
B. S. Dhillon, 1988

Aircraft Landing Gear Design: Principles and Practices
Norman S. Currey, 1988

Gust Loads on Aircraft: Concepts and Applications
Frederic M. Hoblit, 1988

Aircraft Design: A Conceptual Approach
Daniel P. Raymer, 1989

Boundary Layers
A. D. Young, 1989

Aircraft Propulsion Systems Technology and Design
Gordon C. Oates, Editor, 1989

Basic Helicopter Aerodynamics
J. Seddon, 1990

Introduction to Mathematical Methods in Defense Analyses
J. S. Przemieniecki, 1990

Space Vehicle Design
  Michael D. Griffin and James R. French, 1991
Inlets for Supersonic Missiles
  John J. Mahoney, 1991
Defense Analyses Software
  J. S. Przemieniecki, 1991
Critical Technologies for National Defense
  Air Force Institute of Technology, 1991
Orbital Mechanics
  Vladimir A. Chobotov, 1991
Nonlinear Analysis of Shell Structures
  Anthony N. Palazotto and Scott T. Dennis, 1992
Optimization of Observation and Control Processes
  Veniamin V. Malyshev, Mihkail N. Krasilshikov, and Valeri I. Karlov,
  1992
Aircraft Design: A Conceptual Approach
  Second Edition
  Daniel P. Raymer, 1992
Rotary Wing Structural Dynamics and Aeroelasticity
  Richard L. Bielawa, 1992
Spacecraft Mission Design
  Charles D. Brown, 1992
Introduction to Dynamics and Control of Flexible Structures
  John L. Junkins and Youdan Kim, 1993
Dynamics of Atmospheric Re-Entry
  Frank J. Regan and Satya M. Anandakrishnan, 1993
Acquisition of Defense Systems
  J. S. Przemieniecki, Editor, 1993
Practical Intake Aerodynamic Design
  E. L. Goldsmith and J. Seddon, Editors, 1993
Hypersonic Airbreathing Propulsion
  William H. Heiser and David T. Pratt, 1994
Hypersonic Aerothermodynamics
  John J. Bertin, 1994

Published by
American Institute of Aeronautics and Astronautics, Inc., Washington, DC

FOREWORD

This book and its companion volume, *Hypersonic Airbreathing Propulsion* by William H. Heiser and David Pratt, resulted from a series of discussions among faculty members of the Department of Aeronautics at the United States Air Force Academy in 1987. At that time, hypersonic, piloted flight was back in the public eye due to then President Reagan's announcement of a new program to develop and to demonstrate the technology required to operate an airplane-like vehicle which could take off from a normal runway, and use airbreathing engines to climb and accelerate to sufficient altitude and speed to enter Earth orbit. Upon completion of its orbital mission, the vehicle would re-enter the atmosphere and operate as an airplane during descent and landing on a runway. This idea led to the National Aero-Space Plane (NASP) program. The single-stage-to-orbit concept envisioned in the NASP has required, and will continue to require, the best efforts and intellectual talents the nation has available to make it a reality.

The advent of the NASP program was not the only factor that led to these volumes. The last significant hypersonic, manned vehicle program was the Space Shuttle which underwent engineering development in the 1960's. By the late 1980's, much of the talent involved in that program had long since been applied to other areas. The need for a modern treatment of hypersonic aerothermodynamics and airbreathing propulsion analysis and design principles for the academic, industrial and government communities was clear. As a result, the Air Force's Wright Laboratory and the NASP Joint Program Office, both located at Wright-Patterson Air Force Base, Ohio, entered into a cooperative effort with the Department of Aeronautics at the Academy to fund and provide technical and editorial oversight and guidance as these books were developed.

We sincerely hope that these volumes will serve as up-to-date sources of information and insight for the many students, engineers, and program managers involved in the exciting study and application of hypersonic flight in the years ahead.

G. KEITH RICHEY
Chief Scientist
Wright Laboratory

ROBERT R. BARTHELEMY
Director
NASP Joint Program Office

THOMAS M. WEEKS
Acting Chief Scientist
Flight Dynamics Directorate
Wright Laboratory

MICHAEL L. SMITH
Professor and Head
Department of Aeronautics
United States Air Force Academy

PREFACE

As hypersonic flow encounters a vehicle, the kinetic energy associated with hypervelocity flight is converted into increasing the temperature of the air and into endothermic reactions, such as dissociation and ionization of the air near the vehicle surface. The mechanisms for this conversion include adiabatic compression and viscous energy dissipation. Heat is transferred from the high temperature air to the surface. The rate at which heat is transferred to the surface depends upon many factors, including the freestream conditions, the configuration of the vehicle and its orientation to the flow, the difference between the temperature of the air and the temperature of the surface, and the surface catalycity. The determination of the flowfield requires the simultaneous solution of the continuity equation, of the momentum equation, and of the energy equation. Thus, those reponsible for the design of a hypersonic vehicle must determine the aerodynamic heating environment as well as the aerodynamic forces and moments. Hence, the title of this book, *Hypersonic Aerothermodynamics,* reflects the close coupling of the aerodynamic forces with the heating environment.

Scientists and engineers have designed and built hypersonic vehicles since the 1950's. In many of these programs, flight tests have revealed problems that had not been predicted either through analysis/computation or through ground-based testing. These problems exposed hypersonic flow phenomena not previously identified, what we shall call "unknown unknowns"... *Good judgement comes from experience; experience often comes from bad judgement.* The experience gained during earlier programs led to improved analytical techniques and improved ground-test procedures. One of the objectives in writing this book was to document the phenomena that caused the unexpected problems in previous programs. Thus, it is hoped that the lessons learned by previous scientists and engineers through unpleasant surprises can save future generations from the high cost of ignorance.

This book is also intended to be used as a classroom text for advanced undergraduate and for graduate students. It is assumed that the reader of this text understands the basic principles of fluid mechanics, of thermodynamics, of compressible flow, and of heat transfer. In most applications, computer codes of varying degrees of sophistication are required to obtain solutions to hypersonic flowfields. Discussions of the results obtained using computer codes of varying degrees of rigor (and citation of the literature documenting such codes) will be presented throughout the text. However, to work the sample exercises and the homework problems presented in this book, the reader only needs access to NACA Report 1135, "Equations, Tables, and Charts for Compressible Flow," or the equivalent information either through another book or through a computer program. The sample exercises and the homework problems presented in this text are intended to illustrate the basic principles of hypersonic flow and to familiarize the reader with order-of-magnitude values for the various parameters.

To be sure, each course reflects the personality and interests of the instructor. As a result, there are a variety of ways to use this book either as a primary text or as a complement to other resources. The first four chapters present general information characterizing hypersonic flows, discuss numerical formulations of varying degrees of rigor in computational fluid dynamics (CFD) codes, and discuss the strengths and the limitations of the various types of hypersonic experimentation. It is the author's opinion that the majority of the material to be presented in a one-semester, advanced undergraduate course or introductory graduate course in hypersonic flow is contained in Chaps. 1 and 5 through 8. These chapters cover:

1 The general characterization of hypersonic flow;

5 The stagnation-region flowfield, which often provides reference parameters for correlations of the flow parameters;

6 The inviscid flowfield, i.e., the pressure distribution and the conditions at the edge of the boundary layer;

7 The boundary layer (including convective heat transfer) for laminar and turbulent flows and information regarding boundary-layer transition; and

8 The aerodynamic forces and moments.

Following this sequence to cover the course material will take the student from general concepts through the flowfield in a step-by-step fashion.

Viscous/inviscid interactions and shock/shock interactions are discussed in Chap. 9. Review of aerothermodynamic phenomena and their role in the design of a hypersonic vehicle is discussed in Chap. 10.

The third objective in writing this book is to provide young aerospace vehicle designers (not necessarily in chronological age, but in experience) with an appreciation for the complementary role of experiment and of analysis/computation. Those given the challenge of designing a hypersonic vehicle in the 1950's relied considerably on experimental programs. However, the experimental programs used models and partial simulations of the hypersonic environment. Thus, the experimental data were supported by computer codes based on relatively simple theoretical models and by (semi-) empirical correlations developed using simplified analytical models and user experience. Because of dramatic advances in computer hardware and software, the designers of hypersonic vehicles depend more and more on analytical/numerical tools. By the 1990's, the designers of hypersonic vehicles place considerable reliance on computed flowfields. However, even the most sophisticated numerical codes run on the most powerful computers incorporate approximate models for the physical processes and introduce numerical approximations to the differential equations. Thus, even for the most sophisticated numerical codes, experiments are needed to validate the numerical models and to calibrate the range of applicability for the approximations used in the codes. Thus, an efficient process to determine the aerothermodynamic environment of a hypersonic vehicle requires the integration of inputs from the analytical/computational community, from the ground-test community, from the instrumentation community, and from the flight-test community. Even today, the designers of hypersonic vehicles must carefully integrate experimental and numerical expertise.

Because of the diverse audiences for whom this book is written, material is presented from references that probably are not generally available. The author tried to present sufficient information that the reader will benefit from the presentation, even if she/he can not obtain a copy of the cited reference. Similarly, the large number of quotations are intended to pass on the insights and experiences of the original researchers in their own words. Furthermore, the reader is required to use both English units and metric units.

ACKNOWLEDGMENTS

The author is indebted to Col. Michael L. Smith, chairman of the Aeronautics Department (DFAN) at the U.S. Air Force Academy, and to Dr. William Heiser, Distinguished Visiting Professor in the department, for the vision and follow-through that brought this project from concept to reality. The book was written under Air Force Contract No. F0561190D0101, with funds provided by the Flight Dynamics Directorate of the Wright Laboratory and by the National Aero-Space Plane Program Officer. Lt. Col. Tom Yechout served as the Project Officer. Dr. Thomas Weeks supplied valuable editorial comments for the entire text, serving as the principal technial reviewer. Other reviewers for the contracted effort included: Dr. Keith Richey, Dr. Tom Curran, Mr. Ed Gravlin, and Dr. William McClure. As series Editor-in-Chief for the AIAA Education Series, Dr. J. S. Przemieniecki served as the editor of the final draft of this book. Kenneth F. Stetson and E. Vincent Zoby also provided editorial comments for the final draft.

The author is indebted to the large number of hypersonic experts who provided references from their personal libraries, who served as reviewers for specific sections, and who continually offered valuable suggestions as to content and emphasis. They include: Isaiah Blankson, Dennis M. Bushnell, Gary T. Chapman, William D. Escher, Ernst Hirschel, Chen P. Li, Joseph G. Marvin, Richard K. Matthews, Charles E. K. Morris, Jr., Richard D. Neumann, William L. Oberkampf, Jacques Periaux, Joseph S. Shang, Milton A. Silveira, Kenneth F. Stetson, John Wendt, and E. Vincent Zoby. The author wishes to thank each and every one of them, for their help and inspiration were critical to the completion of this project. The original versions of many cited references presented in this text were first published by the Advisory Group for Aerospace Research and Development, North Atlantic Treaty Organisation (AGARD/NATO). The author wishes to thank AGARD and the individual authors for permission to reproduce this material.

Tim Valdez and Jeanie Duvall prepared the text, the figures, and the integrated copy for the many drafts of this book.

This book is dedicated to

The Faculty, Staff, and Students
of the U.S. Air Force Academy

My Parents

My children: Thomas A., Randolph S., Elizabeth A.,
and Michael R.

My wife, Ruth

TABLE
OF
CONTENTS

NOMENCLATURE

Because of the many disciplines represented in this text, one symbol might represent several different parameters, e.g., k and S. The reader should be able to determine the correct parameter from the context in which it is used.

a	speed of sound
A	axial force, i.e., the force acting along the axis of the vehicle
A_{Base}	base area
A_{ref}	reference area
Bf	Breguet factor, Eq. (10-2)
C	Chapman-Rubesin factor, $\rho\mu/(\rho_w\mu_w)$
$\bar{c}$	reference chord length
C_A	axial-force coefficient; mass fraction of atoms
C_d	sectional-drag coefficient
C_D	drag coefficient
C_{D_F}	friction contribution to the total drag
C_{D_N}	drag coefficient of the nose
C_{D_P}	pressure contribution to the total drag
$C_{d,f}$	sectional-drag coefficient, due to friction force
$C_{d,p}$	sectional-drag coefficent, due to pressure force
C_f	skin-friction coefficient
$C_{f\text{inc}}$	incompressible value of the skin-friction coefficient
C_i	mass fraction of species i, Eq. (2-1); mass injection fraction, Eq. (7-23)
C_L	lift coefficient
C_M	pitching-moment coefficient
C_{M_0}	pitching moment about the apex
C_{M_q}	rotary pitching derivative
$C_{M_{\dot{\alpha}}}$	dynamic pitching-moment coefficient
C_N	normal-force coefficient
c_p	specific heat at constant pressure, Eq. (1-13)

C_p pressure coefficient, Eq. (1-5)

c_{pA} specific heat of atoms

c_{pM} specific heat of molecules

$C_{p,\max}$ maximum value of the pressure coefficient

c_v specific heat at constant volume

$C_{p,t2}$ stagnation-point pressure coefficient

c_w specific heat of the model material

C_∞ $\left(\dfrac{\mu_w}{\mu_\infty}\right)\left(\dfrac{T_\infty}{T_w}\right)$

C_1 recombination rate parameter, Fig. 5.11

d diameter of a cylinder

D drag force, i.e., the force acting parallel to the freestream direction

D_i diffusion coefficient of the ith species into a mixture of gases

D_{12} binary diffusion coefficient for a mixture of two gases

e internal energy

E radiation intensity

e_t total energy, Eq. (2-16)

f transformed stream function

g dimensionless total enthalpy, H/He

h static enthalpy;
convective heat-transfer coefficient, $\dot{q}/(T_{aw}-T_w)$, Eq. (4-7)

H total (or stagnation) enthalpy, also H_t

h_A enthalpy of atoms

h_A^o dissociation energy per unit mass of atomic products

h_i metric, or scale factor, Eq. (7-8)

h_M^o heat of formation

h_M enthalpy of molecules

$h_{\alpha=0^\circ}$ heat-transfer coefficient when the configuration is at zero angle-of-attack

h^* Eckert's reference enthalpy, Eq. (7-9b)

$\hat{\imath}$ unit vector in the x-direction

I radiation falling on a unit area

I_{sp} specific impulse, Eq. (10-2)

$\hat{j}$ unit vector in the y-direction

k thermal conductivity;
height of the roughness element;
turbulent kinetic energy

$\hat{k}$ unit vector in the z-direction

K heat-transfer factor, Eq. (5-42);
ratio of the crosswise velocity gradient to the streamwise velocity gradient

Kn Knudsen number (mean-free path/characteristic dimension)

k_v vibrational thermal conductivity

k_w surface reaction-rate parameter, Fig. 7.4

L lift force, i.e., the force acting perpendicular to the freestream direction;
rolling moment;
reference length

Le Lewis number, Eq. (5-7)

$\bar{m}$ mean molecular weight at the conditions of interest

M Mach number;
pitching moment

m_0 molecular weight of the gas in the reference state, Eq. (1-12)

M_0 pitching moment about the apex

MW_{inj} molecular weight of the injectant

MW_{str} molecular weight of the stream (test) gas

$\hat{n}$ unit normal, Fig. 6.3

N normal force, i.e., force acting perpendicular to the vehicle axis;
yawing moment

Nu_x Nusselt number, Eq. (5-31)

p pressure

p_j pressure at the junction

p_{pit} pressure measured with a pitot probe

p_{static} static pressure

p_{upstream} pressure upstream of the interaction

Pr	Prandtl number, Eq. (5-15b)
q	heat flux vector
q	angular velocity about the pitch axis, or pitch rate
$\dot{q}_c$	convective heat-transfer rate (or simply $\dot{q}$)
$\dot{q}_{cond}$	rate at which heat is conducted into the surface
$\dot{q}_r$	radiative heat-transfer rate incident on the surface
$\dot{q}_{rad}$	rate at which heat is radiated from the surface
$\dot{q}_{r,t}$	radiative heat-transfer rate incident on the stagnation point
$\dot{q}_{stored}$	rate at which heat is stored in the surface material
$\dot{q}_{t,ref}$	reference convective heat-transfer rate at the stagnation point
q_1	dynamic pressure, $0.5\rho_1 U_1^2$
r	cross-section radius from the axis-of-symmetry to a point in the boundary layer, Fig. 7.10
r_t	radial distance to the tangency point, Fig. 8.15
R	gas constant; range, Eq. (10-1)
$\hat{R}$	universal gas constant
R_B	base radius
r_c	radial distance to the conical surface, Fig. 6.38
r_{eq}	cross-section radius of the equivalent body which produces the "three-dimensional" flow
R_N	nose radius
r_0	cross-section radius from the axis of symmetry to a point on the body, Fig. 7.10
r_s	radial distance to the shock wave, Fig. 6.38
R_s	radius of curvature of the bow shock
Re	Reynolds number
s	entropy; wetted distance along the model surface
S	entropy, when used in relation to Fig. 1.16 and 1.17; reference area, Eq. (8-41); surface area; transformed x-coordinate, Eq. (5-16b)

SM	static margin, Eq. (8-1)
St	Stanton number
t	time
T	temperature
t_c	characteristic time scale for chemical reactions
t_f	characteristic time scale for fluid motion
t_v	characteristic time scale for vibrational relaxation process
T_v	vibrational temperature
T^*	Eckert's reference temperature, Eq. (7-9a)
u	x-component of velocity
U_{CO}	circular orbit velocity
U_s	velocity of the moving shock wave; velocity of the flow just downstream of the bow shock wave
U_∞	freestream velocity
$U_{\infty,n}$	normal component of the freestream velocity
$U_{\infty,t}$	tangential component of the freestream velocity
v	y-component of velocity
$\mathbf{V}$	(mass-averaged) velocity vector
V	viscous interaction parameter to correlate skin friction and heat-transfer perturbations, Eq. (3-3)
$\mathbf{V}_i$	velocity vector of species i (differs from the mass-averaged velocity)
w	z-component of velocity
W	weight
$\dot{w}_i$	mass-production rate of the ith species
x	independent coordinate variable
x_{cg}	axial coordinate of the center of gravity
x_{cp}	axial coordinate of the center of pressure
x_t	axial distance to the tangency point, Fig. 8.15; "x-position" of transition, Fig. 7.32
y	independent coordinate variable
y_{cg}	y-coordinate of the center of gravity
y_{cp}	y-coordinate of the center of pressure

y_s streamline location for the mass-balance calculation, Fig. 3.21

y^+ dimensionless y-coordinate for a turbulent boundary layer,
$$y^+ = \frac{y}{\nu_w}\sqrt{\frac{\tau_w}{\rho_w}}$$

z independent coordinate variable; compressibility factor, Eq. (1-19)

z_{cg} z-coordinate of the center of gravity, Fig. 8.31

z_i dimensionless species mass-fraction parameter, C_i/C_{1e}, Eq. (5-19c)

α absorptivity of incident radiative heat-transfer rate; angle-of-attack

β ballistic coefficient, $W/(C_D A_{ref})$; angular position of a point on the body, Fig. 6.3

β_p pressure gradient parameter, Eq. (7-19)

β_u upper limits for β

γ ratio of specific heats

δ angle between the surface and the freestream flow, $\theta_c + \alpha$, Fig. 6.30; boundary-layer thickness; deflection angle

δ^* boundary-layer displacement thickness

δ_{BF} deflection angle of the body flap

$\delta_{E/A}$ deflection angle of the elevon/aileron

δ_R deflection angle of the rudder

δ_{SB} deflection angle of the speed brake

δ_T thickness of the thermal boundary layer

ϵ density ratio, Eq. (1-2); emissivity

η angle between the velocity vector and the inward normal, Eq. (6-4); transformed y-coordinate, Eq. (5-16a)

ξ transformed x-coordinate, Eq. (7-10b); roughness induced forward movement of transition parameter, Fig. 7.34

θ_b local inclination of the body

θ_c semi-vertex angle of a cone

λ	mean-free path
λ_σ	mean-free path of the molecules emitted from the surface
μ	viscosity
ν	kinematic viscosity, μ/ρ
ρ	density
σ	Stefan-Boltzmann constant
τ	viscous stress tensor
τ_{chem}	characteristic reaction time, Fig. 4.1
τ_{flow}	characteristic flow time, Fig. 4.1
τ_{ij}	elements of the viscous stress tensor
τ_t	turbulent shear stress
ϕ	complement of θ_b
ϕ_c	angle of the corner
ϕ_t	angle of the tangency point, Fig. 8.15
χ	viscous interaction parameter to correlate the pressure changes, Eq. (3-2)

Subscripts

a	aft cone of a biconic
aw	adiabatic wall conditions
b	blunt-cone value; conditions in the base region of a cone
c	conditions at the surface of a sharp cone, based on inviscid flow
e	conditions at the edge of the boundary layer
f	final value; fore cone of a biconic
fp	flat-plate (or reference) value
ft	flight conditions
FM	free-molecular conditions
i	property of the ith species
j	property of the jth species
k	conditions evaluated at the top of the roughness element, Eq. (7-22)
max	maximum value (often $t2$ conditions)

orig	original conditions, associated with inviscid flow
pk	peak (or maximum) value
r	recovery (or adiabatic-wall) value
ref	reference value
s	sharp-cone value; post-shock value (i.e., conditions in the shock layer)
sat	satellite value
SL	value at standard sea-level conditions
sw	conditions evaluated just downstream of the shock wave
t	transition location; tangency point for a sphere/cone, Fig. 8.15; stagnation-point value; for a turbulent boundary layer
t,ref	reference stagnation-point value
$t2$	conditions at the stagnation point downstream of a normal shock wave
w	conditions at the wall (or surface); conditions at the surface of a wedge, based on inviscid flow
wt	wind-tunnel conditions
1	freestream conditions, or those conditions immediately upstream of a shock wave in a multiple shock-wave flowfield
2	conditions immediately downstream of a shock wave
$2t$	conditions downstream of the shock wave in transformed coordinates, Fig. 4.5
∞	freestream conditions

Superscripts

k	exponent in the continuity equation Eq. (5-10), $k = 0$ for a two-dimensional flow, $k = 1$ for an axisymmetric flow
$*$	see Fig. 5.15; evaluated at Eckert's reference temperature
$'$	fluctuating parameter for a turbulent boundary layer

1
GENERAL CHARACTERIZATION
OF HYPERSONIC FLOWS

1.1 INTRODUCTION

The problems associated with determining the aerothermodynamic environment of a vehicle flying through the atmosphere offer diverse challenges to the designer. First, there is no single environment which characterizes hypersonic flows. The characteristics of the hypersonic environment depend on the mission requirements and vehicle size constraints. Representative re-entry trajectories for several different programs are presented in Fig. 1.1. Presented in Fig. 1.1 are the approximate velocity/altitude parameters for a trajectory for the aeroassisted space transfer vehicle (ASTV), for two trajectories for the Apollo Command Module (an overshoot trajectory and a 20-g limit trajectory), for the best-estimated trajectory for the STS-2 (Shuttle Orbiter) re-entry, for the trajectory of a single-stage-to-orbit vehicle with an airbreathing engine (SSTO), and for the trajectory for a slender (relatively low-drag) re-entry vehicle. For some applications, the vehicle will be very blunt or fly at very high angles-of-attack, so that the drag coefficient is large. Furthermore, the initial flight path angle is relatively small. Thus, the hypersonic deceleration occurs at very high altitudes. Re-entry vehicles from the U.S. manned spacecraft program, e.g., the Apollo and the Shuttle, are characteristic of these applications. The designer of such vehicles may have to account for nonequilibrium thermochemistry, viscous/inviscid interactions, and (possibly) noncontinuum flow models. Other applications require the time of flight to be minimal. For these applications, slender re-entry vehicles enter at relatively large flight path angles so that they penetrate deep into the atmosphere before experiencing significant deceleration. The trajectory for the slender RV, which is presented in Fig. 1.1, illustrates the velocity/altitude environment for a low-drag configuration. As a result, these relatively high ballistic coefficient (i.e., β, which is equal to $W/C_D A_{\text{ref}}$) configurations experience severe heating rates and high dynamic pressures but only for a short period of time. Ablative thermal protection systems are used to shield against the severe, turbulent heating rates.

A hypersonic vehicle with an airbreathing propulsion system must operate at relatively low altitudes to maintain the higher dynamic pressure required for maximum engine performance. For the airbreathing vehicle which flies at hypersonic speeds at relatively low

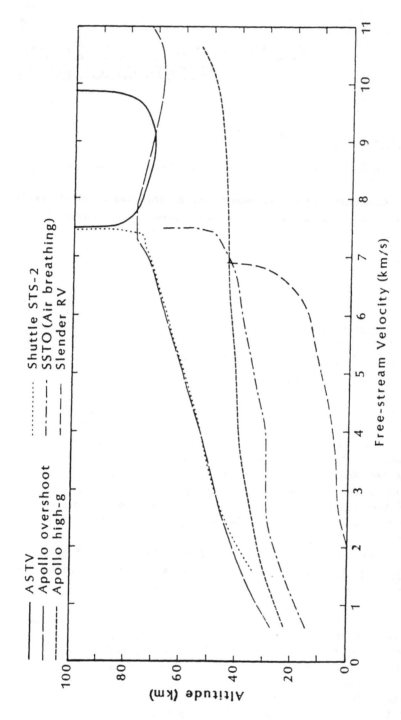

Fig. 1.1 Velocity/altitude parameters for several re-entry vehicles.

altitudes for extended periods of time, high dynamic pressures and high Reynolds numbers cause large aerodynamic loads, uncertainty in boundary-layer transition, and large total surface heating to be major vehicle design drivers. For a detailed presentation of hypersonic airbreathing propulsion systems, the reader is referred to Ref. 1, which is the companion volume to this text.

The aerothermodynamic phenomena which are important to the design of four major classes of hypersonic space-transport vehicles, as presented by Hirschel,[2] are reproduced in Fig. 1.2. The four classes of vehicles for which nominal trajectories are presented include:

1. Winged re-entry vehicles (RV) such as the Space Shuttle Orbiter, the Buran, and Hermes;

2. Hypersonic cruise vehicles (CV) such as the first stage of the Sänger space transportation system;

3. Ascent and re-entry vehicles (ARV) such as the upper stage Horus of the Sänger system; and

4. Aeroassisted orbit transfer vehicles (AOTV), also known as aeroassisted space transfer vehicles (ASTV).

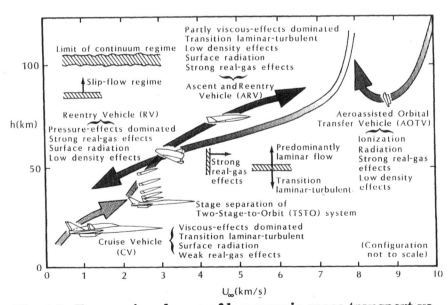

Fig. 1.2 Four major classes of hypersonic space-transport vehicles, and major aerothermodynamic effects, as taken from Ref. 2.

A two-stage-to-orbit (TSTO) launch system, such as the Sänger concept, could consist of an airbreathing-powered first-stage CV and a rocket-powered-second-stage ARV. Staging would occur at approximately Mach 7. Note that the aerothermodynamic phenomena critical to the design differ for the two stages. These differences are due to differences in the trajectories and in the different configurations used in the two stages.

For a freestream velocity of 7 km/s (23 kft/s), the kinetic energy per unit mass, i.e., $U_\infty^2/2$, is 2.45×10^7 J/kg (1.05×10^4 Btu/lbm). As the air particles pass through the shock wave ahead of the vehicle, they slow down, converting kinetic energy into internal energy. For a perfect gas, i.e., one for which the composition is frozen and one for which γ is constant, the increase in internal energy is in the form of random motion of the air particles (temperature). Thus, using a perfect-gas model for the air crossing a normal shock wave, the temperature would increase by approximately 2.4×10^4 K (4.3×10^4 °R). In reality, the change in kinetic energy of the air particles would go into increasing the random thermal motion of the air particles (temperature), exciting the vibrational (internal) energy of the diatomic particles, and dissociating the molecules (producing atomic species). If the density of the air is high enough, there are sufficient collisions between the air particles so that the air downstream of the shock wave quickly achieves the equilibrium state. As will be discussed later in this chapter, the temperature of the air downstream of a normal shock wave is of the order 7×10^3 K (13×10^3 °R) for such flows.

For most cases, the thermodynamic state will be between these limits of frozen flow and of equilibrium flow. Where the actual flow falls between these limits is a function of the density (or, equivalently, the altitude). At higher altitudes, there are fewer air particles and, therefore, fewer collisions between them. Thus, the air particles must travel further to experience the collisions required to excite the vibrational modes and to accomplish the dissociation process.

Having been briefly introduced to the diverse environments that are defined as hypersonic flows, let us now further explore what a hypersonic flow is, and then we can estimate its properties.

1.2   DEFINING HYPERSONIC FLOW

Addressing his graduate class in gas dynamics at Rice University in 1962, H. K. Beckmann said, "Mach number is like an aborigine counting: one, two, three, four, many. Once you reach many, the flow is hypersonic." Although this oversimplifies the problem, the flowfields around blunt bodies begin to exhibit many of the characteristics of hypersonic flows when the Mach number is four, or greater. By definition,

$$M_\infty \equiv \frac{U_\infty}{a_\infty} \gg 1 \qquad (1\text{-}1)$$

is the basic assumption for all hypersonic flow theories. Thus, the internal thermodynamic energy of the freestream fluid particles is small compared with the kinetic energy of the freestream for hypersonic flows. In flight applications, this results because the velocity of the fluid particles is relatively large. The limiting case, where M_∞ approaches infinity because the freestream velocity approaches infinity while the freestream thermodynamic state remains fixed, produces extremely high temperatures in the shock layer.

The high temperatures associated with hypersonic flight are difficult to match in ground-test facilities. Therefore, in wind-tunnel applications, hypersonic Mach numbers are achieved through relatively low speeds of sound. Thus, for the wind tunnel, M_∞ approaches infinity because the speed of sound (which is proportional to the square root of the freestream temperature) goes to "zero" while the freestream velocity is held fixed. As a result, the fluid temperatures in such wind tunnels remain below the levels that would damage the wind tunnel or the model. Ground-based test facilities will be discussed in Chap. 4.

Another assumption common to hypersonic flow is that:

$$\epsilon \equiv \frac{\rho_\infty}{\rho_2} \ll 1 \qquad (1\text{-}2)$$

which is known as the small-density-ratio assumption. Thus, this assumption relates primarily to the properties of the gas downstream of the shock wave. Recall that for a perfect gas,[3]

$$\epsilon = \frac{\gamma - 1}{\gamma + 1} \qquad (1\text{-}3)$$

for a normal shock wave as $M_\infty \to \infty$. Thus, ϵ is $1/6$ for perfect air. Note that typical hypersonic wind tunnels operate at conditions where the test gas can be approximated by the perfect-gas relations. However, during the re-entry flights of the Apollo Command Module, the density ratio approached $1/20$. Thus, the density-ratio simulation in those wind tunnels where the airflow behaves as a perfect gas does not match the flight values. This will have a significant effect on the shock stand-off distance, as well as other parameters. To generate lower-density ratios for wind-tunnel simulations, other gases may be used as the test medium, e.g., hexafluoroethane for which $\gamma = 1.1$ or tetrafluoromethane for which $\gamma = 1.2$. (See Refs. 4 and 5.)

1.2.1 Newtonian Flow Model

As the density ratio across the shock wave (ϵ) becomes very small, the shock layer becomes very thin. As shown in the sketch of Fig. 1.3, the shock layer is the region between the shock wave and the body surface. As a result, one can assume that the speed and the direction of the gas particles in the freestream remain unchanged until they strike the solid surface exposed to the flow. For this flow model (termed *Newtonian flow* since it is similar in character to one described by Newton in the seventeenth century), the normal component of momentum of the impinging fluid particle is wiped out, while the tangential component of momentum is conserved.

Thus, using the nomenclature of Fig. 1.3 and writing the integral form of the momentum equation for a constant-area streamtube normal to the surface,

$$p_\infty + \rho_\infty [U_{\infty,n}]^2 = p_\infty + \rho_\infty [U_\infty \sin \theta_b]^2 = p_w \qquad (1\text{-}4)$$

Rearranging so that the local pressure is written in terms of the pressure coefficient, one obtains:

$$C_p = \frac{p_w - p_\infty}{\frac{1}{2}\rho_\infty U_\infty^2} = 2 \sin^2 \theta_b = 2 \cos^2 \phi \qquad (1\text{-}5)$$

This equation for the pressure coefficient, Eq. (1-5), is based on the Newtonian flow model, where the 2 represents the pressure coefficient at the stagnation point (which is designated $C_{p,t2}$), since $\theta_b = 90$ deg. at the Newtonian stagnation point. Comparisons between Newtonian flow pressures and experimental values will be discussed in detail in Chap. 6.

Pressure coefficients for three "flow models" are presented in Fig. 1.4 as a function of the freestream Mach number for deflec-

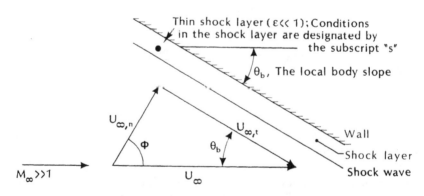

Fig. 1.3 Nomenclature for Newtonian flow model.

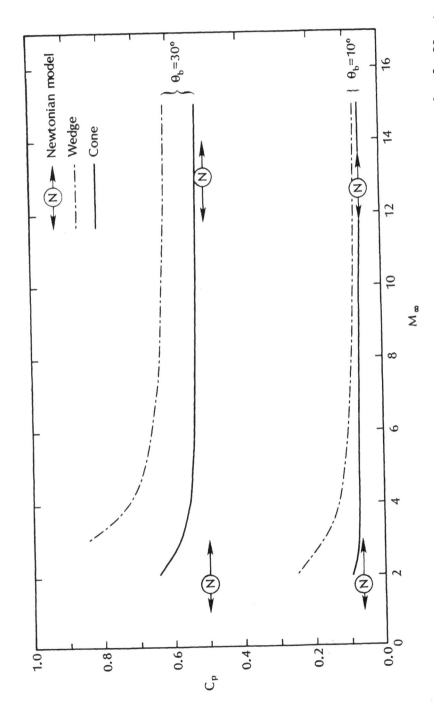

Fig. 1.4 Pressure coefficient for perfect-air ($\gamma = 1.4$) flow past a wedge, past a cone, or for the Newtonian flow model.

tion angles of 10 deg. and of 30 deg. The pressure coefficient for the Newtonian flow model, Eq. (1-5), is independent of Mach number, depending only upon the angle between the freestream flow direction and the surface inclination. For flow past a sharp cone, the pressure coefficients achieve Mach number independence, once the freestream Mach number exceeds 5. Slightly higher freestream Mach numbers are required before the pressure coefficients on a wedge exhibit Mach number independence.

Note that (for both deflection angles) there is relatively close correlation between the pressure coefficients for a sharp cone and those for Newtonian flow. For a sharp cone in a Mach 10 stream of perfect air, the shock wave angle (θ_{sw}) is 12.5 deg. for a deflection angle (θ_b) of 10 deg. and is 34 deg. for $\theta_b = 30$ deg. Thus, the inviscid shock layer is relatively thin for a sharp cone in a hypersonic stream. Since the Newtonian flow model assumes that the shock layer is very thin, the agreement between the pressure coefficients for these two flow models should not be surprising. However, for a wedge located in a Mach 10 stream of perfect air, the shock wave angle is 14.5 deg. when $\theta_b = 10$ deg. and is 38.5 deg. when $\theta_b = 30$ deg. Because the shock wave angle is at a higher inclination angle (relative to the freestream flow), the pressure on the wedge surface is significantly higher, with the differences being greater for the larger deflection angle.

The Newtonian flow model and the various theories for thin shock layers related to the Newtonian approximation are based on the small-density-ratio assumption. The small-density-ratio requirement for Newtonian theory also places implicit restrictions on the body shape in order that the shock layer be thin.

1.2.2 Mach Number Independence Principle

For slender configurations, such as sharp cones and wedges, the strong shock assumption is:

$$M_\infty \sin \theta_b \gg 1 \qquad (1\text{-}6)$$

which is mixed in nature, since it relates both to the flow and to the configuration. The concept termed the "Mach number independence principle" depends on this assumption.

The Mach number independence principle was derived for inviscid flow by Oswatitsch. Since pressure forces are much larger than the viscous forces for blunt bodies or for slender bodies at relatively large angles-of-attack when the Reynolds number exceeds 10^5, one would expect the Mach number independence principle to hold at these conditions. This is demonstrated for the data presented in Figs. 1.5 and 1.6, which are taken from Refs. 6 and 7, respectively. Note that

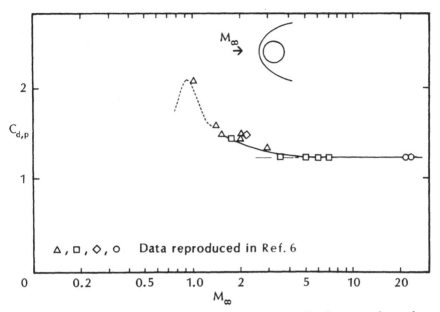

Fig. 1.5 Pressure drag of a right circular cylinder as a function of Mach number, as taken from Ref. 6.

the pressure drag for the blunt, right circular cylinder reaches its limiting value by $M_\infty = 4$ (Fig. 1.5), whereas the limiting values of the aerodynamic coefficients for the "more slender" lifting re-entry body are not achieved until the Mach number exceeds 7 (Fig. 1.6). This is consistent with the requirement that

$$M_\infty \sin \theta_b \gg 1$$

for Mach number independence.

Data presented by Koppenwallner[6] indicate that a significant increase in the total drag coefficient for a right circular cylinder occurs due to the friction drag when the Knudsen number (which is the ratio of the length of the molecular mean-free path to a characteristic dimension of the flowfield) is greater than 0.01. Data presented by Koppenwallner are reproduced in Fig. 1.7. Using the Reynolds number based on the flow conditions behind a normal-shock wave as the characteristic parameter,

$$Re_2 = \frac{\rho_2 U_2 d}{\mu_2} \qquad (1\text{-}7)$$

the friction drag for Re_2 is given by

$$C_{d,f} = \frac{5.3}{(Re_2)^{1.18}} \qquad (1\text{-}8)$$

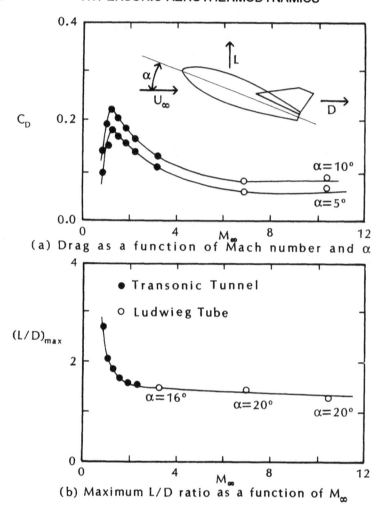

(a) Drag as a function of Mach number and α

(b) Maximum L/D ratio as a function of M_∞

Fig. 1.6 The Mach number independence principle for the MBB-Integral Re-entry Body, as taken from Ref. 7.

The data presented in Figs. 1.5 and 1.7 illustrate the significance of high-altitude effects on the aerodynamic coefficients.

1.3 CHARACTERIZING HYPERSONIC FLOW USING FLUID-DYNAMIC PHENOMENA

If the configuration geometry, the attitude, i.e., the orientation with respect to the freestream, and the altitude at which the vehicle flies or the test conditions of a wind-tunnel simulation are known, one can (in principle) compute the flowfield. When computer-generated flowfield solutions are used to define the aerothermodynamic environment, matching the fluid-dynamic similarity parameters becomes

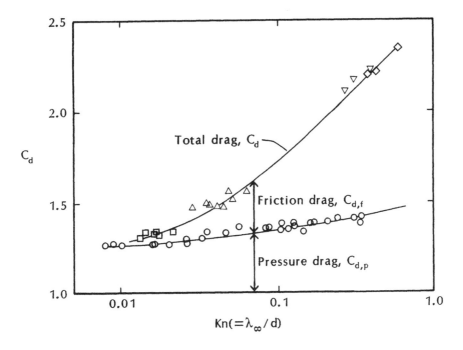

Fig. 1.7   Contributions to the total drag of friction drag and of pressure drag for a right-circular cylinder, as taken from Ref. 6.

a secondary issue. However, it is very important to incorporate "critical" fluid-dynamic phenomena into the computational flow model. What may be a critically important fluid-dynamic phenomenon for one application may be irrelevant to another. For instance, the drag coefficient for a cylinder whose axis is perpendicular to a hypersonic freestream will be essentially constant, independent both of Mach number and of Reynolds number, providing that the Mach number is sufficiently large so that the flow is hypersonic and that the Reynolds number is sufficiently large so that the boundary layer is thin. Refer to Fig. 1.5. Because the configuration is blunt, skin friction is a small fraction of the total drag. For such flows, reasonable estimates of the force coefficients could be obtained from computed flowfields using flow models based on the Euler equations, i.e., neglecting viscous terms in the equations of motion. However, in the lower-density flows encountered at higher altitudes, viscous/inviscid interactions become important and the effects of viscosity can no longer be neglected. Note the Knudsen number dependence of the data presented in Fig. 1.7. For these data, the effect of the viscous/inviscid interactions are correlated in terms of the Reynolds number, Eq. (1-8), or of the Knudsen number.

1.3.1 Noncontinuum Considerations

In fact, at very high altitudes, the air becomes so rarefied that the motion of the individual gas particles becomes important. Rarefied gas dynamics is concerned with those phenomena related to the molecular description of a gas flow (as distinct from continuum) which occur at sufficiently low densities. The parameter which can be used to delineate the flow regimes of interest is the Knudsen number, the ratio of the length of the molecular mean-free path to some characteristic dimension of the flowfield (such as a body dimension). Limits for the high-altitude regimes are presented in Fig. 1.8, as reproduced from Ref. 8. The criterion for free-molecule flow is that the mean-free path λ_σ , i.e., the mean-free path of the molecules emitted from the surface, relative to the incident molecules, is at least 10 times the nose radius. At the other end of the spectrum, the high-altitude extreme of the continuum flow regime is termed the *vorticity interaction regime.* Note that, since the bow shock wave ahead of a blunt body is curved, it generates vorticity in the inviscid flow outside of the boundary layer. The vorticity interaction regime identifies a condition when conventional boundary-layer theory can no longer be used, because the vorticity in the shock layer external to

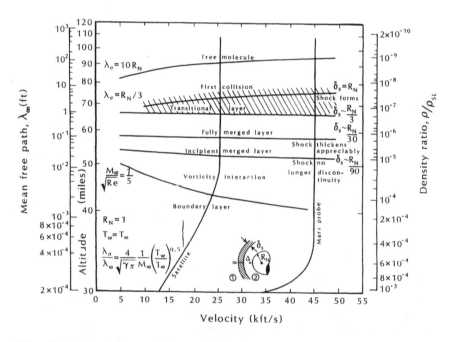

Fig. 1.8 Boundaries for high-altitude flowfield regimes, as taken from Ref. 8.

the viscous region becomes comparable to that within the boundary layer.

The limits of applicability for the continuum-flow model and for a discrete particle model, as presented by Moss and Bird,[9] are reproduced in Fig. 1.9. As the density of a flow is reduced from that of continuum conditions, the assumptions of temperature continuity adjacent to the surface and of zero-surface velocity (i.e., the no-slip condition) are no longer valid. This occurs because the state of molecules adjacent to the surface is affected not only by the surface but also by the flow conditions at a distance of the order of a mean free path from the surface. Consequently, as the flow becomes more rarefied, the spatial region that influences the state of the gas adjacent to the surface increases and gives rise to significant velocity-slip and temperature-jump effects. The calculated temperature jump and velocity slip, as taken from Ref. 9, are presented as a function of the Knudsen number in Figs. 1.10 and 1.11, respectively, for the nose region of the Space Shuttle. The stagnation-point temperature jump is expressed as a fraction of the specified wall temperature, with values ranging from 0.33 to 4.64 for the Knudsen-number range considered. The velocity slip, which is presented in Fig. 1.11 for an axial location of $x = 1.5$ m, indicates that the wall velocity varies from 0.010 U_∞ to 0.028 U_∞ over the range of conditions considered. Moss and Bird[9] concluded that "the results for heating and drag suggest that continuum calculations must be modified to account for slip and temperature jump effects at freestream Knudsen numbers of approximately 0.03."

Thus, there is a portion of the flight environment where the flow can no longer be represented by a continuum. Based on the correla-

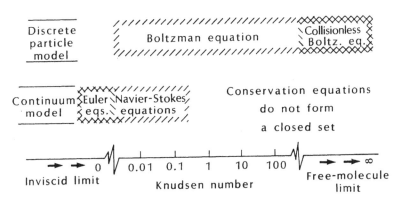

Fig. 1.9 The Knudsen number limits on gas-flow models, as taken from Ref. 9.

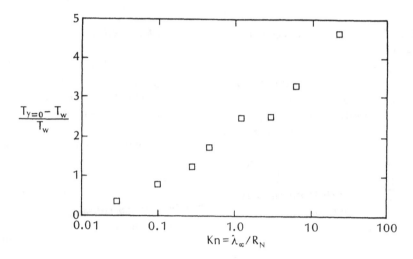

Fig. 1.10 The stagnation-point temperature jump as a function of the Knudsen number for a Space Shuttle, as taken from Ref. 9.

tions presented by Probstein,[8] which are reproduced in Fig. 1.8, the flow should be represented as a free-molecular flow for altitudes above about 90 to 100 miles (145 to 161 km) for a 1-ft radius sphere whose wall temperature is equal to the freestream temperature. Probstein notes that the free-molecular limit based on λ_σ is at a much higher altitude than would be predicted using λ_∞ as the characteristic mean free path. In addition, the correlations given by Probstein indicate that, above 51 miles (82 km), the shock is no longer a discontinuity. Cheng[10] suggests 330,000 ft (100 km) as the maximum altitude for the continuum assumption for a 1-ft sphere.

Flowfield solutions for the nose region of the Space Shuttle Orbiter generated by Moss and Bird[9] indicate that both the direct simulation Monte Carlo (DSMC) and the viscous shock-layer (VSL) methods provide results that are in good agreement with the flight measurements at 92.35 km.

Based on this discussion, it should be clear that there is no single, definitive criterion for an upper-limit altitude above which the continuum model for the flow is no longer valid. Furthermore, since the Knudsen number is a characteristic dimensionless parameter for low-density flows, these limits depend on vehicle size. If the radius were increased by a factor of 10, the corresponding density would be decreased by a factor of 10, and the curve could move upward an amount indicated by the density scale shown at the right of the graph in Fig. 1.8.

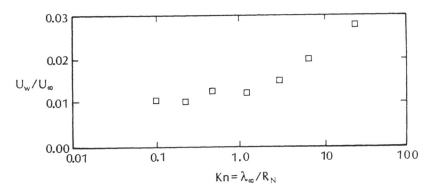

Fig. 1.11 The velocity slip at x = 1.5 m; U_∞ = 7.5 km/s as a function of the Knudsen number for a Space Shuttle, as taken from Ref. 9.

Let us assume that 90 km may be used as an "upper limit for the continuum flow model." The impact that this criterion would have on generating solutions for the aerothermodynamic environment of the trajectories of Fig. 1.1 is illustrated in Fig. 1.12. It is clear that noncontinuum effects are important over much of the trajectory for the aeroassisted space transfer vehicle (ASTV).

1.3.2 Stagnation-Region Flowfield Properties

It is well known that the air particles in the shock layer of a hypervelocity re-entry vehicle undergo vibrational excitation, dissociation, and (possibly) ionization. These chemical phenomena absorb energy, limiting the temperature increase as the kinetic energy of the hypervelocity air particles crossing the bow shock wave is converted to thermal energy. At low altitudes, where the freestream density is sufficiently high, these chemical phenomena tend to reach equilibrium. (See Fig. 1.12.) At higher altitudes, where the freestream density is relatively low, there are not sufficient collisions for the gas to reach an equilibrium state. Although these nonequilibrium states are normally bounded by the equilibrium state and by the frozen state (as stated in Sec. 1.1), the concentrations of individual species and the energy contained in the different internal modes must be calculated by integrating the conservation equations governing the phenomena.

The nomenclature for the flow near the stagnation point of a vehicle in a hypersonic stream is illustrated in Fig. 1.13. The flow passes through the normal portion of the shock wave reaching state 2 and then decelerates isentropically to state $t2$, which constitutes the

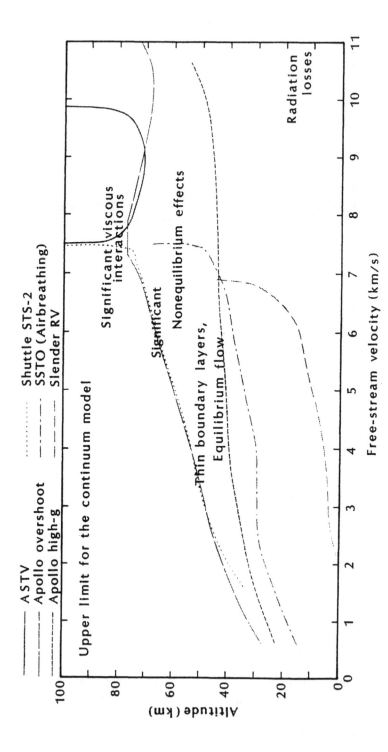

Fig. 1.12 Modeling flow phenomena for hypersonic flight.

outer edge condition for the thermal boundary layer at the stagnation point. The streamline from the shock wave to the stagnation point may be curved for nonaxisymmetric flowfields.

The relations for steady, one-dimensional, inviscid, adiabatic flow in a constant-area streamtube are used to compute the conditions across a normal shock wave.

$$\text{MASS} \quad = \quad \rho_1 U_1 = \rho_2 U_2 \qquad (1\text{-}9)$$

$$\text{PMØM} \quad = \quad p_1 + \rho_1 U_1^2 = p_2 + \rho_2 U_2^2 \qquad (1\text{-}10)$$

$$\text{HTØT} \quad = \quad h_1 + 0.5U_1^2 = h_2 + 0.5U_2^2 = H_t \qquad (1\text{-}11)$$

where H_t is the total (or stagnation) enthalpy of the flow.

If one assumes that the gas is thermally perfect

$$p = \rho RT = \rho \frac{\hat{R}}{m_0} T \qquad (1\text{-}12)$$

where $\hat{R}$ is the universal gas constant and m_0 is the molecular weight of the gas in the perfect (or reference) state. A thermally perfect gas is also known as an *ideal gas*. Introducing the assumption that the gas is calorically perfect, i.e., the specific heat is constant,

$$h = c_p T \qquad (1\text{-}13)$$

If the gas is both thermally perfect and calorically perfect (or, simply, perfect), the ratio of the values of flow properties across the shock wave can be written as a unique function of M_1 (or M_∞, the freestream Mach number) and γ (the ratio of specific heats). The reader should note that a gas can be thermally perfect and calorically "imperfect" but not vice versa. We shall use the term *perfect gas* to describe a gas which is both thermally and calorically perfect. These definitions for a thermally perfect gas, a calorically perfect gas, and a perfect gas are consistent with those of Ref. 3. These relations, which are presented in a number of texts, e.g., Ref. 11, are:

$$\frac{p_2}{p_1} = \frac{2\gamma M_1^2 - (\gamma - 1)}{\gamma + 1} \qquad (1\text{-}14)$$

$$\frac{\rho_2}{\rho_1} = \frac{U_1}{U_2} = \frac{(\gamma + 1)M_1^2}{(\gamma - 1)M_1^2 + 2} \qquad (1\text{-}15)$$

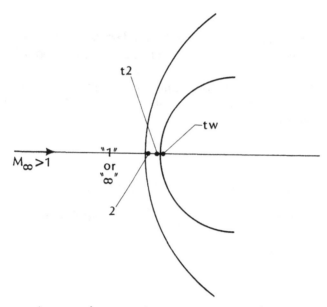

1 or ∞   denotes the free-stream conditions.

2   denotes the conditions immediately downstream of the shock wave.

t2   denotes the conditions at the stagnation point (downstream of the normal portion of the shock wave) but outside of the boundary layer.

tw   denotes conditions at the wall (the inner edge of the boundary layer at the downstream stagnation point).

Fig. 1.13 The nomenclature for the stagnation region.

$$\frac{T_2}{T_1} = \frac{[2\gamma M_1^2 - (\gamma - 1)][(\gamma - 1)M_1^2 + 2]}{(\gamma + 1)^2 M_1^2} \quad (1\text{-}16)$$

If one assumes that the flow decelerates isentropically from the conditions at point 2 (immediately downstream of the normal portion of the shock wave) to the stagnation point outside of the thermal boundary layer (point $t2$):

$$\frac{p_{t2}}{p_1} = \left[\frac{(\gamma + 1)M_1^2}{2}\right]^{\left(\frac{\gamma}{\gamma-1}\right)} \left[\frac{\gamma + 1}{2\gamma M_1^2 - (\gamma - 1)}\right]^{\left(\frac{1}{\gamma-1}\right)} \quad (1\text{-}17)$$

$$\frac{T_{t2}}{T_1} = \frac{T_{t1}}{T_1} = \left(1 + \frac{\gamma - 1}{2}M_1^2\right) \quad (1\text{-}18)$$

Note that, whereas it is generally true that the stagnation enthalpy is constant across a normal-shock wave for an adiabatic flow, as can be seen in Eq. (1-11), the stagnation temperature is constant across a normal shock wave only for the adiabatic flow of a perfect gas. See Eq. (1-18). Note also that, since the ratio of the downstream value to the freestream value for the flow properties (i.e., the ratio of the properties across a normal-shock wave) can be written as a function of γ and of M_1 ($\equiv M_\infty$) only, they do not depend specifically on the altitude.

In reality, for hypersonic flight, the gas molecules that pass through the bow shock wave are excited to higher vibrational and chemical energy modes. This lowers the specific-heat ratio of the gas below the freestream value, if it is assumed that equilibrium exists and that dissociation is not driven to completion. A large amount of the energy that would have gone into increasing the static temperature behind the bow shock wave for a perfect gas is used instead to excite the vibrational energy levels or to dissociate the gas molecules. The dominant chemical reactions that occur in the stagnation region for an equilibrium airflow, as presented in Ref. 12, are presented in Fig. 1.14. At about 3,000 ft/s, the vibrational energy of the air molecules begins to become important. Oxygen dissociation begins when the freestream velocity is between 6,000 ft/s and 8,000 ft/s.

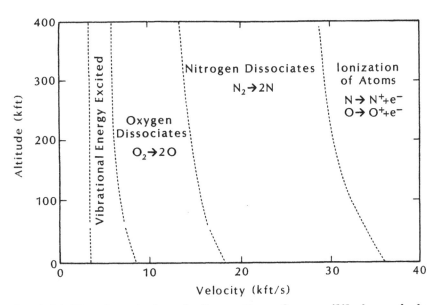

Fig. 1.14 Dominant chemical reactions for equilibrium air in the stagnation region, as taken from Ref. 12.

Nitrogen dissociation occurs at velocities in excess of 15,000 ft/s. Finally, ionization of atoms becomes important at velocities in excess of 30,000 ft/s. Hansen and Heims[12] note that "The dissociation and ionization reactions are pressure dependent because each particle yields two product particles, and such reactions are inhibited by high pressure. Therefore, higher temperature and, consequently, higher velocity are required to produce the reactions at sea level than at high altitudes where much lower pressures occur." The effects of vibration, dissociation, and ionization, etc., will be termed *real-gas effects* in this text.

To account for the departure from the thermally perfect equation of state, Eq. (1-12), due to the chemical reactions in air, the compressibility factor z is introduced. The compressibility factor is the ratio of the molecular weight of the undissociated air to the mean molecular weight at the conditions of interest. Thus,

$$z = \frac{m_0}{\bar{m}} \qquad (1\text{-}19)$$

where $\bar{m}$ is the mean molecular weight of the gas mixture at the conditions of interest. Accounting for the change in the gas composition, the equation of state becomes

$$p = \rho \frac{\hat{R}}{\bar{m}} T = \rho \frac{m_0}{\bar{m}} \frac{\hat{R}}{m_0} T = \rho z R T \qquad (1\text{-}20)$$

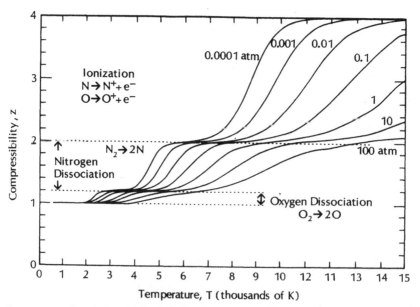

Fig. 1.15 **Compressibility of air as a function of temperature, as taken from Ref. 13.**

The compressibility factor for air, as taken from Ref. 13, is presented as a function of temperature and of pressure in Fig. 1.15. The important reactions for air are also indicated. As Hansen and Heims[12] note, "The ionization reactions occur at very nearly the same temperature and with nearly the same energy changes so that they may be classed together as a single reaction, for the purpose of approximation."

As indicated by the curves presented in Fig. 1.15, the compressibility factor is not influenced by the vibrational excitation and, therefore, is equal to 1.0 until oxygen dissociation begins. Since air contains about 20 percent oxygen, z approaches 1.2 as the oxygen dissociation approaches completion. Similarly, z approaches 2.0 as the nitrogen dissociation approaches completion and all of the molecules have dissociated into atoms. The ionization process produces further increases in z.

As energy is absorbed by the gas molecules entering the shock layer, the conservation laws and the thermophysics dictate certain changes in the forebody flow. The static temperature, the speed of sound, and the velocity in the shock layer are less for the equilibrium, real-gas flow than for a perfect-gas flow. The static pressure computed for air in thermodynamic equilibrium is slightly larger than the perfect-gas value. The density is increased considerably, and, as a result, the shock-layer thickness is reduced significantly.

Based on computations for the property changes across a normal shock wave using thermochemical equilibrium air properties, Huber[14] concluded that "for temperatures of 1,600 K and below (velocities of about 6,000 ft/s and below) the air composition does not change in the shock-compression process. Consequently, ideal-gas relations (that is, relations for a thermally perfect but calorically imperfect gas) can be used in this range." Furthermore, "For temperatures below 800 K (velocities below about 3,500 ft/s) it can be seen... that ideal-gas relations with constant specific-heat ratio of 1.40 may be used."

Equations (1-9) through (1-11) are not restricted to the perfect-gas assumption and can be applied to high-temperature hypersonic flow. We can use the U.S. Standard Atmosphere[15] to define the freestream properties, i.e., p_1, ρ_1, and h_1 at any given altitude. Refer to Table 1.1. Since there are four unknowns in Eqs. (1-9) through (1-11), i.e., p_2, ρ_2, h_2, and U_2, but only three equations, additional relations are needed to obtain a solution. The graphs of Ref. 16, which are presented in Figs. 1.16 and 1.17, are used in tabular form to define the following:

$$\rho(p, h)$$

$$s(p, h)$$

$$T(h, s)$$

Table 1.1a   U.S. Standard Atmosphere, 1976: Metric Units

Geometric Altitude (km)	Pressure (p/p_{SL})	Temperature (K)	Density (ρ/ρ_{SL})	Viscosity (μ/μ_{SL})	Speed of Sound (m/s)
0	1.0000 E+00	288.150	1.0000 E+00	1.00000	340.29
1	8.8700 E−01	281.651	9.0748 E−01	0.98237	336.43
2	7.8461 E−01	275.154	8.2168 E−01	0.96456	332.53
3	6.9204 E−01	268.659	7.4225 E−01	0.94656	328.58
4	6.0854 E−01	262.166	6.6885 E−01	0.92836	324.59
5	5.3341 E−01	255.676	6.0117 E−01	0.90995	320.55
6	4.6600 E−01	249.187	5.3887 E−01	0.89133	316.45
7	4.0567 E−01	242.700	4.8165 E−01	0.87249	312.31
8	3.5185 E−01	236.215	4.2921 E−01	0.85343	308.11
9	3.0397 E−01	229.733	3.8128 E−01	0.83414	303.85
10	2.6153 E−01	223.252	3.3756 E−01	0.81461	299.53
11	2.2403 E−01	216.774	2.9780 E−01	0.79485	295.15
12	1.9145 E−01	216.650	2.5464 E−01	0.79447	295.07
13	1.6362 E−01	216.650	2.1763 E−01	0.79447	295.07
14	1.3985 E−01	216.650	1.8601 E−01	0.79447	295.07
15	1.1953 E−01	216.650	1.5898 E−01	0.79447	295.07
16	1.0217 E−01	216.650	1.3589 E−01	0.79447	295.07
17	8.7340 E−02	216.650	1.1616 E−01	0.79447	295.07
18	7.4663 E−02	216.650	9.9304 E−02	0.79447	295.07
19	6.3829 E−02	216.650	8.4894 E−02	0.79447	295.07
20	5.4570 E−02	216.650	7.2580 E−02	0.79447	295.07
21	4.6671 E−02	217.581	6.1808 E−02	0.79732	295.70
22	3.9945 E−02	218.574	5.2661 E−02	0.80037	296.38
23	3.4215 E−02	219.567	4.4903 E−02	0.80340	297.05
24	2.9328 E−02	220.560	3.8317 E−02	0.80643	297.72
25	2.5158 E−02	221.552	3.2722 E−02	0.80945	298.39
26	2.1597 E−02	222.544	2.7965 E−02	0.81247	299.06
27	1.8553 E−02	223.536	2.3917 E−02	0.81547	299.72
28	1.5950 E−02	224.527	2.0470 E−02	0.81847	300.39
29	1.3722 E−02	225.518	1.7533 E−02	0.82147	301.05
30	1.1813 E−02	226.509	1.5029 E−02	0.82446	301.71
31	1.0177 E−02	227.500	1.2891 E−02	0.82744	302.37
32	8.7743 E−03	228.490	1.1065 E−02	0.83041	303.02
33	7.5727 E−03	230.973	9.4474 E−03	0.83785	304.67
34	6.5473 E−03	233.743	8.0714 E−03	0.84610	306.49
35	5.6708 E−03	236.513	6.9089 E−03	0.85431	308.30
36	4.9200 E−03	239.282	5.9248 E−03	0.86247	310.10
37	4.2758 E−03	242.050	5.0902 E−03	0.87059	311.89
38	3.7220 E−03	244.818	4.3809 E−03	0.87866	313.67
39	3.2452 E−03	247.584	3.7769 E−03	0.88669	315.43
40	2.8338 E−03	250.350	3.2618 E−03	0.89468	317.19

(continued on next page)

Table 1.1a (continued)

Geometric Altitude (km)	Pressure (p/p_{SL})	Temperature (K)	Density (ρ/ρ_{SL})	Viscosity (μ/μ_{SL})	Speed of Sound (m/s)
41	2.4784 E−03	253.114	2.8216 E−03	0.90262	318.94
42	2.1709 E−03	255.878	2.4447 E−03	0.91052	320.67
43	1.9042 E−03	258.641	2.1216 E−03	0.91838	322.40
44	1.6728 E−03	261.403	1.8440 E−03	0.92620	324.12
45	1.4715 E−03	264.164	1.6051 E−03	0.93398	325.82
46	1.2962 E−03	266.925	1.3993 E−03	0.94172	327.52
47	1.1433 E−03	269.684	1.2217 E−03	0.94941	329.21
48	1.0095 E−03	270.650	1.0749 E−03	0.95210	329.80
49	8.9155 E−04	270.650	9.4920 E−04	0.95210	329.80
50	7.8735 E−04	270.650	8.3827 E−04	0.95210	329.80
55	4.1969 E−04	260.771	4.6376 E−04	0.92442	323.72
60	2.1671 E−04	247.021	2.5280 E−04	0.88506	315.07
65	1.0786 E−04	233.292	1.3323 E−04	0.84476	306.19
70	5.1526 E−05	219.585	6.7616 E−05	0.80346	297.06
75	2.3569 E−05	208.399	3.2589 E−05	0.76892	289.40
80	1.0387 E−05	198.639	1.5068 E−05	0.73813	282.54
85	4.3985 E−06	188.893	6.7099 E−06	0.70677	275.52

Reference values: $p_{SL} = 1.01325 \times 10^5$ N/m^2; $T_{SL} = 288.150$ K
$\rho_{SL} = 1.2250$ kg/m^3; $\mu_{SL} = 1.7894 \times 10^{-5}$ kg/s·m

Table 1.1b U.S. Standard Atmosphere, 1976: English Units

Geometric Altitude (kft)	Pressure (p/p_{SL})	Temperature (° R)	Density (ρ/ρ_{SL})	Viscosity (μ/μ_{SL})	Speed of Sound (ft/s)
0	1.0000 E+00	518.67	1.0000 E+00	1.00000 E+00	1116.44
2	9.2981 E−01	511.54	9.4278 E−01	9.8928 E−01	1108.76
4	8.6368 E−01	504.41	8.8811 E−01	9.7849 E−01	1100.98
6	8.0142 E−01	497.28	8.3590 E−01	9.6763 E−01	1093.18
8	7.4286 E−01	490.15	7.8609 E−01	9.5670 E−01	1085.33
10	6.8783 E−01	483.02	7.3859 E−01	9.4569 E−01	1077.40
12	6.3615 E−01	475.90	6.9333 E−01	9.3461 E−01	1069.42
14	5.8767 E−01	468.78	6.5022 E−01	9.2346 E−01	1061.38
16	5.4224 E−01	461.66	6.0921 E−01	9.1223 E−01	1053.31
18	4.9970 E−01	454.53	5.7021 E−01	9.0092 E−01	1045.14
20	4.5991 E−01	447.42	5.3316 E−01	8.8953 E−01	1036.94

(continued on next page)

Table 1.1b (continued)

Geometric Altitude (kft)	Pressure (p/p_{SL})	Temperature (°R)	Density (ρ/ρ_{SL})	Viscosity (μ/μ_{SL})	Speed of Sound (ft/s)
22	4.2273 E−01	440.30	4.9798 E−01	8.7806 E−01	1028.64
24	3.8803 E−01	433.18	4.6462 E−01	8.6650 E−01	1020.31
26	3.5568 E−01	426.07	4.3300 E−01	8.5487 E−01	1011.88
28	3.2556 E−01	418.95	4.0305 E−01	8.4315 E−01	1003.41
30	2.9754 E−01	411.84	3.7473 E−01	8.3134 E−01	994.85
32	2.7151 E−01	404.73	3.4795 E−01	8.1945 E−01	986.22
34	2.4736 E−01	397.62	3.2267 E−01	8.0746 E−01	977.53
36	2.2498 E−01	390.51	2.9883 E−01	7.9539 E−01	968.73
38	2.0443 E−01	389.97	2.7191 E−01	7.9447 E−01	968.08
40	1.8576 E−01	389.97	2.4708 E−01	7.9447 E−01	968.08
42	1.6880 E−01	389.97	2.2452 E−01	7.9447 E−01	968.08
44	1.5339 E−01	389.97	2.0402 E−01	7.9447 E−01	968.08
46	1.3939 E−01	389.97	1.8540 E−01	7.9447 E−01	968.08
48	1.2667 E−01	389.97	1.6848 E−01	7.9447 E−01	968.08
50	1.1511 E−01	389.97	1.5311 E−01	7.9447 E−01	968.08
52	1.0461 E−01	389.97	1.3914 E−01	7.9447 E−01	968.08
54	9.5072 E−02	389.97	1.2645 E−01	7.9447 E−01	968.08
56	8.6402 E−02	389.97	1.1492 E−01	7.9447 E−01	968.08
58	7.8524 E−02	389.97	1.0444 E−01	7.9447 E−01	968.08
60	7.1366 E−02	389.97	9.4919 E−02	7.9447 E−01	968.08
62	6.4861 E−02	389.97	8.6268 E−02	7.9447 E−01	968.08
64	5.8951 E−02	389.97	7.8407 E−02	7.9447 E−01	968.08
66	5.3580 E−02	390.07	7.1246 E−02	7.9463 E−01	968.21
68	4.8707 E−02	391.16	6.4585 E−02	7.9649 E−01	969.55
70	4.4289 E−02	392.25	5.8565 E−02	7.9835 E−01	970.90
72	4.0284 E−02	393.34	5.3121 E−02	8.0020 E−01	972.24
74	3.6651 E−02	394.43	4.8197 E−02	8.0205 E−01	973.59
76	3.3355 E−02	395.52	4.3742 E−02	8.0390 E−01	974.93
78	3.0364 E−02	396.60	3.9710 E−02	8.0575 E−01	976.28
80	2.7649 E−02	397.69	3.6060 E−02	8.0759 E−01	977.62
82	2.5183 E−02	398.78	3.2755 E−02	8.0943 E−01	978.94
84	2.2943 E−02	399.87	2.0761 E−02	8.1127 E−01	980.28
86	2.0009 E−02	400.96	2.7048 E−02	8.1311 E−01	981.63
88	1.9060 E−02	402.05	2.4589 E−02	8.1494 E−01	982.94
90	1.7379 E−02	403.14	2.2360 E−02	8.1677 E−01	984.28
92	1.5850 E−02	404.22	2.0339 E−02	8.1860 E−01	985.60
94	1.4460 E−02	405.31	1.8505 E−02	8.2043 E−01	986.94
96	1.3195 E−02	406.40	1.6841 E−02	8.2225 E−01	988.25
98	1.2044 E−02	407.49	1.5331 E−02	8.2407 E−01	989.57
100	1.0997 E−02	408.57	1.3960 E−02	8.2589 E−01	990.91

(continued on next page)

Table 1.1b (continued)

Geometric Altitude (kft)	Pressure (p/p_{SL})	Temperature (°R)	Density (ρ/ρ_{SL})	Viscosity (μ/μ_{SL})	Speed of Sound (ft/s)
105	8.7691 E−03	411.29	1.1059 E−02	8.3042 E−01	994.19
110	7.0112 E−03	418.38	8.6918 E−03	8.4221 E−01	1002.72
115	5.6288 E−03	425.98	6.8536 E−03	8.5473 E−01	1011.78
120	4.5370 E−03	433.58	5.4275 E−03	8.6715 E−01	1020.77
125	3.6711 E−03	441.17	4.3160 E−03	8.7947 E−01	1029.66
130	2.9815 E−03	448.76	3.4460 E−03	8.9168 E−01	1038.48
135	2.4301 E−03	456.34	2.7620 E−03	9.0379 E−01	1047.21
140	1.9875 E−03	463.92	2.2221 E−03	9.1581 E−01	1055.87
145	1.6311 E−03	471.50	1.7943 E−03	9.2773 E−01	1064.47
150	1.3429 E−03	479.07	1.4539 E−03	9.3955 E−01	1073.00
155	1.1091 E−03	486.64	1.1821 E−03	9.5129 E−01	1081.43
160	9.1763 E−04	487.17	9.7697 E−04	9.5210 E−01	1082.02
165	7.5930 E−04	487.17	8.0840 E−04	9.5210 E−01	1082.02
170	6.2827 E−04	485.17	6.7166 E−04	9.4901 E−01	1079.79
175	5.1877 E−04	477.61	5.6337 E−04	9.3728 E−01	1071.36
180	4.2709 E−04	470.06	4.7126 E−04	9.2547 E−01	1062.86
185	3.5054 E−04	462.51	3.9311 E−04	9.1358 E−01	1054.27
190	2.8681 E−04	454.97	3.2697 E−04	9.0161 E−01	1045.64
195	2.3390 E−04	447.43	2.7114 E−04	8.8955 E−01	1036.94
200	1.9011 E−04	439.89	2.2416 E−04	8.7740 E−01	1028.18
205	1.5398 E−04	432.36	1.8472 E−04	8.6516 E−01	1019.32
210	1.2426 E−04	424.83	1.5172 E−04	8.5283 E−01	1010.40
215	9.9916 E−05	417.30	1.2419 E−04	8.4041 E−01	1001.41
220	8.0026 E−05	409.78	1.0129 E−04	8.2790 E−01	992.36
225	6.3840 E−05	402.26	8.2316 E−05	8.1529 E−01	983.20
230	5.0716 E−05	394.74	6.6639 E−05	8.0259 E−01	973.98
235	4.0117 E−05	387.23	5.3735 E−05	7.8978 E−01	964.67
240	3.1608 E−05	381.62	4.2959 E−05	7.8015 E−01	957.64
245	2.4822 E−05	376.26	3.4218 E−05	7.7089 E−01	950.92
250	1.9428 E−05	370.90	2.7169 E−05	7.6158 E−01	944.09

Reference values: $p_{SL} = 2116.22 \text{ lbf/ft}^2$
$T_{SL} = 518.67°\text{R}$
$\rho_{SL} = 0.002377 \text{ slugs/ft}^3$
$\mu_{SL} = 1.2024 \times 10^{-5} \text{ lbm/ft·s}$
$= 3.740 \times 10^{-7} \text{ lbf·s/ft}^2$

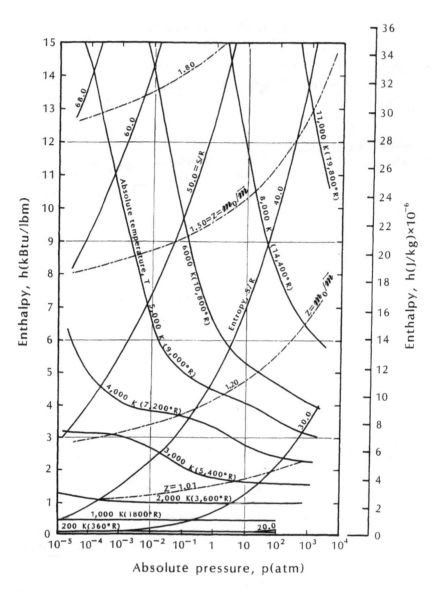

Fig. 1.16 Thermodynamic properties of air in chemical equilibrium, as taken from Ref. 16.

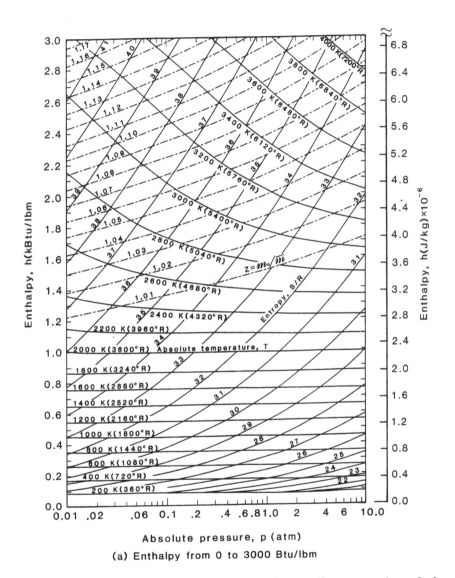

(a) Enthalpy from 0 to 3000 Btu/lbm

Fig. 1.17 Detailed charts for thermodynamic properties of air in chemical equilibrium, as taken from Ref. 16.

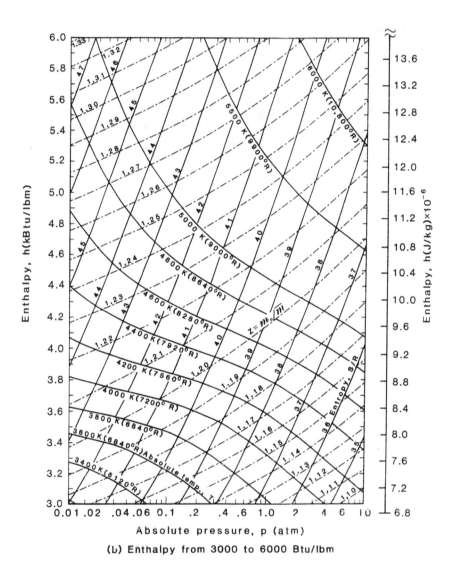

(b) Enthalpy from 3000 to 6000 Btu/lbm

Fig. 1.17 Continued.

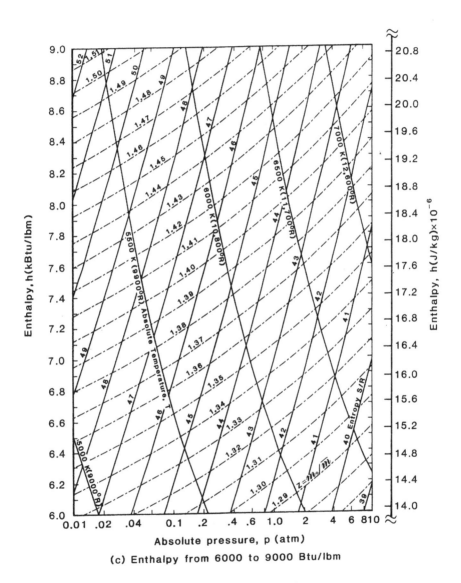

(c) Enthalpy from 6000 to 9000 Btu/lbm

Fig. 1.17 Continued.

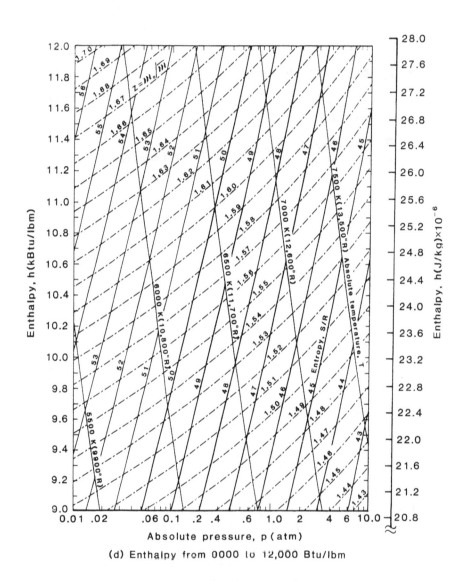

(d) Enthalpy from 0000 to 12,000 Btu/lbm

Fig. 1.17 Continued.

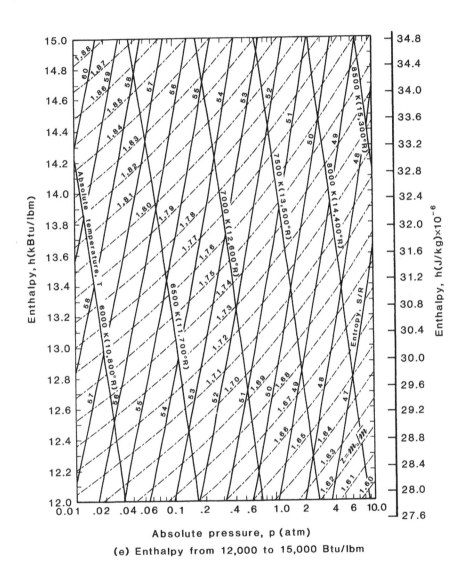

Absolute pressure, p (atm)

(e) Enthalpy from 12,000 to 15,000 Btu/lbm

Fig. 1.17 Concluded.

Exercise 1.1:

Calculate the flow downstream of a normal shock wave, when $U_1 = 15,000$ ft/s at an altitude of 150,000 ft.

Solution:
Using Table 1.1(b),

$$p_1 = (1.3429 \times 10^{-3})(2116.22) = 2.8419 \text{ lbf/ft}^2$$
$$T_1 = 479.07°\text{R}$$
$$\rho_1 = (1.4539 \times 10^{-3})(0.002377) = 3.4559 \times 10^{-6} \text{ slugs/ft}^3$$
$$a_1 = 1073.00 \text{ ft/s}$$

Thus,

$$M_1 = \frac{U_1}{a_1} = 13.979$$

There are a variety of approaches to solve this system of equations. Let us rearrange Eqs. (1-10) and (1-11).

$$p_2 = p_1 + \rho_1 U_1^2 - \rho_2 U_2^2 = \text{PMØM} - \rho_2 U_2^2$$
$$h_2 = h_1 + 0.5U_1^2 - 0.5U_2^2 = \text{HTØT} - 0.5U_2^2$$

Once we find values of p_2 and h_2, we can use Figs. 1.16 and 1.17 to find T_2 and z_2. Then we use Eq. (1-20) to calculate ρ_2,

$$\rho_2 = \frac{p_2}{z_2 R T_2}$$

Rearrange Eq. (1-9) to solve for U_2,

$$U_2 = \frac{\rho_1 U_1}{\rho_2}$$

Compare the values of ρ_2 as calculated in successive iterations until the desired accuracy is achieved.

Since the freestream air can be modeled using perfect-gas relations,

$$h_1 = c_p T_1 = (0.2404)(479.07) = 115.168 \text{ Btu/lbm}$$

Note: For the enthalpy, the conversion factor is

$$0.5U^2 \left[\frac{\text{ft}^2}{\text{s}^2}\right] \bigg/ \left[\left(32.174\frac{\text{ft lbm}}{\text{lbf s}^2}\right)\left(778.2\frac{\text{ft lbf}}{\text{Btu}}\right)\right]$$

Thus,

$$PM\emptyset M = p_1 + \rho_1 U_1^2 = 780.424 \text{ lbf/ft}^2$$

$$HT\emptyset T = h_1 + 0.5U_1^2 = 4608.373 \text{ Btu/lbm}$$

For Iteration 1: Assume $U_2 = 0$ ft/s

$$p_2 = 780.424 \text{ lbf/ft}^2 = 0.3688 \text{ atm}$$

$$h_2 = 4608.373 \text{ Btu/lbm}$$

Using Fig. 1.17b: $T_2 = 8970°$R

$$z_2 = 1.229$$

$$\rho_2 = \frac{780.424 \text{ lbf/ft}^2}{(1.229)(1716.16\frac{\text{ft}^2}{\text{s}^2\text{°R}})(8970°R)} = 4.1250 \times 10^{-5}\frac{\text{slugs}}{\text{ft}^3}$$

$$U_2 = \frac{\rho_1 U_1}{\rho_2} = 1256.69\,\frac{\text{ft}}{\text{s}}$$

For Iteration 2: $U_2 = 1256.69$ ft/s

$$p_2 = 715.279 \text{ lbf/ft}^2 = 0.3380 \text{ atm}$$

$$h_2 = 4576.836 \text{ Btu/lbm}$$

Using Fig. 1.17b: $T_2 = 8880°$R
$$z_2 = 1.227$$

$$\rho_2 = \frac{715.279}{(1.227)(1716.16)(8880)} = 3.8253 \times 10^{-5} \text{ slugs/ft}^3$$

$$U_2 = \frac{\rho_1 U_1}{\rho_2} = 1355.17 \text{ ft/s}$$

For Iteration 3: $U_1 = 1355.17$ ft/s

$$
\begin{aligned}
p_2 &= 710.173 \text{ lbf/ft}^2 = 0.3356 \text{ atm} \\
h_2 &= 4571.699 \text{ Btu/lbm}
\end{aligned}
$$

Using Fig. 1.17b: $T_2 = 8880°\text{R}$
$z_2 = 1.227$

The values from successive iterations are unchanged within the accuracy of the chart. Thus,

$$
\rho_2 = 3.8253 \times 10^{-5} \text{ slugs/ft}^3
$$

To compute the stagnation-point properties, note that the flow decelerates isentropically from the conditions immediately downstream of the normal-shock wave, i.e., 2, to the stagnation point, i.e., t2, as defined in Fig. 1.13. Using Fig. 1.17b and the air properties from iteration 3, the entropy (S_2/R) is equal to 41.0.

$$
\frac{S_{t2}}{R} = \frac{S_2}{R} = 41.0
$$

Furthermore, since t2 is the stagnation point, $h_{t2} = H_{t2} = \text{HTØT} = 4608.373$ Btu/lbm.

$$
\begin{aligned}
p_{t2} &= 0.34 \text{ atm} = 719.515 \text{ lbf/ft}^2 \\
T_{t2} &= 8950 \text{ °R} \\
z_{t2} &= 1.229
\end{aligned}
$$

Furthermore, the pressure coefficient at the stagnation point is

$$
C_{p,t2} = \frac{p_{t2} - p_1}{0.5\rho_1 U_1^2} = 1.8433
$$

Expressing our computed values in terms of some commonly used (dimensionless) ratios,

$$
\frac{\rho_2}{\rho_1} = 11.069 \; ; \quad \frac{p_{t2}}{p_1} = 253.18 \; ; \quad \frac{T_{t2}}{T_1} = 18.682 \; ; \quad \frac{H_{t2}}{h_1} = 40.014
$$

Wittliff and Curtis[17] computed the property changes across a normal-shock wave for real air in thermodynamic equilibrium. The 1959 ARDC model atmosphere[18] was used to specify the ambient conditions ahead of the shock wave. Note that there are differences between the 1959 ARDC atmosphere and the 1976 U.S. Standard Atmosphere. The corresponding values computed by Wittliff and Curtis[17] are

$$M_1 = 13.68 ; \quad z_2 = 1.227$$

$$\frac{\rho_2}{\rho_1} = 10.86 ; \quad \frac{p_{t2}}{p_1} = 250.9 ; \quad \frac{T_{t2}}{T_1} = 18.07 ; \quad \frac{H_{t2}}{h_1} = 38.59$$

Note that these computations have been made for the equilibrium assumptions, i.e., infinite reaction rates. The parameters will not be applicable to finite-rate nonequilibrium problems.

1.3.2.1 Stagnation pressure downstream of a normal-shock wave. Comparisons of the stagnation pressure downstream of a normal shock wave as computed for equilibrium air with the corresponding perfect-gas value are presented in Figs. 1.18 and 1.19. The values computed for Mach numbers from 4 through 24 at an altitude of 150,000 ft are presented in terms of the dimensionless ratio, p_{t2}/p_∞, and of the pressure coefficient at the stagnation point, $C_{p,t2}$. The comments made

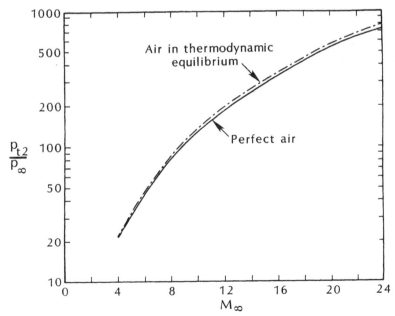

Fig. 1.18 The pressure at the stagnation point downstream of a normal-shock wave at an altitude of 150,000 ft.

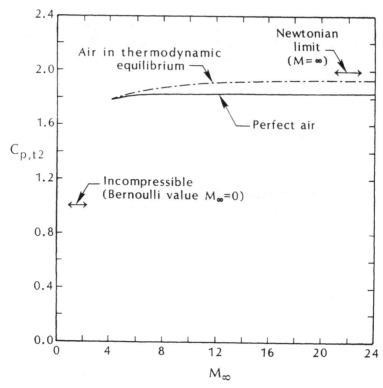

Fig. 1.19 Pressure coefficient at the stagnation point down-stream of a normal-shock wave at an altitude of 150,000 ft.

earlier regarding the qualitative differences between the perfect-gas values and those for equilibrium air, i.e., "The static pressure computed for air in thermodynamic equilibrium is slightly larger than the perfect-gas value," are illustrated in Figs. 1.18 and 1.19. Note that, at $M_\infty = 4$, the stagnation-pressure coefficient ($C_{p,t2}$) is approximately 1.8 both for perfect air and for air in thermodynamic equilibrium. At $M_\infty = 24$, $C_{p,t2}$ is 1.932 for the equilibrium air model as compared with 1.838 for perfect air and 2 for Newtonian flow.

Candler and MacCormack[19] generated numerical solutions for a two-dimensional hypersonic flowfield that is ionized and in thermochemical nonequilibrium. Comparing the surface pressures for a seven-species reacting gas flow model with those for a perfect-gas flow model for $M_\infty = 25.9$ at 71 km, they found: "The surface pressure is almost identical for each case." Actually, the stagnation point pressure computed using the reacting-gas flow model was slightly higher than the perfect-gas value. Thus, the stagnation-point pressure is essentially the same whether or not equilibrium is achieved. Furthermore, the *real-gas* value is only slightly greater than the *perfect-gas* value for a given freestream condition.

Referring to Eq. (1-10), note that, for hypersonic flow,

$$p_1 \ll \rho_1 U_1^2 \quad \text{and} \quad p_2 \gg \rho_2 U_2^2$$

As a result, we obtain the approximation that:

$$p_2 \approx p_{t2} \approx \rho_1 U_1^2 = 2q_1$$

for flow across the normal portion of the shock wave. Thus, the stagnation-point pressure for hypersonic flow is (to first order) independent of the flow chemistry and is approximately twice the dynamic pressure, q_1.

Experimental pressure measurements from the Apollo program support the previous comments, which are based on computations. For the environment of the Apollo Command Module during atmospheric entry at near-lunar-return velocities, Lee and Goodrich[20] note that "wind-tunnel measurements provide a good description of the pressure on the entry face during hypersonic flight."

The reader should note that this relative insensitivity of the pressure applies specifically to the pressures downstream of the normal portion of the bow shock wave near the stagnation point. Significant differences may exist away from the stagnation point. Although relatively simple techniques can provide first-order estimates of the pressures away from the stagnation point, e.g., refer to Chap. 6, there are phenomena that introduce large uncertainties in the pressure. The pressure distributions for flows involving shock/shock interactions, i.e., multiple shock waves, shock/boundary-layer interactions, and flow separations, such as base flows, are very sensitive to a number of parameters, as will be discussed in Chaps. 6 and 9.

The ratio of T_{t2}/T_∞ is presented in Fig. 1.20. As noted earlier, the energy absorbed by the dissociation process causes the real-gas equilibrium temperature to be markedly lower than the perfect-gas values. The specific heat correlations that were presented by Hansen[13] as a function of pressure and of temperature can be used to identify conditions where the dissociation of oxygen and of nitrogen affect the properties. Hansen notes that, at all pressures, the dissociation of oxygen is essentially complete before the dissociation of nitrogen begins. Based on Hansen's correlations, the oxygen dissociation reaction occurs near Mach 7 and the nitrogen reaction occurs near Mach 18 for the equilibrium air model at 150,000 feet. The computed values presented in Fig. 1.20 are in reasonable agreement with those presented by Wittliff and Curtis[17] using the 1959 ARDC atmosphere

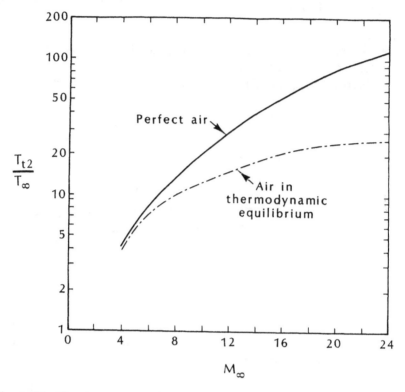

Fig. 1.20 The temperature at the stagnation point downstream of a normal-shock wave at an altitude of 150,000 ft.

and the thermodynamic properties of air reported by Hilsenrath and Beckett.

The static temperatures downstream of a normal-shock wave as computed by Witliff and Curtis[17] for velocities up to 26,000 ft/s and by Marrone[21] are presented in Fig. 1.21. For these calculations, it is assumed that the flow is uniform, steady, adiabatic, and (except for the shock wave) inviscid. Furthermore, it is assumed that the air is a homogeneous mixture that can be treated as a continuum and that complete thermochemical equilibrium exists downstream of the shock wave. Since these calculations are made for velocities up to 50,000 ft/s and for altitudes in excess of 300 kft, the reader should question the validity of these assumptions. Nevertheless, these maps of the equilibrium-air normal-shock temperatures may be used as a quick reference for investigating various flight trajectories, since they provide a bound (corresponding to the limiting case of infinitely fast reaction rates) for the range of possible temperatures.

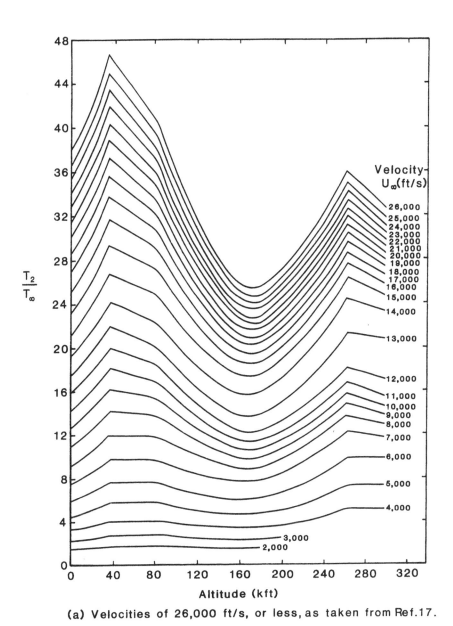

(a) Velocities of 26,000 ft/s, or less, as taken from Ref.17.

Fig. 1.21 The equilibrium air static temperature downstream of a normal-shock wave.

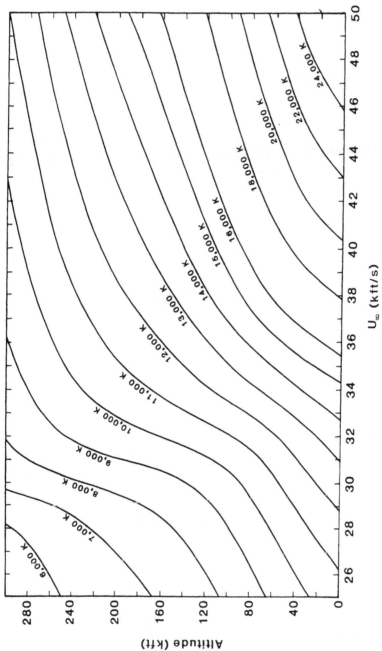

(b) Velocities of 25,000 ft/s and greater, as taken from Ref. 21.

Fig. 1.21 Concluded.

1.3.2.2   Additional comments regarding the stagnation-region properties.
Dimensional values of the stagnation pressure and of the stagnation
temperature, as calculated for equilibrium air using the techniques
and references of Exercise 1.1, are presented in Fig. 1.22. Included for
reference in Fig. 1.22 are the (approximate) velocity/altitude pair-
ings for the re-entry trajectories of a slender RV and of a Space
Shuttle Orbiter and a nominal trajectory for an SSTO vehicle with
an airbreathing engine. Note that, although the flow is hypersonic
for all these vehicles, the environment is radically different. For the
Space Shuttle Orbiter, the re-entry angle (or path angle or angle of
descent) is relatively small, i.e., less than 5 deg. Much of the deceleration occurs at such high altitudes that the total pressure at the
stagnation point is less than one-fourth of the standard atmospheric
pressure at sea level. For a slender RV, the re-entry angle is relatively large, i.e., 20 deg. to 30 deg., and the vehicle rapidly descends
deep into the atmosphere maintaining hypersonic velocity. As a result, the stagnation pressure can be roughly 50 times the standard
sea-level atmospheric value. Not only will these differences have a
dramatic impact on the loads acting on the vehicles, but they will
impact the Reynolds number, which in turn influences the time at
which boundary-layer transition occurs and, therefore, the convective heat transfer to which the vehicle is subjected. However, the
times of flight are very different also.

From the previous discussions, it should be clear that the changes
in a fluid property across a normal-shock wave are a function of M_1
and γ only for a perfect gas [see Eq. (1-14)], i.e.,

$$\frac{p_2}{p_1} = f(M, \gamma)$$

However, for a reacting gas in chemical equilibrium, three freestream
parameters are necessary to obtain the ratios of properties across a
normal shock (the freestream velocity and two thermodynamic properties),

$$\frac{p_2}{p_1} = f(U_1, p_1, T_1)$$

As noted earlier, computations made by the present author, such
as those presented in Figs. 1.18, 1.19, 1.20, and 1.21 which use the
1976 U.S. Standard Atmosphere[15] to define the freestream properties, differ from those of Refs. 17 and 21 which use the 1959 ARDC
atmosphere. Some of the differences can be traced directly to the
freestream properties. Furthermore, at the higher altitudes, the
1976 U.S. Standard Atmosphere differs from the 1962 U.S. Standard
Atmosphere.[22] Data from Shuttle flights have indicated the existence
of so-called *potholes in the sky.* As illustrated in Fig. 1.23, large den-

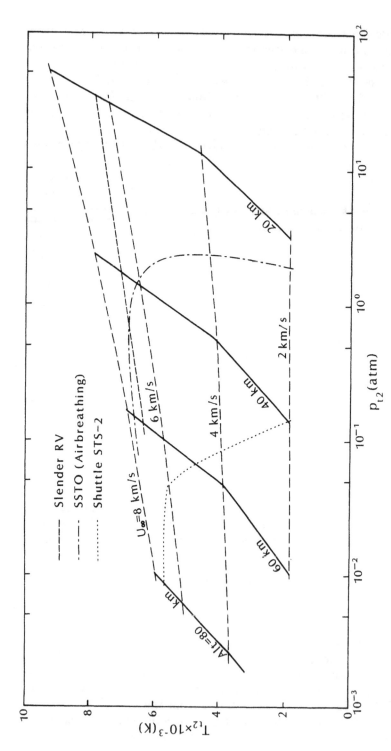

Fig. 1.22 Dimensional values of the stagnation temperature and stagnation pressure for equilibrium air.

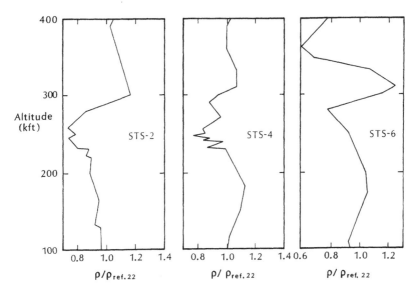

Fig. 1.23 Shuttle-derived densities compared to the 1962 U.S. Standard Atmosphere, which is Ref. 22, as taken from Ref. 23.

sity excursions occur over relatively narrow altitude ranges. Talay et al.[23] computed simulations of aerobraking trajectories of aeroassisted space transfer vehicles (ASTV's) returning from geosynchronous orbit in which the effects of off-nominal atmospheres were examined. They found that:

> None of the vehicles which safely negotiated passage through the 1962 standard atmosphere, would survive a pass through a 25 percent higher density situation. Too much energy is dissipated during the pass initiated under nominal entry conditions and the vehicle deorbited. Conversely, all the vehicles exited the 25 percent lower density atmosphere, but ended up in high apogee orbits with the higher L/D vehicles missing the required plane changes by wide margins. Significantly more propulsive maneuvers than the nominal case are then required to return these vehicles to the Shuttle.

Walberg[24] noted that the large density excursions posed a significant challenge to ASTV flight control systems.

By now, it should be clear that "real-gas" effects have a significant effect on the temperature downstream of the shock wave. This will clearly impact the density downstream of the shock wave. The density ratio for equilibrium air is presented as a function of altitude in Fig. 1.24. These values are taken from Witliff and Curtis,[17] who note that the crossing of the curves for velocities in the range of

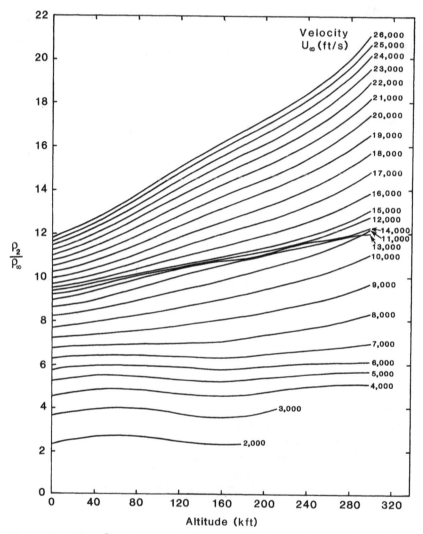

Fig. 1.24 The density ratio for equilibrium air as a function of altitude, as taken from Ref. 17.

11,000 ft/s to 14,000 ft/s is related to oxygen dissociation occurring downstream of the shock wave. Note also that the limiting value of the density ratio based on the perfect-gas model, i.e., $\rho_2 = 6\rho_1$, is achieved with a freestream velocity of 6,000 ft/s.

1.4 CONCLUDING REMARKS

As indicated in Fig. 1.12, the problems confronting the designer of a hypersonic vehicle may differ significantly from one application to another. One project, e.g., an aeroassisted space transfer vehicle, may be dominated by concerns about noncontinuum flows, complicated by nonequilibrium chemistry. For another project, e.g., hypervelocity anti-armor projectiles, which operate at speeds approaching 3 km/s near sea level, the flow is primarily equilibrium with a thin, turbulent boundary layer. Developing a reliable boundary-layer transition criterion could be a major problem for still other projects, e.g., a single-stage-to-orbit vehicle. In this chapter we have briefly characterized the nature of hypersonic flows and discussed the evaluation of the properties of air.

REFERENCES

[1] Heiser, W. H., and Pratt, D., *Hypersonic Airbreathing Propulsion,* AIAA, Washington, DC, 1993.

[2] Hirschel, E. H., "Viscous Effects," In Space Course 1991, Aachen, Feb. 1991, pp. 12-1 to 12-35.

[3] Ames Research Staff, "Equations, Tables, and Charts for Compressible Flow," NACA Rept. 1135, 1953.

[4] Hunt, J. L. , Jones, R. A., and Smith, K. A., "Use of Hexafluoroethane to Simulate the Inviscid Real-Gas Effects of Blunt Re-entry Vehicles," NASA TN D-7701, Oct. 1974.

[5] Miller, C. G., "Experimental Investigation of Gamma Effects on Heat Transfer to a 0.006 Scale Shuttle Orbiter Fluids at Mach 6," AIAA Paper 82-0826, St. Louis, MO, June 1982.

[6] Koppenwallner, G., "Experimentelle Untersuchung der Druckverteilung und des Widerstands von querangestroemten Kreiszylindern bei hypersonischen Machzahlen in Bereich von Kontinuums— bis freier Molekularstromung," *Zeitschrift für Flugwissenschaften,* Vol. 17, No. 10, Oct. 1969, pp. 321–332.

[7] Krogmann, P., "Aerodynamische Untersuchungen an Raumflugkoerpern in Machzahlbereich $Ma_\infty = 3$ bis 10 bei hohen Reynold-

szahlen," *Zeitschrift für Flugwissenschaften*, Vol. 21, No. 3, Mar. 1973, pp. 81–88.

[8] Probstein, R. F., "Shock Wave and Flow Field Development in Hypersonic Re-Entry," *ARS Journal*, Vol. 31, No. 2, Feb. 1961, pp. 185–194.

[9] Moss, J. N., and Bird, G. A., "Direct Simulation of Transitional Flow for Hypersonic Re-Entry Conditions," H. F. Nelson (ed.), *Thermal Design of Aeroassisted Orbital Transfer Vehicles*, Vol. 96 of Progress in Astronautics and Aeronautics, AIAA, New York, 1985, pp. 338–360.

[10] Cheng, H. K., "Recent Advances in Hypersonic Flow Research," *AIAA Journal*, Vol. 1, No. 2, Feb. 1963, pp. 295–310.

[11] Bertin, J. J., *Engineering Fluid Mechanics*, 2nd Ed., Prentice-Hall, Englewood Cliffs, NJ, 1987.

[12] Hansen, C. F., and Heims, S. P., "A Review of Thermodynamic, Transport, and Chemical Reaction Rate Properties of High Temperature Air," NACA TN-4359, July 1958.

[13] Hansen, C. F., "Approximations for the Thermodynamic and Transport Properties of High-Temperature Air," NACA TR R-50, Nov. 1957.

[14] Huber, P. W., "Hypersonic Shock-Heated Flow Parameters for Velocities to 46,000 Feet Per Second and Altitudes to 323,000 Feet," NASA TR R-163, 1963.

[15] Staff, *U.S. Standard Atmosphere, 1976,* Government Printing Office, Washington, DC, Dec. 1976.

[16] Moeckel, W. E., and Weston, K. C., "Composition and Thermodynamic Properties of Air in Chemical Equilibrium," NACA TN-4265, Aug. 1958.

[17] Wittliff, C. E., and Curtis, J. T., "Normal Shock Wave Parameters in Equilibrium Air," Cornell Aeronautical Lab. Rept. No. CAL-111, Nov. 1961.

[18] Minzner, R. A., Champion, K. S. W., and Pond, H. L., "The ARDC Model Atmosphere, 1959," Air Force Cambridge Research Center Rept. AFCRC TR-59-267, Aug. 1959.

[19] Candler, G., and MacCormack, R., "The Computation of Hypersonic Ionized Flows in Chemical and Thermal Nonequilibrium," AIAA Paper 88-0511, Reno, NV, Jan. 1988.

[20] Lee, D. B., and Goodrich, W. D., "The Aerothermodynamic Environment of the Apollo Command Module During Superorbital Entry," NASA TND-6792, Apr. 1972.

[21] Marrone, P. V., "Normal Shock Waves in Air: Equilibrium Composition and Flow Parameters for Velocities from 26,000 to 50,000 ft/sec," Cornell Aeronautical Lab. Rept. No. CAL AG-1729-A-2, Aug. 1962.

[22] Staff, *U.S. Standard Atmosphere, 1962*, Government Printing Office, Washington, DC, Dec. 1962.

[23] Talay, T. A., White, N. H., and Naftel, J. C., "Impact of Atmospheric Uncertainties and Viscous Interaction Effects on the Performance of Aeroassisted Orbit Transfer Vehicles," H. F. Nelson (ed.)., *Thermal Design of Aeroassisted Orbital Transfer Vehicles*, Vol. 96 of Progress in Astronautics and Aeronautics, AIAA, New York, 1985, pp. 198–229.

[24] Walberg, G. D., "A Survey of Aeroassisted Orbit Transfer," *Journal of Spacecraft and Rockets*, Vol. 22, No. 1, Jan. – Feb. 1985, pp. 3–18.

PROBLEMS

The following four points have been selected from the Space Shuttle re-entry trajectory. They are to be used in Problems 1.1 through 1.3.

(a) $U_\infty = 26,400$ ft/s;   Altitude $= 246,000$ ft

(b) $U_\infty = 16,840$ ft/s;   Altitude $= 199,000$ ft

(c) $U_\infty = 10,268$ ft/s;   Altitude $= 162,000$ ft

(d) $U_\infty = 3,964$ ft/s;   Altitude $= 100,000$ ft

1.1 For the four points of the Shuttle trajectory, use the freestream properties presented in Table 1.1

(a) the freestream Mach number, U_∞/a_∞

(b) the freestream Reynolds number, $\rho_\infty U_\infty L/\mu_\infty$ where the characteristic length $L = 1290$ in. ($L = 107.5$ ft)

1.2 For the four points of the Shuttle trajectory, calculate the conditions downstream of the normal portion of the shock wave using the perfect-gas relations ($\gamma = 1.4$).

Calculate:

p_2 the static pressure downstream of the shock wave

T_2 the static temperature downstream of the shock wave

z_2 the compressibility factor downstream of the shock wave

ϵ ρ_∞/ρ_2

p_{t2} the stagnation pressure downstream of the normal-shock wave

T_{t2} the stagnation pressure downstream of the normal-shock wave

$C_{p,t2}$ the stagnation-point pressure coefficient,
$(p_{t2} - p_\infty)/(0.5\rho_\infty U_\infty^2)$

1.3   Repeat Problem 1.2 assuming that the air is in thermodynamic equilibrium. Thus, use Figs. 1.16 and 1.17 to calculate the thermodynamic properties.

1.4   Consider a sphere 1.0 m in diameter flying at 7 km/s at an altitude of 70 km.

(a) What is the freestream Mach number of this flow?

(b) What is the freestream Reynolds number of this flow?

(c) Assuming that the air is in thermodynamic equilibrium, calculate p_2, T_2, p_{t2}, and T_{t2}. Use metric units.

1.5   If the deflection angle is 15 deg., calculate C_p as a function of M_∞ for:

(a) tangent wedge (NACA Report 1135)

(b) tangent cone (NACA Report 1135)

(c) Newtonian flow

For Problems 1.6 and 1.7, consider a "flat-plate," delta-wing configuration, shown in the sketch of Fig. 1.25, in a hypersonic stream at an angle-of-attack (α). Neglect viscous effects and use the Newtonian flow model to calculate the pressures acting on the wing. Thus, $C_p = 0$ on the upper (or leeward) surface.

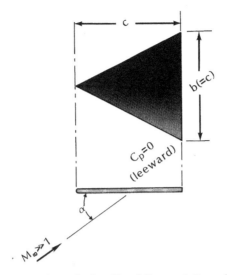

Fig. 1.25   Sketch for Problems 1.6 and 1.7.

1.6   The normal force (N) acts perpendicular to the surface and the axial force (A) is tangent to the surface. Obtain expressions for the normal force coefficient:

$$C_N = \frac{N}{q_\infty A_{\text{ref}}} = \frac{N}{q_\infty (0.5bc)}$$

and the axial force coefficient:

$$C_A = \frac{A}{q_\infty A_{\text{ref}}} = \frac{A}{q_\infty (0.5bc)}$$

as a function of the angle-of-attack.

1.7   The lift force (L) acts perpendicular to the freestream velocity, and the drag force (D) acts parallel to the freestream velocity. Obtain expressions for the lift coefficient and for the drag coefficient as a function of the angle-of-attack. At what angle-of-attack does the maximum lift coefficient occur? What is $C_{L_{\text{max}}}$? Develop an expression for L/D.

2
BASIC EQUATIONS
OF MOTION

2.1 INTRODUCTION

In order to obtain solutions of the flowfield between the bow shock wave and the surface of a vehicle traveling at hypersonic speeds, it is necessary to develop the governing equations of motion and the appropriate flow models. For the development of the basic governing equations, the flow is assumed to be a continuum.

As shown in the sketch of Fig. 2.1, the bow shock wave is curved. As noted in Fig. 2.1, the freestream conditions will be designated by the symbol 1 or by the symbol ∞. These symbols will be used interchangeably. Recall that the entropy change across a shock wave depends on the freestream Mach number and the shock inclination angle. Thus, the entropy change across the bow shock wave depends on where the flow crosses the shock wave. As a result, the flow in the inviscid portion of the shock layer, i.e., the flow between the bow shock wave and the boundary layer, is rotational. The inviscid flow in region 2, i.e., the flow downstream of the nearly normal portion of the bow shock wave, is subsonic. The flow downstream of that portion of the shock wave where the inclination angle is relatively low or where the flow has accelerated from region 2, i.e., the flow in region 3, is supersonic.

Consider the processes which may occur as the air particles move through the flowfield. The process could be a chemical reaction or the exchange of energy among the various modes, e.g., translational, rotational, vibrational, and electronic, of the atoms and of the molecules. The transfer of energy between the various energy modes is accomplished through collisions between the molecules, the atoms, and the electrons within the gas. The accommodation time is determined by the frequency with which effective collisions occur. At high altitudes, where the air density is low, chemical states do not necessarily reach equilibrium. Nonequilibrium processes occur in a flow when the time required for a process to accommodate itself to the local conditions within a particular region is of the same order as the time it takes the air particles to cross that region, i.e., the transit time. If the accommodation time is very short compared to the transit time, the process is in equilibrium. If the accommodation time is long with respect to the transit time, the chemical composition is "frozen" as the flow proceeds around the vehicle.

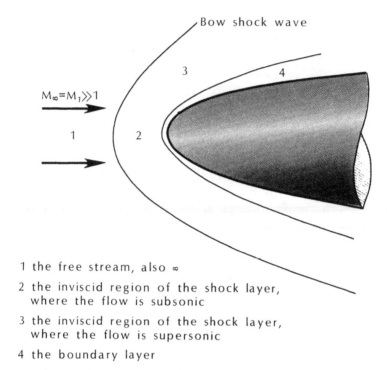

$M_\infty = M_1 \gg 1$

1 the free stream, also ∞

2 the inviscid region of the shock layer,
  where the flow is subsonic

3 the inviscid region of the shock layer,
  where the flow is supersonic

4 the boundary layer

Fig. 2.1 Hypersonic flowfield.

The shapes of the bow shock wave, of the shear layer, and of
the recompression shock wave as determined by Park[1] using optical
visualization techniques are reproduced in Fig. 2.2. Note that the
stand-off distance and the shape of the bow shock wave are sensitive
functions of the chemical state. The shock shape is affected by chemical reactions, because chemical reactions affect the temperature and,
therefore, the density. Furthermore, the effective isentropic exponent
of the gas is changed which, in turn, affects the pressure distribution
over the vehicle. Although the magnitude of the pressure difference
at a specific location may be small (see the discussion at the end of
Chap. 4), the integrated effect on the pitching moment and on the
stability of the vehicle may be significant.

2.1.1 Equilibrium Flows

When the density is sufficiently high so that there are sufficient collisions between particles to allow the equilibration of energy transfer
between the various modes, the flow is in thermochemical equilibrium. For an equilibrium flow, any two thermodynamic properties,
e.g., p and T, can be used to uniquely define the state. As a re-

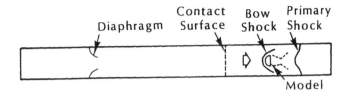

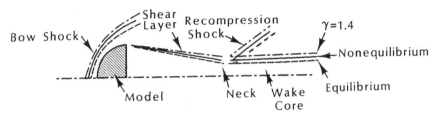

Fig. 2.2 The effect of gas chemistry on the flowfield around a blunt body, as taken from Ref. 1. Copyright © 1990 by John Wiley & Sons, Inc. Reprinted by permission of John Wiley & Sons, Inc.

sult, the remaining thermodynamic properties and the composition of the gas can be determined. The thermodynamic properties of air in thermochemical equilibrium were presented in Figs. 1.16 and 1.17.

2.1.2 Nonequilibrium Flows

A nonequilibrium gas state may result when the flow particles pass through a strong shock wave (dissociation nonequilibrium) or undergo a rapid expansion (recombination nonequilibrium). In either case, the nonequilibrium state occurs because there have not been sufficient collisions to achieve equilibrium during the time characteristic of the fluid motion. If the rate at which air particles move through the flowfield is greater than the chemical and thermodynamic reaction rates, the energy in the internal degrees of freedom (and, therefore, the energy that would be released if the gas were reacting chemically) is frozen within the gas. Calculations by Scott[2] have shown that both dissociation nonequilibrium and recombination nonequilibrium would occur at Shuttle entry conditions. Typical Shuttle trajectories are presented in Fig. 2.3. According to Rakich et al.,[3] nonequilibrium effects can occur above an altitude of 40 km (1.312×10^6 ft) and at velocities greater than 4 km/s (13.12 kft/s).

When a nonequilibrium reacting flow is computed, the dynamic behavior of the flow is significantly affected by the chemical reactions. Classical translational temperature is defined in terms of the average translational kinetic energy per particle. This temperature is classically associated with the "system temperature" in the "one-temperature" models. Park[4] notes that the use of a one-temperature model in the computation of a nonequilibrium reacting flow leads to a substantial overestimation of the rate of equilibration. Because of the slow equilibration rate of vibrational energy, multiple-temperature models are used to describe a flow which is out of equilibrium. Lee[5] recommends a three-temperature model.

> Rotational temperature tends to equilibrate very fast with translational temperature and, hence, can be considered to be equal to heavy-particle translational temperature T. Electron temperature T_e deviates from heavy-particle translational temperature T because of the slow rate of energy transfer between electrons and heavy particles caused by the large mass disparity between electrons and heavy particles. Vibrational temperature T_v departs from both electron temperature T_e and heavy-particle translational temperature T because of the slow equilibration of vibrational energy with electron and translational energies.

However, Park[4] notes that the three-temperature chemical-kinetic model is complex and requires many chemical rate parameters. As a compromise between the three-temperature model and the conventional one-temperature model, Park recommends a two-temperature chemical-kinetic model. One temperature, T, is used to characterize both the translational energy of the atoms and molecules and the rotational energy of the molecules. A second temperature, T_v, is used to characterize the vibrational energy of the molecules, the translational energy of the electrons, and the electronic excitation energy of atoms and molecules.

According to Park[1]:

> Without accounting for the nonequilibrium vibrational temperature, there is little chance that a CFD calculation can reproduce the experimentally observed phenomena. ...In a one-temperature model, the temperature at the first node point behind a shock wave is very high, and so the chemical reaction rates become very large. ...In a two-temperature model, the vibrational temperature is very low behind the shock, and so chemical reaction rates are nearly zero there. Chemical reaction rates become large only after a few node points behind the shock.

When a multiple-temperature model is used, an independent conservation equation must be written for each part of energy characterized by that temperature.

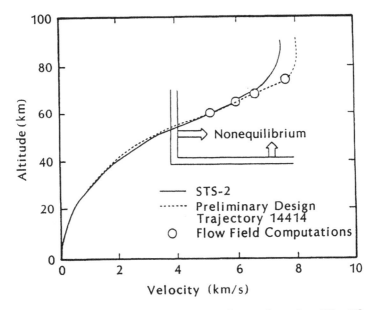

Fig. 2.3 Region of nonequilibrium flows for the Shuttle, as taken from Ref. 3.

2.1.3 Equilibrium Conditions: Thermal, Chemical, and Global

Let us follow the work of Mitcheltree and Gnoffo[6] to define the conditions for thermal equilibrium, chemical equilibrium, and global equilibrium. A mixture of gases at a point is in local thermal equilibrium when the internal energies of each species form a Boltzmann distribution for the heavy-particle translational temperature T across each of their respective energy spectra. However, when the distribution of vibrational energies does not fit the Boltzmann distribution for temperature T, thermal nonequilibrium effects are present at that point. For the two-temperature model, the distribution of vibrational energies still forms a Boltzmann distribution but at a different temperature T_v. Thermal equilibrium exists when $T = T_v$.

A mixture of gases at a point is in local chemical equilibrium when the chemical species concentrations at that point are a function of the local pressure and of the local temperature alone. Chemical equilibrium occurs when the chemical reaction rates are significantly faster than the time scales of the local fluid motion, so that the species conservation equations reduce to a balance between production and destruction of the species due to chemical reactions. If the effects of convection and of diffusion affect local species concentrations, the flow is in chemical nonequilibrium.

Combining these two definitions, we can define the conditions for local thermochemical equilibrium at a point in the flowfield. The characteristic time scales for the fluid motion, the vibrational relaxation process, and the chemical reactions are denoted by t_f, t_v, and t_c, respectively. Thermal equilibrium requires $t_v \ll t_f$, everywhere. Similarly, chemical equilibrium requires $t_c \ll t_f$, everywhere.

2.2 DEPENDENT VARIABLES

Although the various formulations of the equations of motion may employ different dependent variables, the variables common to the formulations are two thermodynamic properties and either the velocity vector (if the vector form of the momentum equation is used) or the three velocity components (if the scalar components of the momentum equation are used). For a chemically reacting flow, one must determine the composition of the gas. As noted earlier, for equilibrium processes, once any two thermodynamic properties are known, the remaining properties can be uniquely determined. Thus, once the two thermodynamic properties that are used as dependent variables have been determined, the equilibrium composition can be determined. However, such is not the case for nonequilibrium processes. For nonequilibrium flows, the mass density (or number density) for each of the NI species must be determined.

In the formulations of nonequilibrium hypersonic flows, the number of chemical species varies from four,[7] i.e., N_2, O_2, N, and O, which allows air to be approximately modeled by two species: molecules and atoms, to eleven,[5] i.e., N_2, O_2, N, O, NO, N^+, O^+, N_2^+, O_2^+, NO^+, and e^-, which is used to approximate high-temperature air (greater than 9000 K). When the gas mixture is assumed to contain electron and positive ion species, charge neutrality is assumed.

The mass fraction of species i is defined as:

$$C_i = \frac{\rho_i}{\rho} = \frac{\rho_i}{\sum \rho_i} \qquad (2\text{-}1)$$

Note that the overall mass density is given by:

$$\rho = \sum_{i=1}^{NI} \rho_i \qquad (2\text{-}2)$$

Instead of solving every one of the individual species conservation equations for all of the C_i's, the expression for the density of the mixture can be rearranged:

$$\rho_{NI} = \rho - \sum_{i=1}^{NI-1} \rho_i$$

or:

$$C_{NI} = 1 - \sum_{i=1}^{NI-1} C_i$$

Widhopf and Wang[8] recommend this approach because it allows a strong coupling, since ρ is directly a dependent variable evaluated from the governing conservation equations.

Summing the mass production rates (per unit volume) over all of the NI species, one obtains:

$$\sum_{i=1}^{NI} \dot{w}_i = 0 \tag{2-3}$$

The chemical source terms are determined from the reactions that occur between the components of the gas. Since the magnitudes of the published reaction-rate coefficients vary greatly from one source to another, the interested reader should refer to the current literature to evaluate the present state-of-the-art in this area.

2.3 TRANSPORT PROPERTIES

Before the equations of motion can be applied for a given application, it is necessary to develop appropriate expressions for the transport coefficients which appear in the momentum, energy, and mass flux terms in these equations. Gradients of physical properties in the flowfield produce a molecular transport which is directly proportional to the gradient but in the opposite direction. The transport of the momentum of the flow which is proportional to the velocity gradient, as represented by the laminar shear stress, gives rise to the coefficient of viscosity. The transport of thermal energy which is proportional to the temperature gradient gives rise to the thermal conductivity. In self-diffusion, the transport of tracer molecules in proportion to the concentration gradient of the tracer molecules gives rise to the diffusion coefficient. Thus, the coefficient of viscosity, the thermal conductivity, and the diffusion coefficient are known as *transport properties*.

Since the transport of momentum, energy, and chemical species is due to collision processes among the particles that make up the flowfield, the theoretical prediction of the transport properties requires a knowledge of the potential energy curves which describe the interactions between the various colliding particles. Thus, in order to calculate the transport coefficients once the relative populations of the constituents have been determined, one must develop models that describe the collision cross section and the dynamics of the collisions between the particles that make up the flowfield. Because the researchers who attempt to calculate the transport properties use a variety of interaction models, the values of the transport properties may vary from source to source. To paraphrase a comment often made by R. D. Neumann: "Even basic fluid properties may involve approximate models that have a limited range of applicability."

2.3.1 Coefficient of Viscosity

Using a model described physically as a rigid, impenetrable sphere surrounded by an inverse-power attractive force, Sutherland developed an equation to calculate the coefficient of viscosity:

$$\mu = 1.458 \times 10^{-6} \frac{T^{1.5}}{T + 110.4} \qquad (2\text{-}4a)$$

where T is the temperature (in K) and the units for μ are kg/s·m, or

$$\mu = 2.27 \times 10^{-8} \frac{T^{1.5}}{T + 198.6} \qquad (2\text{-}4b)$$

where T is the temperature (in $°R$) and the units for μ are lbf·s/ft^2. This model is qualitatively correct in that the molecules attract one another when they are far apart and exert strong repulsive forces upon one another when they are close together.

Chapman and Cowling[9] note that Eqs. (2-4a) and (2-4b) closely represent the variation of μ with temperature over a "fairly" wide range of temperatures. They caution, however, that the success of Sutherland's equation in representing the variation of μ with temperature for several gases does not establish the validity of Sutherland's molecular model for those gases.

> In general it is not adequate to represent the core of a molecule as a rigid sphere, or to take molecular attractions into account to a first order only. The greater rapidity of the experimental increase of μ with T, as compared with that for non-attracting rigid spheres, has to be explained as due partly to the 'softness' of the repulsive field at small distances, and partly to attractive forces which have more than a first-order effect. The chief value of Sutherland's formula seems to be as a simple interpolation formula over restricted ranges of temperature.

The Lennard-Jones model for the potential energy of an interaction, which takes into account both the softness of the molecules and

Table 2.1 Comparison of the coefficient of viscosity for air as tabulated by Svehla[10] and as calculated using Sutherland's equation, Eq. (2-4a).

T (K)	$\mu \times 10^5$ (kg/m·s)[a]	$\mu \times 10^5$ (kg/m·s)[b]
200	1.360	1.329
400	2.272	2.285
600	2.992	3.016
800	3.614	3.624
1000	4.171	4.152
1200	4.695	4.625
1400	5.197	5.057
1600	5.670	5.456
1800	6.121	5.828
2000	6.553	6.179
2200	6.970	6.512
2400	7.373	6.829
2600	7.765	7.132
2800	8.145	7.422
3000	8.516	7.702
3200	8.878	7.973
3400	9.232	8.234
3600	9.579	8.488
3800	9.918	8.734
4000	10.252	8.974
4200	10.580	9.207
4400	10.902	9.435
4600	11.219	9.657
4800	11.531	9.874
5000	11.838	10.087

[a] From Ref. 10
[b] Calculated using Eq. (2-4a)

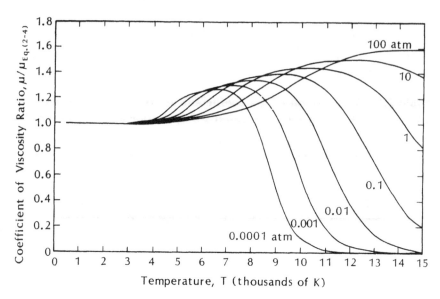

Fig. 2.4 Ratio of the coefficient of viscosity for air to the reference coefficient, Eq. (2-4), as a function of temperature, as taken from Ref. 11.

their mutual attraction at large distances, has been used by Svehla[10] to calculate the viscosity and the thermal conductivity of gases at high temperatures. The coefficients of viscosity for air as tabulated by Svehla are compared with the values calculated using Eq. (2-4a) in Table 2.1. Note that there are significant differences between the values of viscosity based on these two "models."

For temperatures below 3000 K (5400 °R), the viscosity of air is independent of pressure. The transport properties of air at high temperatures should take into account the collision cross sections for atom/atom and atom/molecule collisions. The coefficient of viscosity, as taken from Ref. 11, is presented as the ratio $\mu/\mu_{Eq. (2-4)}$ in Fig. 2.4. The ratio is unity until dissociation of molecules becomes appreciable. Then the mean free path between molecular collisions becomes larger because the collision diameters for the atoms are smaller than for the molecules. Thus, the momentum exchange takes place between more widely separated planes in the gas, and the viscosity increases.

The coefficient of viscosity for equilibrium air at high temperatures as presented by Bruno[12] is compared in Fig. 2.5 with values presented by Yos.[13] Differences are small for temperatures of 3000 K and below. However, at higher temperatures, the models assumed in the calculations of the collision integrals have a sizeable effect on the calculated values for the viscosity coefficients.

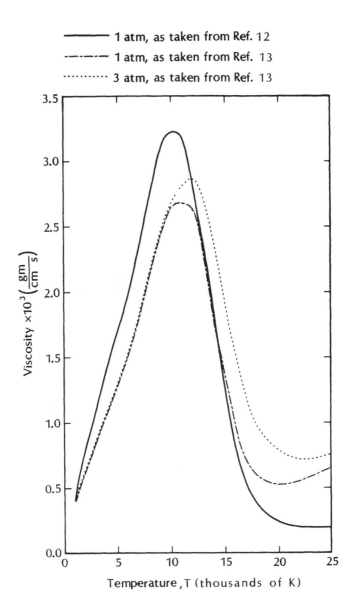

Fig. 2.5   Coefficient of viscosity of equilibrium air at high temperatures.

Exercise 2.1:

What is the coefficient of viscosity at the stagnation point, downstream of the normal portion of the bow shock wave, when $U_1 = 15,000$ ft/s at an altitude of 150,000 ft, i.e., the flow of Exercise 1.1.

Solution:

Using the results from Exercise 1.1,

$$T_{t2} = 8950 \, °R \text{ and } p_{t2} = 0.34 \text{ atm} = 719.515 \text{ lbf/ft}^2$$

In order to use Fig. 2.4, we need to convert the temperature to degrees Kelvin. Thus,

$$T_{t2} = 4972 \text{ K}$$

Using Fig. 2.4,

$$\frac{\mu_{t2}}{\mu_{Eq.(2\text{-}4b)}} \approx 1.045$$

From Eq. (2-4b),

$$\mu_{Eq.(2\text{-}4b)} = 2.27 \times 10^{-8} \frac{T^{1.5}}{T + 198.6}$$

$$= 2.10 \times 10^{-6} \frac{\text{lbf·s}}{\text{ft}^2}$$

Thus,

$$\mu_{t2} = 2.20 \times 10^{-6} \frac{\text{lbf·s}}{\text{ft}^2}$$

To compare this coefficient of viscosity with the values presented in Fig. 2.5, let us calculate the coefficient of viscosity at the stagnation point in metric units. From Eq. (2-4a),

$$\mu_{Eq.(2\text{-}4a)} = 1.458 \times 10^{-6} \frac{T^{1.5}}{T + 110.4}$$

$$= 1.0057 \times 10^{-4} \frac{\text{kg}}{\text{s·m}} = 0.0010057 \frac{\text{gm}}{\text{cm·s}}$$

Thus,

$$\mu_{t2} = 0.001051 \frac{\text{gm}}{\text{cm·s}}$$

From Fig. 2.5, the coefficient of viscosity, as taken from Yos,[13] is approximately 0.0013 gm/(cm·s), that presented by Bruno[12] is approximately 0.00166 gm/(cm·s), and that presented by Svehla[10] (see Table 2.1) is 0.00118 gm/(cm·s).

2.3.2 Thermal Conductivity

For multicomponent gas mixtures, the heat flux "vector" which appears in the energy equation includes contributions resulting from the transport of energy due to heat conduction, the direct transport of enthalpy (sensible and formation) by species whose velocity V_i differs from the bulk (mass-averaged) velocity, the energy transferred by gas radiation, and higher-order effects. Radiative heat flux will be discussed briefly in Chap. 5.

The thermal conductivity, as taken from Ref. 11, is presented as a function of temperature for selected pressures in Fig. 2.6. Again, the coefficient is referenced to a coefficient of the Sutherland form:

$$k = 1.993 \times 10^{-5} \frac{T^{1.5}}{T + 112} \tag{2-5a}$$

where T is the temperature (in K) and the thermal conductivity is in W/(cm·K) or

$$k = 2.39 \times 10^{-7} \frac{T^{1.5}}{T + 202} \tag{2-5b}$$

where T is the temperature (in °R) and the thermal conductivity is in Btu/(ft·s·°R). To determine the values for k presented in Fig. 2.6, Hansen[11] divided the energy transfer through the gas into two independent parts. One part is the energy transferred by collisions as in the ordinary thermal conductivity of nonreacting gases. The other part is the energy transferred by the diffusion of the gas particles and the reactions which occur to reestablish chemical equilibrium. The second mode of energy transfer, which takes place whenever the gas undergoes a chemical reaction, is due to the diffusion of chemical species. These particles then react with one another, giving off or absorbing the heat of reaction and causing heat transfer which may be considerably larger than the ordinary heat transfer due to molecular collisions.

Note that local maxima appear in the thermal conductivity correlations presented in Fig. 2.6. The equilibrium thermal conductivity coefficients for nitrogen, oxygen, and air at 1 atm, as presented by Yos,[13] are presented in Fig. 2.7. The local maxima in the coefficient

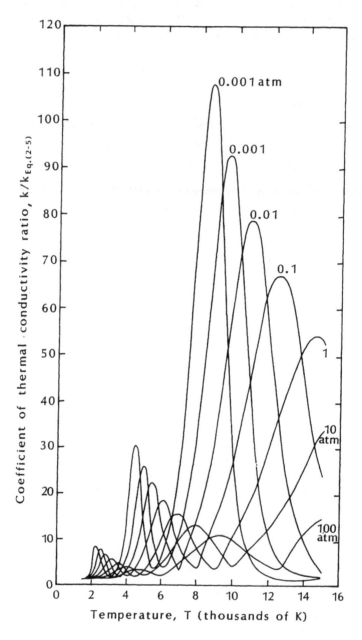

Fig. 2.6 Ratio of the coefficient of thermal conductivity of air to the reference coefficient, $k_{Eq.\ (2-5)}$ as a function of temperature, as taken from Ref. 11.

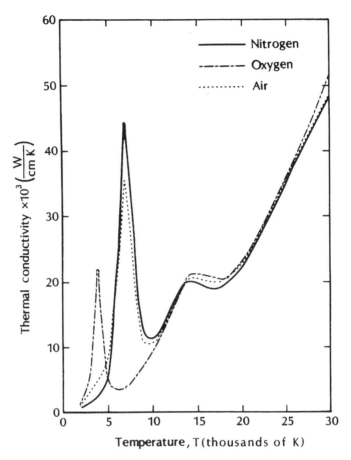

Fig. 2.7 A comparison of the (equilibrium) coefficients of thermal conductivity of air at 1 atm, as taken from Ref. 13.

of thermal conductivity for oxygen which occurs at 4000 K correspond to the temperatures associated with the dissociation reaction for oxygen. Refer to Fig. 1.15. Similarly, local maxima evident in the coefficient of thermal conductivity for nitrogen which occurs at 7000 K corresponds to the temperatures associated with the dissociation reaction for nitrogen. Since the collision cross sections for nitrogen and for oxygen are roughly equal (up to the point where ionization occurs), air can be treated essentially as a two-component mixture of atoms and molecules. Dorrance[14] presents relations for calculating the transport properties of polyatomic gas mixtures.

For polyatomic molecules, one must account for the internally stored energy transport just as one accounts for the translational energy. Bruno[12] notes that, at moderate temperatures (on the order of

2000 K), the translational and the internal contributions are almost equally divided. At higher temperatures, additional contributions are due to chemical heat release transport and ionization. Citing better collision integrals and the inclusion of the heat-of-reaction transport, Bruno[12] presents coefficients of thermal conductivity which are considerably different from other values found in the literature. The coefficients of thermal conductivity as presented by Yos[13] for equilibrium air at 1 atmosphere are compared in Fig. 2.8 with the values reproduced in Bruno[12] for 1 atm.

Brun[15] suggests that the thermal conductivity is divided into terms representing translation, rotation, and vibration. Brun notes that the translational component of the coefficient of thermal conductivity depends only weakly on nonequilibrium effects. However, the vibrational component depends strongly on nonequilibrium effects.

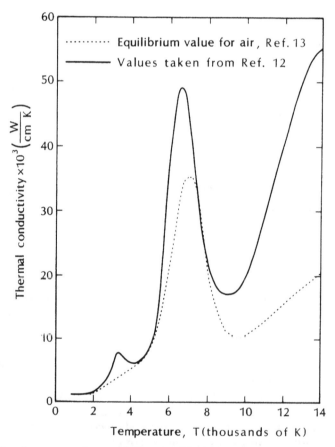

Fig. 2.8   A comparison of the coefficient of thermal conductivity of air at 1 atm.

Exercise 2.2:

What is the coefficient of thermal conductivity at the stagnation point, downstream of the normal portion of the bow shock wave, when $U_1 = 15,000$ ft/s at an altitude of 150,000 ft, i.e., the flow of Exercise 1.1.

Solution:

Using the results from Exercise 1.1,

$$T_{t2} = 8950 \,°\mathrm{R} \text{ and } p_{t2} = 0.34 \text{ atm} = 719.515 \text{ lbf/ft}^2$$

In order to use Fig. 2.6, we need to convert the temperature to degrees Kelvin. Thus,

$$T_{t2} = 4972 \text{ K}$$

Using Fig. 2.6,

$$\frac{k_{t2}}{k_{\mathrm{Eq.(2\text{-}5)}}} \approx 5.6$$

The reference thermal conductivity, i.e., $k_{\mathrm{Eq.\ (2\text{-}5a)}}$, is

$$k_{\mathrm{Eq.(2\text{-}5a)}} = 1.993 \times 10^{-5} \frac{T^{1.5}}{T+112} = 1.375 \times 10^{-3} \frac{\mathrm{W}}{\mathrm{cm\ K}}$$

Thus,

$$k_{t2} = 7.70 \times 10^{-3} \frac{\mathrm{W}}{\mathrm{cm\ K}}$$

In Fig. 2.8, the value presented by Bruno[12] is approximately 8.7×10^{-3} W/(cm·K), while that of Yos[13] is approximately 8.5×10^{-3} W/(cm·K). Note that the values presented in Fig. 2.8 are for a pressure of 1 atm. Since the thermal conductivity is pressure-dependent at these conditions, the values of Fig. 2.8 are not directly applicable.

2.3.3 Diffusion Coefficient

In a multicomponent flow in which there are concentration gradients, the net mass flux of species i is the sum of the fraction of the mass transported by the overall fluid motion and that transported by diffusion. The diffusional velocity of a particular species is due to molecular diffusion, to pressure gradients, and to thermal diffusion (also known as the Soret effect). The Soret effect is a second-order collisional transport effect that moves lighter molecules to hotter gas regions.

For most applications, the principal contributor to the diffusional transport is modeled in terms of the concentration gradients. According to Fick's law of diffusion:

$$C_i \, V_i = -D_i \, \nabla C_i \qquad (2\text{-}6)$$

for each of the i species. Since C_i represents the mass fraction of the ith species, i.e.,

$$C_i = \frac{\rho_i}{\rho} = \frac{\rho_i}{\Sigma \rho_i} \qquad (2\text{-}1)$$

Eq. (2-6) can be rewritten as:

$$\rho_i \, V_i = -\rho \, D_i \, \nabla C_i \qquad (2\text{-}7)$$

Note that the conservation of mass requires that:

$$\sum_{i=1}^{N} C_i \, V_i = 0$$

Bruno[12] cites two aspects of Fick's law which violate physical reality: (1) chemical potential gradients, not concentration gradients, drive diffusion, and (2) when C_i goes to zero, diffusional velocities go to infinity, according to Eq. (2-6). Nevertheless, a diffusion coefficient can be a useful approximation in modeling the flow.

In practical applications, a multicomponent diffusion coefficient D_{ij} is calculated where collisions occur simultaneously among all species. Values for D_{ij} are usually found from binary collision theory, i.e., considering as many binary mixtures as there are gas pairs in the multicomponent gas mixture. The use of binary diffusion coefficients is obviously an approximation, which is most justifiable because of its convenience. Bartlett et al.[16] suggest an approximation of the form:

$$D_{ij} = \frac{\bar{D}}{F_i F_j} \qquad (2\text{-}8a)$$

where

$$F_i = \left(\frac{m_i}{26} \right)^{0.461} \qquad (2\text{-}8b)$$

$$F_j = \left(\frac{m_j}{26} \right)^{0.461} \qquad (2\text{-}8c)$$

$$\bar{D} = \frac{C T^{1.5}}{p} \qquad (2\text{-}8d)$$

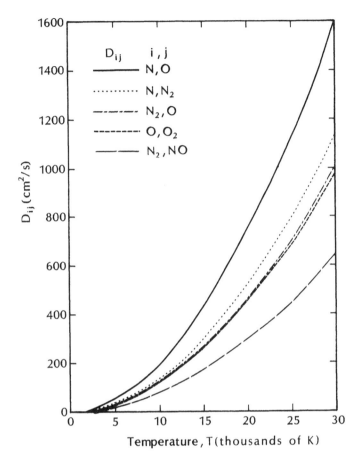

Fig. 2.9   The diffusion coefficients for air components, as taken from Ref. 13.

m_i and m_j are the molecular weights of the ith and of the jth species, respectively, and C is a constant.

Representative binary diffusion coefficients, as tabulated by Yos,[13] are presented in Fig. 2.9.

2.3.4 Additional Comments

Hansen and Heims[17] note that, over a range of temperature, the thermal conductivity ratio $[k/k_{\text{Eq. (2-5)}}]$ is nearly proportional to the specific heat. Thus, until ionization occurs, the Prandtl number for equilibrium air falls within the range from 0.6 to 1.0. The Prandtl number, as taken from Ref. 11, is presented as a function of temperature for selected pressures in Fig. 2.10.

The fact that the Prandtl number is essentially 1 for air over a wide range of temperatures and pressures is significant in computing the flow. For a Prandtl number of 1, Reynolds analogy indicates that the heat-transfer coefficient is directly proportional to the skin-friction coefficient for an incompressible flow, as discussed in Ref. 18. The correlation between skin friction and heat transfer holds approximately for compressible attached boundary layers. This is evident in the two photographs of the Apollo Command Module that are combined in Fig. 2.11. On the windward side (the right-hand side) where flow is attached, oil flow patterns from the wind-tunnel model indicate significant shear forces (i.e., skin friction) at the surface, and the char patterns on the flight vehicle indicate relatively high heating. In addition, on the leeward side where the flow is separated, the oil on the surface of the wind tunnel model is not moved by shear, and the recovered Command Module shows no effect of heating.

The transport properties for reactive-gas nonequilibrium flows are the subject of numerous investigations. The interested reader should refer to the literature for developments in this area, e.g., Refs. 15 and 19.

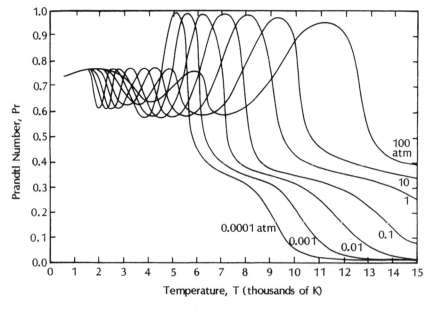

Fig. 2.10 Prandtl number for air as a function of temperature, as taken from Ref. 11.

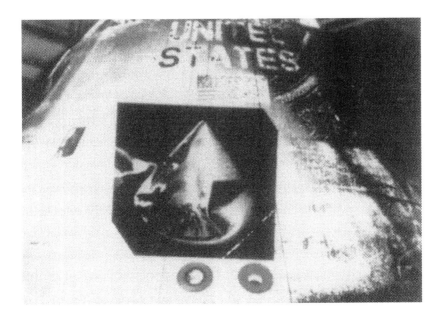

Fig. 2.11 A comparison between the oil-flow pattern (indicating skin friction) obtained in the wind tunnel and the char patterns on a recovered Apollo Command Module.

2.4 CONTINUITY EQUATION

The continuity equation represents the physical requirement that the rate of change of mass within a control volume plus the net outflow of the mass through the surface surrounding the control volume equals the rate at which mass is produced in the control volume. When this relation is applied to each chemical species, one obtains

$$\frac{\partial \rho_i}{\partial t} + \nabla \cdot [\rho_i(\boldsymbol{V} + \boldsymbol{V}_i)] = \dot{w}_i \qquad (2\text{-}9)$$

In Eq. (2-9), $\boldsymbol{V}$ is the mass-averaged velocity and $\boldsymbol{V}_i$ is the diffusion velocity of component i of the gas mixture. Specifically, the diffusion velocity $\boldsymbol{V}_i$ represents the relative velocity of species i with respect to the averaged local gas motion. Thus, the net mass flux of species i is the sum of the fraction of the mass transported by the overall fluid motion and that transported by diffusion. Based on Fick's law of diffusion, the transport of species i by diffusion is proportional to the negative of the species concentration. Thus,

$$\frac{\partial \rho_i}{\partial t} + \nabla \cdot (C_i \rho \boldsymbol{V}) - \nabla \cdot (\rho D_i \nabla C_i) = \dot{w}_i \qquad (2\text{-}10)$$

The four terms in Eq. (2-10) represent (in the order in which they appear): the rate of change of the mass of species i per unit volume in the cell, the flux of mass of species i convected across the cell walls with the mixture velocity, the diffusion of species i across the cell walls, and the mass production of species i due to chemical reactions.

As noted in Sec. 2.2, instead of solving for every one of the individual species for ρ_i, the overall (mixture) mass conservation equation is used to replace the NI-th species equation. The overall mass conservation equation is obtained by summing Eq. (2-10) over all of the species. Combining Eq. (2-2) and Eq. (2-3) and noting that the overall mass flow owing to diffusion is zero, one obtains the overall mass conservation equation:

$$\frac{\partial \rho}{\partial t} + \nabla \cdot (\rho V) = 0 \qquad (2\text{-}11)$$

Park[20] recommends the use of a formulation that retains a redundant species equation, i.e., the solution algorithm employs a conservation equation for each of the N species, Eq. (2-10), and the overall mass conservation equation, Eq. (2-11). Park recommends the redundancy for the following reasons. If one of the species variables, e.g., O, is expressed as a combination of the others, then the numerical errors acquired by each of the species variables during a computing step will be compounded in the value of O. If the concentration of O remains relatively large, then the resulting error would be small. However, in a reacting flow where the value of O becomes small, the compounded error could be a large fraction of the time value of O. By retaining the redundant equation, if the sum of all species concentration does not satisfy the overall mass because of numerical errors, the individual species concentration values are multiplied by the same adjustment factor to insure that the sum equals the total mass. (The reader is reminded that the value of one formulation over another may be user-dependent or dependent on the numerical technique used.)

The differential form of the overall continuity equation in Cartesian coordinates is:

$$\frac{\partial \rho}{\partial t} + \frac{\partial}{\partial x}(\rho u) + \frac{\partial}{\partial y}(\rho v) + \frac{\partial}{\partial z}(\rho w) = 0 \qquad (2\text{-}12)$$

2.5   MOMENTUM EQUATION

The momentum equation is obtained from Newton's law that the sum of the forces acting on a system of fluid particles is equal to the time rate of change of linear momentum. If one represents the forces and the velocity field as vector quantities, the result will be a single momentum equation. However, in our development of the momentum equation, we will consider the components of the forces and of the velocity field. Thus, we will have three scalar equations representing the component momentum equations.

The formulation includes both the pressure forces and the viscous forces acting on a surface in space (both of which are surface forces) but neglects the weight of the particles within the control volume (which is a body force). To relate the stresses to the fluid motion, it is assumed that the stress components may be expressed as a linear function of the components of the rate-of-strain. When all the velocity gradients are zero, i.e., the shear stresses vanish, the stress components must reduce to the hydrostatic pressure, p. Using index notation, the component momentum equations may be written in Cartesian coordinates as:

$$\frac{\partial}{\partial t}[\rho u_i] + \frac{\partial}{\partial x_j}[\rho u_i u_j] = -\frac{\partial p}{\partial x_i}$$

$$+ \frac{\partial}{\partial x_j}\left[\mu\left(\frac{\partial u_i}{\partial x_j} + \frac{\partial u_j}{\partial x_i}\right) - \frac{2}{3}\mu\frac{\partial u_k}{\partial x_k}\delta_{ij}\right] \qquad (2\text{-}13)$$

The four terms in Eq. (2-13) represent, respectively: the time rate of change of the ith component of momentum per unit volume in a cell, the flux of the ith component of momentum convected across the cell walls with the mixture velocity u_i, the pressure forces acting on the cell walls in the i direction, and the viscous forces acting on the cell walls in the i direction.

Taking $i = 1$ and summing over j in Eq. (2-13), we obtain the x-component of the momentum equation:

$$\frac{\partial}{\partial t}(\rho u) + \frac{\partial}{\partial x}(\rho u^2) + \frac{\partial}{\partial y}(\rho uv) + \frac{\partial}{\partial z}(\rho uw)$$

$$= -\frac{\partial p}{\partial x} + \frac{\partial}{\partial x}\left[2\mu\frac{\partial u}{\partial x} - \frac{2}{3}\mu\nabla\cdot V\right]$$

$$+ \frac{\partial}{\partial y}\left[\mu\left(\frac{\partial u}{\partial y} + \frac{\partial v}{\partial x}\right)\right] + \frac{\partial}{\partial z}\left[\mu\left(\frac{\partial u}{\partial z} + \frac{\partial w}{\partial x}\right)\right] \qquad (2\text{-}13a)$$

Taking $i = 2$ and summing over j in Eq. (2-13), we obtain the y-component of the momentum equation:

$$\frac{\partial}{\partial t}(\rho v) + \frac{\partial}{\partial x}(\rho vu) + \frac{\partial}{\partial y}(\rho v^2) + \frac{\partial}{\partial z}(\rho vw)$$

$$= -\frac{\partial p}{\partial y} + \frac{\partial}{\partial x}\left[\mu\left(\frac{\partial v}{\partial x} + \frac{\partial u}{\partial y}\right)\right]$$

$$+ \frac{\partial}{\partial y}\left[2\mu\frac{\partial v}{\partial y} - \frac{2}{3}\mu\nabla\cdot V\right] + \frac{\partial}{\partial z}\left[\mu\left(\frac{\partial v}{\partial z} + \frac{\partial w}{\partial y}\right)\right] \qquad (2\text{-}13b)$$

The z-component of the momentum equation is:

$$\frac{\partial}{\partial t}(\rho w) + \frac{\partial}{\partial x}(\rho w u) + \frac{\partial}{\partial y}(\rho w v) + \frac{\partial}{\partial z}(\rho w^2)$$

$$= -\frac{\partial p}{\partial z} + \frac{\partial}{\partial x}\left[\mu\left(\frac{\partial w}{\partial x} + \frac{\partial u}{\partial z}\right)\right]$$

$$+ \frac{\partial}{\partial y}\left[\mu\left(\frac{\partial w}{\partial y} + \frac{\partial v}{\partial z}\right)\right] + \frac{\partial}{\partial z}\left[2\mu\frac{\partial w}{\partial z} - \frac{2}{3}\mu\nabla\cdot V\right] \quad (2\text{-}13c)$$

The conservation of momentum is unaffected by chemical reactions or ionization. Even with ionization, there is no electrostatic force on the flow for approximate charge neutrality. However, as discussed in Sec. 2.3, the multicomponent nature of the reacting flow affects the value of the viscosity, as well as the other transport coefficients (the coefficient of thermal conductivity and the diffusion coefficient). The viscosity μ is affected by the multicomponent nature of the reacting flow, but is not affected by the existence of multitemperature.

For many applications, the terms involving the product of the coefficient of viscosity times a velocity gradient are negligible over extensive regions of the flowfield. We will use the term *inviscid flow* to describe those portions of the flowfield where the viscous stresses are negligibly small. By using the term *inviscid flow*, we emphasize that we are assuming that the combined product of viscosity and the relevant velocity gradients (both normal stresses and shear stresses) has a small effect on the flowfield and not that the fluid viscosity is zero.

In regions of the flowfield where the viscous shear stresses are negligibly small, i.e., in regions where the flow is inviscid, the components of the momentum equation become:

$$\frac{\partial}{\partial t}(\rho u) + \frac{\partial}{\partial x}(\rho u^2) + \frac{\partial}{\partial y}(\rho u v) + \frac{\partial}{\partial z}(\rho u w) = -\frac{\partial p}{\partial x} \quad (2\text{-}14a)$$

$$\frac{\partial}{\partial t}(\rho v) + \frac{\partial}{\partial x}(\rho v u) + \frac{\partial}{\partial y}(\rho v^2) + \frac{\partial}{\partial z}(\rho v w) = -\frac{\partial p}{\partial y} \quad (2\text{-}14b)$$

$$\frac{\partial}{\partial t}(\rho w) + \frac{\partial}{\partial x}(\rho w u) + \frac{\partial}{\partial y}(\rho w v) + \frac{\partial}{\partial z}(\rho w^2) = -\frac{\partial p}{\partial z} \quad (2\text{-}14c)$$

These equations, derived in 1755 by Euler, are called the Euler equations.

In reality, most flowfields (for which the continuum assumption is valid) will contain regions where the flow is inviscid and regions where the viscous forces cannot be neglected. In order to generate solutions for such flows, one may divide the flow into two layers: the viscous boundary layer adjacent to the wall and the inviscid flow outside of the boundary layer. Even for computational techniques which treat the flowfield as a single layer, computational time may be saved by employing the Euler equations away from the wall.

2.6 ENERGY EQUATION

Before we develop the total energy equation, let us introduce a few definitions. For the present discussion, the temperature of air is assumed to be 9000 K (16,200 °R), or less. Thus, the ionization of the atomic species will be negligible. Furthermore, transfer of energy by radiation is negligible. The overall thermodynamic energy per unit mass, i.e., the specific internal energy, is defined as:

$$e = \sum C_i e_i \qquad (2\text{-}15)$$

The specific internal energy is the sum of the energy in heavy particle translation, the energy in rotation, the energy in vibration, and the latent chemical energy of the species. The total energy per unit mass is the sum of the specific internal energy and the kinetic energy of the directed motion per unit mass:

$$e_t = e + \frac{1}{2}(\boldsymbol{V} \cdot \boldsymbol{V}) = e + \frac{1}{2}V^2 \qquad (2\text{-}16)$$

Note that the kinetic energy is interpreted as the energy associated with the observable fluid motion. The overall specific enthalpy is defined as:

$$h = \sum C_i h_i = \sum C_i e_i + \frac{p}{\rho} \qquad (2\text{-}17)$$

Using the notation of Back[21] combined with the energy equation presented by Bird et al.,[22] one obtains:

$$\sum \rho_i \frac{\partial h_i}{\partial t} + \sum \left[\rho_i (\boldsymbol{V} + \boldsymbol{V}_i) \cdot \nabla \right] h_i + \sum \dot{w}_i h_i$$
$$= \frac{dp}{dt} - \nabla \cdot (\boldsymbol{q}) + \tau : (\nabla \boldsymbol{V}) \qquad (2\text{-}18)$$

where $\boldsymbol{q}$ is the heat-flux vector and τ is the viscous stress tensor. Multiplying the continuity equation for each individual species, i.e.,

Eq. (2-10) by h_i and summing over i, we obtain:

$$\sum \dot{w}_i h_i = \sum h_i \frac{\partial \rho_i}{\partial t} + \sum h_i \nabla \cdot (\rho_i V)$$

$$- \sum h_i \nabla \cdot (\rho D_i \nabla C_i) \qquad (2\text{-}19)$$

Substituting this expression for $\sum \dot{w}_i h_i$ into Eq. (2-18), we obtain:

$$\frac{\partial}{\partial t}(\rho h) + \nabla \cdot \left(\sum \rho_i h_i V \right) - \nabla \cdot \left\{ \sum \rho h_i D_i \nabla C_i \right\}$$

$$= \frac{dp}{dt} - \nabla \cdot (q) + \tau : (\nabla V) \qquad (2\text{-}20)$$

Noting that

$$\sum \rho_i h_i = \rho \sum C_i h_i = \rho h$$

we can write Eq. (2-20)

$$h\frac{\partial \rho}{\partial t} + \rho \frac{\partial h}{\partial t} + h \left\{ \nabla \cdot (\rho V) \right\} + \rho (V \cdot \nabla) h - \nabla \cdot \left\{ \sum h_i \rho D_i \nabla C_i \right\}$$

$$= \frac{\partial p}{\partial t} + (V \cdot \nabla)p - \nabla \cdot q + \tau : (\nabla V) \qquad (2\text{-}21)$$

Multiplying the overall continuity equation by h, we obtain:

$$h[\frac{\partial \rho}{\partial t} + \nabla \cdot (\rho V)] = h[0] = 0 \qquad (2\text{-}22a)$$

Let us analyze four terms from Eq. (2-21), as follows:

$$\rho\frac{\partial h}{\partial t} + \rho(V \cdot \nabla)h - \frac{\partial p}{\partial t} - (V \cdot \nabla)p$$

$$= \rho\frac{\partial e}{\partial t} - \frac{p}{\rho}\frac{\partial \rho}{\partial t} + \frac{\partial p}{\partial t} + \rho(V \cdot \nabla)e - \frac{p}{\rho}\left[(V \cdot \nabla)\rho\right]$$

$$+ (V \cdot \nabla)p - \frac{\partial p}{\partial t} - (V \cdot \nabla)p$$

$$= \rho\frac{\partial e}{\partial t} + \rho(V \cdot \nabla)e - \frac{p}{\rho}\left[\frac{\partial \rho}{\partial t} + (V \cdot \nabla)\rho\right] \qquad (2\text{-}22b)$$

Multiplying the overall continuity equation by e, we obtain:

$$e\frac{\partial \rho}{\partial t} + e\nabla \cdot (\rho V) = 0 \qquad (2\text{-}22c)$$

If the momentum equation

$$\rho \frac{dV}{dt} = -\nabla p + \nabla \cdot \tau$$

is multiplied by the dot product of V, we obtain:

$$V\rho \cdot \frac{dV}{dt} = -V \cdot \nabla p + V \cdot (\nabla \cdot \tau)$$

Using the vector/tensor identity presented by Bird et al.,[22]

$$\tau : \nabla V \equiv \nabla \cdot (\tau \cdot V) - V \cdot (\nabla \cdot \tau)$$

Combining these last two equations, we obtain:

$$\tau : \nabla V = \nabla \cdot (\tau \cdot V) - V \cdot \nabla p - V\rho \cdot \frac{dV}{dt} \qquad (2\text{-}22\text{d})$$

and

$$\rho V \cdot \frac{dV}{dt} = \rho \frac{d}{dt} \left(\frac{V^2}{2} \right) \qquad (2\text{-}22\text{e})$$

Substituting Eqs. (2-22a) through (2-22e) into Eq. (2-21), we obtain:

$$\frac{\partial(\rho e)}{\partial t} + \nabla \cdot (\rho V e) + \rho \frac{\partial}{\partial t} \left(\frac{V^2}{2} \right) + \rho(V \cdot \nabla)\frac{V^2}{2} + V \cdot \nabla p$$

$$- \frac{p}{\rho} \left[\frac{\partial \rho}{\partial t} + (V \cdot \nabla)\rho \right] - \nabla \cdot (\tau \cdot V)$$

$$+ \nabla \cdot q - \nabla \cdot \left\{ \sum h_i \rho D_i \nabla C_i \right\} = 0 \qquad (2\text{-}23)$$

Let us use the vector definition (with the third term multiplied by the factor ρ/ρ):

$$V \cdot \nabla p \equiv \nabla \cdot (pV) - \frac{p}{\rho}(\rho \nabla \cdot V) \qquad (2\text{-}24\text{a})$$

Multiplying the overall continuity equation by $V^2/2$, we obtain:

$$\frac{V^2}{2} \left[\frac{\partial \rho}{\partial t} + \nabla \cdot (\rho V) \right] = 0 \qquad (2\text{-}24\text{b})$$

Substituting Eqs. (2-24a) and (2-24b) into Eq. (2-23), we obtain:

$$\frac{\partial(\rho e)}{\partial t} + \nabla \cdot (\rho V e) + \rho \frac{\partial}{\partial t}\left(\frac{V^2}{2}\right) + \frac{V^2}{2}\frac{\partial \rho}{\partial t} + \rho V \cdot \nabla \left(\frac{V^2}{2}\right)$$

$$+ \frac{V^2}{2}\nabla \cdot (\rho V) + \nabla \cdot (pV) - \frac{p}{\rho}\left[\frac{\partial \rho}{\partial t} + (V \cdot \nabla)\rho + \rho\nabla \cdot V\right]$$

$$- \nabla \cdot (\tau \cdot V) + \nabla \cdot q - \nabla \cdot \left\{\sum h_i \rho D_i \nabla C_i\right\} = 0 \qquad (2\text{-}25)$$

Using the overall continuity equation to eliminate terms and the definition for the total energy (e_t) as given by Eq. (2-16) to combine others, we can write:

$$\frac{\partial}{\partial t}(\rho e_t) + \nabla \cdot (\rho V e_t) + \nabla \cdot (pV) - \nabla \cdot (\tau \cdot V) + \nabla \cdot q$$

$$- \nabla \cdot \left\{\sum h_i \rho D_i \nabla C_i\right\} = 0 \qquad (2\text{-}26)$$

The terms represent (in sequence): (1) the rate of change of total energy per unit volume in a cell, (2) the flux of total energy through the cell walls, (3) the work done by the pressure forces, (4) the work done by the viscous forces, (5) the conduction of energy through the cell walls due to temperature gradients, and (6) the diffusion of enthalpy through the cell walls due to concentration gradients. In developing Eq. (2-26), the flux of energy through a multicomponent gas is a result of temperature gradients and diffusion.

For a two-dimensional flow in Cartesian coordinates,

$$e_t = e + 0.5(u^2 + v^2) = \sum C_i e_i + 0.5(u^2 + v^2)$$

Equation (2-26) then becomes:

$$\frac{\partial}{\partial t}(\rho e_t) + \frac{\partial}{\partial x}[u(\rho e_t + p)] - \frac{\partial}{\partial x}[u\tau_{xx} + v\tau_{xy} - q_x]$$

$$- \frac{\partial}{\partial x}\left[\sum h_i \rho D_i \frac{\partial C_i}{\partial x}\right] + \frac{\partial}{\partial y}[v(\rho e_t + p)]$$

$$- \frac{\partial}{\partial y}[u\tau_{yx} + v\tau_{yy} - q_y] - \frac{\partial}{\partial y}\left[\sum h_i \rho D_i \frac{\partial C_i}{\partial y}\right] = 0 \qquad (2\text{-}27)$$

In these relations,

$$q_x = -k\frac{\partial T}{\partial x} - k_v\frac{\partial T_v}{\partial x} \tag{2-28a}$$

and

$$q_y = -k\frac{\partial T}{\partial y} - k_v\frac{\partial T_v}{\partial y} \tag{2-28b}$$

For a two-temperature model, there are two temperature gradients: one in T (the translational/rotational temperature), and one in T_v (the vibrational temperature). The unsubscripted symbol k designates the thermal conductivity relating to the translational/rotational temperature T, while k_v designates the thermal conductivity relating to the vibrational temperature T_v.

2.7 GENERAL FORM OF THE EQUATIONS OF MOTION IN CONSERVATION FORM

Let us establish the following nomenclature for the general form of the equations of motion in conservation form:

$$\frac{\partial U}{\partial t} + \frac{\partial(E_i - E_v)}{\partial x} + \frac{\partial(F_i - F_v)}{\partial y} + \frac{\partial(G_i - G_v)}{\partial z} = S \tag{2-29}$$

In Eq. (2-29), the subscript i denotes the terms that appear in the equations of motion for an inviscid flow, and the subscript v denotes the terms of the viscous flux vectors, i.e., those terms that are unique to the equations of motion when the viscous and the heat-transfer effects are included. Although many authors split the flux vectors into inviscid and viscous components as indicated by Eq. (2-29) and as given in the following subsections, many do not. The vector of the chemical source terms is represented by S.

2.7.1 Overall Continuity Equation

The reader should be readily able to identify the terms in the general form of the equations of motion in conservation form which describe the overall continuity equation, since it is ordinarily written in conservation form. Comparing Eq. (2-12) with Eq. (2-29),

$$U = \rho \; ; \quad E_i = \rho u \; ; \quad F_i = \rho v \; ; \quad G_i = \rho w \; ;$$
$$\text{and } E_v = F_v = G_v = S = 0$$

Note that even though the overall continuity equation applies both to viscous regions and to inviscid regions, there are no terms relating uniquely to the viscous flow.

2.7.2 Momentum Equation

The reader may be more familiar with the x-momentum equation (which neglects the body forces) written as:

$$\rho\frac{\partial u}{\partial t} + \rho u\frac{\partial u}{\partial x} + \rho v\frac{\partial u}{\partial y} + \rho w\frac{\partial u}{\partial z} =$$

$$-\frac{\partial p}{\partial x} + \frac{\partial}{\partial x}\left[2\mu\frac{\partial u}{\partial x} - \frac{2}{3}\mu\nabla\cdot V\right]$$

$$+\frac{\partial}{\partial y}\left[\mu\left(\frac{\partial u}{\partial y} + \frac{\partial v}{\partial x}\right)\right] + \frac{\partial}{\partial z}\left[\mu\left(\frac{\partial w}{\partial x} + \frac{\partial u}{\partial z}\right)\right] \quad (2\text{-}30)$$

or

$$\rho\frac{\partial u}{\partial t} + \rho u\frac{\partial u}{\partial x} + \rho v\frac{\partial u}{\partial y} + \rho w\frac{\partial u}{\partial z}$$

$$= -\frac{\partial p}{\partial x} + \frac{\partial(\tau_{xx})}{\partial x} + \frac{\partial(\tau_{yx})}{\partial y} + \frac{\partial(\tau_{zx})}{\partial x} \quad (2\text{-}31)$$

where:

$$\tau_{xx} = 2\mu\frac{\partial u}{\partial x} - \frac{2}{3}\mu\nabla\cdot V$$

$$\tau_{yx} = \mu\left(\frac{\partial u}{\partial y} + \frac{\partial v}{\partial x}\right)$$

$$\tau_{zx} = \mu\left(\frac{\partial u}{\partial z} + \frac{\partial w}{\partial x}\right)$$

Equations (2-30) and (2-31), forms of the x-momentum equation which are commonly used in the analysis of boundary layers (e.g., Ref. 18), are not in conservation form. If we multiply the overall continuity equation, Eq. (2-11), by u and add it to Eq. (2-31), we obtain the conservation-law form of the x-momentum equation.

$$\frac{\partial(\rho u)}{\partial t} + \frac{\partial}{\partial x}\left[(p + \rho u^2) - (\tau_{xx})\right] + \frac{\partial}{\partial y}\left[(\rho uv) - (\tau_{yx})\right]$$

$$+\frac{\partial}{\partial z}\left[(\rho uw) - (\tau_{zx})\right] = 0 \quad (2\text{-}32)$$

Comparing Eq. (2-32) with Eq. (2-29), it is clear that:

$$U = \rho u \; ; \quad E_i = p + \rho u^2 \; ; \quad F_i = \rho uv \; ; \quad G_i = \rho uw \; ;$$

$$E_v = \tau_{xx} \; ; \quad F_v = \tau_{yx} \; ; \quad G_v = \tau_{zx} \; ; \quad \text{and} \quad S = 0$$

2.7.3 Energy Equation

The vector expressions in Eq. (2-26) were expanded for a two-dimensional flow to provide Eq. (2-27). Extending that technique for a three-dimensional flow:

$$U = \rho e_t$$

$$E_i = u(\rho e_t + p)$$

$$E_v = u\tau_{xx} + v\tau_{xy} + w\tau_{xz} + q_x + \sum h_i \rho D_i \frac{\partial C_i}{\partial x}$$

$$F_i = v(\rho e_t + p)$$

$$F_v = u\tau_{yx} + v\tau_{yy} + w\tau_{yz} + q_y + \sum h_i \rho D_i \frac{\partial C_i}{\partial y}$$

$$G_i = w(\rho e_t + p)$$

$$G_v = u\tau_{zx} + v\tau_{zy} + w\tau_{zz} + q_z + \sum h_i \rho D_i \frac{\partial C_i}{\partial z}$$

$$S = 0$$

2.7.4 Continuity Equation for an Individual Species

Comparing Eq. (2-10) with Eq. (2-29):

$$U = \rho C_i$$

$$E_i = \rho C_i u \; ; \quad F_i = \rho C_i v \; ; \quad G_i = \rho C_i w$$

$$E_v = \rho D_i \frac{\partial C_i}{\partial x} \; ; \quad F_v = \rho D_i \frac{\partial C_i}{\partial y} \; ; \quad G_v = \rho D_i \frac{\partial C_i}{\partial z}$$

$$S = \dot{w}_i$$

2.7.5 The Vectors

Thus, when the fundamental equations governing the unsteady flow of a gas, without body forces or external heat addition, are written in conservation form, the terms U, E_i, F_i, F_v, G_i, and G_v can be written by the following vectors:

$$
U = \begin{bmatrix} \rho \\ \rho u \\ \rho v \\ \rho w \\ \rho e_t \\ \rho C_1 \\ \vdots \\ \rho C_i \end{bmatrix}
$$

$$
E_i = \begin{bmatrix} \rho u \\ p + \rho u^2 \\ \rho u v \\ \rho u w \\ (\rho e_t + p)u \\ \rho C_1 u \\ \vdots \\ \rho C_i u \end{bmatrix}
\qquad
E_v = \begin{bmatrix} 0 \\ \tau_{xx} \\ \tau_{xy} \\ \tau_{xz} \\ u\tau_{xx} + v\tau_{xy} + w\tau_{xz} + q_x + \sum h_i \rho D_i \frac{\partial C_i}{\partial x} \\ \rho D_1 \frac{\partial C_1}{\partial x} \\ \vdots \\ \rho D_i \frac{\partial C_i}{\partial x} \end{bmatrix}
$$

$$
F_i = \begin{bmatrix} \rho v \\ \rho v u \\ p + \rho v^2 \\ \rho v w \\ (\rho e_t + p)v \\ \rho C_1 v \\ \vdots \\ \rho C_i v \end{bmatrix}
\qquad
F_v = \begin{bmatrix} 0 \\ \tau_{yx} \\ \tau_{yy} \\ \tau_{yz} \\ u\tau_{yx} + v\tau_{yy} + w\tau_{yz} + q_y + \sum h_i \rho D_i \frac{\partial C_i}{\partial y} \\ \rho D_1 \frac{\partial C_1}{\partial y} \\ \vdots \\ \rho D_i \frac{\partial C_i}{\partial y} \end{bmatrix}
$$

$$
G_i = \begin{bmatrix} \rho w \\ \rho w u \\ \rho w v \\ p + \rho w^2 \\ (\rho e_t + p)w \\ \rho C_1 w \\ \vdots \\ \rho C_i w \end{bmatrix}
\qquad
G_v = \begin{bmatrix} 0 \\ \tau_{zx} \\ \tau_{zy} \\ \tau_{zz} \\ u\tau_{zx} + v\tau_{zy} + w\tau_{zz} + q_z + \sum h_i \rho D_i \frac{\partial C_i}{\partial x} \\ \rho D_1 \frac{\partial C_1}{\partial z} \\ \vdots \\ \rho D_i \frac{\partial C_i}{\partial z} \end{bmatrix}
$$

$$S = \begin{bmatrix} 0 \\ 0 \\ 0 \\ 0 \\ 0 \\ \dot{w}_1 \\ \vdots \\ \dot{w}_i \end{bmatrix}$$

where

$$e_t = u_e + \frac{1}{2}(u^2 + v^2 + w^2)$$

and

$$\tau_{xx} = 2\mu\frac{\partial u}{\partial x} - \frac{2}{3}\mu\nabla \cdot V$$

$$\tau_{yy} = 2\mu\frac{\partial u}{\partial y} - \frac{2}{3}\mu\nabla \cdot V$$

$$\tau_{zz} = 2\mu\frac{\partial u}{\partial z} - \frac{2}{3}\mu\nabla \cdot V$$

$$\tau_{xy} = \tau_{yx} = \mu\left(\frac{\partial u}{\partial y} + \frac{\partial v}{\partial x}\right)$$

$$\tau_{xz} = \tau_{zx} = \mu\left(\frac{\partial u}{\partial z} + \frac{\partial w}{\partial x}\right)$$

$$\tau_{yz} = \tau_{zy} = \mu\left(\frac{\partial v}{\partial z} + \frac{\partial w}{\partial y}\right)$$

$$q_x = k\frac{\partial T}{\partial x} ; \qquad q_y = k\frac{\partial T}{\partial y} ; \qquad \text{and} \quad q_z = k\frac{\partial T}{\partial z}$$

2.8 CONCLUDING REMARKS

In Chap. 1, we saw that there are a wide variety of environments that are classified as hypersonic flows. Similarly, one might introduce any one of a wide variety of assumptions (depending upon the application), when developing flow models for solutions to the equations of motion. For instance, the gas properties (and composition) may be computed using the perfect-gas model, using the assumption that the gas is in equilibrium, or using the finite-rate chemistry of a nonequilibrium flow. As noted earlier, different analysts employ different temperatures when representing a reacting gas which is not in equilibrium. The rigor of the flow model should be consistent with the application.

REFERENCES

[1] Park, C., *Nonequilibrium Hypersonic Aerothermodynamics*, John Wiley & Sons, New York, 1990.

[2] Scott, C. D., "Space Shuttle Laminar Heating with Finite-Rate Catalytic Recombination," T. E. Horton (ed.), *Thermophysics of Atmospheric Entry*, Vol. 82 of Progress in Astronautics and Aeronautics, AIAA, New York, 1982, pp. 273–289.

[3] Rakich, J. V., Stewart, D. A., and Lanfranco, M. J., "Results of a Flight Experiment on the Catalytic Efficiency of the Space Shuttle Heat Shield," AIAA Paper 82-0944, St. Louis, MO, June 1982.

[4] Park, C., "Assessment of Two-Temperature Kinetic Model for Ionizing Air," AIAA Paper 87-1574, Honolulu, HI, June 1987.

[5] Lee, J. H., "Basic Governing Equations for the Flight Regimes of Aeroassisted Orbit Transfer Vehicles," H. F. Nelson (ed.), *Thermal Design of Aeroassisted Orbit Transfer Vehicles*, Vol. 96 of Progress in Astronautics and Aeronautics, AIAA, New York, 1985, pp. 3–53.

[6] Mitcheltree, R., and Gnoffo, P., "Thermochemical Nonequilibrium Issues for Earth Reentry of Mars Missions Vehicles," AIAA Paper 90-1698, Seattle, WA, June 1990.

[7] Shang, J. S., and Josyula, E., "Numerical Simulations of Non-Equilibrium Hypersonic Flow Past Blunt Bodies," AIAA Paper 88-0512, Reno, NV, Jan. 1988.

[8] Widhopf, G. F., and Wang, J. C. T., "A TVD Finite-Volume Technique for Nonequilibrium Chemically Reacting Flows," AIAA Paper 88-2711, San Antonio, TX, June 1988.

[9] Chapman, S., and Cowling, T. G., *The Mathematical Theory of Non-Uniform Gases*, Cambridge University Press, Cambridge, England, 1960.

[10] Svehla, R. A., "Estimated Viscosities and Thermal Conductivities of Gases at High Temperatures," NASA TR R-132, 1962.

[11] Hansen, C. F., "Approximations for the Thermodynamic and Transport Properties of High-Temperature Air," NACA TN R-50, Nov. 1957.

[12] Bruno, C., "Real Gas Effects," J. J. Bertin, R. Glowinski, and J. Periaux (eds.), published in *Hypersonics, Volume 1: Defining the Hypersonic Environment*, Birkäuser Boston, Boston, MA, 1989.

[13] Yos, J. M., "Transport Properties of Nitrogen, Hydrogen, Oxygen, and Air to 30,000 K," AVCO Research and Advanced Development Division RAD-TM-63-7, Mar. 1963.

[14] Dorrance, W. H., *Viscous Hypersonic Flow*, McGraw-Hill, New York, 1962.

[15] Brun, R., "Transport Properties in Reactive Gas Flows," AIAA Paper 88-2655, San Antonio, TX, June 1988.

[16] Bartlett, E. P., Kendall, R. M., and Rindal, R. A., "An Analysis of the Coupled Chemically Reacting Boundary Layer and Charring Ablator, Part IV—A Unified Approximation for Mixture Transport Properties for Multicomponent Boundary-Layer Applications," NASA CR-1063, June 1968.

[17] Hansen, C. F., and Heims, S. P., "A Review of Thermodynamic, Transport, and Chemical Reaction Rate Properties of High Temperature Air," NACA TN-4359, July 1958.

[18] Bertin, J. J., *Engineering Fluid Mechanics*, 2nd Ed., Prentice-Hall, Englewood Cliffs, NJ, 1987.

[19] Jaffe, R. L., "The Calculation of High-Temperature Equilibrium and Nonequilibrium Specific Heat Data for N_2, O_2, and NO," AIAA Paper 87-1633, Honolulu, HI, June 1987.

[20] Park, C., "Convergence of Computation of Chemical Reacting Flows," J. N. Moss and C. D. Scott (eds.), *Thermophysical Aspects of Re-entry Flows*, Vol. 103 of Progress in Astronautics and Aeronautics, AIAA, New York, 1985, pp. 478–513.

[21] Back, L. H., "Conservation Equations of a Viscous, Heat-Conducting Fluid in Curvilinear Orthogonal Coordinates," Jet Propulsion Lab. TR-32-1332, Sep. 1968.

[22] Bird, R. B., Stewart, W. E., and Lightfoot, E. N., *Transport Phenomena*, John Wiley & Sons, New York, 1960.

PROBLEMS

2.1 For a blunt vehicle flying at $M_1 = 16.35$ at an altitude of 60.65 km, the stagnation conditions downstream of the normal shock wave are:
$$p_{t2} = 0.070 \text{ atm}; \quad T_{t2} = 5244 \text{ K}$$

Using the correlations of Figs. 2.4 and 2.6, what are the coefficient of viscosity (μ_{t2}) and the coefficient of thermal conductivity (k_{t2})?

2.2 For a blunt vehicle flying at 3964 ft/s at an altitude of 100,000 ft, the stagnation conditions downstream of the normal shock wave are:
$$p_{t2} = 0.2317 \text{ atm}; \quad T_{t2} = 1716 \text{ °R}$$

Using the correlations of Figs. 2.4 and 2.6, what are the coefficient of viscosity (μ_{t2}) and the coefficient of thermal conductivity (k_{t2})?

2.3 From our study of steady, compressible, inviscid flow in a variable-area streamtube, we know that a subsonic flow accelerates in the converging section and decelerates in the diverging section. Using the Euler equation for a one-dimensional, steady flow, how does the static pressure vary in the converging section? That is, does the pressure increase or decrease in the streamwise direction (or, equivalently, is it an adverse pressure gradient or a favorable pressure gradient)? How does the static pressure vary in the diverging section?

2.4 A supersonic flow decelerates in the converging section and accelerates in the diverging section. Using the Euler equation for a one-dimensional, steady flow, how does the static pressure vary in the converging section? How does the static pressure vary in the diverging section?

3
DEFINING THE
AEROTHERMODYNAMIC
ENVIRONMENT

3.1 INTRODUCTION

The aerothermodynamic environment (i.e., the aerodynamic forces and moments and the heating distributions) of a vehicle that will fly at hypersonic speeds may include some, if not all, of the following phenomena: boundary-layer transition and turbulence, viscous/inviscid interactions, separated flows, nonequilibrium chemistry and the effects of surface catalycity, ablation, and noncontinuum effects. The analysis is further complicated if the vehicle contains an airbreathing, scramjet propulsion system, such as shown in Fig. 3.1. Since the vehicle's forebody serves as a compression surface for the inlet flow, one must be able to describe the transition process and subsequent three-dimensional turbulent boundary layer with reasonable accuracy. Since the afterbody serves as a nozzle, one must be able to describe a complex, three-dimensional flowfield where chemistry and viscous/inviscid interactions are important. Obviously, a tip-to-tail analysis of the flowfield over the vehicle of Fig. 3.1 in hypersonic flight presents tremendous challenges. What tools are available to the designer?

One may wish to build and test a scale model in a hypersonic wind tunnel or in another type of ground-based facility. For a relatively simple shape, properly nondimensionalized wind-tunnel data can be readily applied to the flight environment. For example, Lee et al.[1] note that, for the Apollo Command Module:

1. Flight data verify that wind-tunnel measurements provide a good description of pressure variations on the aft conical section of the Command Module during entry.

2. Heating rates measured with asymptotic calorimeters on the conical section agreed with wind-tunnel measurements extrapolated on the basis of laminar theory.

A key phrase in this quote is "extrapolated on the basis of laminar theory." As we will see in Chap. 4, all ground-based tests are only partial simulations of the hypersonic flight environment. Furthermore, the extrapolation process becomes more complicated for more complex shapes. As will be discussed further in this and other chapters, the Space Shuttle experienced a pitch-up moment in flight that

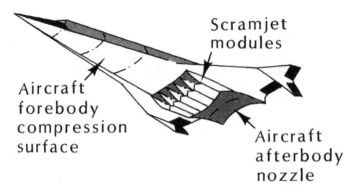

Fig. 3.1 Hypersonic vehicle with an airbreathing propulsion system.

was not predicted by the wind-tunnel data base, requiring a body-flap deflection almost twice that predicted based on the Aerodynamic Design Data Book (ADDB).

Progress in computer hardware and software made during the 1960s and 1970s led people to predict that the computer would completely replace the wind tunnel in the design process. Chapman[2] envisioned the day when turbulent eddy simulations "could be conducted from essentially first principles of the flow over practical aircraft configurations." Indeed, because of continued improvements in computer hardware and software, computational fluid dynamics (CFD) plays an ever greater role in the design process. Note that the term computational fluid dynamics implies the integration of two disciplines: fluid dynamics and computation. Thus, in developing a CFD code to generate flowfield solutions, we must realize that approximations are made both in modeling fluid-dynamic phenomena, e.g., turbulence, and in the numerical formulation, e.g., developing grids to define the configuration and to define the flowfield points where dependent variables are to be computed.

Li[3] states:

> The CFD code development and application may follow seven steps.
>
> 1) Select the physical processes to be considered.
> 2) Decide upon mathematical and topological models.
> 3) Build body geometry and space grid.
> 4) Develop a numerical solution method.
> 5) Incorporate the above into a computer code.
> 6) Calibrate and validate the code against benchmark data.
> 7) Predict aerodynamic coefficients, flow properties, and aeroheatings.

Li warns that, "In any case, one must carefully weigh consistency among levels of physical modeling, computer resources available, and simulation requirements when employing CFD methodology."

Neumann[4] notes:

> The codes are NOT an end in themselves... They represent engineering tools; tools that require engineering judgement to use and critical appraisal to understand. Hypersonics must be dominated by an increased understanding of fluid mechanic reality and an appreciation of the difference between that reality and the modeling of that reality. CFD represents the framework for that modeling study and experiments represent the technique for introducing physical reality into the modeling process. Finally, classical analytical theory and the trend information produced by theory gives us the direction with which to assemble the point data from these numerical solutions in an efficient and meaningful way to produce understanding.

Mehta[5] notes:

> In a broad sense, computational requirements for design performance estimates are the following: credibility of computations, the acceptable level of accuracy, the uncertainty of test data, safety factors, and consistency in determination. The key characteristics of a majority of these requirements is sensitivity. The presented vision of the CFD design technology development triad is essential in fulfilling these requirements. The triad brings together different technologies, CFD, design, and testing (ground-based and flight). The process of establishing the credibility of a CFD code is called code certification.

The discussion in this section underscores the fact that the hypersonic design process must employ the synergistic integration of ground-based testing, basic analytic technologies, computational fluid dynamics, and flight testing. In some instances, data from ground-based tests are used to develop correlations for use at the design flight conditions. In other situations, the primary objective of a ground-test program is to generate high-quality data that can be used to develop, to validate, and to calibrate numerical codes. In either case, one should always compare experimental measurements to the corresponding computed values. Furthermore, when developing a CFD code, one should compare computed flowfield parameters with the corresponding measured values to validate the numerical models for the fluid-dynamic phenomena. One should exercise the complete code over a range of flow conditions, comparing the computations with measurements (both ground-based and flight) to calibrate the code and to establish the limits of its applicability.

3.2 EMPIRICAL CORRELATIONS COMPLEMENTED BY ANALYTICAL TECHNIQUES

As noted in Sec. 3.1, wind-tunnel data were used in dimensionless form to define the pressure distributions, the convective heat-transfer distributions, and the aerodynamic forces and moments for the Apollo re-entry vehicle. For example, h (the convective heat-transfer coefficient measured at a specific location) was divided by $h_{t,\text{ref}}$ (a reference, stagnation-point heat-transfer coefficient), measured at the same flow conditions in the wind tunnel (wt), producing the dimensionless ratio $(h/h_{t,\text{ref}})_{wt}$. To determine the local convective heat-transfer rate at flight conditions (ft)

$$\dot{q}_{ft} = \left(\frac{h}{h_{t,\text{ref}}} \right)_{wt} (\dot{q}_{t,\text{ref}})_{ft} \qquad (3\text{-}1)$$

where the relations of Fay and Riddell[6] or of Detra et al.[7] (see Ch. 5) could be used to determine the reference, stagnation-point heating rate at flight conditions, $(\dot{q}_{t,\text{ref}})_{ft}$. This approach was pioneered at the Air Force Flight Dynamics Laboratory for the ASSET program of the 1960s.[8]

As discussed in the previous section, such wind-tunnel-based empirical correlations complemented by analytical solutions provided reasonable estimates of the actual flight environment for the Apollo Command Module (at least until radiative heat transfer became a factor at superorbital velocities). Neumann[4] notes, "These systems are dominated by a small contribution of viscous effects on the aerodynamics and aerodynamic heating of the vehicle. A direct result of being pressure dominated is that developmental tests can, for the most part, be conducted in 'representative' hypersonic wind tunnels without the need for more exact duplication of flight characteristics."

3.2.1 Correlation Techniques for Shuttle Orbiter Heating

Although the Shuttle Orbiter flowfield contains complexities such as shock/shock interactions, extensive use was made of wind tunnels in determining the aerodynamics of the Space Shuttle Orbiter, e.g., Refs. 9, 10, and 11. Haney[12] stated that the Space Shuttle Orbiter's thermal protection system (TPS) was designed mainly on the basis of wind-tunnel data. For the windward (or lower) surface of the Orbiter, the empirical correlations complemented by analytical techniques were developed as follows. As shown in Fig. 3.2a, the Orbiter was divided into a combination of simple shapes, i.e., cones, cylinders, flat plates, spheres, and wedges, for which analytical solutions are available. For example, the wing leading edge was represented

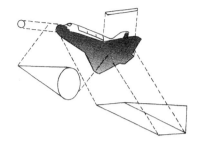

(a) Representative flow models

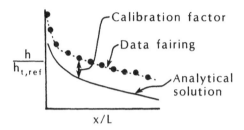

(b) Fuselage lower centerline

Fig. 3.2   Design methodology for the windward (lower) surface of the Orbiter, as taken from Ref. 12.

by a swept cylinder in order to obtain estimates for the leading-edge heating rates (outside of the shock/shock interaction region). Shock/shock interactions will be discussed in Chap. 9. The heating to the windward fuselage was calculated using the standard Eckert reference enthalpy flat plate solutions,[13] designated *Analytical solution* in Fig. 3.2b. Because the flow over the simple shape is only a crude approximation of the actual flow, there are significant differences in edge flow properties and in streamline patterns, which produce significant differences between the theoretical heat transfer and that measured in the wind tunnel on the Orbiter model. Adjustments, designated *Calibration factor* in Fig. 3.2b, were used to take into account such variations as streamline divergence and flow running lengths. When the analytical solutions for the simple geometric shapes are multiplied by position-dependent calibration factors, one can match the wind-tunnel measurements, designated *Data fairing* in Fig. 3.2b. These calibration factors, developed both for laminar and for turbulent boundary layers, were held constant in extrapolations to flight conditions.

For the upper (or leeward) surface, the wind-tunnel data were nondimensionalized as the ratio of the local heat-transfer coefficient divided by a reference, stagnation-point heat transfer, i.e.,

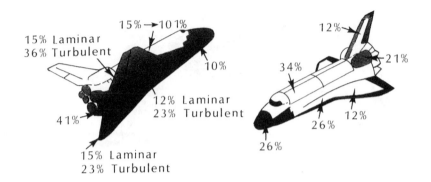

Fig. 3.3   Uncertainties in pre-flight heating estimates for the Shuttle Orbiter based on empirical correlations complemented by analytical solutions, as taken from Ref. 12.

$(h/h_{t,\text{ref}})_{wt}$. See Eq. (3-1). Using correlations of this dimensionless heat-transfer ratio as a function of angle-of-attack, of angle-of-sideslip (yaw), of the freestream Mach number, and of the freestream Reynolds number, the wind-tunnel data were applied directly to flight.

The uncertainties inherent in extrapolating wind-tunnel heat-transfer data to flight conditions, including wind-tunnel data scatter, scaling to flight, uncertainty in local flow properties, and extrapolating data from regions where multiple shock waves interact with the boundary layer, force the introduction of conservatism into the estimates of the heating environment prior to the first flight. Estimates of the heating uncertainties are reproduced in Fig. 3.3. The magnitude of these uncertainties may seem large, but the determination of the heat-transfer rates represent a severe challenge both to the experimentalist and to the analyst.

3.2.2 Correlations for Viscous Interactions with the External Flow

The effect of the boundary layer on the inviscid flowfield may be represented by displacing the actual surface by the boundary-layer displacement thickness. Since the boundary layer displaces the external stream, this stream deflection will change the shape of the shock wave and, therefore, the flowfield. The growth of the boundary layer is determined by the pressure distribution, the flow properties at the edge of the boundary layer, etc., while the values of these parameters, themselves, depend on the magnitude of the displacement effect. For instance, the lower the Reynolds number, the thicker the boundary layer and the greater the viscous-interaction-induced effect on the flowfield. Thus, this viscous/inviscid interaction is a complex phenomena in which the boundary-layer "history" plays an important role. Note also that this phenomenon is more significant for slender bodies, such as slender cones, since the changes in the effective geometry due to boundary-layer growth will be proportionally larger. Furthermore, the shock wave will lie closer to the body, the higher the Mach number.

To correlate the viscous/inviscid-interaction-induced flowfield perturbations, Koppenwallner[14] identified two different parameters: one for the pressure, another for the skin friction and the heat transfer. To correlate the pressure changes, the hypersonic viscous interaction parameter is:

$$\chi = \frac{M_\infty^3 \sqrt{C_\infty}}{\sqrt{Re_{\infty,x}}} \tag{3-2}$$

where

$$C_\infty = \left(\frac{\mu_w}{\mu_\infty}\right)\left(\frac{T_\infty}{T_w}\right)$$

To correlate the viscous/inviscid-induced perturbations in the skin friction or the heat transfer, Koppenwallner recommends:

$$V = \frac{M_\infty \sqrt{C_\infty}}{\sqrt{Re_{\infty,x}}} \tag{3-3}$$

as the viscous interaction parameter.

Pressure data obtained by Talbot et al.[15] on slender, sharp cones (having semi-vertex angles of 3 deg. and 5 deg.) are reproduced in Fig. 3.4. The induced pressure increase $(p_2 - p_c)$ divided by the pressure for inviscid flow past a sharp cone (p_c) is correlated in terms of the viscous-interaction parameter:

$$\chi_c = \frac{M_c^3 \sqrt{C_c}}{\sqrt{Re_{c,x}}} \tag{3-4}$$

where the subscript c denotes the inviscid flow properties at the surface of a sharp cone. Note the use of the unperturbed, sharp-cone values of the edge properties in the correlation for the induced pressure, e.g., Eq. (3-4), rather than the use of the freestream values,e.g., Eq. (3-2), is consistent with the analysis of Hayes and Probstein.[16]

The pre-flight predictions of the aerodynamics for the Shuttle Orbiter relied on an extensive wind-tunnel data base. As noted by Romere and Young,[9] the traditional freestream Reynolds number was selected as the flowfield scaling parameter below Mach 15, while the viscous interaction parameter

$$V_\infty = \frac{M_\infty \sqrt{C_\infty}}{\sqrt{Re_{\infty,L}}} \tag{3-5}$$

was used at higher Mach numbers. In Eq. (3-5), L is the length of the Shuttle model. However, above Mach 10 the predictions of the trim characteristics based on the wind-tunnel data were significantly different from the flight results. Features of the flow model that account for the differences between the tunnel data and the flight data will be discussed at the end of this chapter. Viscous interactions are discussed further in Sec. 6.6.

Exercise 3.1:

A sharp cone with a semi-vertex angle of 5 deg. (i.e., $\theta_c = 5$ deg.) is placed in the 96-in. Hypersonic Shock Tunnel at Calspan. The freestream test conditions are:

$$M_\infty = M_1 = 16$$
$$p_\infty = p_1 = 14.21 \text{ N/m}^2$$
$$T_\infty = T_1 = 56.4 \text{ K} \quad (\text{Thus, } T_t = 2944 \text{ K})$$

Use the experimental data presented in Fig. 3.4 to obtain the viscous-induced pressure rise (p_2/p_c) at two points on the conical surface: (1) 30 cm from the apex and (2) 1.0 m from the apex. Assume that the air behaves as a perfect gas and that $C_c = 1.0$.

Solution:

Use the charts in NACA Report 1135[17] to find the inviscid (reference) properties at the surface of the cone:

$$C_{p_c} = \frac{p_c - p_1}{q_1} = 0.017 \quad \text{and} \quad M_c = 12 \text{ (approx.)}$$

Thus,

$$\frac{p_c}{p_1} = 1 + \frac{\gamma}{2} M_1^2 C_{p_c} = 4.046 \; ; \quad p_c = 57.49 \text{ N/m}^2$$

For $M_c = 12$ and for adiabatic flow of perfect air (the stagnation temperature is unchanged across the shock wave):

$$\frac{T_c}{T_t} = 0.03356 \; ; \quad T_c = 98.80 \text{ K}$$

Thus,

$$a_c = 20.047\sqrt{T_c} = 199.26 \text{ m/s} \; ; \quad U_c = M_c a_c = 2391.12 \text{ m/s}$$

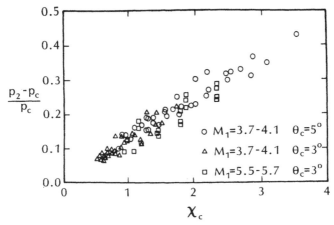

Fig. 3.4 Induced pressure increment as a function of the parameter χ_c, as taken from Ref. 15.

Using Sutherland's equation, Eq. (2-4a), to calculate the viscosity:

$$\mu_c = 1.458 \times 10^{-6} \frac{T_c^{1.5}}{T_c + 110.4} = 6.84 \times 10^{-6} \text{ kg/s·m}$$

and

$$\rho_c = \frac{p_c}{RT_c} = \frac{57.49 \frac{\text{N}}{\text{m}^2}}{\left(287.05 \frac{\text{N·m}}{\text{kg·K}}\right)(98.80 \text{ K})} = 0.00203 \text{ kg/m}^3$$

Thus, the unit Reynolds number at the surface of the cone for the unperturbed flow is:

$$Re_c = \frac{\rho_c U_c}{\mu_c} = 0.7092 \times 10^6 \text{ /m}$$

For $x_c = 30$ cm $= 0.30$ m,

$$Re_{c,x} = 212,760 \quad \text{and} \quad \chi_c = 3.75$$

From the data presented in Fig. 3.4,

$$\frac{p_2 - p_c}{p_c} = 0.45 ; \quad \frac{p_2}{p_c} = 1.45$$

or, in terms of the freestream static pressure,

$$\frac{p_2}{p_1} = \left(\frac{p_2}{p_c}\right)\left(\frac{p_c}{p_1}\right) = 5.867$$

For $x_c = 1.00$ m,

$$Re_{c,x} = 709,200 \quad \text{and} \quad \chi_c = 2.05$$

$$\frac{p_2 - p_c}{p_c} = 0.25 \; ; \quad \frac{p_2}{p_c} = 1.25$$

Again,

$$\frac{p_2}{p_1} = \left(\frac{p_2}{p_c}\right)\left(\frac{p_c}{p_1}\right) = 5.058$$

Note that the viscous-induced pressure perturbation is much greater at the first station (at $x_c = 30$ cm, $p_2 = 1.45p_c$) than at the second station (at $x_c = 1.0$ m, $p_2 = 1.25p_c$). Because a laminar boundary layer grows as $(x_c)^{0.5}$, while the reference shock wave, which is straight, grows as $(x_c)^{1.0}$, the boundary layer becomes a smaller fraction of the shock layer with distance from the apex. Thus, the viscous-induced pressure perturbations (for these high Mach number / low Reynolds number flows) are greatest nearest the apex of the cone.

3.3 GENERAL COMMENTS ABOUT CFD

Computational fluid dynamics provides a valuable tool for the designer of a hypersonic vehicle. Continuing and rapid developments in computer hardware and software insure an ever-increasing role for CFD in the design process. A large fraction of the readers of this text will participate in the development of their own organization's code. An equally large fraction of the readers will use a code developed by workers from other organizations. Often multiple organizations participate for several years in the development of a single code. Whether you are developing your own code or using a code developed elsewhere, it is important that you understand the grid scheme used to represent the body and the flowfield, the numerical algorithms used to obtain the flowfield solution, and the models used to represent fluid mechanic phenomena, thermochemical phenomena, and flow properties. The material presented in this chapter is intended to help the reader develop that understanding.

For the reader desiring more detailed information, there is a wealth of material in the open literature. Several textbooks describing the fundamentals of CFD have been written, e.g., Refs. 18 and 19. The American Institute of Aeronautics and Astronautics (AIAA) organizes CFD conferences at two-year intervals. Similar CFD conferences are periodically held at NASA centers. The AIAA and NASA publish proceedings from the conferences. Researchers from Europe and from the United States meet periodically to exchange information about modeling grid schemes and computing hypersonic flows. Collections of the papers describing the state-of-the-art of CFD technology in Europe and in the United States are presented in Refs. 20 and 21. Since CFD is a rapidly developing technology, the interested reader is encouraged to use the literature, of which Refs. 18 through 21 represent an infinitesimal fraction.

3.3.1 Introductory Comments

At this point, it is appropriate to reproduce the statement of Deiwert et al.[22]

> Computational fluid dynamics involves the numerical solution of the equations of motion which describe the conservation of mass, momentum, and energy. The most general forms of these equations are the compressible Navier-Stokes equations for continuum flight regimes and the Boltzmann equation for rarefied flight regimes. Many continuum flowfields have been well simulated for a variety of shapes and flow conditions where strong viscous/inviscid interactions and/or flow separation are important by advancing these equations in time until a steady state is asymptotically achieved. When there is no flow reversal and the flow in the streamwise direction is supersonic, these equations can be simplified by neglecting the streamwise viscous terms. The solution to these simplified equations, referred to as the parabolized Navier-Stokes equations, can be found by efficient streamwise marching techniques. Further simplification can be achieved, when viscous/inviscid interactions are weak, by decoupling the viscous and the inviscid dominated regions from one another and simulating the regions separately in an iterative manner. Here the inviscid Navier-Stokes equations, termed the Euler equations, are solved in the inviscid region away from body surfaces. Near the body surface, the viscous dominated boundary-layer equations are solved. A fourth simplification which can be used for strong viscous/inviscid interactions is the viscous shock layer approximation. This method is used for the stagnation region of hypersonic blunt bodies between the bow shock and the body surface.

> Real gas effects include thermochemical nonequilibrium, where finite rate processes for chemical and energy ex-

change phenomena occur, and radiative transport is a coupled process. To account for chemical reactions, conservation equations for each chemical species must be added to the flow-field equation set. There are five flow-field equations (one continuity, three momentum, and one energy equation). For dissociating and ionizing air, there are typically 11 species (N_2, O_2, N, O, NO, O^+, N^+, NO^+, N_2^+, O_2^+, and e^-). The inclusion of conservation equations for each of these species nearly triples the number of equations to be solved. When there are combustion processes or gas/surface interactions or ablation products, the number of species increases dramatically. To account for thermal nonequilibrium and radiative transport, there are additional energy equations to describe the energy exchange between the various energy modes (such as translational, rotational, vibrational, and electronic). To further complicate the analysis, the range of time scales involved in thermochemical process is many orders of magnitude wider than the mean flow time scale. This is the single most complicating factor in computational aerothermodynamics. Coupled radiative transport results in a system of integrodifferential equations which are exceedingly difficult to solve. Simplifying assumptions (such as either optical transparency, or gray gas, or tangent slab models) are generally used to reduce this level of complexity.

Numerical solutions of the flowfield around a complete vehicle model, the hypersonic research aircraft X24C-10D, in a Mach 5.95 stream have been generated using the Reynolds-averaged Navier-Stokes equations.[23] However, even with current capabilities in computer hardware and software, the development and the use of Navier-Stokes codes to generate flowfield solutions for complex three-dimensional reacting-gas flows are costly and time consuming. At the time of this writing, codes modeling the full Navier-Stokes equation find limited use in design studies. Such codes (possibly incorporating selected simplifications) can be used to provide benchmark solutions.

Monnoyer et al.[24] state, "The coupled solution of second-order boundary layer and the Euler equations provides an efficient tool for the calculation of hypersonic viscous flows. It is restricted, however, to flows with weak viscous interaction, i.e., no shock/boundary interaction nor streamwise separation can be considered."

One approach to reducing the computational efforts is the zonal method, i.e., one in which the flowfield is divided into zones according to the local flow characteristics. Thus, to calculate the flow over the nose region of the Shuttle Orbiter, Li[3] divided the flowfield into three

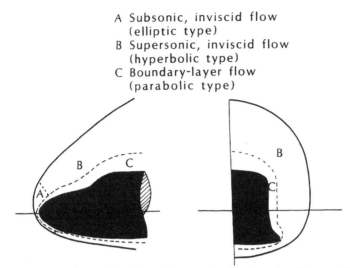

A Subsonic, inviscid flow
   (elliptic type)
B Supersonic, inviscid flow
   (hyperbolic type)
C Boundary-layer flow
   (parabolic type)

Fig. 3.5   A zonal method for flow calculation for Space Shuttle Orbiter, as taken from Ref. 3.

zones, as shown in Fig. 3.5, as follows:

A. A time-iterative Euler code is used to solve the elliptic equations characteristic of the subsonic, inviscid flow in the nose region.

B. For the inviscid, supersonic flow outside of the boundary layer, a space-stepping Euler code is used for the hyperbolic equations.

C. Because boundary-layer theory is based on the assumptions of negligible diffusion in the direction parallel to the surface and of very little convection in the direction perpendicular to the wall, the boundary-layer equations are of the parabolic type and space marching techniques can be applied.

This approach is reliable only if the boundary layer has little interaction with the inviscid flow. It breaks down, therefore, for the leeside of the Orbiter at high angles-of-attack.

The level of simplifications that must be introduced into the modeling in order to generate numerical solutions for the design process depends on the complexity of the application, on the time available, on the computer available, and on the accuracy required. There is no simple, general answer.

3.3.2 Grid Considerations

Every code that is used to compute the aerothermodynamic environment requires a subprogram to define the geometric characteristics

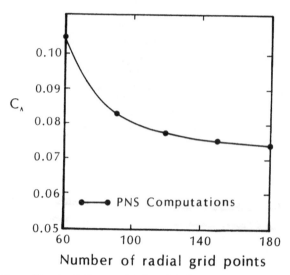

Fig. 3.6   The effect of the radial grid density on the axial force coefficient for a triconic in a Mach 20 stream of perfect air with a laminar boundary layer, as taken from Ref. 25.

of the vehicle and a grid scheme to identify points or volumes within the flowfield. Although there will be no further discussion of the techniques to describe the vehicle geometry, the reader should not minimize the importance of the geometry subprogram.

There are a number of techniques that have been developed for generating the computational grids that are required in the finite-difference, finite-volume, and finite-element solutions of the partial differential equations for arbitrary regions. A poorly chosen grid-scheme may cause the results to be erroneous or may fail to reveal critical aspects of the true solution. Neumann and Patterson[25] discussed Parabolized Navier-Stokes solutions of the flow over a blunted triconic (conical angles of 5 deg., 7.5 deg., and 10 deg.) at zero angle-of-attack in a Mach 20 stream of perfect air. The grid density in the radial direction was found to have a significant effect on the axial force coefficient, both for an entirely laminar boundary layer and for one in which transition occurred. The effect of radial grid density on the values of C_A for a laminar boundary layer are reproduced in Fig. 3.6.

Degani and Schiff[26] note, "The choice of a computational grid for any flowfield computation is dependent on the flow structures that must be resolved." A sketch of the flow structures in the cross-flow plane for a pointed body at a large angle-of-attack, as taken from Ref. 26, is reproduced in Fig. 3.7. The radial stretching was chosen to give a value of y^+ at the first point above the body surface equal to 5, since a close radial resolution is necessary to obtain accurate turbulent flow results. Degani and Schiff state that:

> The radial stretching was also adjusted to give adequate radial resolution of the leeward vortices. In addition, computational grids consisting of 36, 72, and 144 points equispaced around the body ($\Delta\phi$ = 10 deg., 5 deg., and 2.5 deg., respectively) were used for each cone flow case. It was observed that, for the cases considered, the coarsest circumferential resolution ($\Delta\phi$ = 10 deg.) was inadequate to properly resolve even the primary leeward side vortices. As a result the secondary separation points were not observed in the computed solutions, even though the modified turbulence model was employed. For those cases computed with $\Delta\phi$ = 5 deg. (72 circumferential points) and the modified turbulence model the primary vortices were resolved, and both the primary and secondary cross-flow separation points were observed. The locations of the primary separation points were found to be in good agreement with the experimental ones. However, the small secondary vortex structure was not adequately resolved. A further increase in circumferential resolution to $\Delta\phi$ = 2.5 deg. changed the surface-pressure distributions and location of the primary vortices and primary separation points only slightly, but resulted in increased resolution of the

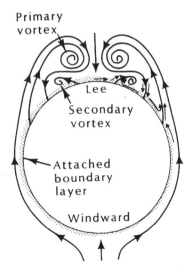

Fig. 3.7 Flow structure in the cross-flow plane, as taken from Ref. 26.

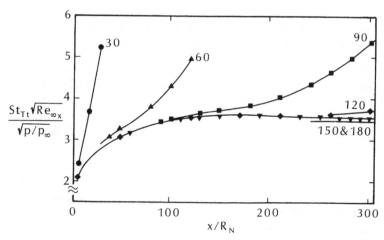

Fig. 3.8 The effect of grid density on the solution quality for
M_∞ **= 20 flow past a 5 deg. semi-vertex angle cone,** R_N **= 1.0 in.,**
T_w **= 2000 °R,** Re/**in. = 10^4, as taken from Ref. 27.**

secondary vortices and closer agreement of the computed
secondary separation points with experiment.

Grids may be structured or unstructured. There are no *a priori*
requirements on how grids are to be oriented. However, in some cases,
the manner in which the flow-modeling information is formulated
may influence the grid structure. For instance, since the turbulence
models are often formulated in terms of the distance normal to the
surface, the grid scheme utilized for these turbulent boundary layers
employ surface-oriented coordinates where one of the coordinate axes
is locally perpendicular to the body surface.

The computation of the heat transfer or of the skin friction requires detailed resolution of the flow very near the surface. The effect
of grid density on the computed heat-transfer rates is illustrated in
Fig. 3.8, which is taken from Ref. 27. Six solutions represent grid
densities from 30 to 180 points between the body surface and the
shock wave. As the computation proceeds downstream from the
nose and the distance between the body and the shock wave grows,
the physical distance between adjacent nodes will also grow. For this
particular application, the heat-transfer rates computed at a given x-
station approach a limiting value as the number of nodes is increased.
As noted by Neumann,[27] "There comes a time in the solution when
adding additional points has no effect on the solution quality. In
fact, solutions with 150 and 180 points clearly generate the same
heat transfer."

Siddiqui et al.[28] investigated the grid-dependency effects on
Navier-Stokes solvers for varying amounts of damping (or artificial

viscosity) and different differencing schemes. It was concluded, "The various algorithms investigated herein are found to be grid dependent, i.e., the resolution of the grid directly affects the quality of the solution. Improper mesh sizes can result in underprediction of the heating rates by orders of magnitude. The degree of accuracy depends on the dissipative nature of the algorithm. The more dissipative a scheme, the finer the grid resolution that will be required for an accurate estimation of the heating loads."

How does one approach questions relating to grid spacing? Lomax and Inouye[29] state, "It is not always wise to demand larger computing capacity merely to push ahead a few more steps before exponentially growing instabilities started by numerical truncations swamps the first few significant figures in the calculations. A more sensible approach is to face the problem with analysis and attempt to suppress nonessential instabilities by appropriate numerical methods." They state further, "The important distinction between an essential and a nonessential instability is that the latter would *not* occur in an exact analytical solution. What we refer to as nonessential instabilities are started by round-off, truncation or end-of-array inaccuracies due entirely to the fact that numerical methods are employed." Finally, "The complexity of the problem makes complete mathematical rigor in these studies practically impossible, and one is forced to rely on experience with linearized equations and familiarity with the physical problem for help in making the arguments plausible."

In the discussion of Fig. 3.7, it was noted that the first point above the body surface was at y^+ of 5. Accurate solutions of a turbulent boundary layer require resolution of the dependent variables down to the laminar sublayer. Thus, sophisticated grid strategies may be required. There are a variety of coordinate transformations that can be used. The specific transformation employed is usually tailored to the application of interest. The grid distribution normal to the body surface used by Waskiewicz and Lewis[30] is reproduced in Fig. 3.9. They note that a large number of points were placed near the wall to give more accurate predictions of the normal derivatives used in calculating surface quantities, such as skin friction and heat transfer. An exponential fit was used to give a nearly linear distribution of points at the shock which permitted more accurate predictions in the inviscid region near the shock wave.

Other grid-generation strategies exist, such as multigrid sequencing, which is used when solving elliptic problems by means of a time-iterative approach. Coarse grids might be used to obtain initial solutions, while a finer grid is used for final results. The procedures require interpolation of flow variables in evolving from the coarse grid to the finer grid.

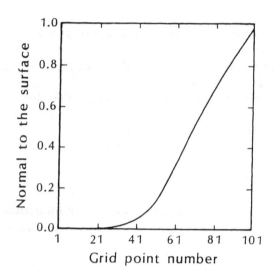

Fig. 3.9 Grid-point distribution normal to body surface, as taken from Ref. 30.

Another grid-generation strategy is the use of adaptive grid embedding for solving nonequilibrium hypersonic flow. The bow shock wave processes a substantial portion of the air within the shock layer of high Mach number, blunt-body flows. At high altitudes, nonequilibrium phenomena occur so that the characteristic chemical time scale (or, equivalently, the relaxation length) may differ by several orders of magnitude from the characteristic convection (fluid motion) time scale. Thus, accurate numerical simulations require the resolution of events occurring within extremely small distances. Typically, such resolution requirements lead to the use of very fine grids and lengthy computations. Aftosmis and Baron[31] state, "Adaptive grid embedding provides a promising alternative to more traditional clustering techniques. This method locally refines the computational mesh by sub-dividing existing computational cells based on information from developing solutions. By responding to the resolution demands of chemical relaxation, viscous transport, or other features, adaptation provides additional mesh refinement only where actually required by the developing solution."

3.4 COMPUTATIONS BASED ON A TWO-LAYER FLOW MODEL

Many of the computer codes that are used to define the aerothermodynamic environment divide the shock-layer flowfield (i.e., that portion of the flowfield between the curved, bow shock wave and the vehicle's surface) into two regions: (1) a rotational, inviscid flow

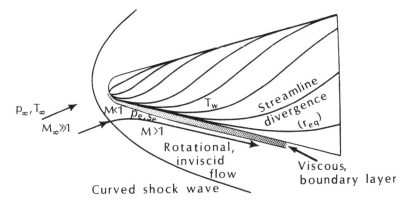

Fig. 3.10 Characterization of a flow around a re-entry vehicle at hypersonic speed.

where viscous effects are negligible; and (2) the thin, viscous, boundary layer adjacent to the surface.

A sketch illustrating the characteristics of the two-layer flow model is presented in Fig. 3.10. The pressure distribution and the properties at the edge of the boundary layer are determined from the solution of the inviscid flow. Assuming the boundary layer is thin, the pressure is constant across the boundary layer. As a result, the pressure at the surface is equal to the pressure from the inviscid solution at the edge of the boundary layer. The skin friction and the convective heat transfer are determined from the boundary-layer solution.

Chapter 6 will be devoted to a discussion of the pressure distributions for various hypersonic flows. Chapter 7 will be devoted to a discussion of the boundary-layer characteristics. Thus, in this chapter, pressures and boundary layers will be discussed only as necessary to illustrate the effects of various flow model assumptions on the computed flow.

The principal purpose of the codes described in this section is to provide tools which are efficient to use and which have suitable accuracy over a range of applications. Therefore, the code developers introduce approximate flow models that are appropriate for the application of interest and that simplify the problem to be solved. However, the approximate flow models employed in these computer codes reflect a wide range of rigor (or sophistication) in simulating the actual flow. Thus, the codes must be exercised to determine the range of conditions for which the flow models provide reasonable approximations of reality and for which the computed aerothermodynamic parameters are valid.

For purposes of the discussion in this section, two categories will be considered:

1. Those codes offering a menu of pressure options with relatively simple models for the boundary-layer parameters, termed (for convenience) *conceptual design codes,* and

2. Those codes in which simplifications to the governing equations are introduced so that the Euler equations are solved to define the inviscid flow and the boundary-layer equations are used to describe the inner region.

The term *two-layer CFD code* will be used for codes in the second category. Within each category, an individual code will include a number of approximations, representing a wide range of rigor in flow modeling.

3.4.1 Conceptual Design Codes

During the initial phases of a development program, the designer is faced with evaluating the performance of various configurations which might satisfy the mission requirements. Thus, the designer has need for computational tools that are capable of predicting the aerodynamic characteristics and the heat-transfer distributions, i.e., the aerothermodynamic environment, for a wide variety of configurations. The desired code should be economical and easy-to-use, employing engineering methods that represent realistic modeling of the actual flow about the candidate configurations over an entire trajectory. Since the candidate configurations may contain wings, fins, body flaps, and flat surfaces, the code should be able to model strong shock waves with associated viscous/inviscid interactions.

Because of the need for conceptual design codes, several organizations have developed such codes. One of the most widely used of these codes is the Supersonic/Hypersonic Arbitrary Body Program (S/HABP).[32] In the first step, the vehicle geometry is divided into flat elemental panels.

The most frequently used techniques for computing the inviscid pressure acting on a panel employ one of the simple impact or expansion methods. These methods require the impact angle (or the change in the angle of an element from a previous point) and, in some cases, the freestream Mach number. McCandless and Cruz[33] used such techniques to compute the pressures acting on an advanced aerospace plane (see Fig. 3.11). "Modified Newtonian theory was used for the nose region and the leading edge areas of the vehicle, and the tangent-cone method was applied to the top, wedge sides, and bottom ramp impact surfaces. For the shadow regions,

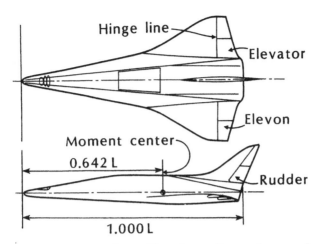

Fig. 3.11 Planform and profile view of an advanced aerospace plane, as taken from Ref. 33.

either Prandtl-Meyer or Newtonian theory was used, depending on the freestream Mach number." (Modified Newtonian theory will be discussed in Chap. 6.)

Using the S/HABP code to predict the forces and moments acting on a thin ellipse cross-section vehicle with monoplane fins (see Fig. 3.12) at a Mach number of 4.63 for angles-of-attack up to 14 deg., Fisher[34] notes, "Difficulty was encountered when trying to predict C_M accurately. Although some methods approximate the data trend, the magnitude is inaccurate. Perhaps the Modified Newtonian and Prandtl-Meyer method is a poor choice to apply on the body of this configuration, since this method should not be applied to a nose impact angle less than that for shock detachment."

The previous two paragraphs indicate that the same pressure model provides suitable results for one application but not for an-

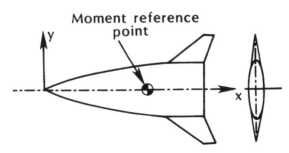

Fig. 3.12 Lifting body with an elliptic cross section, as taken from Ref. 34.

other. The reader should be aware that the choice of which of these simple impact methods provides the best estimates of the forces and moments depends on many parameters, including the configuration geometry, the Mach number, and the angle-of-incidence. Thus, although a conceptual design code may be highly flexible in that it can be applied both to simple shapes and to very complicated shapes, the prediction accuracy depends on the user's experience with the code.

Another approach for determining the local pressures includes the calculation of the interference effects from one component on another. This capability employs some of the simple impact methods but using the change in angle of an element from a previous point and the local flow conditions for that point determined from the flowfield subroutine.

The most challenging aspect of analyzing the flow over a complex shape is the calculation of the viscous flow due to difficulties in developing simple, yet realistic models for turbulence, viscous/inviscid interactions, etc. A detailed knowledge of the local flow properties along surface streamlines is required for a realistic boundary-layer solution. The developers of the S/HABP wanted an engineering approach for calculating the viscous forces that was simple yet retained the essential characteristics of the boundary-layer problem. The skin friction was calculated either using the relations for incompressible flow over a flat plate with correlation factors to account for compressibility and for the heating (or cooling) to the wall, or using an integral boundary-layer method.

The detailed distributions for the pressure and for the skin friction that are computed using the approximate methods of a conceptual code may differ from the actual distributions. Nevertheless, once the user has gained sufficient experience in which option(s) to select for a given application, suitable estimates of the overall force and moment coefficients can be obtained for a wide range of applications (e.g., Refs. 34 and 35).

Using the experience gained from S/HABP, an improved conceptual design code, the Missile Aerodynamic Design Method (MADM),[36] was developed. In addition to updating and to improving existing programs and techniques for the S/HABP, additional options were developed for MADM to calculate flowfield parameters, including an axisymmetric Euler solver and a two-dimensional Euler solver. Although solutions of the Euler equations provide a more rigorous definition of the inviscid flowfield, they require significantly more effort in code development and in computer resources.

3.4.2 Characteristics of Two-Layer CFD Models

The blunted leading edge of a vehicle designed for hypersonic flight in the earth's atmosphere produces a detached bow shock wave, as

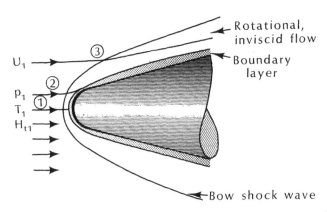

Fig. 3.13 Nomenclature for the two-layer flow model.

shown in Fig. 3.13. Since the entropy increase across the shock wave depends on the shock inclination angle, the entropy downstream of the shock wave is greatest for Streamline 1 and least for Streamline 3. Because all of the fluid particles originate in the freestream where the total enthalpy is uniform, the flow remains isoenergetic outside of the boundary layer. The assumption that the flow outside of the boundary layer is isoenergetic, i.e., the total enthalpy is constant, is reasonable, unless the temperatures in the inviscid shock-layer flow are sufficiently high so that gas radiates out through the shock layer, a phenomenon that occurs only at very high velocities, i.e., in excess of 12 km/s. Since the entropy is constant along a streamline for a steady, inviscid adiabatic flow, the entropy will vary continuously through the shock layer. A fluid particle having crossed the curved, bow shock wave at a particular location will retain the entropy and the stagnation pressure associated with the shock inclination at the point where it crossed the shock layer. The entropy (or the stagnation pressure) will be constant along a streamline until it enters the boundary layer. As the gas flows over the body, all of the high entropy gas initially processed by the bow shock wave, which is termed the entropy layer, is eventually entrained, or swallowed, by the boundary layer (e.g., Streamlines 1 and 2 in Fig. 3.13). Additional gas subsequently entrained by the boundary layer is a cooler, lower-entropy gas which has traversed only the weaker portions of the shock wave (e.g., Streamline 3). As will be pointed out in Sec. 3.4.2.2, one must include entropy-layer analysis to have accurate predictions of quantities such as the edge Mach number, the skin friction, etc.

At low Reynolds number / high Mach number conditions, the interaction between the rotational external flow and the boundary layer would invalidate the two-layer approach. However, at moderate to high Reynolds numbers, a coupled Euler/boundary-layer approach with features specific to hypersonic flows, e.g., gas chemistry, surface catalycity, and entropy swallowing, provides an efficient tool for the calculation of hypersonic viscous flows over a wide range of conditions. The two-layer CFD models are restricted to flows with weak viscous/inviscid interactions. Thus, unless special compensation is introduced to the flow model, flows with streamwise boundary-layer separation or with strong shock/boundary-layer interactions should not be considered.

The basic principle of a coupled inviscid region/boundary layer for the flowfield is that the flows are matched at their interface. Although there are a variety of ways to compute the two-layer flowfield, the classic approach is to solve the inviscid flowfield first for the actual configuration. Using the inviscid flowfield solution at the body surface provides the conditions at the edge of the boundary layer, e.g., p_e, s_e, and U_e in Fig. 3.10. Subject to the boundary conditions at the edge and at the wall, a boundary-layer solution can be obtained. The spreading of adjacent streamlines can be used to define a scale factor or "equivalent cross-section radius" (r_{eq}), which is analogous to the cross-section radius for an axisymmetric flow. See Fig. 3.10. The equivalent cross-section radius determined from the divergence of the streamlines allows a three-dimensional flow to be modeled by the axisymmetric analogue, or small cross-flow assumption.[37] As the boundary layer grows, more and more inviscid flow is entrained into the boundary layer. This has two effects on the flow modeling. The effect of the boundary layer on the inviscid flow is represented by displacing the wall by the displacement thickness, i.e., the distance the external streamlines are shifted to account for the mass-flow deficit due to the presence of the boundary layer. Boundary-layer growth also changes the conditions at the edge of the boundary layer. The inviscid flowfield might then be computed for the equivalent configuration, i.e., the actual configuration plus the displacement thickness.

In a close-coupling strategy, an interactive procedure would be set up, continuing until successive computations yielded a flow that is unchanged within some tolerance. However, Monnoyer et al.[24] found convergence problems with their close-coupling procedures. It was found that "This effect is avoided by using only one coupling iteration step (perturbation coupling) and the differences observed on the final results of each of these coupling methods are small."

3.4.2.1 Two-layer flowfields.

A variety of investigators have developed computer codes that can be used to generate solutions of steady-state flowfields over complex three-dimensional bodies. In the treatment of Marconi et al.,[38] all shock waves within the flowfield are followed and the Rankine-Hugoniot relations are satisfied across them. Since the conservation form of the Euler equations is used in the Weilmuenster

formulation,[39] which is given the acronym HALIS (High Alpha Inviscid Solution), shocks which lie within the computational domain are
captured. To simplify the geometry, the upper surface of the Shuttle
wing was filled in with an elliptic-curve segment which eliminated
the need to deal with the complex viscous-dominated flowfield on
the leeside of the wing. With this approximation, Weilmuenster was
able to compute the pressure field in the vicinity of the interaction
between the fuselage-generated shock wave and the wing-generated
shock wave. Since the code is limited to inviscid flows, it does not
provide information about the heat transfer to the surface. Nevertheless, the flow phenomena identified in the computed pressure fields
provide insights into the heating environment, including streaks of locally high heating. Weilmuenster[39] notes, "Namely, the streaks seem
to originate at a point near the wing/body shock intersection and
the severity of the heating decreases with increasing angle-of-attack
which corresponds to a decreased strength of the interior wing shock
as determined by the HALIS code."

At relatively high Reynolds numbers, the presence of the boundary layer has a second-order effect on the static pressures acting on
the windward surface. The coupled Euler/boundary-layer method of
Monnoyer et al.[24] has been used to calculate the flow over the hyperboloid shown in Fig. 3.14 in a Mach 10 stream of equilibrium air. The
outer grid line is fitted to the bow shock wave and the thick line in the

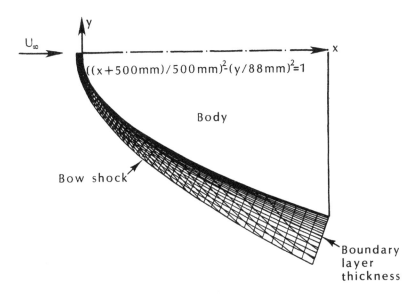

**Fig. 3.14 Shock-fitted Euler grid and axisymmetric hyper-
boloid, as taken from Ref. 24.**

computational domain is the boundary-layer thickness. Pressure distributions, as computed using an uncoupled model, a close-coupled model, and the Navier-Stokes model, are reproduced in Fig. 3.15. The uncoupled method underpredicts the pressure, as represented by the lower curve, because the displacement effect of the viscous, boundary layer is not simulated. Adding the displacement thickness to the actual body effectively makes the configuration blunter and, thus, increases the values of C_p. Note that the pressures computed by the close-coupled model agree very well with those computed using the Navier-Stokes formulation.

A finite-volume algorithm has been used[40] to compute the inviscid flowfield of the assured crew return vehicle (ACRV), as shown in Fig. 3.16. Since the computed aerodynamic coefficients were to be compared with data from the Langley Research Center's 31-Inch Mach 10 Wind Tunnel, perfect-gas relations ($\gamma = 1.4$) were used. As evident in Fig. 3.17, the computed values of the normal force coefficient are in reasonable agreement with the data. For the two Reynolds numbers for which data were obtained, viscous effects had little, if any, effect on the measured values of C_N. The values of C_A, computed by Weilmuenster et al.,[40] which are relatively constant over this angle-of-attack range, are reproduced in Fig. 3.18. As one would expect, the computed values of C_A based on the Euler equations are significantly below the experimental values. Note, however, that the values of C_A measured at the higher Reynolds number are closer to the computed inviscid values.

At those angles-of-attack where the flow on the leeside of a slender body separates and forms a vortex pattern, Euler-based flow models

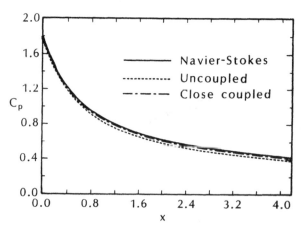

Fig. 3.15 Pressure distribution along the hyperboloid, $M_\infty = 10$, $T_\infty = 220$ K, $\rho_\infty = 7.7 \times 10^{-4}$ kg/m^3, $Re_{\infty R_N} = 2.47 \times 10^3$, adiabatic wall, perfect air, as taken from Ref. 24.

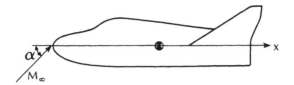

Fig. 3.16 Assured crew return vehicle (ACRV) of Ref. 40.

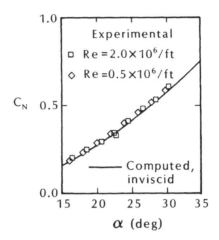

Fig. 3.17 Computed and measured normal-force coefficients, M_∞ = 10, and γ = 1.4, as taken from Ref. 40.

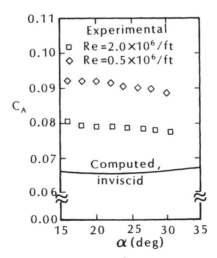

Fig. 3.18 Computed and measured axial force coefficients, M_∞ = 10, and γ = 1.4, as taken from Ref. 40.

usually fail to represent the actual flow. When such a configuration is treated with an inviscid code, a cross-flow shock develops and the surface pressure ahead of the shock wave becomes unrealistically low. A more realistic leeside flow pattern can be achieved with an Euler code by employing a semi-empirical model for defining the separation point on the body and the change in the inviscid flow properties at separation. This technique effectively changes the body surface to simulate flow separation.

3.4.2.2 Evaluating the flow properties at the edge of the boundary layer.
Vehicles designed for hypersonic flight through the atmosphere often use some degree of nose bluntness to reduce the convective heat transfer and to alleviate asymmetric vortex effects associated with the subsonic portion of the flight. As a result, the bow shock wave is curved. The entropy increase (or, equivalently, the total pressure decrease) is proportional to the local inclination of the shock wave and to the freestream Mach number. For perfect-gas flow, the entropy increase across the shock wave (where the local inclination is θ_{sw}) is given by:

$$(\gamma - 1)\frac{s_2 - s_1}{R} = \ell n \left[\frac{2\gamma M_1^2 \sin^2 \theta_{sw} - (\gamma - 1)}{\gamma + 1}\right]$$

$$-\gamma \ell n \left[\frac{(\gamma + 1)M_1^2 \sin^2 \theta_{sw}}{(\gamma - 1)M_1^2 \sin^2 \theta_{sw} + 2}\right] \qquad (3\text{-}6)$$

Thus, as noted when discussing Fig. 3.13, the streamline passing through the nearly normal portion of the bow shock wave (e.g., Streamline 1) experiences a larger entropy increase when crossing the shock than does the streamline passing through the more oblique portion of the shock wave (e.g., Streamline 3). Since the entropy is constant along a streamline for an inviscid, adiabatic and steady flow, the entropy will vary continuously through the shock layer, depending upon where the streamline crossed the shock wave. Fluid particles, having crossed the curved shock wave at a particular inclination, retain the entropy and the stagnation pressure associated with that shock inclination as they move through the inviscid shock layer.

Proceeding downstream from the stagnation point, the boundary layer grows into the rotational, inviscid flow. The flow entering the boundary layer is initially the hot, high-entropy gas that has been stagnated, or nearly so, after passing through the nearly normal portion of the bow shock wave. As the gas flows over the body, all of the high-entropy gas is eventually entrained, or swallowed, by the boundary layer. Additional gas subsequently entrained by the boundary

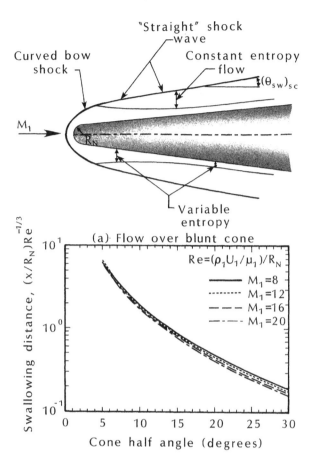

(a) Flow over blunt cone

(b) Correlation developed by Blottner, as taken from Ref. 43.

Fig. 3.19 Correlation for the entropy-swallowing length for perfect-air flow past a spherically blunted cone.

layer is a cooler, lower-entropy gas which has passed through the weaker portions of the shock wave. The effect of streamline swallowing on the local edge conditions has been long recognized, e.g., Refs. 41 and 42.

Perfect-air flowfields have been computed[43] for Mach 8 flow over a spherically blunted cone ($\theta_c = 5.25$ deg. and $R_N = 1.699$ in.) flying at an altitude of 125 kft. As shown in the sketch of Fig. 3.19a, the shock wave eventually becomes "straight" at its limiting value, i.e., θ_{sw} achieves the value corresponding to the value for a sharp cone, $(\theta_{sw})_{sc}$. The word *straight* is in quotes because weak waves crossing the shock layer produce second-order flowfield perturbations.

The entropy (or total pressure) is constant between the streamline crossing the bow shock wave at the tangency point and the shock wave. However, the entropy gradients in the inviscid shock-layer flow that are produced when the flow passes through the curved bow shock wave persist for a considerable length. A correlation developed by Blottner[43] (and reproduced in Fig. 3.19b) shows that it may take hundreds of nose radii before the variable-entropy flow associated with the curved bow shock wave is swallowed by the boundary layer.

The length affected by variable-entropy edge conditions is discussed in Exercise 3.2 and the impact of the rotational flows on the boundary-layer edge conditions is discussed in Exercise 3.3.

Exercise 3.2:

A spherically blunted cone ($\theta_c = 5.25$ deg. and $R_N = 1.699$ in) flies at Mach 8 at 125 kft. Using the correlation of Fig. 3.19b, determine the length required before the variable entropy flow is swallowed.

Solution:

Since the vehicle is flying at 125 kft, use Table 1.1b to determine ρ_1, μ_1, and a_1.

$$\rho_1 = \left(\frac{\rho_1}{\rho_{SL}}\right)\rho_{SL} = (4.3160 \times 10^{-3})(0.002377)$$

$$= 1.026 \times 10^{-5} \frac{\text{lbf s}^2}{\text{ft}^4}$$

$$\mu_1 = \left(\frac{\mu_1}{\mu_{SL}}\right)\mu_{SL} = (8.7947 \times 10^{-1})(3.740 \times 10^{-7})$$

$$= 3.29 \times 10^{-7} \frac{\text{lbf s}}{\text{ft}^2}$$

$$a_1 = 1029.66 \frac{\text{ft}}{\text{s}}$$

Thus,

$$U_1 = M_1 a_1 = 8237.28 \frac{\text{ft}}{\text{s}}$$

and

$$Re_{\infty,R_N} = \frac{\rho_1 U_1 R_N}{\mu_1}$$

$$= \frac{(1.026 \times 10^{-5})(8237.28)(0.1416)}{3.29 \times 10^{-7}} = 3.638 \times 10^4$$

Using Fig. 3.19b, it is seen that for $\theta_c = 5.25$ deg.,

$$\left(\frac{x}{R_N}\right)(Re_{\infty,R_N})^{-1/3} = 6$$

so that

$$\left(\frac{x}{R_N}\right) = 6(33.135) = 198.8$$

Thus, one must go almost 200 nose radii before the entropy layer is swallowed for this slender cone. The swallowing length decreases rapidly as the conical half-angle is increased.

As has been stated, the static pressure acting across the boundary layer is relatively insensitive to the flow model. However, as evident in Fig. 3.19, significant entropy gradients can persist for significant distances. Thus, when entropy layer swallowing is being considered, the location of the outer edge of the boundary layer must be defined. Because of the rotational character of the inviscid flow, the velocity and the temperature gradients are not zero at the outer edge of the boundary layer but have values associated with the rotational inviscid flow. The boundary-layer edge is often defined as the location where the total-enthalpy gradient goes to zero since the inviscid part of the shock layer is usually assumed to be adiabatic. This is the definition used by Adams et al.[44] However, as will be illustrated in Exercise 3.4, the total enthalpy within the boundary layer can exceed the inviscid value of the total enthalpy in certain cases. In these cases, there will be at least two locations where the total-enthalpy gradient goes to zero.

Exercise 3.3:

A sphere is to be placed in Tunnel B at AEDC where $M_1 = 8.0$, $p_{t1} = 850$ psia, and $T_{t1} = 1300\,°R$. For this wind-tunnel flow, the air will be assumed to behave as a perfect gas with $\gamma = 1.4$. Nevertheless, it will also be assumed that $\rho_2 \gg \rho_1$, so that the shock wave is parallel to the surface. In reality, for a normal shock wave in a Mach 8 stream of perfect air, $\rho_2 = 5.565\rho_1$. This inconsistency in the basic assumptions will not significantly detract from the objective of this problem, which is to illustrate the effect that the value of the entropy has on the local edge conditions.

Consider the two streamlines shown in the sketch of Fig. 3.20.

(1) For Streamline 1-3, the freestream air (designated by the subscript 1) passes through the normal portion of the shock wave

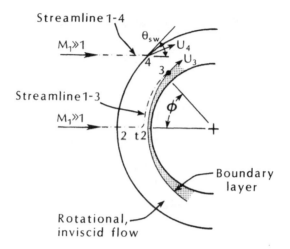

Fig. 3.20   Nomenclature for Exercise 3.3.

before coming to rest at $t2$, the stagnation point external to the boundary layer. The pressure increases suddenly as the air particles pass through the shock wave (from Condition 1 to Condition 2), then increases slowly as the air downstream of the shock wave approaches the stagnation point ($t2$). The air particles accelerate along this streamline as the pressure decreases from the stagnation pressure (p_{t2}) to the local pressure at Condition 3(p_3) as defined by the pressure distribution. The entropy increases suddenly as the air particles pass through the shock wave, then remains constant along the streamline downstream of the shock wave. Thus,

$$s_3 = s_{t2} = s_2$$

In essence, the air has passed through a normal-shock wave (NS), which fixes the entropy, and has undergone an isentropic expansion (IE) from p_{t2} to the local pressure p_3.

(2) For Streamline 1-4, the air particles pass from the freestream through the oblique portion of the shock wave. Thus, the conditions at Point 4 are those immediately downstream of a shock wave whose inclination angle, because it is parallel to the body surface, is $\theta_{sw} = 90$ deg. $- \phi$. This flow model will be termed *parallel shock entropy* (PSE).

    (a) Assume that the surface-pressure distribution is given by the modified Newtonian flow model. Calculate U_3, M_3, and the unit Reynolds number ($\rho_3 U_3/\mu_3$) for $\phi_3 = 45$ deg., i.e., where Point 3 is 45 deg. from the axis-of-symmetry (or the stagnation point).

(b) Assume that the oblique shock wave relations for perfect air can be used to calculate the conditions at Point 4. Calculate U_4, M_4, and the unit Reynolds number ($\rho_4 U_4 / \mu_4$) for $\phi_4 = 45$ deg.

Note that both Points 3 and 4 are located 45 deg. from the axis-of-symmetry. Although different models are used to calculate the static pressures at Points 3 and 4, the calculated static pressures are within 13 percent of each other. However, the entropy at Point 3 is much greater than that at Point 4. The difference in entropy has a dramatic effect on the local velocity, on the Mach number, and on the unit Reynolds number.

Solution:

(a) Using the tabulated values of Ref. 17 for $M_1 = 8.0$:

$$\frac{p_1}{p_{t1}} = 0.1024 \times 10^{-3} \; ; \; \frac{p_{t2}}{p_{t1}} = 0.8488 \times 10^{-2} \; ; \; \frac{T_1}{T_{t1}} = 0.7246 \times 10^{-1}$$

Thus, the freestream static properties are:

$$p_1 = \left(\frac{p_1}{p_{t1}}\right) p_{t1} = (0.1024 \times 10^{-3})850 = 0.08704 \text{ psia}$$

$$T_1 = \left(\frac{T_1}{T_{t1}}\right) T_{t1} = (0.7246 \times 10^{-1})1300 = 94.198 \text{ °R}$$

$$U_1 = M_1 a_1 = 8.0 \left[49.02\sqrt{94.198}\right] = 3806.13 \text{ ft/s}$$

The conditions at the stagnation point downstream of the normal portion of the shock wave are calculated as follows:

$$p_{t2} = \left(\frac{p_{t2}}{p_{t1}}\right) p_{t1} = (0.8488 \times 10^{-2})850 = 7.2148 \text{ psia}$$

Since Streamline 1-3 is entirely outside of the boundary layer, the total enthalpy remains constant along the streamline (in fact, through the entire region outside of the boundary layer). For a perfect gas, the total temperature also remains constant. Thus,

$$T_{t3} = T_{t2} = T_{t1} = T_t = 1300 \text{ °R}$$

Using the modified Newtonian flow relation to define the surface-pressure distribution and that along Streamline 1-3 between Points $t2$ and 3,

$$C_p = C_{p,t2} \cos^2 \phi$$

or,

$$\frac{p - p_1}{q_1} = \frac{p_{t2} - p_1}{q_1} \cos^2 \phi$$

Rearranging:

$$\frac{p}{p_{t2}} = \cos^2 \phi + \frac{p_\infty}{p_{t2}} \sin^2 \phi$$

Thus, for $\phi = 45$ deg.,

$$\frac{p}{p_{t2}} = \frac{p_e}{p_{t2}} = \frac{p_3}{p_{t2}} = 0.5060 \quad ; \quad p_3 = 3.6507 \text{ psia}$$

The use of the subscripts e and 3, indicate that the static pressure at Point 3 is the static pressure at the edge of the boundary layer. It is also the static pressure at the wall, since

$$\frac{\partial p}{\partial n} \approx 0$$

across the boundary layer. Since the flow along Streamline 1-3 has expanded isentropically from $t2$ to 3, and since

$$\frac{p_3}{p_{t2}} = 0.5060$$

therefore,

$$M_3 = 1.036$$

Further, once the Mach number is known,

$$\frac{T_3}{T_t} = 0.8231 \; ; \quad T_3 = 1070 \text{ °R}$$

$$a_3 = 49.02\sqrt{T_3} = 1603.51 \; \frac{\text{ft}}{\text{s}}$$

$$U_3 = M_3 a_3 = 1661.23 \; \frac{\text{ft}}{\text{s}}$$

$$\rho_3 = \frac{p_3}{RT_3} = \frac{\left(3.6507 \; \frac{\text{lbf}}{\text{in}^2}\right) \left(144 \; \frac{\text{in}^2}{\text{ft}^2}\right)}{\left(1716.16 \; \frac{\text{ft}^2}{\text{s}^2 \text{ °R}}\right) (1070 \text{ °R})}$$

$$\rho_3 = 2.863 \times 10^{-4} \; \frac{\text{lbf s}^2}{\text{ft}^2} \quad \left(\text{or} \quad \frac{\text{slugs}}{\text{ft}^3}\right)$$

Using Sutherland's equation, Eq. (2-4b), for the viscosity:

$$\mu_3 = 2.27 \times 10^{-8} \, \frac{T_3^{1.5}}{T_3 + 198.6} = 6.3 \times 10^{-7} \, \frac{\text{lbf s}}{\text{ft}^2}$$

Thus, the unit Reynolds number for Point 3 is:

$$Re_3/\text{ft} = \frac{\rho_3 U_3}{\mu_3} = \frac{\left(2.863 \times 10^{-4} \, \frac{\text{lbf s}^2}{\text{ft}^4}\right)\left(1661.23 \, \frac{\text{ft}}{\text{s}}\right)}{\left(6.3 \times 10^{-7} \, \frac{\text{lbf s}}{\text{ft}^2}\right)}$$

$$Re_3/\text{ft} = 7.549 \times 10^5 \, /\text{ft}$$

(b) The conditions upstream of the oblique shock wave are the freestream conditions. They were calculated for part (a):

$$p_1 = 0.08704 \text{ psia} ; \quad T_1 = 94.198 \, °\text{R}$$

The property changes across an oblique shock wave in perfect air are given in Ref. 17. However, to be consistent with the nomenclature of Fig. 3.20, the subscript 4 will be used to designate conditions immediately downstream of the oblique shock wave.

$$\frac{p_4}{p_1} = \frac{7M_1^2 \sin^2 \theta_{sw} - 1}{6} = 37.1667 ; \quad p_4 = 3.2350 \text{ psia}$$

$$\frac{U_4^2}{U_1^2} = 1 - \frac{5}{36} \frac{(M_1^2 \sin^2 \theta_{sw} - 1)(7M_1^2 \sin^2 \theta_{sw} + 5)}{M_1^4 \sin^2 \theta_{sw}}$$

$$U_4 = 0.7201 U_1 = 2740.86 \, \frac{\text{ft}}{\text{s}}$$

$$\frac{T_4}{T_1} = \frac{(7M_1^2 \sin^2 \theta_{sw} - 1)(M_1^2 \sin^2 \theta_{sw} + 5)}{36 M_1^2 \sin^2 \theta_{sw}} = 7.1623$$

Thus,

$$T_4 = 674.677 \, °\text{R}$$

$$M_4 = = \frac{U_4}{a_4} = \frac{2740.86}{49.02\sqrt{674.677}} = 2.153$$

The local density and the local viscosity are needed to calculate the unit Reynolds number,

$$\rho_4 = \frac{p_4}{RT_4} = \frac{\left(3.2350 \ \frac{\text{lbf}}{\text{in}^2}\right)\left(144 \ \frac{\text{in}^2}{\text{ft}^2}\right)}{\left(1716.16 \ \frac{\text{ft}^2}{\text{s}^2 \ ^\circ\text{R}}\right)(674.677 \ ^\circ\text{R})}$$

$$\rho_4 = 4.023 \times 10^{-4} \ \frac{\text{lbf s}^2}{\text{ft}^4}$$

Using Sutherland's equation for the viscosity,

$$\mu_4 = 2.27 \times 10^{-8} \ \frac{T_4^{1.5}}{T_4 + 198.6} = 4.6 \times 10^{-7} \ \frac{\text{lbf s}}{\text{ft}^2}$$

Thus,

$$\cdot Re_4/\text{ft} = \frac{\rho_4 U_4}{\mu_4} = 2.397 \times 10^6 \ /\text{ft}$$

Comparing the results from part (a), the NS/IE model for Streamline 1-3, and part (b), the PSE model for Streamline 1-4:

	NS/IE	PSE
Static pressure	3.6507 lbf/in^2	3.2350 lbf/in^2
Velocity	1661.23 ft/s	2740.86 ft/s
Mach number	1.036	2.153
Unit Reynolds number	0.755 $\times$ 10^6/ft	2.397 $\times$ 10^6/ft

The entropy of the streamline that has passed through the normal portion of the bow shock wave, i.e., Streamline 1-3, is much greater than the entropy of the streamline that passed through the shock wave where the inclination is 45 deg., i.e., Streamline 1-4. The value of the entropy has a significant effect on the flow properties. The properties, in turn, would affect skin-friction calculations, heat-transfer calculations, and estimates of where boundary-layer transition would occur. Thus, for a two-layer flow model, it is important to know where the streamline at the edge of the boundary layer crossed the bow shock wave and the strength of the shock wave at this point, since that will define the edge entropy.

Mayne and Adams[45] have evaluated the effects of streamline swallowing for perfect-air flow over a 22.5 deg. asymptotic half-angle hyperboloid in a Mach 10 flow over a large range of Reynolds number. Two flow models were used. For the simpler model, the pressure distribution was determined using an inviscid method. The local flow conditions are determined by isentropically expanding the flow from the stagnation conditions downstream of the normal shock wave to the known local pressure. This is essentially the normal-shock wave (NS)/isentropic expansion (IE) model of Exercise 3.3. Mayne and Adams[45] note that:

Assuming the pressure along the outer edge of the boundary layer on a blunt body to have the inviscid surface value, the determination of the local edge flow conditions may be improved by taking into consideration the inclination of the bow shock where the flow crossed the shock. The point at which the flow along the edge of the boundary layer crossed the shock can be determined by matching the mass flow in the boundary layer at a given location to the freestream mass flow in a cylinder with radius extending out to the location to be determined. Referring to Fig. 3.21, this may be expressed as

$$\rho_\infty U_\infty \pi y_s^2 = \int_0^{y_e} 2\pi \, r \, \rho u \, dy$$

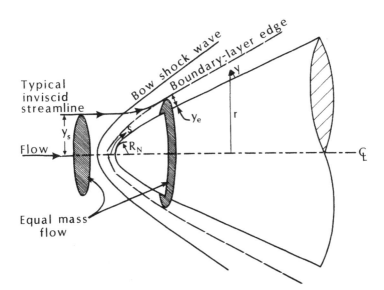

Fig. 3.21 Geometry of 22.5 deg. asymptotic half-angle hyperboloid, showing inviscid streamline swallowing.

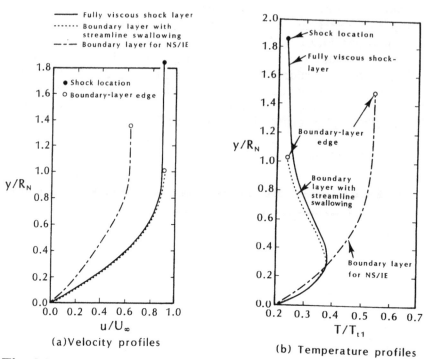

Fig. 3.22 Velocity and temperature profiles at $s = 20R_N$ for a 22.5 deg. hyperboloid in a Mach 10 stream of perfect air, $T_w = 0.2\,T_{t1}$, as taken from Ref. 45.

After y_s is found, the shock inclination at that point can be determined and the flow conditions along the boundary layer at the corresponding body location can be computed by crossing the oblique shock at y_s with the freestream flow and allowing the flow to expand isentropically to the known local boundary-layer edge pressure. Naturally, a swallowing analysis such as described above requires that the shape of the bow shock be known in addition to the body surface pressure.

These parameters were obtained from the blunt-body and the method-of-characteristic solutions of Inouye et al.[46]

Velocity and temperature profiles for the lowest Reynolds number considered,[45] i.e., $Re_{\infty R_N} = 400$, are reproduced in Fig. 3.22. The wall-to-total-temperature ratio for the computations was 0.2. Note that the edge velocity is much higher and the edge temperature is much lower when one uses the mass-balancing technique to locate where the edge streamline crossed the bow shock wave. The results computed using the mass-balancing flow model are designated *bound-ary layer with streamline swallowing* in Fig. 3.22. The relatively low edge velocity and high edge temperature associated with the high-entropy streamline of the NS/IE model is consistent with the results of Exercise 3.3. Included for comparison are the results for a fully viscous shock layer, as computed using the technique described by Davis.[47] Because the approach of Davis considers the entire shock layer to be viscous with the shock stand-off distance determined as part of the solution, there are none of the matching problems associated with boundary-layer analysis.

Even though the edge temperature is much lower, the temperature profile computed properly accounting for the entropy swallowing has a relative maximum (probably due to viscous dissipation associated with the higher velocities). See Fig. 3.22b. As a result, there is a large temperature gradient at the wall for the entropy-swallowing solution. Therefore, the solution for the boundary layer without streamline swallowing underpredicts the heat transfer. With respect both to surface quantities, i.e., heat flux and skin friction, and to profiles, the streamline-swallowing boundary-layer analysis agrees well with the results of the fully viscous shock-layer analysis.

Goodrich et al.[42] developed two-layer flowfield solutions for the windward surface of the Shuttle Orbiter at four flight conditions, ranging from the time at which peak heating occurs to the time corresponding to the beginning of boundary-layer transition. They used a two-layer model to provide "a relatively economical description of the complete Orbiter flowfield." The general objective of their study "was to examine the sensitivity of the Orbiter heating rates and boundary-layer edge conditions to considerations of: (1) the inviscid flowfield 'entropy layer'; (2) equilibrium air versus chemically and vibrationally frozen flow; and (3) nonsimilar terms in the boundary-layer computations." Nonsimilar terms will be discussed in Chap. 7. In their two-layer approach, Goodrich et al. used local inviscid flow conditions to establish surface streamlines and the local edge conditions needed for the boundary-layer computations. The inviscid solution patches one technique for the nose region and another for the downstream, supersonic region of the windward surface of the Shuttle.

To assess the effects of finite-rate chemistry, flowfield solutions were presented by Goodrich et al.[42] for equilibrium air and for chemically and vibrationally frozen air (i.e., a perfect gas with $\gamma = 1.4$). The local entropy at the edge of the boundary layer was assumed to be either the normal shock value (s_{t2}) or the local inviscid entropy (s_δ) at the edge conditions in the inviscid flow as established through the growth of the boundary layer into the inviscid flow. Note that this technique differs from the mass-balancing procedure (such as

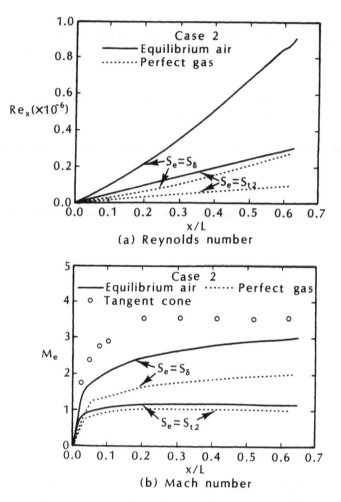

Fig. 3.23   Effects of gas chemistry and entropy-layer swallowing on centerline boundary-layer edge conditions for Shuttle Orbiter, Case 2, as taken from Ref. 42.

that used by Mayne and Adams[45]). The Reynolds number and the edge Mach number distributions for the windward pitch plane of the Shuttle Orbiter are presented in Fig. 3.23 for $\alpha = 40.2$ deg., altitude $= 226$ kft, and velocity $= 21.7$ kft/s (termed Case 2). These results indicate a strong influence of the entropy on the Reynolds number and on the local Mach number. Using the normal-shock/isentropic-expansion model (NS/IE), the flow is barely supersonic ($M_e \cong 1.2$) along most of the centerline, whereas the local edge Mach number approaches 3 at $x = 0.6L$, if one accounts for the entropy variations with the equilibrium-air assumption. Gas chemistry affects the edge conditions significantly, but less than entropy-layer swallowing. Further-more, Goodrich et al.[42] note that "Heating predictions downstream of $x = 0.04L$ were strongly dependent on entropy-layer swallowing. Gas chemistry had a relatively small influence on heating predictions with entropy layer swallowing and fully catalytic surface."

These improved techniques for computing the inviscid flowfield allow one to investigate streamlines away from the plane-of-symmetry. However, even these more rigorous two-layer approaches are not applicable to the flows with strong viscous/inviscid interactions.

3.5   COMPUTATIONAL TECHNIQUES THAT TREAT THE ENTIRE SHOCK LAYER IN A UNIFIED FASHION

The correlations presented thus far indicate that, for many flows, solutions of suitable accuracy can be obtained for the windward flow-field using two-layer models, e.g., dividing the flowfield into an inviscid rotational flow between the shock wave and the edge of the boundary layer and a thin, viscous boundary layer adjacent to the wall. This is especially true if the numerical model for inviscid flow incorporates the Euler equations. Furthermore, the flowfield can be computed in a relatively short time. However, solutions obtained using the two-layer approach may require considerable interaction by the user, since it may be necessary to process externally the output from the inviscid solution and prepare it as input for the viscous formulation. However, in regions where there are strong viscous/inviscid interactions, such as occur near the leeward plane-of-symmetry for vehicles at relatively high angles-of-attack, two-layer models may not be adequate. Because of these difficulties (and those identified in the previous discussion), it is desirable to develop solution techniques that treat the entire shock layer in a unified fashion.

A variety of assumptions and simplifications are introduced into the computational models that are used to treat the shock layer in a unified fashion. For instance, the "parabolized" Navier-Stokes (PNS) equations are obtained from the complete Navier-Stokes equations by neglecting the unsteady terms and the streamwise viscous derivative terms. (This assumption is usually valid for large Reynolds number flows over bodies which do not experience severe geometric variations in the streamwise direction). If it is assumed that the viscous, streamwise derivative terms are small compared with the viscous, normal and circumferential derivatives, a tremendous reduction in computing time and in storage requirements is possible over that required for the time-dependent approaches. Since the equations are parabolic in the streamwise direction, a marching-type numerical-solution technique can be used.

A major objective of the presentation in this section is to provide the reader with a brief introduction to some of the characteristics of unified models for the shock-layer flowfield. The reader who is not familiar with terms such as *shock fitting, explicit formulation,* etc., should refer to texts such as Refs. 18 and 19.

As noted by Prabhu and Tannehill,[48] "The PNS equations are hyperbolic-parabolic in the ξ direction if the inviscid region of the flowfield is completely supersonic, if there is no streamwise flow separation within the domain of interest, and if the streamwise pressure gradient, $dp/d\xi$, in the subsonic portion of the boundary layer is treated in such a manner as to suppress departure or exponentially growing solutions." As noted by Rakich et al.[49] the pressure gradient term in the streamwise momentum equation is a main factor that differentiates the parabolic approximation from boundary-layer approximations. If $dp/d\xi$ is retained, then a marching method of solution is not well posed. Thus, the numerical treatment of the streamwise pressure gradient becomes an important aspect of the computational flow model.

A PNS code requires a starting solution on an initial data surface where the inviscid flow is supersonic. The presence of an embedded subsonic region over the blunted nose requires the use of an axisymmetric or three-dimensional time-dependent Navier-Stokes code to provide a "starter solution," i.e., an initial "plane" of data. Mach number contours in the blunt-nose region of a Space Shuttle at 30 deg. angle-of-attack have been taken from Ref. 50 and are reproduced in Fig. 3.24. Note that the feasible region for PNS marching is downstream of the $M = 1.0$ contours, which is consistent with the requirement stated in the first sentence of this paragraph. The initial solution surface from which Rakich et al.[50] began marching the PNS solutions is further downstream. A second code, capable of generating solutions in the mixed subsonic/supersonic region, is needed to provide the initial solution surface.

At the outer boundary, the Rankine-Hugoniot relations are enforced to solve for the shock wave that separates the shock-layer flow from the undisturbed freestream. Since the flow properties at the node immediately downstream of the shock wave are often computed using an explicit formulation whereas the properties at the other internal points are computed using implicit techniques, there may be significant step-size considerations implicit in the numerical-solution procedure. In those cases where the shock wave boundary conditions for the system of equations are evaluated using an explicit formulation, stability restrictions may be introduced such that the unrestricted step size in the marching direction associated with the "unconditional stability of implicit techniques" is no longer valid. Such is the case in the PNS formulation, when the bow shock wave

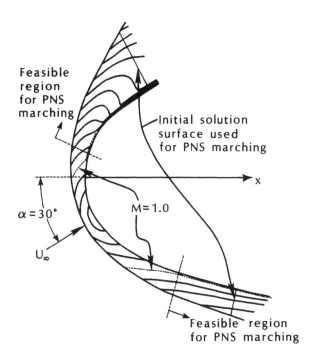

Fig. 3.24 Mach number contours in the blunt-nose region, as taken from Ref. 50.

is fitted using an explicit technique, e.g., Ref. 51. The stability condition takes the form of a modified CFL number given as

$$\frac{\lambda_{\max}\Delta x}{\Delta y} \leq 1$$

where Δx is the streamwise marching step and Δy corresponds to the mesh size in the radial direction. The eigenvalue ($\lambda_{\max}$) is determined from the matrix structure of the numerical algorithm as applied at the shock boundary. If the stability condition given above is violated when marching the PNS equations, oscillations in the shock boundary will occur, leading to inaccurate predictions of the shock wave shape and/or instabilities in the shock-fitting algorithm.

Using pressure distributions and oil-flow patterns, Stetson[52] identified the relevant characteristics of the flow over a blunted cone when $\alpha \geq \theta_c$. A sketch of the flow is presented in Fig. 3.25. The flow model contains symmetrical supersonic helical vortices with an attachment line on the most leeward ray. The vortices are in contact with the surface (at least up to $\alpha = 18$ deg. for $\theta_c = 5.6$ deg.), and there is no subsonic reverse flow or singular points associated with the vortex

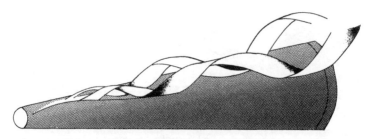

Fig. 3.25 Sketch of the hypersonic flow over a blunted cone when $\alpha \geq \theta_c$, as taken from Ref. 52.

pattern. The reattaching vortex produces a "feather" pattern in the oil flow near the leeward pitch plane and creates relatively high heating, as will be discussed subsequently. "Stream ribbons" are used to illustrate the vortex flow in Fig. 3.25. The separation pattern is a free-vortex type of separation (as opposed to the separation bubble) and is basically a cross-flow pattern.

Pitot-pressure surveys were made of the leeward region at $x = 0.87L$ for a sharp cone whose semi-vertex angle is 5.6 deg. at an angle-of-attack of 10 deg. in a Mach 14.2 stream. In order to make approximate Mach number calculations for the leeward flowfield, Stetson[52]

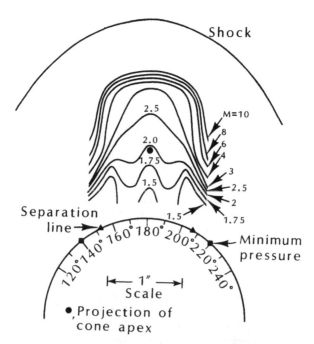

Fig. 3.26 Approximate Mach number contours at $x = 0.87L$ for sharp cone ($\theta_c = 5.6$ deg.), at $\alpha = 10$ deg., as taken from Ref. 52.

assumed that the static pressure in the region where the vortices lie is constant and equal to the surface value. The Mach number contours which were calculated using the ratio of p_{pit}/p_{static} are reproduced in Fig. 3.26. As indicated in the Mach number contours, the flow is supersonic throughout the region. Stetson[53] notes, "Even with an allowance for a reasonable error in static pressure, this general conclusion of supersonic flow is not altered." Stetson[52] concluded that the fact that the leeward flow was wholly supersonic could be the reason why there was no communication between the vortices of the leeward cone flowfield and the base region. The fact that the leeward flow remains supersonic (even though $\alpha > \theta_c$) allows us to use a variety of techniques to compute the flowfield, including the parabolized Navier-Stokes approach which includes both viscous and inviscid regions in a single formulation.

Lubard and Rakich[54] used the parabolized Navier-Stokes formulation to compute the hypersonic flow ($M_\infty = 10.6$) past a blunted cone ($\theta_c = 15$ deg. and $\alpha = 15$ deg.). The computed flowfields indicate that the circumferential separation zone first appears on the leeward side for $x = 8R_N$. The computed cross-plane velocity-vector distribution for $\beta \geq 150$ deg. (as taken from Ref. 54) is reproduced in Fig. 3.27. The recirculating flow associated with the leeward vortex which forms at a circumferential location of approximately 155 deg. can be clearly discerned. The corresponding pressure distributions (both computed and measured) are presented in Fig. 3.28. The increased pressure in the leeward plane-of-symmetry, i.e., $\beta = 180$ deg., is due to the viscous/inviscid interaction associated with the re-

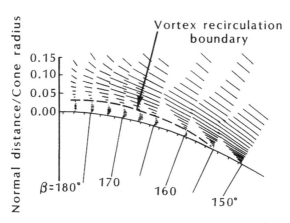

Fig. 3.27 Cross plane velocity vector distribution for a blunt cone (θ_c =15 deg., α = 15 deg.) at x =14.8R_N, as taken from Ref. 54.

attaching vortex pattern. The successful calculation of the pressure
distribution and of cross-flow separation occurs because the calculation is based on a single-layer system of three-dimensional parabolic
equations which are approximations to the full, steady Navier-Stokes
equations.

The importance of viscous/inviscid interactions is evident in the
flow patterns presented in Figs. 3.25 and 3.27 and in the pressure
distributions presented in Fig. 3.28. Photographs of the oil-flow patterns on models where a free-vortex layer separation occurs exhibit
a featherlike pattern as the vortices reattach in the leeward plane-
of-symmetry. Note that there are locally high heating rates near
the leeward plane-of-symmetry of a blunt cone at 15 deg. angle-of-
attack ($\alpha = \theta_c$). The heating rates computed using the parabolized
Navier-Stokes analysis of Lubard and Rakich,[54] which are presented
in Fig. 3.29, are in good agreement with the data of Cleary[55] for this
relatively simple configuration. Thus, by treating the entire flowfield as a single layer, "the effects of the viscous/inviscid interaction
and the entropy gradients due to both the curved bow shock and
the angle-of-attack effects are automatically included," as stated by
Lubard and Rakich.[54] Conversely, these flow phenomena will not be

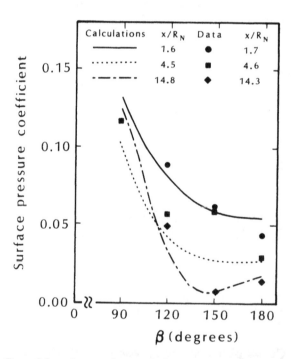

**Fig. 3.28 Leeside circumferential pressure distribution for a
blunt cone θ_c = 15 deg., α = 15 deg., M_∞ = 10.6, as taken from
Ref. 54.**

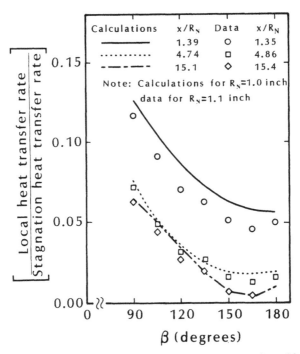

Fig. 3.29 Leeside circumferential heat-transfer distribution for a blunt cone θ_c = 15 deg., α = 15 deg., M_∞ = 10.6, compared with data from Ref. 55, as taken from Ref. 54.

modeled when the flowfield is divided into two regions: an inviscid region and the viscous boundary layer.

Exercise 3.4:

The increased drag and increased convective heating associated with a turbulent boundary layer have a significant impact on the design of slender vehicles like the NASP, the Saenger, or a hypersonic, airbreathing cruiser. Thus, it is important to reduce uncertainties in the location of boundary-layer transition for these vehicles. A program integrating CFD, ground-based testing, and flight testing would contribute to understanding the effect of various parameters on the boundary-layer transition process for a slender vehicle. Solutions of a slender (θ_c = 5.25 deg.), slightly blunted (R_N = 1.699 in.= $0.068 R_B$) vehicle in a Mach 8 stream at 125 kft have been computed using a PNS code for perfect air (Ref. 43). Partial output from an axial station, $x = 150 R_N$ from the apex, is presented in Table 3.1.

Table 3.1 PNS-computed values used as input for Exercise 3.4.

$\tilde{y}$	$\dfrac{p}{p_\infty}$	$\dfrac{\rho}{\rho_\infty}$	$\dfrac{u}{a_\infty}$	$\dfrac{v}{a_\infty}$
0.00000	1.95613	1.72597	0.00000	0.00000
0.00057	1.95619	1.19557	0.40561	0.03726
0.00132	1.95603	0.92523	0.83269	0.07643
0.00222	1.95600	0.76823	1.26713	0.11627
0.00492	1.95627	0.57597	2.36704	0.21696
0.00858	1.95627	0.49670	3.61540	0.33080
0.02023	1.95594	0.58371	6.50353	0.59338
0.03558	1.95579	1.08217	7.72745	0.70228
0.07843	1.95374	1.55812	7.89029	0.69384
0.12471	1.94833	1.58536	7.89986	0.67029
0.18512	1.93772	1.59238	7.90604	0.64097
0.28834	1.91172	1.58255	7.91241	0.59473
0.46279	1.85050	1.54918	7.92169	0.52370
1.00000	1.32541	1.22211	7.97095	0.20347

The y-axis is normal to the axis-of-symmetry of the cone, which is the x-axis. In the table, $\tilde{y}$ is a dimensionless y-coordinate:

$$\tilde{y} = \frac{y - y_w}{y_{sw} - y_w}$$

where y_w is the coordinate of the conical surface (or wall) and y_{sw} is that of the shock wave.

Since the transition criteria will most likely make use of conditions at the edge of the boundary layer, determine the static temperature (T_e/T_∞) and the Mach number (M_e) at the edge of the boundary layer.

Solution:

In Sec. 3.4.2.2, it was noted that the boundary-layer edge can be defined as the location where the total enthalpy gradient goes to zero. Based on this criterion, the edge of the boundary layer is the location where $H(y)$ first equals H_∞ within some increment, e.g., $0.99H_\infty$.

To convert the output presented in the table in the problem statement to the required parameters:

$$\frac{T}{T_\infty} = \frac{(p/p_\infty)}{(\rho/\rho_\infty)} \tag{3-7}$$

$$\frac{U}{U_\infty} = \sqrt{\left(\frac{u}{a_\infty}\right)^2 + \left(\frac{v}{a_\infty}\right)^2} \, \frac{a_\infty}{U_\infty} \tag{3-8}$$

$$M = \frac{U}{49.02\sqrt{T}} \tag{3-9}$$

$$H = c_p T + \frac{U^2}{2(778.2)(32.174)} \tag{3-10}$$

Using Table 1.1 for the freestream properties of air at 125 kft,

$$T_\infty = 441.17\,^\circ\text{R} \quad \text{and} \quad a_\infty = 1029.66 \text{ ft/s}$$

Thus,

$$U_\infty = M_\infty a_\infty = 8237.28 \text{ ft/s}$$

and

$$H_\infty = c_p T_\infty + 0.5 U_\infty^2$$

$$H_\infty = \left(0.2404 \, \frac{\text{Btu}}{\text{lbm}\,^\circ\text{R}}\right)(441.17\,^\circ\text{R}) + \frac{\left(8237.28 \, \frac{\text{ft}}{\text{s}}\right)^2}{2\left(778.2 \, \frac{\text{ft lbf}}{\text{Btu}}\right)\left(32.174 \, \frac{\text{ft lbm}}{\text{lbf s}^2}\right)}$$

$$H_\infty = 1461.064 \, \frac{\text{Btu}}{\text{lbm}}$$

Using Eqs. (3-7) through (3-10), the values shown in Table 3.2 were obtained.

These four parameters are presented as a function of $\tilde{y}$ in Fig. 3.30. Based on the criteria that the boundary-layer edge corresponds to $H = 0.99 H_\infty$, $\tilde{y}_e = 0.031$. At this point, $M = 5.4$ and $T = 2.12 T_\infty$. However, the static temperature continues to decrease with distance from the wall beyond this point. Using the static temperature profile, $\tilde{y}_e = 0.072$, where $M = 7.05$ and $T = 1.26 T_\infty$. This value for the Mach number compares more favorably with the sharp-cone value. Using the charts of Ref. 17, $M_c = 7.14$ for a sharp

Table 3.2 Derived values.

$\tilde{y}$	$\dfrac{T}{T_\infty}$	$\dfrac{U}{U_\infty}$	$\dfrac{H}{H_\infty}$	M
0.00000	1.1334	0.0000	0.08226	0.0000
0.00057	1.6362	0.0509	0.12117	0.3184
0.00132	2.1141	0.1045	0.16359	0.5751
0.00222	2.5461	0.1591	0.20828	0.7975
0.00492	3.3965	0.2971	0.32842	1.2898
0.00858	3.9385	0.4538	0.47689	1.8294
0.02023	3.3509	0.8163	0.86124	3.5677
0.03558	1.8073	0.9699	1.00363	5.7720
0.07843	1.2539	0.9901	1.00014	7.0738
0.12471	1.2290	0.9910	1.00005	7.1520
0.18512	1.2169	0.9915	1.00004	7.1908
0.28834	1.2080	0.9918	1.00003	7.2197
0.46279	1.1945	0.9924	1.00003	7.2642
1.00000	1.0845	0.9967	1.00001	7.6568

cone for which $\theta_c = 5.25$ deg. in a Mach 8 stream. However, recall from Exercise 3.2 that it was not until an x of 198.8 R_N was reached that the variable-entropy inviscid flow would have been swallowed for this condition. Since these profile calculations are for $x = 150$ R_N (which is the end of the vehicle), the rotational character of the inviscid flow (which still exists) would create temperature gradients in the flow external to the boundary layer, similar to those evident in Fig. 3.30. Thus, there is an uncertainty as to the exact location of the boundary-layer edge. Although the PNS technique employed herein produces a solution for the entire shock layer, this uncertainty may introduce problems, if one wishes to use transition correlations based on boundary-layer techniques.

Note that $H(\tilde{y})$ actually exceeds H_∞, reaching a maximum value of $1.00363 H_\infty$ for the points calculated in this example. A possible explanation for $H(\tilde{y})$ being greater than H_∞ is that heat transferred outward from the peak temperature (caused by viscous dissipation of the high-speed local flow) adds to the already high contribution due to kinetic energy. The point to be made is that, if correlations involving the edge properties are to be used and if one, therefore, needs to locate precisely the coordinate of the edge of the boundary

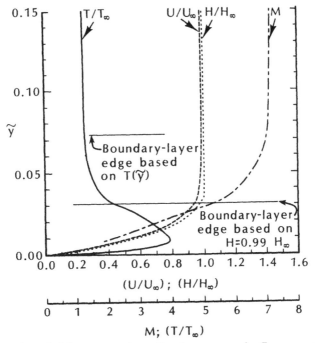

Fig. 3.30   Flowfield parameters for hypersonic flow past a slender cone of Exercise 3.4.

layer, care must be taken in establishing the criteria for defining the edge of the boundary layer.

3.6   CALIBRATION AND VALIDATION OF THE CFD CODES

Bradley[56] defines the concepts of code validation and code calibration as follows.

> CFD code validation implies detailed surface and flow-field comparisons with experimental data to verify the code's ability to accurately model the critical physics of the flow. Validation can occur only when the accuracy and limitations of the experimental data are known and thoroughly understood and when the accuracy and limitations of the code's numerical algorithms, grid-density effects, and physical basis are equally known and understood over a range of specified parameters.

> CFD code calibration implies the comparison of CFD code results with experimental data for realistic geometries that

are similar to the ones of design interest, made in order to provide a measure of the code's ability to predict specific parameters that are of importance to the design objectives without necessarily verifying that all the features of the flow are correctly modeled.

Because CFD codes are being applied to the design of complex configurations, carefully designed test programs with innovative instrumentation are required for use in code-calibration and in code-validation exercises.

Using the definitions presented in the quoted paragraphs, both for code validation and for code calibration, computed values of a specific parameter are compared with the corresponding measured values. Some code developers prefer to compare the results computed using the code under development with the results computed using an established, reference (benchmark) CFD code. Comparisons with parameters computed using an established code play a useful role during the development of a CFD code. Furthermore, in some instances, comparison of two sets of computations is the only possible way of "evaluating" the code under development, since the necessary data for validating the models are not available.

Huebner at al.[57] compared perfect-air flowfield solutions from a PNS code that used an implicit, upwind algorithm with data for a sharp 10 deg. half-angle cone in a Mach 7.95 stream.[58] Because of the completeness of these data and the simplicity of the configuration, the Tracy data have been used in numerous validation/calibration studies. In the numerical algorithm, the influence of the streamwise pressure gradient on the subsonic portion of the boundary layer was treated using the technique that was described earlier.[49] Surface pressures and heat-transfer rates for a laminar boundary layer ($Re_{\infty,L} = 4.2 \times 10^5$) for $\alpha = 4$ deg., 12 deg., and 20 deg. are reproduced in Figs. 3.31 and 3.32, respectively. Huebner et al.[57] noted that there was very good agreement between the computed and the measured pressures on the leeward side for all three angles-of-attack. The computational results underpredict the experimental pressures on the windward side. Huebner et al.[57] cite other studies that attribute this discrepancy to experimental errors due to the relatively large pressure-tap size as compared with the small boundary-layer thickness on the windward side. The heat-transfer results are presented as the ratio $h/h_{\alpha=0^\circ}$, which is the local heat-transfer coefficient divided by the heat-transfer coefficient for the same location with the model at zero angle-of-attack. The computational results agree with the measurements for all three angles-of-attack. The computational solutions model the increased heating near $\beta = 150$ deg. associated with the reattachment of the vortices in the leeward plane-of-symmetry for the higher angles-of-attack.

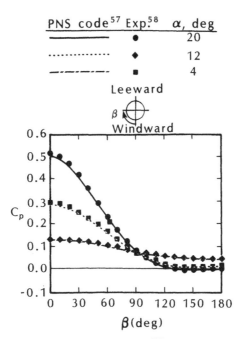

Fig. 3.31 Comparison of computed[57] and experimental[58] circumferential pressure distributions, M_∞ = 7.95, $Re_{\infty,L}$ = 4.2 × 10^5, as taken from Ref. 57.

Huebner et al.[57] compared the computed flowfield solutions with pitot-pressure measurements at a location where $Re_{\infty,L} = 3.6 \times 10^5$. For the comparisons, which are reproduced for $\alpha = 12$ deg. in Fig. 3.33, the experimental flowfield data are presented on the left-half plane and the computed total Mach number contours on the right-half plane. The pitot-pressure measurements were used to determine the shock location and to define the edge of the boundary-layer region. The location of the minimum pitot pressure defines the inner boundary of the shear layer. Inside the line of minimum pitot pressure is a region of small cross-flow recirculation. Note that there is a change in the cross-flow velocity direction near the body at a circumferential angle between 156 deg. and 160 deg., which corresponds to the minimum heat-transfer location on the body.

Deiwert et al.[22] and Marvin[59] discuss a calibration experiment that was intended to determine the applicability of the air chemistry model used in a PNS code. Drag data for a 10 deg. half-angle, sharp cone (the model length was 2.54 cm along the axis-of-symmetry) fired down a ballistic range are reproduced in Fig. 3.34. For these test conditions the flow is laminar, any viscous/inviscid interaction is small, and the temperature in the viscous layer is sufficiently high to

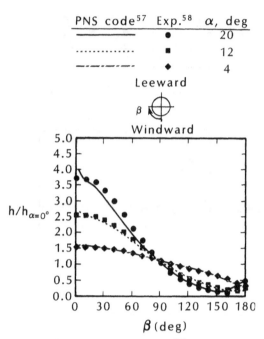

Fig. 3.32   Comparison of computed[57] and experimental[58] circumferential heat-transfer distributions, $M_\infty = 7.95$, $Re_{\infty,L} = 4.2 \times 10^5$, as taken from Ref. 57.

cause dissociation of the air. The angle-of-attack range represents the variation (uncertainty) in launch and flight path angle of the cones from various firings done nominally at zero angle-of-attack. The measurement accuracy for any ballistic range data is a function of the model size and shape, the clarity of the photographs, and the skill of the film reader. Included in the figure are computed values of the drag coefficient from (1) a PNS code including a nonequilibrium-air model, (2) a PNS code incorporating ideal-gas relations, and (3) the sharp-cone tables for perfect air. For these conditions, the pressure drag contributes about 40 percent of the total drag at zero angle-of-attack.

Deiwert et al.[22] note:

> During the course of comparing the computer solutions with experimental results, we made several observations relating to code calibration and validation. When comparing absolute values, such as drag coefficient, all sources of experimental, as well as computational, error must be evaluated. For example, since drag is sensitive to Reynolds number, the measurement accuracy of that number as well

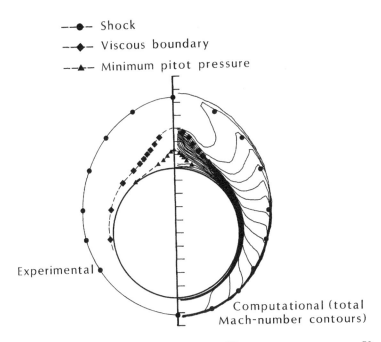

—•— Shock
—◆— Viscous boundary
—▲— Minimum pitot pressure

Experimental

Computational (total
Mach-number contours)

Fig. 3.33   Comparison of computed[57] and experimental[58] flow-field geometries for sharp cone (θ_c = 10 deg.), M_∞ = 7.95, $Re_{\infty,L}$ = 3.6×10⁵, α = 12 deg., as taken from Ref. 57.

as that of other input parameters should be evaluated. In one case a 10 percent change in calculated Reynolds number resulted in a 6.5 percent change in the drag coefficient. This is outside of the experimental error range and could make comparison between theory and experiment meaningless. Since all initial and boundary conditions used in a computation cannot or have not been measured, code sensitivity to these conditions should be explored.   All sources of error in the experimental data should be documented. The sensitivity of the computer code to grid size and shape, and to initial and boundary conditions, should be well documented. This is especially true when people other than the code developers are running the code.

3.7   DEFINING THE SHUTTLE PITCHING MOMENT — A HISTORICAL REVIEW

The aerodynamic characteristics of the Space Shuttle Orbiter were determined based on an extensive wind-tunnel test and analysis program. As noted by Romere and Whitnah,[60] "In general, wind

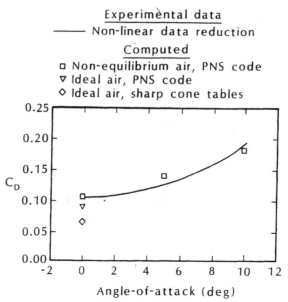

Experimental data
——— Non-linear data reduction
Computed
□ Non-equilibrium air, PNS code
▽ Ideal air, PNS code
◇ Ideal air, sharp cone tables

Fig. 3.34   A calibration experiment used to evaluate a real-gas chemistry model in a PNS code, sharp cone, θ_c = 10 deg., M_∞ = 15, $Re_{\infty,L}$ = 0.4 × 10^6, as taken from Refs. 22 and 59.

tunnel data cannot be used directly for prediction; the most valid set of wind tunnel results must be adjusted for unsimulated conditions. The major adjustments applied to the Space Shuttle wind tunnel data base involved corrections for nonsimulation of structural deformation, flowfield parameters, and the profile drag due to thermal protection system roughness and minor protuberances." The traditional freestream Reynolds number was selected for the flowfield scaling parameter below Mach 15, while a viscous interaction parameter [see Eq. (3-5) for the definition of V_∞] was utilized at higher Mach numbers.

Hoey[61] noted, "A significant discrepancy in the pitch trim predictions has been observed on all flights (Fig. 3.35). Elevon pulses, bodyflap sweeps, and pushover-pullup maneuvers have isolated the individual pitching moment contributions from the elevon, bodyflap, and angle of attack and determined that they are all close to predictions. The trim prediction error has thus been isolated to the basic pitching moment." Reviewing the data from the pullup/pushover maneuvers, from the body-flap pulse maneuvers, and from the elevon maneuvers, Romere and Whitnah[60] came to the same conclusion. "The ratio of the change in elevon deflection to the change in body flap deflection from the flight data is as was predicted. This lends additional strength to the inference that both body flap and elevon

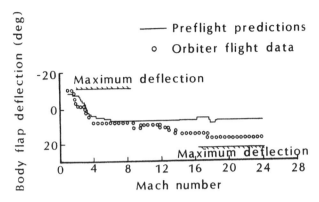

Fig. 3.35   A comparison of body flap defection as a function of Mach number, as taken from Ref. 61.

effectiveness were well predicted. Therefore, one would conclude that the most probable cause for the hypersonic trim discrepancy would be an error in the predicted basic pitching moment of the vehicle."

The fact that a body-flap deflection twice its nominal value (i.e., 15.0 deg. instead of 7.5 deg.) was required in order to trim the Orbiter during the STS 1 flight is equivalent to a change in pitching moment of 0.03. With the L/D ratio predicted very well over most of the angle-of-attack range, what would cause this large discrepancy between the pitching moment predicted prior to the first flight and that extracted from the flight data?

3.7.1 Model Proposed by Maus and His Co-workers

Maus and his co-workers conducted a study of the Shuttle Orbiter flowfield in order to determine why body-flap deflections twice those predicted prior to the first flight were required to maintain trim. The study focused on the effects of Mach number, of gas chemistry (specifically, equilibrium air or perfect air), and of the boundary layer on the Orbiter aerodynamics. The reference conditions were those of a Mach 8, perfect-air condition characteristic of the wind-tunnel flow in Tunnel B at AEDC.

Numerical codes for predicting steady supersonic/hypersonic inviscid flows were used to compute flowfields over a modified Orbiter geometry.[62,63] The major differences between the computational model geometry (see Fig. 3.36) and the actual Orbiter geometry include: (1) the wing-sweep back angle is increased from 45 deg. (the value for the actual Orbiter) to 55 deg.; (2) the wing thickness of the model is about twice that of the Orbiter; (3) the computational geometry is squared off at the body-flap hinge line; and (4) the rudder and OMS pods are not included in the model geometry.

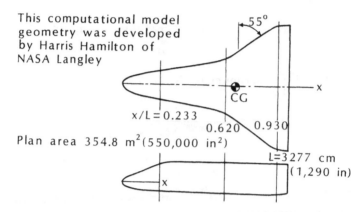

Fig. 3.36 Computational model geometry used in the flowfield computations of Griffith et al., as taken from Ref. 63.

Real-gas pressures (specifically those for an equilibrium-air computation) are slightly higher than the perfect-air values in the nose region and somewhat lower on the afterbody. As a result, the effect of the real-gas assumption is to produce a more positive (nose-up) pitching moment relative to the perfect-gas model. The equivalent effects of Mach number and of real gas on the center-of-pressure locations, as it moves forward from the reference conditions are reproduced in Fig. 3.37. Based on these computations, the center of pressure moves forward approximately 12 in. for an angle-of-attack of 35 deg. and approximately 20 in. for an angle-of-attack of 20 deg.

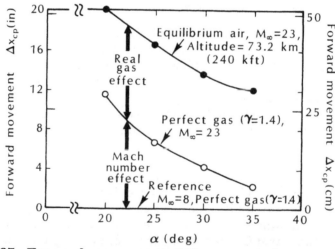

Fig. 3.37 Forward movement of the center-of-pressure location due to Mach number change (from 8 to 23) and due to gas model (perfect air to equilibrium air), as taken from Ref. 63.

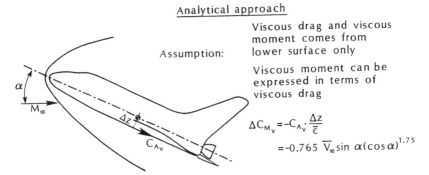

Analytical approach

Assumption:

Viscous drag and viscous moment comes from lower surface only

Viscous moment can be expressed in terms of viscous drag

$$\Delta C_{M_v} = -C_{A_v} \cdot \frac{\Delta z}{\bar{c}}$$

$$= -0.765 \, \overline{V}_\infty \sin \alpha (\cos \alpha)^{1.75}$$

Fig. 3.38 Viscous contribution to pitching moment, as taken from Ref. 63.

The analysis of the viscous effects on the Space Shuttle Orbiter aerodynamics was pursued using two complementary approaches: (1) fully viscous computations for a modified Orbiter geometry using a parabolized Navier-Stokes code and (2) the development of simple analytical expressions for the viscous contribution to C_A and to C_M from theoretical considerations and experimental data. As shown in Fig. 3.38, Griffith et al.[63] assume that the viscous drag acts only over the windward surface of the Orbiter, producing a negative contribution to the pitching moment.

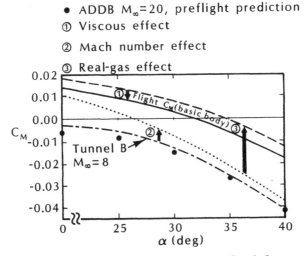

- ADDB $M_\infty = 20$, preflight prediction
① Viscous effect
② Mach number effect
③ Real-gas effect

Fig. 3.39 Buildup of flight C_M using methodology model for $M_\infty = 23$, at an altitude of 73.1 km (240kft), $\delta_{EV} = \delta_{BF} = 0$, $x_{cg} = 0.650L$, as taken from Ref. 63.

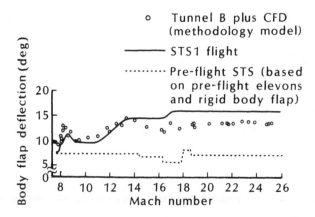

Fig. 3.40 The body-flap deflection for pre-flight predictions, for STS 1 flight data, and for buildup computations, as taken from Ref. 64.

An example of the buildup of C_M for the basic configuration is presented in Fig. 3.39, which is for $M_\infty = 23$ flight at 240,000 ft. The correlation illustrates the incremental changes in C_M due to Mach number, to equilibrium-air model, and to the boundary layer in going from the Mach 8 Tunnel B conditions to the flight conditions for $\delta_{EV} = \delta_{BF} = 0$. Note that the values of C_M based on the computed increments using the buildup technique are 0.02, and more, greater than the preflight predictions. Using the CFD-based corrections, such as those depicted in Figs. 3.38 and 3.39, the rigid-body-flap-deflection requirements as a function of Mach number are presented in Fig. 3.40. Updating the simulation model with flight data results in excellent agreement between the predicted body-flap deflections and the flight data.[64]

3.7.2 Model Proposed by Koppenwallner

Koppenwallner has observed that the pitching moment difference between the preflight predictions and the flight data for the Shuttle Orbiter could be explained by viscous-rarefaction effects. For the Shuttle Orbiter at Mach 12 at 53 km, the value of the viscous parameter V_∞ is 0.005. Koppenwallner notes that the severe discrepancies between preflight predictions and flight data occur in the hypersonic high-altitude region where $M_\infty \geq 12$ and $V_\infty \geq 0.005$.

Koppenwallner[65] notes that the viscous shear of the cross flow in the nose region cannot be neglected. "This viscous shear is large due to the small Reynolds number at the nose region and acts on a large lever arm. The nose region shear shall contribute a nose-up

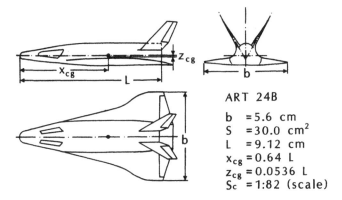

ART 24B

b = 5.6 cm
S = 30.0 cm^2
L = 9.12 cm
x_{cg} = 0.64 L
z_{cg} = 0.0536 L
S_c = 1:82 (scale)

Fig. 3.41 Wind-tunnel model of the re-entry configuration ART24B, as taken from Ref. 65.

pitching moment contrary to the lower surface shear. To neglect the nose region seems therefore questionable."

Aerodynamic data on the ART24B (Fig. 3.41) were obtained in the DFVLR hypersonic vacuum tunnel and in the Göttingen Ludwig Tube Tunnel and cover a range of V_∞ from 0.006 to 0.1. The viscous rarefaction effects strongly influence the longitudinal stability, as indicated in the data reproduced in Fig. 3.42. With V_∞ increasing from 0.006 to 0.1, the slope $dC_M/d\alpha$ decreases and a nose up change in the pitching moment $(+\Delta C_M)$ is observed. At α = 30 deg., the viscous-induced change in C_M amounts to ΔC_M = +0.03.

The very strong influence of the high-altitude viscous effects on the aerodynamic stability and control is shown in Fig. 3.43. Flap effectiveness is reduced by 50 percent, and a nose-up pitching moment change of 0.03 is induced at $(M_\infty/\sqrt{Re_{\infty,L}})$ = 0.1. Thus, Koppenwallner[65] concludes, "Wind tunnel experiments and qualitative arguments show that the aerodynamic performance of re-entry vehicles is strongly influenced by viscous low density effects. It is of importance to note that the nose up pitching moment change of the Shuttle at high altitudes can be explained by viscous rarefaction effects."

3.7.3 Final Comments About the Pitching Moment

The body-flap deflections required to trim the Space Shuttle Orbiter for the higher Mach numbers were twice those expected based on preflight predictions. One team of analysts, using CFD codes with varying degrees of rigor, concluded that the discrepancies could be attributed to the differences between the Mach number, the gas chemistry, and the character of the windward boundary layer of the

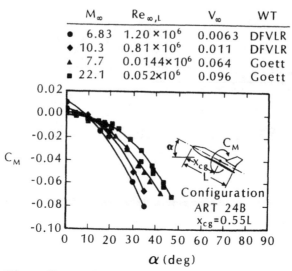

Fig. 3.42 The effect of high-altitude viscous effects on the pitching moment of the ART24B, as taken from Ref. 65.

flight environment relative to the reference conditions. Another analyst, using ground-based test data, obtained at relevant values of the viscous simulation parameter, attributed the discrepancies to a failure to model properly the low-density effects. Each approach (biased by the experiences of the analysts) was supported by analysis and by a data base.

The point of this discussion is not to determine which approach is correct. In fact, at the high Mach number/high-altitude flight conditions where the discrepancies were observed, elements of both

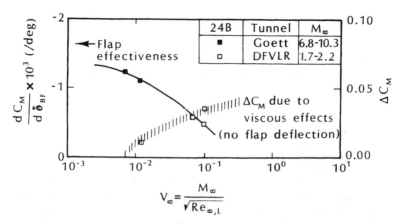

Fig. 3.43 Reduction of flap effectiveness and change of pitching moment ΔC_M due to viscous effects, as taken from Ref. 65.

analyses are present. The important point is that, when the available tools are used to define the hypersonic aerothermodynamic environment, we must carefully evaluate the approximations, the models, and the simulations inherent in the tools. As noted by Neumann,[27] "An accepted model may well explain trends for entirely wrong reasons and, used in a slightly different context, may be totally invalid."

3.8 CONCLUDING REMARKS

The designers of a hypersonic vehicle must make complementary use of the available tools, both experimental and analytical. Although neither ground-test facilities nor CFD provides the complete answer to a designer's needs, each offers certain advantages. As noted by Korkegi[66]:

> CFD can, in principle, give considerably greater detail of a flowfield than is possible in any wind tunnel as all aerodynamic parameters are computed at each grid point. CFD provides a capability for configuration optimization or for determining the effect of configuration changes before commitment to model construction is made. In this respect, CFD helps in making more effective use of ground-test facilities. On the other hand, when integrated forces and moments are desired, CFD is subject to the inherent mathematical inaccuracies associated with small differences of large numbers.

The designers should use the best possible mix of the available tools, with rigor consistent with the application.

REFERENCES

[1] Lee D. B., Bertin, J. J., and Goodrich, W. D., "Heat-Transfer Rate and Pressure Measurements Obtained During Apollo Orbital Entries," NASA TND-6028, Oct. 1970.

[2] Chapman, D. R., "Computational Aerodynamics Development and Outlook," AIAA Paper 79-0129, New Orleans, LA, Jan. 1979.

[3] Li, C. P., "Computation of Hypersonic Flow Fields," J. J. Bertin, R. Glowinski, and J. Periaux (eds.), *Hypersonics, Volume II: Computation and Measurement of Hypersonic Flows*, Birkhäuser Boston, Boston, MA, 1989.

[4] Neumann, R. D., "Missions and Requirements," AGARD Rept. 761 "Special Course on Aerothermodynamics of Hypersonic Vehicles," Neuilly sur Seine, France, 1988.

[5] Mehta, U. B., "Computational Requirements for Hypersonic Flight Performance Estimates," *Journal of Spacecraft and Rockets*, Vol. 27, No. 2, Mar.–Apr. 1990, pp. 103–112.

[6] Fay, J. A., and Riddell, F. R., "Theory of Stagnation Point Heat Transfer Dissociated Air," *Journal of the Aeronautical Sciences*, Vol. 25, No. 2, Feb. 1958, pp. 73–85, 121.

[7] Detra, R. W., Kemp, N. H., and Riddell, F. R., "Addendum to Heat Transfer to Satellite Vehicles Reentering the Atmosphere," *Jet Propulsion*, Vol. 27, No. 12, Dec. 1957, pp. 1256–1257.

[8] Weeks, T. M., private communication, July 1991.

[9] Romere, P. O., and Young, J. C., "Space Shuttle Entry Longitudinal Aerodynamic Comparisons of Flight 2 with Preflight Predictions," *Journal of Spacecraft and Rockets*, Vol. 20, No. 6, Nov.–Dec. 1983, pp. 518–523.

[10] Whitnah, A. M., and Hillje, E. R., "Space Shuttle Wind Tunnel Testing Program Summary," NASA Reference Publication 1125, 1984.

[11] Surber, T. E., and Olsen, D. C., "Space Shuttle Orbiter Aerodynamic Development," *Journal of Spacecraft and Rockets*, Vol. 15, No. 1, Jan.–Feb. 1978, pp. 40–47.

[12] Haney, J. W., "Orbiter Entry Heating Lessons Learned from Development Flight Test Program," *Shuttle Performance: Lessons Learned*, NASA CP-2283, Part 2, Mar. 1983.

[13] Eckert, E. R. G., "Engineering Relations for Friction and Heat Transfer to Surfaces in High Velocity Flow," *Journal of the Aeronautical Sciences*, Vol. 22, No. 8, Aug. 1955, pp. 585–587.

[14] Koppenwallner, G., "Hypersonic Aerothermodynamics," von Karman Institute for Fluid Dynamics, Lecture series 1984-01, Feb. 1984.

[15] Talbot, L., Koga, T., and Sherman, P. M., "Hypersonic Viscous Flow Over Slender Cones," *Journal of the Aerospace Sciences*, Vol. 26, No. 11, Nov. 1959, pp. 723–730.

[16] Hayes, W. D., and Probstein, R. F., *Hypersonic Flow Theory*, Academic Press, New York, 1959.

[17] Ames Research Staff, "Equations, Tables, and Charts for Compressible Flow," NACA Rept. 1135, 1953.

[18] Hirsch, C., *Numerical Computation of Internal and External Flows, Volume 2: Computational Methods for Inviscid and Viscous Flows*, John Wiley & sons, Chichester, England, 1990.

[19] Hoffmann, K. A., *Computational Fluid Dynamics for Engineers*, Engineering Education System, Austin, TX, 1989.

[20] Bertin, J. J., Glowinski, R., and Periaux, J. (eds.), *Hypersonics, Volume II: Computing Hypersonic Flows*, Birkhäuser Boston, Boston, MA, 1989.

²¹ Bertin, J. J., Periaux, J., and Ballmann, J. (eds.), *Advances in Hypersonics, Volume 3: Computing Hypersonic Flows*, Birkhäuser Boston, MA, Boston, 1992.

²² Deiwert, G. S., Strawa, A. W., Sharma, S. P., and Park, C., "Experimental Program for Real Gas Flow Code Validation at Ames Research Center," Paper No. 21, AGARD Symposium on Validation of Computational Fluid Dynamics, Lisbon, Portugal, May 1988.

²³ Shang, J. S., and Scherr, S. J., "Navier-Stokes Solution for a Complete Re-Entry Configuration," *Journal of Aircraft*, Vol. 23, No. 12, Dec. 1986, pp. 881–888.

²⁴ Monnoyer, F., Mundt, C., and Pfitzner, M., "Calculation of the Hypersonic Viscous Flow Past Reentry Vehicles with an Euler-Boundary Layer Coupling Method," AIAA Paper 90-0417, Reno, NV, Jan. 1990.

²⁵ Neumann, R. D., and Patterson, J. L., "Results of an Industry Representative Study of Code to Code Validation of Axisymmetric Configurations at Hypervelocity Flight Conditions," AIAA Paper 88-2691, San Antonio, TX, June 1988.

²⁶ Degani, D., and Schiff, L. B., "Computation of Supersonic Viscous Flows Around Pointed Bodies at Large Incidence," AIAA Paper 83-0222, Reno, NV, Jan. 1983.

²⁷ Neumann, R. D., "Defining the Aerothermodynamic Methodology," J. J. Bertin, R. Glowinski, and J. Periaux (eds.), *Hypersonics, Volume I: Defining the Hypersonic Environment*, Birkäuser Boston, Boston, MA, 1989.

²⁸ Siddiqui, M. S., Hoffmann, K. A., Chiang, S. T., and Rutledge, W. H., "A Comparative Study of the Navier-Stokes Solvers with Emphasis on the Heat Transfer Computations of High Speed Flows," AIAA Paper 92-0835, Reno, NV, Jan. 1992.

²⁹ Lomax, H., and Inouye, M., "Numerical Analysis of Flow Properties about Blunt Bodies Moving at Supersonic Speeds in an Equilibrium Gas," NASA TR R-204, July 1964.

³⁰ Waskiewicz, J. D., and Lewis, C. H., "Hypersonic Viscous Flows over Sphere-Cones at High Angles of Attack," AIAA Paper 78-64, Huntsville, AL, Jan. 1978.

³¹ Aftosmis, M. J., and Baron, J. R., "Adaptive Grid Embedding in Nonequilibrium Hypersonic Flow," AIAA Paper 89-1652, Buffalo, NY, June 1989.

³² Gentry, A. E., Smyth, D. N., and Oliver, W. R., "The Mark IV Supersonic-Hypersonic Arbitrary-Body Program," AFFDL-TR-73-159, Nov. 1973.

³³ McCandless, R. S., and Cruz, C. I., "Hypersonic Characteristics of an Advanced Aerospace Plane," AIAA Paper 85-0346, Reno, NV, Jan. 1985.

[34] Fisher, C. M. E., "Experiences Using the Mark IV Supersonic Hypersonic Arbitrary Body Program," Paper 31 in AGARD Conference Proceedings No. 428, *Aerodynamics of Hypersonic Lifting Vehicles,* Nov. 1987.

[35] Draper, A. C., and Buck, M. L., "Lifting Bodies—An Attractive Aerodynamic Configuration Choice for Hypervelocity Vehicles," Paper 30 in AGARD Conference Proceedings No. 428, *Aerodynamics of Hypersonic Lifting Vehicles,* Nov. 1987.

[36] Williams, J. E., Matthews, B. L., Adiasor, M. I., Casey, L. E., Deusinger, C. D., Donley, D. W., Goetz, P. R., McCotter, F., Moore, M. E., and Slutzky, S. L., "Missile Aerodynamic Design Method (MADM)," AFWAL TR-87-3108, Feb. 1988.

[37] Cooke, J. C., "An Axially Symmetric Analogue for General Three-Dimensional Boundary Layers," British Aeronautical Research Council, R&M No. 3200, 1961.

[38] Marconi, F., Salas, M., and Yeager, L., "Development of a Computer Code for Calculating the Steady Super/Hypersonic Flow Around Real Configurations, Volume I—Computational Technique," NASA CR-2675, Apr. 1976.

[39] Weilmuenster, K. J., "Comparison of Inviscid Flow Computations with Flight Data for the Shuttle Orbiter," *Journal of Spacecraft and Rockets,* Vol. 22, No. 3, May–June 1985, pp. 297–303.

[40] Weilmuenster, K., Smith, R., and Greene, F., "Assured Crew Return Vehicle Flowfield and Aerodynamic Characteristics," AIAA 90-0229, Reno, NV, Jan. 1990.

[41] Rotta, N. R., and Zakkay, V., "Effects of Nose Bluntness on the Boundary-Layer Characteristics of Conical Bodies at High Speeds," *Astronautica Acta,* Vol. 13, 1968, pp. 507–516.

[42] Goodrich, W. D., Li, C. P., Houston, C. K., Chiu, P. B., and Olmedo, L., "Numerical Computations of Orbiter Flowfields and Laminar Heating Rates," *Journal of Spacecraft and Rockets,* Vol. 14, No. 5, May 1977, pp. 257–264.

[43] Oberkampf, W. L., Wong, C. C., and Blottner, F. R., private transmittal, Feb. 1991.

[44] Adams, J. C., Jr., Martindale, W. R., Mayne, A. W., Jr., and Marchand, E. O., "Real-Gas Scale Effects on Shuttle Orbiter Laminar Boundary-Layer Parameters," *Journal of Spacecraft and Rockets,* Vol. 14, No. 5, May 1977, pp. 273–279.

[45] Mayne, A. W., Jr., and Adams, J. C., Jr., "Streamline Swallowing by Laminar Boundary Layers in Hypersonic Flow," AEDC-TR-71-32, Mar. 1971.

[46] Inouye, M., Rakich, J. V., and Lomax, H., "A Description of Numerical Methods and Computer Programs for Two-Dimensional and

Axisymmetric Flow Over Blunt-Nosed and Flares Bodies," NASA TN D-2970, Aug. 1965.

[47] Davis, R. T., "Numerical Solution of the Hypersonic Viscous Shock-Layer Equations," *AIAA Journal,* Vol. 8, No. 5, May 1970, pp. 843–851.

[48] Prabhu, D. K., and Tannehill, J. C., "Numerical Solutions of Shuttle Orbiter Flowfield Including Real-Gas Effects," *Journal of Spacecraft and Rockets,* Vol. 23, No. 3, May–June 1986, pp. 264–272.

[49] Rakich, J. V., Vigneron, Y. C., and Agarwal, R., "Computation of Supersonic Viscous Flows Over Ogive-Cylinders at Angle of Attack," AIAA Paper 79-0131, New Orleans, LA, Jan. 1979.

[50] Rakich, J. V., Venkatapathy, E., Tannehill, J. C., and Prabhu, D., "Numerical Solution of Space Shuttle Orbiter Flowfield," *Journal of Spacecraft and Rockets,* Vol. 21, No. 1, Jan.–Feb. 1984, pp. 9–15.

[51] Rizk, Y. M., Chaussee, D. M., and McRae, D. S., "Computation of Hypersonic Viscous Flow Around Three-Dimensional Bodies at High Angles of Attack," AIAA Paper 81-1261, Palo Alto, CA, June 1981.

[52] Stetson, K. F., "Experimental Results of Laminar Boundary Layer Separation on a Slender Cone at Angle of Attack at $M_\infty = 14.2$," Aerospace Research Laboratories, ARL 71-0127, Aug. 1971.

[53] Stetson, K. F., "Boundary-Layer Separation on Slender Cones at Angle of Attack," *AIAA Journal,* Vol. 10, No. 5, May 1972, pp. 642–648.

[54] Lubard, S. C., and Rakich, J. V., "Calculation of the Flow on a Blunt Cone at High Angle of Attack," AIAA Paper 75-149, Pasadena, CA, Jan. 1975.

[55] Cleary, J. W., "Effects of Angle of Attack and Bluntness on Laminar Heating-Rate Distributions of a 15° Cone at a Mach Number of 10.6," NASA TN D-5450, Oct. 1969.

[56] Bradley, R. G., "CFD Validation Philosophy," Paper No. 1, AGARD Symposium on Validation of Computational Fluid Dynamics, May 1988, Lisbon, Portugal.

[57] Huebner, L. D., Pittman, J. L., and Dilley, A. D., "Computational Validation of a Parabolized Navier-Stokes Solver on a Sharp Cone at Hypersonic Speeds," AIAA Paper 88-2566CP, Williamsburg, VA, June 1988.

[58] Tracy, R. R., "Hypersonic Flow Over a Yawed Circular Cone," Graduate Aeronautical Laboratories California Institute of Technology Memorandum No. 69, Aug. 1963.

[59] Marvin, J. G., "Accuracy Requirements and Benchmark Experiments for CFD Validation," Paper No. 2, AGARD Symposium on Validation of Computational Fluid Dynamics, May 1988, Lisbon, Portugal.

[60] Romere, P. O., and Whitnah, A. M., "Space Shuttle Entry Longitudinal Aerodynamic Comparisons of Flights 1–4 with Preflight Predictions," *Shuttle Performance: Lessons Learned*, NASA CP-2283, Part 1, Mar. 1983.

[61] Hoey, R. G., "AFFTC Overview of Orbiter-Reentry Flight-Test Results," *Shuttle Performance: Lessons Learned*, NASA CP-2283, Part 2, Mar. 1983.

[62] Maus, J. R., Griffith, B. J., Szema, K. Y., and Best, J. T., "Hypersonic Mach Number and Real Gas Effects on Space Shuttle Orbiter Aerodynamics," AIAA Paper 83-0343, Reno, NV, Jan. 1983.

[63] Griffith, B. J., Maus, J. R., and Best, J. T., "Examination of the Hypersonic Longitudinal Stability Problem – Lessons Learned," *Shuttle Performance: Lessons Learned*, NASA CP-2283, Part 1, Mar. 1983.

[64] Griffith, B. J., Maus, J. R., Majors, B. M., and Best, J. T., "Addressing the Hypersonic Simulation Problem," *Journal of Spacecraft and Rockets*, Vol. 24, No. 4, July–Aug. 1987, pp. 334–341.

[65] Koppenwallner, G., "Low Reynolds Number Influence on Aerodynamic Performance of Hypersonic Lifting Vehicles," Paper 11, AGARD Conference Proceedings No. 428, *Aerodynamics of Hypersonic Lifting Vehicles*, Nov. 1987.

[66] Korkegi, R. H., "Impact of Computational Fluid Dynamics on Development Test Facilities," *Journal of Aircraft*, Vol. 22, No. 3, Mar. 1985, pp. 182–187.

PROBLEMS

3.1 Pressure data are to be obtained on a sharp cone with a semi-vertex angle of 5 deg. to be tested in the low-density hypersonic wind tunnels at the DLR in Göttingen, Germany. The test gas is cold nitrogen, for which:

$\gamma = 1.4$, $R = 297$ N·m/kg·K, and

$$\mu = 1.39 \times 10^{-6} \frac{T^{1.5}}{T + 102} \text{ kg/s·m}$$

The test conditions include:

$U_\infty = 1340$ m/s
$\rho_\infty = 5.14 \times 10^{-5}$ kg/m^3
$T_\infty = 8.3$ K

The reference length (x_c) is 71.4 mm.

What is M_∞? V, as given by Eq. (3-3)?

Using the correlation presented in Fig. 3.4, determine the ratio p_2/p_1 (i.e., p_2/p_∞).

3.2 Repeat Problem 3.1, but assume the test conditions to be:

$$U_\infty = 1740 \text{ m/s}$$
$$\rho_\infty = 39.12 \times 10^{-5} \text{ kg/m}^3$$
$$T_\infty = 15.6 \text{ K}$$

3.3 A spherically blunted cone ($\theta_c = 5$ deg., $R_N = 5$ cm) is to be tested in a low-density hypersonic wind tunnel at the DLR in Göttingen, Germany. The test gas is cold nitrogen, for which:

$\gamma = 1.4$, $R = 297$ N·m/kg·K, and

$$\mu = 1.39 \times 10^{-6} \frac{T^{1.5}}{T + 102} \text{ kg/s·m}$$

The test conditions include:

$$U_\infty = 1340 \text{ m/s}$$
$$\rho_\infty = 5.14 \times 10^{-5} \text{ kg/m}^3$$
$$T_\infty = 8.3 \text{ K}$$

What is the swallowing distance (x/R_N)?

3.4 A sphere is to be tested in Tunnel B at AEDC where $M_1 = 8$, $p_{t1} = 850$ psia, and $T_{t1} = 1300$ °R. For this wind-tunnel flow, the air will be assumed to behave as a perfect gas with $\gamma = 1.4$. Consider the two streamlines shown in the sketch of Fig. 3.20.

(a) For Streamline 1-3, the freestream air passes through the normal portion of the shock wave, stagnating at $t2$, before accelerating isentropically to Point 3, for which $\phi_3 = 60$ deg. Assume that the surface-pressure distribution is given by the modified Newtonian flow model. Calculate U_3, M_3, and the unit Reynolds number ($\rho_3 U_3/\mu_3$) for $\phi_3 = 60$ deg.

(b) For Streamline 1-4, the air particles pass from the freestream through the oblique portion of the shock wave. Thus, the conditions at Point 4 are those immediately downstream of a shock wave whose inclination angle, because it is parallel to the body surface, is $\theta_{sw} = 90$ deg. $- \phi$. For $\phi_4 = 60$ deg., calculate U_4, M_4, and the unit Reynolds number ($\rho_4 U_4/\mu_4$).

4
EXPERIMENTAL MEASUREMENTS
OF HYPERSONIC FLOWS

4.1 INTRODUCTION

The vehicle design process integrates experimental data obtained from tests conducted in ground-based facilities with computed flowfield solutions and with data obtained from flight tests to define the aerothermodynamic environments which exist during the vehicle's mission. Only flight tests of the full-scale vehicle provide an uncompromising representation of the vehicle's environment. Of course, flight tests of the full-scale vehicle are very expensive and can occur only after an extensive design, development, and fabrication program has been completed. However, there have been numerous research-oriented flight-test programs employing elementary shapes, e.g., blunted cones, that were designed to obtain data of fundamental interest, e.g., boundary-layer transition measurements at real-gas conditions obtained during the Re-entry F Flight Test. Nevertheless, the majority of the experimental information about the flowfield is obtained in ground-based test facilities.

Since complete simulations of the flowfield cannot be obtained in a ground-based facility, the first and most important step in planning a ground-based test program is establishing the test objectives. As stated by Matthews et al.,[1] "A precisely defined test objective coupled with comprehensive pretest planning are essential for a successful test program."

There are many reasons for conducting ground-based test programs. They include the following:

1. Obtain data to define the aerodynamic forces and moments and/or heat-transfer distributions for complete configurations whose complex flowfields resist computational modeling.

2. Use partial configurations to obtain data defining local flow phenomena, such as the inlet flowfield for a hypersonic air-breathing engine or the shock/boundary-layer interaction using a fin or a wing mounted on a plane surface.

3. Obtain detailed flowfield data to be used in developing flow models for use in a computational algorithm (code validation).

4. Obtain measurements of parameters, such as the heat transfer and the drag, to be used in comparison with computed flowfield solutions over a range of configuration geometries and of flow conditions (code calibration).

5. Document aerodynamic effects of aerosurface settings, failures, etc.

6. Certify airbreathing engines.

Because reliance on CFD as a design tool continues to increase, an ever-increasing number of tests are conducted for reasons 3 and 4. Experimental programs are needed to validate the numerical models used to represent physical processes and flow chemistry and to calibrate the code to determine the range of conditions for which the values of the computed parameters are of suitable accuracy. As noted by Martellucci,[2] "The process of CFD code validation is one that must be ever ongoing as long as there are new systems and vehicle configurations, or new applications of current systems.... In approaching code validation, one must be critical of both the working tenets of the code and the quality, thoroughness, and applicability of the data [on] which the validation will be based."

4.2 GROUND-BASED SIMULATION OF HYPERSONIC FLOWS

The parameters that can be simulated in ground-based facilities include:

1. The freestream Mach number,
2. The freestream unit Reynolds number (and its influence on the character of the boundary layer),
3. The freestream velocity,
4. The pressure altitude,
5. The total enthalpy of the flow,
6. The density ratio across the shock wave,
7. The test gas,
8. The wall-to-total temperature ratio, and
9. The thermochemistry of the flowfield.

Note that some of the parameters are interrelated, e.g., the freestream velocity and the total enthalpy of the flow, and the density ratio across the shock wave and the test gas. Furthermore, in many instances where two parameters are related, e.g., the freestream Mach number and the freestream velocity, one may simulate one parameter (the Mach number) but not the other (the velocity, since neither the total enthalpy nor the speed of the sound are matched).

Complete simulations of the flowfield cannot be obtained in any one ground-based facility. In a statement attributed to J. Leith Potter in Ref. 3, "Aerodynamic modeling is the art of partial simulation." Thus, one must decide which parameters are critical to accomplishing the objectives of the test program. In fact, during the development of a particular vehicle, the designers will most likely utilize many different facilities with the run schedule, the model, the instrumentation, and the test conditions for each program tailored to answer specific questions.

Neumann[4] notes that, "For configurations, such as the Space Shuttle, which operate at high angles of attack dominated by a blunt nosed entropy layer, Mach number is not a significant parameter." However, Mach number and viscous effects become important to the Shuttle Orbiter design at lower angles-of-attack. Miller[5] studied the effect of gamma (using air and tetrafluoromethane as the test gas) and of the Reynolds number on the Shuttle Orbiter over an angle-of-attack range from 15 deg. to 45 deg. The heat-transfer distribution was essentially insensitive to the test gas or to the Reynolds number at an angle-of-attack of 45 deg., where the large effective bluntness dominates the windward flowfield. However, at 20 deg. angle-of-attack, the heat-transfer distribution is sensitive both to the test gas and to the Reynolds number. Thus, Neumann[4] concluded, "the physical phenomena which dominate the design and which dictate the required capabilities of experimentation to be conducted change as the characteristics of the configuration change. Test facilities and developmental philosophy appropriate for one class of configurations must be re-thought as a different design is analyzed."

In addition to the nine parameters described above, additional factors must be considered when developing a test plan. The additional factors include:

1. Model scale,
2. Test time,
3. Types of data available, and
4. Flow quality (including uniformity, noise, cleanness, and steadiness).

Consider a test program where a primary objective is to obtain experimental data to verify a code's ability to accurately model the physics and/or the chemistry of the flow, i.e., code validation data. Measurements at the model surface, e.g., heat-transfer data and surface pressures, are not sufficient. Code validation data must include information about the flowfield away from the surface, e.g., flow visualization data defining shock waves and density contours, velocity measurements, and gas-chemistry measurements. In any test program, one must be able to define the freestream flow, since it is the upstream boundary condition for the flow around the model. Defining the freestream flow is critical for code validation test-programs. Flow nonuniformity, unsteadiness, or noise may affect the validity of the data and make them unusable for code validation applications.

The test time is also an important test parameter. As shown in Fig. 4.1, the test time may vary from fractions of a millisecond for shock-tube flows to hours for "conventional" wind tunnels. Continuous-flow wind tunnels using air as the test gas have maximum stagnation temperatures slightly greater than 1000 K. If the flow involves nonequilibrium chemistry, the model scale affects the relation between the characteristic flow time and the characteristic reaction time. Furthermore, the specification of the instrumentation must reflect the test time and the model size. Thus, these many varied factors provide considerable challenge to the pre-test planning effort.

For many applications, experimental data provide needed information to develop realistic computer flow models and provide design data for flowfields whose complexity resists computer flow models. Thus, in the design of a hypersonic vehicle, CFD and ground-based testing play complementary roles.

In a paper presented in 1983, Romere and Young[6] note that 27,000 occupancy hours of wind-tunnel testing were used to develop the pre-flight aerodynamic predictions for the Shuttle. To extrapolate the wind-tunnel results to the flight environment, adjustments were made to correct for nonsimulation of structural deformation, flowfield parameters, and the profile drag due to thermal protection system roughness and minor protuberances. The traditional freestream Reynolds number was selected as the flowfield scaling pa-

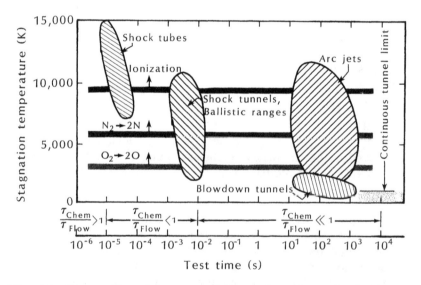

Fig. 4.1 Stagnation temperature as a function of test time for a variety of hypersonic facilities.

rameter below Mach 15, while a viscous interaction parameter was used at higher Mach numbers.

In a summary of the total wind-tunnel testing program for the Phase C/D of the Space Shuttle design development, Whitnah and Hillje[7] reported that a total of 493 aerodynamics tests were conducted using 52,993 hours, 139 heating tests using 12,138 hours, and 77 structures tests using 6,166 hours. The difference between the specific number of wind-tunnel hours used to define the Space Shuttle aerodynamics as taken from the literature probably reflects different accounting bases (supersonic only as opposed to all Mach numbers, excluding in-house NASA tests, etc.).

4.3 GROUND-BASED HYPERSONIC FACILITIES

Since there is no single ground-based facility capable of duplicating the hypersonic flight environment, different facilities are used to address various aspects of the design problems associated with hypersonic flight. Not all of the numerous special-purpose facilities will be discussed here. Instead, we will focus our attention on four types of facilities that are used to define the aerothermodynamic environment, i.e., the aerodynamic forces and moments, the heat-transfer

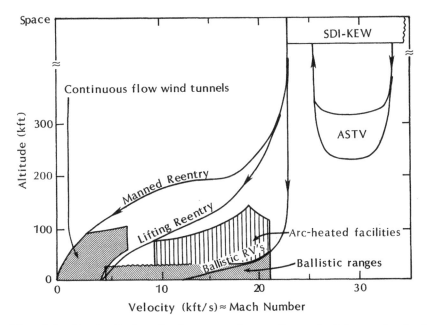

Fig. 4.2 Typical simulation envelopes for wind tunnels, ballistic ranges, and arc-heated facilities as compared with flight conditions, as taken from Ref. 3.

distribution, and the surface-pressure distribution. In this chapter, we will discuss four types of facilities:

1. Shock tubes,

2. Arc-heated test facilities,

3. Hypersonic wind tunnels, and

4. Ballistic free-flight ranges.

The capability of continuous-flow wind tunnels, ballistic ranges, and arc heaters to "simulate" hypersonic flows, as taken from Ref. 3, is reproduced in Fig. 4.2. Both the flight and the simulated conditions are presented in forms of altitude/velocity combinations. As will be discussed later, the simulation of altitude in a continuous flow wind tunnel actually means that the freestream static pressure in the test section (p_1) corresponds to the value at that altitude for the standard atmosphere, e.g., Table 1.1. However, the freestream static temperature (T_1) in the tunnel will rarely correspond to the atmospheric value. Thus, there is only a partial simulation of the atmospheric conditions. Similar differences exist in the velocity simulation. Note the statement in Fig. 4.2 that the velocity in kft/s is approximately equal to the Mach number is true for the atmosphere but is not true for facilities where the static temperature is not equal to the atmospheric value. As stated by Neumann,[8] "At best, hypersonic facilities currently available are partial simulation facilities." Note also that, in Fig. 4.2, nothing is given about other parameters such as model scale, which in turn affects the value of the simulated Reynolds number.

4.3.1 Shock Tubes

A shock tube is a facility in which a shock wave passes through a test gas, creating a high-pressure/high-temperature environment that can be used to study chemical kinetics and heat transfer in a dissociated gas. In principle, a shock-tube facility consists of two sections, initially divided by a diaphragm. As shown in Fig. 4.3, one of the sections is filled with a "driver" gas at high pressure; the other is filled with a "driven" gas (which is the test gas) at low pressure. After the diaphragm bursts, the piston-like motion of the driver gas (to the right) causes a shock wave to move through the test (driven) gas at the shock velocity (U_s). Refer to Fig. 4.4. The moving shock wave compresses ($p_2 > p_1$), accelerates, and heats the test gas. The pressure across the contact surface (the interface between the driver gas and the driven gas) is constant, i.e., $p_2 = p_3$. At any point in the shock tube, the flow useful for testing is that between the shock wave and the interface. Referring to Fig. 4.1, the test time for shock-tube flows is of the order of fractions of a millisecond.

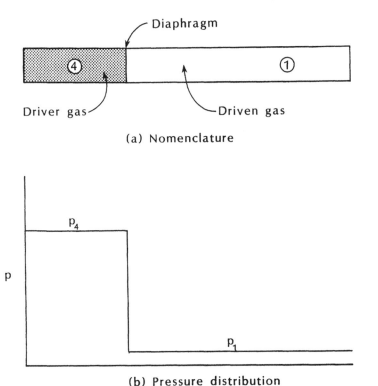

(a) Nomenclature

(b) Pressure distribution

Fig. 4.3 Shock-tube nomenclature and pressure distribution prior to diaphragm burst.

In order to study the high-temperature phenomenon associated with hypersonic flight, the speed of the primary shock must be large. Sharma et al.[9] note that increased shock speeds can be obtained by using a light gas, such as hydrogen or helium, as the driver gas and by heating the driver gas. Of course, high shock velocities create test times measured in microseconds.

Exercise 4.1:

A shock tube can be used to obtain stagnation-point heat-transfer measurements in high-temperature, high-pressure air. An example of a program where a shock tube has been used to obtain measurements of the stagnation-point heat-transfer rates in dissociated air is presented in Ref. 10. As shown in Fig. 4.5a, the primary shock wave travels (U_s) at 7000 ft/s into still air where static pressure (p_1) is 0.1 atm and where static temperature (T_1) is 540 °R.

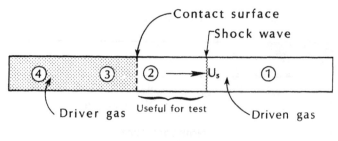

(a) Nomenclature

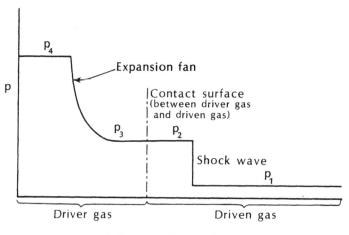

(b) Pressure distribution

Fig. 4.4 Shock-tube nomenclature and pressure distribution after diaphragm burst but prior to wave reflection.

1. The primary shock wave accelerates, compresses, and heats the air. Calculate the velocity, the static pressure, and the static temperature of the shocked air, i.e., U_2, p_2, and T_2.

2. The air behind the primary shock wave constitutes the "freestream" for the hemisphere cylinder mounted in the shock tube. Because the shocked flow is supersonic, a bow shock wave forms, as shown in Fig. 4.5c. Calculate the conditions at the stagnation point downstream of the bow shock wave, i.e., the pressure and the temperature at $t2$.

Use perfect-gas relations to calculate the conditions ahead of the primary shock wave, i.e., region 1. Downstream of the shock waves, use the charts for equilibrium air, e.g., Figs. 1.16 and 1.17.

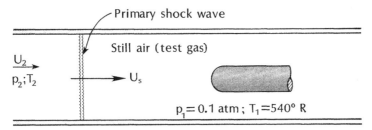

(a) Flow with primary shock wave upstream of the model

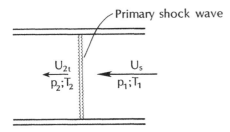

(b) Transformed coordinates so that the primary shock wave is stationary

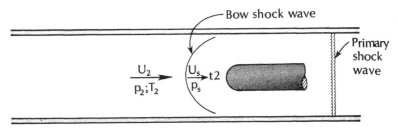

(c) Flow around hemi-sphere cylinder

Fig. 4.5 Use of shock tube to study stagnation-point heat-transfer rate.

Solution:

1. Applying the perfect-gas relations to the air in region 1,

$$a_1 = \sqrt{\gamma R T_1} = 1139 \text{ ft/s}$$

Therefore, $M_s = \dfrac{U_s}{a_1} = 6.145$

$$\rho_1 = \frac{p_1}{R T_1} = \frac{211.622 \frac{\text{lbf}}{\text{ft}^2}}{(1716.16 \frac{\text{ft}^2}{\text{s}^2 \, ^\circ\text{R}})(540 \, ^\circ\text{R})} = 0.000228 \frac{\text{lbf s}^2}{\text{ft}^4}$$

For the shock wave moving into still air, as shown in Fig. 4.5a, the flow is unsteady. Let us transform the flow into a coordinate system where the shock wave is fixed and the test gas moves through the shock wave from the right, as shown in Fig. 4.5b. The flow is steady in the coordinate system of Fig. 4.5b. Thus, we can use Eqs. (1-9) through (1-11), with $U_s = U_1$:

$$\rho_2 U_{2t} = \rho_1 U_1 = 1.5960 \; \frac{\text{lbf s}}{\text{ft}^3}$$

$$p_2 + \rho_2 U_{2t}^2 = p_1 + \rho_1 U_1^2 = 11,384 \; \frac{\text{lbf}}{\text{ft}^2}$$

$$h_2 + 0.5 U_{2t}^2 = h_1 + 0.5 U_1^2 = 27.750 \times 10^6 \; \frac{\text{ft}^2}{\text{s}^2}$$

Using the iterative procedure of Exercise 1.1 and the thermo-dynamic properties of equilibrium air in Fig. 1.17,

$$\rho_2 = 0.00143 \; \frac{\text{lbf s}^2}{\text{ft}^4}$$

$$p_2 = 9610.69 \; \frac{\text{lbf}}{\text{ft}^2} = 4.541 \text{ atm}$$

$$T_2 = 3895 \; °R$$

and $\quad\quad U_{2t} = 1120 \text{ ft/s}$

But U_{2t} is the velocity in the transformed coordinate system. The actual velocity of the shocked air (relative to the shock-tube walls) is:

$$U_2 = 7000 - 1120 = 5880 \text{ ft/s}$$

Using the figures in Ref. 11, the speed of sound of the shocked air is 2900 ft/s. Thus,

$$M_2 = \frac{5880}{2900} = 2.027$$

So the flow is supersonic, as stated earlier. As a result, a bow shock wave will form ahead of the sphere/cylinder. See Fig. 4.5c.

2. To calculate the conditions downstream of the bow shock wave, we again use Eqs. (1-9) through (1-11). The properties of air are found in Fig. 1.17 or in the original reference, Ref. 11.

$$\rho_s U_s = \rho_2 U_2 = 8.4084 \, \frac{\text{lbf s}}{\text{ft}^3}$$

$$p_s + \rho_s U_s^2 = p_2 + \rho_2 U_2^2 = 59,052 \, \frac{\text{lbf}}{\text{ft}^2}$$

$$h_s + 0.5 U_s^2 = h_2 + 0.5 U_2^2 = 44.410 \times 10^6 \, \frac{\text{ft}^2}{\text{s}^2}$$

The subscript 2 denotes the effective "freestream" conditions for the sphere/cylinder model, which have been generated by the passing of the primary shock wave. See Fig. 4.5c. The static conditions downstream of the bow shock wave are designated by the subscript s, and the stagnation-point conditions by the subscript $t2$. Note that U_s in Fig. 4.5c denotes the velocity downstream of the bow shock wave and differs from the velocity of the primary shock wave (also denoted U_s) in Fig. 4.5a.

The static conditions behind the bow shock wave are:

$$U_s = 2147 \text{ ft/s}$$
$$p_s = 40,999 \text{ lbf/ft}^2 = 19.373 \text{ atm}$$
$$T_s = 5520 \, ^\circ\text{R}$$

Decelerating the air isentropically from these conditions to the stagnation point, we find:

$$p_{t2} = 52,905 \text{ lbf/ft}^2 = 25 \text{ atm}$$
$$T_{t2} = 5760 \, ^\circ\text{R}$$

The results in Exercise 4.1 demonstrate that, although an extremely high velocity can be generated behind a normal-shock wave (U_2) in a constant-area duct, the ambient temperature is correspondingly high. As a result, although the Mach number of the flow behind the shock wave is supersonic, it is limited. The variation of the flow Mach number behind a normal-shock wave moving into still air is presented as a function of the shock Mach number in Fig. 4.6. The results presented in Fig. 4.6 (including the points representing experimental data) are taken from Ref. 12. Since the excitation of the higher vibrational states of the molecules and dissociation absorb considerable amounts of energy, the static temperature behind the normal-shock wave is not as high as that which would be achieved if the gas were perfect. Consequently, the flow Mach number behind the normal-shock wave is higher when the real-gas effects are taken into account.

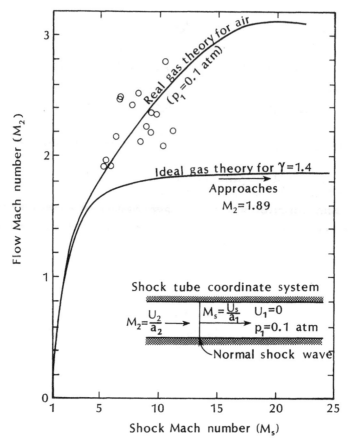

Fig. 4.6 Flow Mach number obtainable in a constant-area shock tube, as taken from Ref. 12.

As noted by Hertzberg,[12] "The most important limitation of the shock tube, aside from the Mach number limitation, is the extremely brief period during which high temperatures can be sustained. The conventional shock tube is, nonetheless, a valuable tool for the investigation of high temperature phenomenon in air when it is not necessary to simulate the flow Mach number."

Consider now a situation in which there is no model in the shock tube, i.e., the tube is as shown in Fig. 4.4. When the shock wave reaches the end of the driven tube, all of the test (driven) gas will have been compressed and will have a velocity in the direction of the shock wave's travel. Upon striking the end of the tube, the shock wave will be reflected as a shock wave. As the shock wave reflects back through the driven gas, the driven gas is brought to rest, satisfying the boundary condition that there is no flow through

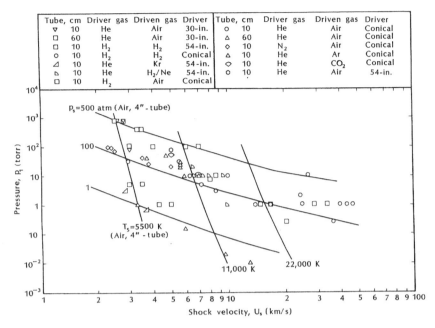

Tube, cm		Driver gas	Driven gas	Driver	Tube, cm		Driver gas	Driven gas	Driver
▽	10	He	Air	30-in.	○	10	He	Air	Conical
□	60	He	Air	30-in.	△	60	He	Air	Conical
□	10	H₂	H₂	54-in.	◇	10	N₂	Air	Conical
○	10	H₂	H₂	Conical	△	10	He	Ar	Conical
△	10	He	Kr	54-in.	○	10	He	CO₂	Conical
◺	10	He	H₂/Ne	54-in.	○	10	He	Air	54-in.
□	10	H₂	Air	Conical					

Fig. 4.7   Measured shock velocity vs. initial driven-tube charging pressure p_1, as taken from Ref. 9.

the right-end wall. The reflected shock wave produces additional compression and heating of the test gas. The conditions downstream of the reflected shock wave (which are designated by the subscript 5), as taken from Ref. 9, are reproduced in Fig. 4.7. As discussed previously, light driver gases and increasing the temperature of the driver gas by heating (in this case by an electrical arc discharge) produces high shock speeds and high temperatures downstream of the reflected shock wave.

4.3.2 Arc-Heated Test Facilities (Arc Jets)

As indicated in Fig. 4.1, facilities using an electric-arc heater provide relatively high-stagnation temperatures for relatively long test times. As noted by Anfimov,[13] "Electric-arc heating of gas in wind tunnels is used for different purposes: (a) to prevent condensation during expansion of air or another test gas to hypersonic Mach numbers; (b) to duplicate air physical/chemical transformations for study of vehicle aerogasdynamics and heat transfer at hypersonic velocities and also for study of plasma envelopes around vehicles, their radiation and radio wave transmission through them; and (c) to duplicate the real thermal protection ablation process due to material internal phys-

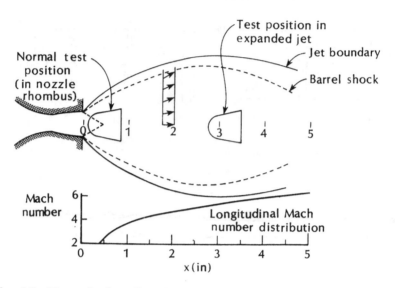

Fig. 4.8   Descriptive sketch of plume from HEAT (H-2), as taken from Ref. 14.

ical/chemical changes as well as due to material physical/chemical interaction with a high temperature environment." In this section, the discussion will focus on the ability to evaluate the Thermal Protection System (TPS) in ground-based facilities. To evaluate the TPS in a ground-based facility, one must be able to generate surface temperatures on the order of a few thousand degrees Fahrenheit for extended periods of time, while subjecting the model to a representative fluid mechanic environment, i.e., representative pressure and shear forces. These arc-heated facilities are often called *arc jets*.

In an arc-heated test facility, the test gas is passed through a high-power electric arc inside an elongated pressure vessel and then through a converging/diverging nozzle to produce a supersonic flow. Although the stagnation (impact) pressure and the stagnation en thalpy are at the re-entry value, the freestream Mach number is usually relatively low. This is illustrated in Fig. 4.8, which is taken from Ref. 14. A sketch of the plume and the longitudinal Mach-number distribution for the High Enthalpy Ablation Test (HEAT) Facility at the Arnold Engineering Development Center (AEDC) indicates a maximum Mach number less than six. However, as noted earlier in this chapter, there are many configurations for which the flow is independent of the Mach number. As shown in Fig. 4.9, as the freestream Mach number increases, the stagnation-point heat transfer in the expanded plume decreases.[14] There is a similar streamwise decrease in the total (impact) pressure.

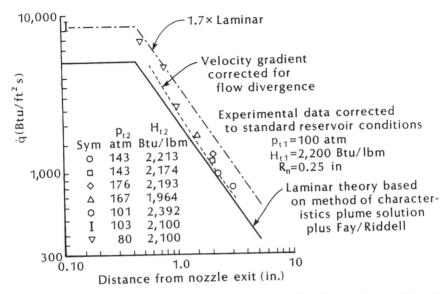

Fig. 4.9 Stagnation-point heat transfer in plume from M = 2 nozzle, as taken from Ref. 14.

A concern when testing in an arc-jet facility is the definition of the freestream. Water-cooled copper electrodes used in high-power electric-arc heaters may be subjected to partial erosion, resulting in gas-flow pollution. Furthermore, the mixing process between the test gas and the electric arc in the pressure vessel may cause unsteadiness and nonuniformity of the gas in the test section. However, the degree of flow nonuniformity and contamination can be reduced to very acceptable levels through careful arc-heater design.[15]

Arc-heated test facilities are unique in their capability to provide the necessary environment to evaluate TPS materials. Typical data by which the performance of the TPS material is evaluated include surface recession rates, shape changes which result from ablation, and temperature distributions.

Laganelli and Martellucci[16] studied the interaction heating effects that accompany regions of discontinuous ablation which exist for heat shields made of dissimilar materials. For instance, the differential ablation of a model composed of a graphite nose and a carbon phenolic frustrum produced the shape shown in Fig. 4.10. Laganelli and Martellucci[16] report that data obtained at General Electric "indicate that higher heat-transfer rates occur in the immediate downstream region, which results from the fluid-mechanical interaction of the step change in the injection with the local flow. This larger heat transfer implies increased recession, which relates to the starting process in achieving local necking."

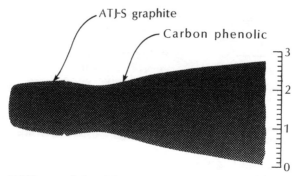

Fig. 4.10 Differential ablation resulting in necking of the model, as taken from Ref. 16.

Arc jets can be used to determine the effect of the gas composition on the performance of an ablative thermal protection system. Bertin et al.[17] note that, " The external degradation parameter, i.e., surface recession, indicates that air is the most severe environment, followed by N_2/CO_2 with inert N_2 the least severe... The composition of the external environment is an important factor in the internal degradation process, since the chemical reactions at the surface and in the boundary layer determine the thickness of char layer and, to some degree, the surface temperature. Since the temperature at the char/pyrolysis interface depends on the degradation properties of the material and is usually assumed to be a constant, the thermal gradient (and, hence the rate at which energy is supplied to the pyrolysis zone) across the char is dependent upon the composition of the external environment."

4.3.3 Hypersonic Wind Tunnels

We will use the term *hypersonic wind tunnel* to describe those facilities in which the test gas is accelerated from a reservoir (or still-

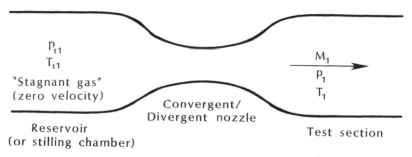

Fig. 4.11 Nomenclature for a hypersonic wind tunnel.

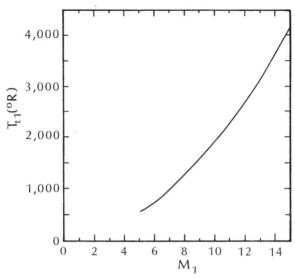

Fig. 4.12 Reservoir temperature required to maintain perfect air at T_1 = 90 °R when expanded to M_1.

ing chamber) where the gas is essentially at rest, through a convergent/divergent nozzle, achieving hypersonic Mach numbers in the test section. The nomenclature for a hypersonic wind tunnel is illustrated in Fig. 4.11. Except for the thin boundary layer near the wall, the acceleration may be assumed to be an isentropic process. However, in some facilities, there are weak waves and nonequilibrium effects which create departures from isentropic flow in the central part of the tunnel.

4.3.3.1 Characterization of hypersonic wind tunnels.   As the test gas accelerates, the static temperature decreases. Typical hypersonic wind tunnels operate such that the static temperature in the test section (T_1) approaches the liquefaction limit. As a result, the test section temperature for air is approximately 90 °R (50 K). The reservoir temperature (T_{t1}) that is required to maintain the static temperature in the test section at 90 °R when perfect air is accelerated to M_1 is presented in Fig. 4.12. Note that, if the air in the stilling chamber is at room temperature, one can achieve a test section Mach number of 5. To avoid liquefaction of air as it expands to the test section conditions where the Mach number is 10, the stagnation temperature, i.e., the temperature in the stilling chambers (T_{t1}), should be approximately 1900 °R. Thus, because of the limitations of heater capability, the maximum Mach number for a continuous-flow wind tunnel using air as the test gas is approximately 10. However, using better insulation and heater buildup methods, the Wright Laboratory 20-inch tunnel operates at Mach numbers of 12 and 14.

Of concern to the hypersonic wind-tunnel aerothermodynamicist is the combination of maximum Mach number and of minimum stagnation temperature for which condensation-free flow can be generated. Based on an experimental study of air condensation, Daum and Gyarmathy[17A] concluded that, in a rapidly expanding nozzle flow at low stream pressures (less than about 0.05 mm Hg), significant supercooling of the air can be achieved, since the onset of condensation was due to the spontaneous condensation of nitrogen. At these low pressures, an approximately constant experimental supercooling value of about 22 K was obtained.

To achieve higher Mach number flows, the stagnation temperature must be in excess of 1900 °R. There are various ways to generate the required high-pressure/high-temperature condition in the reservoir. An example is the use of the flow conditions downstream of the reflected shock wave in a shock tube, i.e., region 5 in Sec. 4.3.1, as the reservoir for a shock tunnel. However, the test time for such facilities is very short. The short run time reduces the energy requirements and alleviates tunnel and model thermal/structural interactions.

Using run time to classify hypersonic wind tunnels, there are three types:

1. Impulse facilities, which have run times of 1 s or less;

2. Intermittent tunnels (blowdown or indraft), which have run times from a few seconds to several minutes; and

3. Continuous tunnels, which can operate for hours.

As indicated in Fig. 4.1, the facilities with the shortest test times have the higher stagnation temperatures. Arc discharge or reflected shock waves in a shock tube are used to generate the short-duration, high-temperature stagnation conditions.

As noted at the start of this chapter, the flow quality (including uniformity, noise, cleanness, and steadiness) can affect the results obtained in a ground-test program. As noted by Pate,[18] disturbance modes in the supersonic tunnels include vorticity (turbulence fluctuations), entropy fluctuations (temperature spottiness) which are traceable to the stilling chamber, and pressure fluctuations (radiated aerodynamic noise). These disturbances can affect the results of boundary-layer transition studies conducted in hypersonic wind tunnels.

Trimmer et al.[3] note that even the high-enthalpy, short-duration facilities operate on the borderline between perfect-gas and real-gas flows. Referring to Fig. 4.12, it can be seen that, in order to maintain a test section temperature of 90 °R when the Mach number

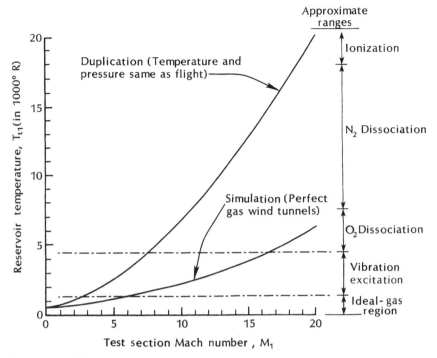

Fig. 4.13 Energetic species in a wind-tunnel reservoir, as taken from Ref. 19.

is 8.5, the stagnation temperature for perfect air must be 1400 °R. However, Boudreau[19] notes that "the vibrational state is excited beginning approximately at 1400 °R." As shown in Fig. 4.13, which is taken from Ref. 19, the reservoir gas of many hypersonic wind tunnels is excited to various energetic states. Thus, for the higher Mach number facilities, vibrational excitation occurs in the stilling chamber, followed by vibrational freezing downstream of the throat, and subsequent rapid relaxation in the downstream section of the nozzle. According to Boudreau,[19] the "improper characterization of hypersonic flowfields manifests itself as an error in the Mach number." As indicated in Fig. 4.14, AEDC's Tunnel C, which is heated by conventional clean air heaters, exhibits a Mach-number error of as much as 1.5 percent compared to that predicted by isentropic flow using the ratio of freestream pitot pressure to reservoir pressure. These relatively small errors in Mach numbers can produce significant errors in the nondimensionalized data, if they are ignored. Properly characterized wind tunnels, such as Tunnel C, will yield excellent experimental data..

Based on the discussion in this section, it should be clear that an integral part of any test program should be the analysis of the data

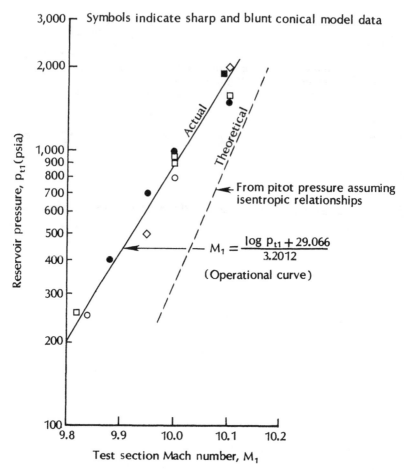

Fig. 4.14 Suggested correction to Mach numbers for Tunnel C (AEDC), as taken from Ref. 19.

and comparisons with theoretical/computational solutions. Conversely, computed flows should be compared with experimental data. An important part of this process is the facility calibration, since freestream flow nonuniformities can affect the experimental measurements.

4.3.3.2 Simulation parameters. Once the objectives of the wind-tunnel program have been established, numerous questions remain to be answered in developing the test plan. One must establish what parameters are to be simulated, define the model configuration(s), and establish the data requirements. In the next sections, we will discuss some of the simulation parameters available to the test planner.

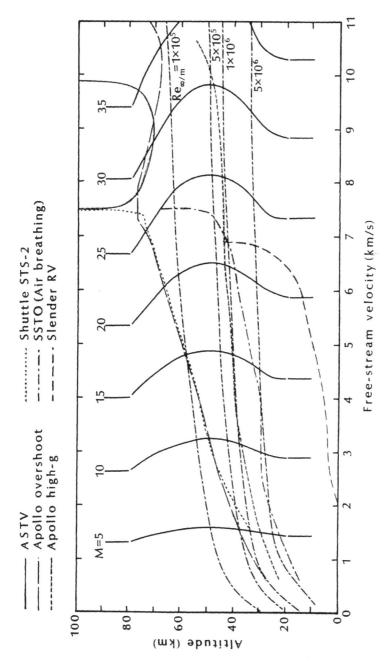

Fig. 4.15 Lines of constant Mach number and of constant unit Reynolds number.

4.3.3.3 Mach number/Reynolds number test conditions.   Let us assume
that we should match the Mach number and the Reynolds number
in order to achieve similarity between the flowfield of the wind-tunnel
simulation and that for the prototype vehicle. The freestream Mach
number and the freestream unit Reynolds number, i.e., the Reynolds
number per meter, are presented in Fig. 4.15. Because the Shuttle
Orbiter enters the atmosphere at a relatively small path angle and
operates at relatively high angles-of-attack, considerable deceleration
occurs at relatively low unit Reynolds numbers. The Mach number
is 15, when the unit Reynolds number reaches 1×10^5. As a result,
boundary-layer transition for the Shuttle Orbiter occurs at an altitude of approximately 50 km when the Mach number was 10. After
reviewing data from numerous Shuttle Orbiter re-entries, Bouslog
et al.[194] reported that, in extreme cases, boundary-layer transition
occurred as early as Mach 17. "During the STS-28, protruding tile
gap fillers apparently caused early transition." For a slender RV,
which enters the atmosphere at a steep path angle and operates at
relatively small angles-of-attack, the unit Reynolds number is very
large even when the Mach number exceeds 20. Thus, boundary-layer
transition occurs at altitudes below 40 km for slender RVs, when
the Mach number is in excess of 20. This comment is intended to
underscore the fact that there are no universal boundaries of flow
phenomena.

As has been noted, typical hypersonic wind tunnels operate such
that the static temperature in the test section approaches the liquefaction limit. Thus, hypersonic Mach numbers are achieved with relatively low freestream velocities (which relate to the kinetic energy),
because the speed of sound (which relates to the static temperature)
is relatively low.

Exercise 4.2:

In May 1987, a Mach 8 free jet nozzle was installed in Tunnel C
(AEDC). In addition to providing a third Mach number nozzle (the
other two nozzles provide Mach numbers of 4 and of 10), it considerably extended the Reynolds number envelope. The operating
envelope, as taken from Ref. 20, is reproduced in Fig. 4.16.

The pressure and the temperature in the stilling chamber are
1200 psia and 1390 °R, respectively. If the air is expanded to a
Mach number of 8, calculate: (a) the freestream static temperature,
(b) the freestream static pressure, (c) the freestream velocity, (d) the
freestream unit Reynolds number, and (e) the freestream dynamic
pressure.

Assume an isentropic expansion of perfect air.

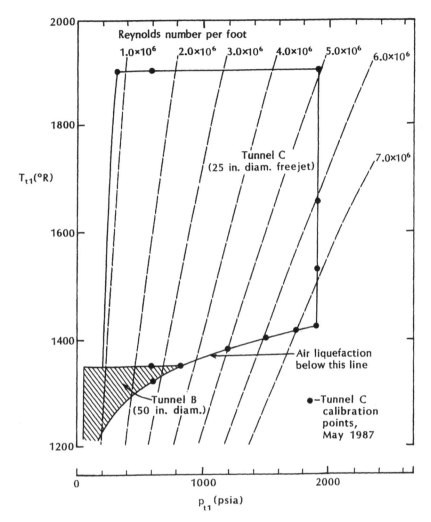

Fig. 4.16 The operating envelope for the Mach 8 nozzle in Tunnel C (AEDC), as taken from Ref. 20.

Solution:

Using the relations for isentropic flow of perfect air or the corresponding tables, e.g., Ref. 21, for $M_1 = 8$,

$$\frac{T_1}{T_{t1}} = 0.07246 \quad ; \quad \frac{p_1}{p_{t1}} = 0.0001024$$

(a) $T_1 = 100.72 \,°R$;   (b) $p_1 = 0.12288$ psia

To calculate the velocity, we first need to calculate the speed of sound.

$$a_1 = \sqrt{\gamma R T_1} = 49.017\sqrt{T_1} = 491.93 \text{ ft/s}$$

(c) $U_1 = M_1 a_1 = 3935.4 \text{ ft/s}$

To calculate the unit Reynolds number, we first need to calculate the density and the viscosity. The density will be calculated using Eq. (1-12).

$$\rho_1 = \frac{p_1}{R T_1} = \frac{(0.12288)(144) \frac{\text{lbf}}{\text{ft}^2}}{\left(1716.16 \frac{\text{ft}^2}{\text{s}^2 \, {}^\circ\text{R}}\right)(100.72 \, {}^\circ\text{R})}$$

$$\rho_1 = 1.024 \times 10^{-4} \frac{\text{lbf s}^2}{\text{ft}^4} \left(\text{or} \, \frac{\text{slugs}}{\text{ft}^3}\right)$$

Using Sutherland's equation, Eq. (2-4b), to calculate the viscosity:

$$\mu_1 = 2.27 \times 10^{-8} \frac{T_1^{1.5}}{T_1 + 198.6} = 7.67 \times 10^{-8} \frac{\text{lbf s}}{\text{ft}^2}$$

(d) Thus, the unit Reynolds number is:

$$Re/\text{ft} = \frac{\rho_1 U_1}{\mu_1} = 5.255 \times 10^6 \; /\text{ft}$$

This compares reasonably well with the value of 5.00×10^6, which was presented by Cooper and Eaves[20] (see Fig. 4.16). Referring to Table 2.1, one would expect that the actual viscosity is 3 percent to 5 percent greater than Sutherland's value at 56 K. Using the "improved" value for the viscosity would bring our calculation of the Reynolds number into line with the value presented by Cooper and Eaves. AEDC personnel recognize this. That is why they use the equation of Hirschfelder instead of Sutherland's equation to calculate the viscosity at temperatures below 100 K (180 °R).

The dynamic pressure is given by:

(e) $q_1 = \frac{1}{2}\rho_1 U_1^2 = \frac{\gamma}{2} p_1 M_1^2 = 5.505 \text{ psia}$

The calibration of the Mach 8 nozzle for Tunnel C of AEDC indicated that the tunnel could be operated over a range of reservoir pressure from 300 to 1900 psia and a range of total temperatures from 1300 to 1900 °R. As a result, the unit Reynolds number ranged from 0.8×10^6 to 7.6×10^6 per foot. This supplements the Mach 8 capabilities of Tunnel B at AEDC, which can be operated over a range of reservoir pressures from 50 to 850 psia at a maximum stagnation temperature of 1350 °R. The resulting unit Reynolds number for Tunnel B ranges from 0.3×10^6 to 4.7×10^6 per foot. Other factors may enter into the choice of the facility, such as tunnel size, flow quality (if known), instrumentation capabilities, etc.

The range of conditions for the Calspan Hypersonic Shock Tunnel, the freestream Mach number, and the freestream unit Reynolds number (i.e., $Re_{\infty/\text{ft}} = \rho_\infty U_\infty / \mu_\infty$) has been taken from Ref. 22 and reproduced in Fig. 4.17. The corresponding altitudes at which these

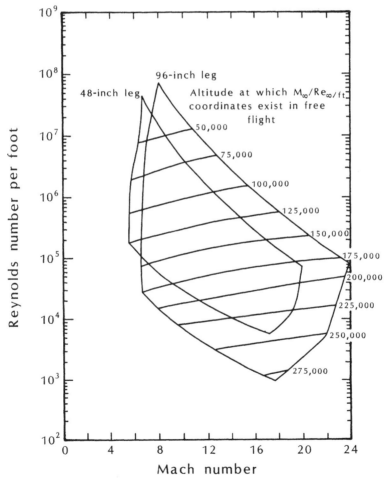

Fig. 4.17 Calspan Hypersonic Shock Tunnel performance, as taken from Ref. 22.

$M_\infty/Re_{\infty/\text{ft}}$ coordinates occur in flight is also indicated in Fig. 4.17. The reader should note that the relatively low freestream temperatures typical of wind tunnels (note that $T_{\infty,\text{min}}$ is assumed to be equal to $1.10\ T_{LOX}$) influence the correlation between $Re_{\infty/\text{ft}}$ and the altitude correlations.

Exercise 4.3:

What is the pressure altitude, the temperature altitude, and the density altitude simulated by the freestream test conditions of Exercise 4.2?

Solution:

Nondimensionalizing the pressure and the density, as done in Table 1.1b, the results from Exercise 4.2 are:

$$p_1 = 0.12288 \text{ psia} = 8.361 \times 10^{-3} p_{SL},$$

$$T_1 = 100.72^\circ\text{R, and}$$

$$\rho_1 = 1.024 \times 10^{-4}\ \frac{\text{slugs}}{\text{ft}^3} \doteq 4.308 \times 10^{-2} \rho_{SL}$$

Referring to Table 1.1b, the pressure altitude is approximately 105 kft and the density altitude is approximately 76 kft. Clearly, there is no altitude in Table 1.1b at which the ambient temperature for the 1976 U.S. Standard Atmosphere is even close to the static temperature in the test section. As a result, in order to achieve a Mach number of 8 in flight, the velocity would be much greater than that in the tunnel simulation.

Early manned spaceflight programs made extensive use of wind-tunnel data to determine the re-entry convective heating environment. The range of flow conditions for that portion of the wind-tunnel investigations for the Apollo Command Module directed by North American Aviation/Manned Spacecraft Center (NASA)[23] is presented in Fig. 4.18. The wind-tunnel flow conditions are given in terms of the freestream Mach number and the freestream Reynolds number (based on the maximum body diameter). As a result of the low freestream static temperature in the wind tunnel, the tunnel flow conditions should not be related to the velocity/altitude coordinate system. The extreme flight conditions that might be experienced by the Apollo capsule upon re-entry are indicated in Fig. 4.18 by

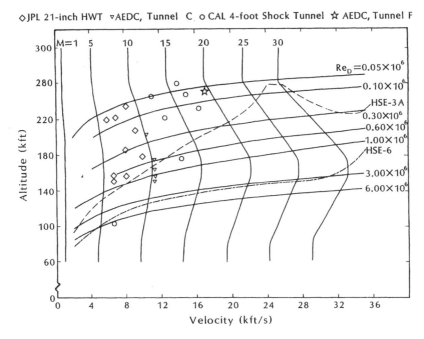

Fig. 4.18 The test conditions for the Apollo Command Module heat-transfer, wind-tunnel program, as taken from Ref. 23.

the HSE-3A overshoot trajectory and by the HSE-6, 20 g emergency re-entry trajectory. Either the velocity/altitude coordinates or the Mach-number/Reynolds-number coordinates may be used to define a given flight condition for these trajectories.

Lee et al.[24] found that "Heating rates measured with asymptotic calorimeters on the conical section (of the Apollo Command Module) agreed with wind-tunnel measurements extrapolated on the basis of laminar theory" during re-entry from orbital velocities. For re-entries at superorbital speeds, one had to include the effects of blowing (due to ablation) for the convective heating rates to agree with the theoretical prediction.[25] This successful extrapolation from the wind-tunnel environment to the flight environment may reflect the character of this blunt-body flowfield, as well as the nondimensionalization procedures used. As evident in Fig. 4.18, generating wind-tunnel simulations of the high-Reynolds-number flows required to study boundary-layer transition is limited to relatively low, hypersonic Mach numbers.

Trimmer et al.[3] note:

> The concept of Mach number independence above Mach 10 or 12 is frequently employed to mitigate the need for test data even though the vehicle may fly at Mach 20 or

above. Many high Mach number facilities have produced
data that are less accurate than the lower speed, more
conventional facilities. Thus, some Mach number effects
have been obscured by facility-related effects. High Mach
number effects do exist, but they tend to be configuration
dependent. For example, the Apollo Command Module
aerodynamics were insensitive to Mach numbers above 8.
The Space Shuttle Orbiter, however, demonstrated important effects above Mach 8.

This last conclusion was based on the analysis of the Shuttle aerodynamic coefficients by Maus et al.[26] However, in a subsequent study by
the same group, Griffith et al.[27] noted that "Ground test data above
Mach 10 were not of sufficiency to define the moment characteristics
of the Orbiter."

For a final comment on Mach number effects, let us recall the
statement of Boudreau,[19] "In the past, many hypersonic facilities
were reputed to produce data of inferior quality when, in fact, it was
poor characterization of the flowfield which was principally at fault."

Reynolds number simulation may present considerable difficulties
to the test planner. Neumann[8] notes:

... the data to generate is that data which can be extrapolated to flight, while the data to minimize is that data
which is dominated by the details of the selected flow facility. Methodology exists by which both laminar and turbulent boundary layer data can be extrapolated to flight.
The extrapolation of transitional data, on the other hand,
is extremely difficult. This would suggest that one should
focus on the generation of either fully laminar or fully turbulent data (or both as appropriate) and minimize the
generation of transitional data which cannot easily be extrapolated to flight and which, in any case, is dominated
by the details of the flow facility. Such a tactic is imple-
mented by testing at the extremes of the Reynolds number
capability in the selected test facility where both fully laminar data and, conversely, rapidly transitioned turbulent
data can be achieved.

However, in many cases, the test conditions in the wind tunnel
are such that the natural state of the boundary layer on the model
would be in the transitional state. Thus, in cases where we need to
simulate turbulent boundary layers, roughness elements are placed on
the model to artificially promote boundary-layer transition. See Fig.
4.19. For hypersonic flow, roughness-induced boundary-layer transition depends on (1) the local Mach number, (2) a Reynolds number
parameter based on the roughness-element dimension, (3) the wall-
to-total temperature ratio (T_w/T_{t1}), (4) the pressure gradient, (5) the

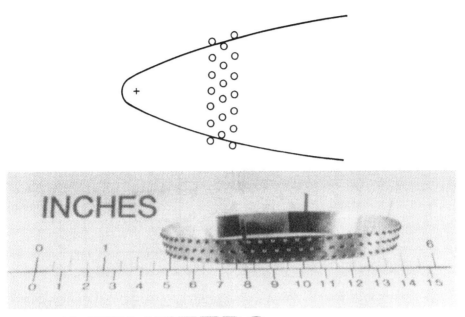

Fig. 4.19 Sketch of rows of spheres which serve as boundary-layer trips.

model configuration, (6) the type of roughness, and (7) the spacing of the roughness elements. Experiments into the effects of the trip geometry, the trip size, and the location of the roughness elements on the position of transition for hypersonic flows have been reported by Morrisette et al.[28] and Nestler and McCauley.[29]

As suggested by Braslow,[30] correlations for the effect of roughness elements on the transition location often include the Reynolds number based on fluid conditions at the top of the roughness elements, i.e., ρ_k, U_k, and μ_k, and the height of the roughness elements, i.e., k.

$$Re_k = \frac{\rho_k U_k k}{\mu_k} \qquad (4\text{-}1)$$

Sterrett et al.[31] note that the size of the roughness elements necessary to move transition close to the trips for the higher Mach numbers becomes very large. For hypersonic flow, the trip sizes are so large that they can create significant spanwise variations in the boundary layer, pressure drag, and local increases in heating. A schlieren photograph of the trip-induced flowfield perturbations that result when a wedge with trips is exposed to the Mach number 8 stream of Tun-

Fig. 4.20   The effect of boundary-layer trips on hypersonic flow past a sharp wedge, $M_\infty = 8$, $Re_{\infty/ft} = 3.6 \times 10^6$, as taken from Ref. 32.

nel B (AEDC), as taken from Ref. 32, is reproduced in Fig. 4.20. Three rows of spherical trips, 0.039 inches in diameter, were located two inches from the leading edge of the wedge. A strong interaction between the shock wave off the trips and the bow shock wave is evident in the photograph. Furthermore, viscous/inviscid interactions, vortical flows, and entropy layers persist downstream.[32]

Sterrett et al.[31] recommend "whether roughness should be used to promote turbulent flow in hypersonic wind-tunnel tests depends on the particular purpose of any experiment." Ericsson and Reding[33] note:

> When only the static aerodynamics are sought, it is often possible to simulate full-scale aerodynamics in subscale tests. In the case of attached flow, one can use a boundary layer trip to cause transition to occur at the proper location, thereby assuring that the relative extent of laminar and turbulent flow is simulated. ...The location of boundary layer transition has significant influence on attached flow unsteady aerodynamics at hypersonic speeds and on separated flow dynamics at all speeds. Any us-

age of boundary layer tripping devices is prohibited in dynamic tests because it prevents simulation of this coupling between vehicle motion and transition location.

4.3.3.4 High-altitude similarity parameters.
For high Mach number / low Reynolds number conditions associated with hypersonic flight at high altitudes, the boundary layer becomes thick relative to the total dimension of the shock layer. The resultant interaction between the boundary layer and the inviscid flow external to the boundary layer can significantly affect the distributions of pressure and of heat transfer. The viscous interaction, i.e., the mutual modification of the boundary-layer growth and the body-supported pressure field, can be related to the ratio of c (the effective velocity at which the vorticity spreads) to a (the acoustic propagation velocity, i.e., the speed of sound). Liepmann and Roshko[34] introduce the approximation that

$$c \simeq \sqrt{\frac{\nu}{t}} = \sqrt{\frac{\nu U}{x}}$$

so that

$$c = \frac{U}{\sqrt{Re_x}}$$

Since

$$a = \frac{U}{M}$$

the ratio of (c/a) is given by:

$$\frac{c}{a} = \frac{M}{\sqrt{Re_x}} \qquad (4\text{-}2)$$

which serves as a viscous interaction parameter for low-density, hypersonic flows. In an effort to better account for varying wall-temperature ratio, Potter[35] converts from a Reynolds number based on μ_∞ to one based on some characteristic viscosity. Thus, the viscous interaction parameter (sometimes referred to as the *slip parameter*) is:

$$\bar{V}_\infty = M_\infty \left(\frac{C}{Re_{\infty L}} \right)^{0.5} \qquad (4\text{-}3)$$

where C is the Chapman-Rubesin coefficient relating viscosity and temperature, i.e.,

$$C = \left(\frac{\mu^*}{\mu_\infty} \right) \left(\frac{T_\infty}{T^*} \right) \qquad (4\text{-}4)$$

where the starred quantities correspond to the chosen reference condition.

Hayes and Probstein[36] note that, for a boundary-layer induced interaction, the disturbance to the external flowfield is entirely due to the distribution of the displacement thickness of the boundary layer itself and that, without the viscous effects, the flowfield would be completely undisturbed. Describing boundary-layer induced interactions as "self-induced" interactions, Hayes and Probstein develop a hypersonic viscous interaction parameter for induced pressure perturbations on wedges as:

$$\chi_{\text{orig}} = \frac{M_{\text{orig}}^3 \sqrt{C_{\text{orig}}}}{\sqrt{Re_{x,\text{orig}}}} \qquad (4\text{-}5)$$

and

$$C_{\text{orig}} = \left(\frac{\mu_b}{\mu_{\text{orig}}}\right)\left(\frac{T_{\text{orig}}}{T_b}\right) \qquad (4\text{-}6)$$

where the subscript "orig" denotes the original conditions associated with the uniform, inviscid flow at the wedge surface for a "no-boundary-layer" flow model, and the subscript b refers to properties evaluated at the surface of the body. Holden[37] used a variation of this viscous interaction parameter to correlate the dependence of incipient separation on a curved compression surface. The flow conditions in Holden's investigation were such that the boundary layer was completely laminar and the wall highly cooled.

Love[38] noted that the possibility of significant viscous effects at high altitude was recognized in the early analytical studies of entry, and experimental studies of complete configurations were begun in the early 1960's. Reliable methods for correlating data are essential to proper interpretation if one is to minimize wind-tunnel programs for a vehicle that experiences a wide spectrum of Mach number and Reynolds number. Thus, in 1973, Love[38] cautioned that one must carefully choose the correlation parameter for high-altitude viscous effects.

Woods et al.[39] reviewed the investigation conducted at the Langley Research Center in 1974 to examine the Mach number / Reynolds number effects on Shuttle Orbiter aerodynamics for $5 \le M_\infty \le 20$. The results for various correlation parameters are summarized in Fig. 4.21, which is taken from Ref. 39. The classic hypersonic viscous interaction parameter, $\bar{\chi}'_\infty = M_\infty^3 \sqrt{C'_\infty}/\sqrt{Re_{\infty L}}$, was used to correlate local effects (pressure, heating, skin friction, etc.). In this expression, C'_∞ is the Chapman-Rubesin viscosity coefficient evaluated in the freestream based on reference temperature conditions, i.e., $(\mu_\infty T')/(\mu' T_\infty)$. When the total integrated axial-force

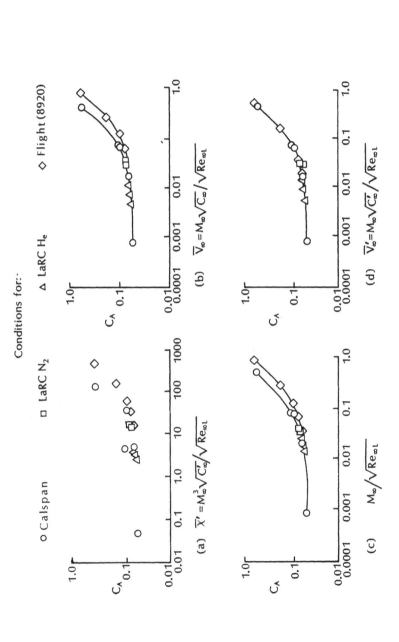

Fig. 4.21 Correlations of the estimated Shuttle axial-force coefficient based on a 15 deg. cone axial force per unit surface area nondimensionalized by Orbiter reference conditions, as taken from Ref. 39.

coefficient is presented as a function of $\bar{\chi}'_\infty$, a separate trend appears to be established for each facility and for the entry condition. The large variation indicated for Calspan conditions can be attributed to the extreme range of operating conditions and the use of different nozzles to obtain these conditions. The slip parameter, $\bar{V}_\infty = M_\infty \sqrt{C_\infty}/\sqrt{Re_{\infty L}}$, does correlate results for the wind-tunnel test conditions. However, for entry flight conditions, a different trend is established. The Chapman-Rubesin viscosity coefficient, C_∞, is evaluated in the freestream based on wall conditions, i.e., $(\mu_\infty T_w)/(\mu_w T_\infty)$, and is nearly 1.0 in most air facilities. Therefore, in many instances, it is assumed to be 1.0, and data are then correlated by $M_\infty/\sqrt{Re_{\infty L}}$. The correlations of Ref. 39 presented in this manner show the same trend as obtained with $\bar{V}_\infty$. Ground facility results establish one trend and entry conditions another. The parameter $\bar{V}'_\infty = M_\infty \sqrt{C'_\infty}/\sqrt{Re_{\infty L}}$, where C'_∞ is based on a reference temperature, T', correlates the present results for both ground facility and entry conditions (ideal gas). Therefore, for this simplistic analysis, $\bar{V}'_\infty$ appeared capable of correlating hypersonic data for wind-tunnel and flight conditions.

Wilhite et al.[40] noted that the maximum L/D degrades with the increasing altitude. Thus, in order to predict realistically the aerodynamic performance of vehicles for which the high-altitude portion of their re-entry trajectory has a significant impact on their design, one must be able to characterize the flow in various flight regimes. Such vehicles include the ASTV, the Space Shuttle Orbiter, and the Apollo Command Module. The variations in similarity parameters for the re-entry of the Space Shuttle Orbiter are presented in Fig. 4.22. Wilhite et al.[40] divide the flow into four separate flight regimes: (simple) hypersonic, a viscous-interaction regime, the noncontinuum transition flow, and the free-molecule flow (FMF). It is assumed that a continuum-flow model can be used both for the (simple) hypersonic regime and the viscous-interaction regime, but not for the other two. As noted in Sec. 1.3, the Knudsen number (Kn) is a dimensionless parameter associated with these two regimes, i.e., the noncontinuum transition flow and the free-molecular flow. Because the viscous-interaction regime identifies flow conditions where mutual interactions occur between the boundary layer and the inviscid flow, it occurs at high Mach numbers and low Reynolds numbers. In the (simple) hypersonic regime, the flowfield can be divided into two regions that can be analyzed separately: the inviscid flow between the shock envelope and the boundary layer, and the viscous boundary layer adjacent to the surface.

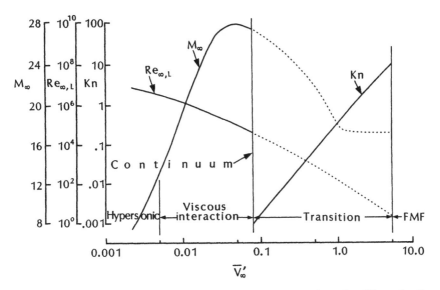

Fig. 4.22 Variation of similarity parameters for the Shuttle Orbiter re-entry, as taken from Ref. 40.

Exercise 4.4:

An experimental program was conducted in the Calspan 96-Inch Hypersonic Shock Tunnel to investigate the effect that the windward surface temperature has on the heat transfer to the leeward surface of the Space Shuttle Orbiter.[41] The model was 1.075 ft long, i.e., 0.01 scale. Test conditions included:

	M_∞	$Re_{\infty L}$ $(\times 10^{-6})$	p_{t2} (psia)	T_{t2} (°R)
(i)	10.05	0.525	5.340	4641
(ii)	11.82	1.696	13.57	4776
(iii)	15.70	0.630	2.825	4651
(iv)	15.81	0.121	0.681	5299
(v)	16.01	0.945	4.094	4657
(vi)	18.26	0.145	0.442	4423

(a) What is the range for the following viscous-interaction parameter?

$$\frac{M_\infty}{\sqrt{Re_{\infty L}}}$$

Compare these values with those appearing in Fig. 4.21c.

(b) What is the total enthalpy at the stagnation point for flow condition (v)? Estimate the freestream velocity in the tunnel at this condition.

(c) Assume that the vehicle is flying at $M_\infty = 16.01$ at 200,000 ft. What would be the corresponding freestream velocity and total enthalpy?

Solution:

(a) The viscous-interaction parameter ranges from 0.0091 to 0.0480. Specifically,

Case	$M_\infty/\sqrt{Re_{\infty L}}$
(i)	0.0139
(ii)	0.0091
(iii)	0.0198
(iv)	0.0455
(v)	0.0165
(vi)	0.0480

(b) Using Fig. 1.17, for

$$
\begin{aligned}
p_{t2} &= 4.094 \text{ psia} = 0.278 \text{ atm} \\
T_{t2} &= 4657 \text{ }^\circ R \\
h_{t2} &= 1400 \text{ Btu/lbm} = 35,053 \times 10^6 \text{ ft}^2/\text{s}^2
\end{aligned}
$$

Using the approximation for hypersonic flow that

$$H = h_{t2} = h_\infty + \frac{U_\infty^2}{2} \approx \frac{U_\infty^2}{2}$$

the freestream velocity is approximated as:

$$U_\infty \approx \sqrt{2H} = \sqrt{2h_{t2}} = 8373 \text{ ft/s}$$

(c) From the 1976 U.S. Standard Atmosphere presented in Table 1.1b, the speed of sound at 200,000 ft is 1028.18 ft/s. Thus,

$$U_\infty = M_\infty a_\infty = 16,461 \text{ ft/s}$$

and

$$H \approx \frac{U_\infty^2}{2} = 135.485 \times 10^6 \text{ ft}^2/\text{s}^2 = 5411 \text{ Btu/lbm}$$

4.3.3.5 Matching the freestream Mach numbers and the stagnation-point conditions.   In Figs. 1.18 through 1.20, the pressure and the temperature of equilibrium air at the stagnation point (outside of the thermal boundary layer) were presented as a function of the freestream Mach number.   Assume that we are to develop a wind-tunnel program that simulates these parameters at specific points along the trajectory. We have chosen a design trajectory for the re-entry of the Space Shuttle Orbiter which is presented in Fig. 4.23. Some of the aerothermodynamic flow features associated with the trajectory, e.g., where maximum heating occurs, are indicated in Fig. 4.23. Also included is a summary of the aerosurface functional requirements, i.e., the legend across the top of the figure provides information about the role of the speed brakes (δ_{SB}), the body flap (δ_{BF}), the rudder (δ_R), and the elevon/aileron ($\delta_{E/A}$). By comparing the position (and length) of the heading, information can be obtained about the functions of a particular aerosurface, e.g., the rudder is not used for velocities in excess of 4,000 ft/s (the yaw jets are), and the speed brake is kept closed for velocities in excess of 10,000 ft/s because of the thermal seal problems. A sketch of the Shuttle Orbiter, showing the relative size and locations of these aerosurfaces, is presented in Fig. 4.24.

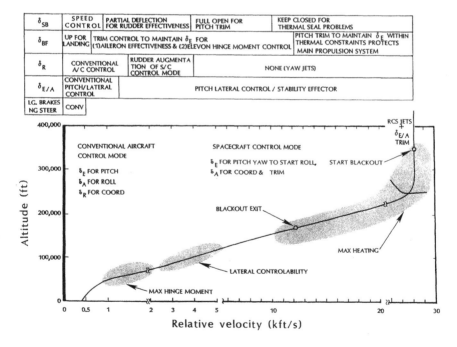

Fig. 4.23   Design trajectory for the re-entry of the Space Shuttle Orbiter.

As presented in Table 4.1, four flight conditions have been selected for simulation in our hypothetical wind-tunnel program. They are:

1. Velocity = 26,400 ft/s and altitude = 246,000 ft, a flight condition near the peak convective heating environment;

2. Velocity = 16,840 ft/s and altitude = 199,000 ft, a flight condition in the black-out region and near the high Mach number end of wind-tunnel capabilities;

3. Velocity = 10,268 ft/s and altitude = 162,000 ft, a flight condition near the onset of boundary-layer transition; and

4. Velocity = 3,964 ft/s and altitude = 100,000 ft, a flight condition where lateral controllability is a concern.

Note that there is no existing wind tunnel using air as a test gas that provides Mach numbers in excess of 25. Of the other three conditions, simulation both of the stagnation pressure and of the stagnation temperature at the desired Mach number can be achieved only at the lowest Mach number, i.e., in the Aerothermal Tunnel C at the Arnold Engineering Development Center (AEDC). Aerothermal Tunnel C is a specially designed modification (using existing Tunnel C support equipment and instrumentation) to fill the need for supersonic aerothermal test capability.[42] The reader should note that the selection of facilities given in Table 4.1 reflects the author's

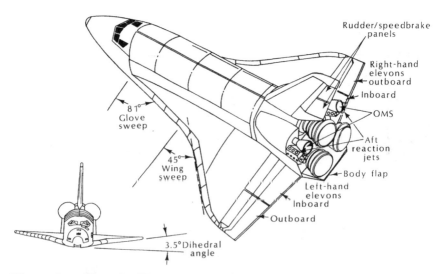

Fig. 4.24 Sketch illustrating the aerosurfaces of the Space Shuttle Orbiter.

Table 4.1   Simulating the Mach number and pressure-altitude for the Shuttle Orbiter entry trajectory in a hypersonic wind tunnel

		— — — — — F L I G H T — — — — —				— — — — S I M U L A T I O N — — — →			
	Velocity (ft/s)	Altitude (ft)	M_∞ (—)	p_{t2} (lbf/ft²)	T_{t2} (°R)	M_∞ (—)	p_{t2} (lbf/ft²)	T_{t2} (°R)	Facility*
1	26,400	246,000	28.36	57.3	11,257	No wind tunnel available			
2	16,840	199,000	16.05	148.8	9,385	16	142.9	4,000	C-96*
3	10,268	162,000	9.49	215.8	5,508	10	209.6	1,911	AEDC-C1
4	3,964	100,000	4.00	490.2	1,716	4	490.2	1,660	AEDC-C2

* C-96: Calspan 96-Inch Hypersonic Shock Tunnel, Ref. 22

* AEDC-C1: Arnold Engineering Development Center (AEDC) Tunnel C

* AEDC-C2: Arnold Engineering Development Center (AEDC) Aerothermal Tunnel C

experience and background. There are other wind-tunnel facilities that could be used to simulate these flows. The interested reader is referred to Ref. 43, which presents a brief summary of the operating characteristics of hypersonic facilities.

4.3.3.6 Use of test gases other than air.
For some applications, the test planners may want to obtain data at conditions not available in wind tunnels where air is the test gas. By using four different test gases, i.e., air, tetrafluoromethane, helium, and nitrogen, the Hypersonic Facilities Complex (HFC) at the Langley Research Center (NASA) provides a wide range of hypersonic simulation parameters. As noted by Miller,[44] the nine hypersonic wind tunnels of the HFC "provide a range of Mach number from 6 to 22, unit Reynolds number from 0.03 to 40 million per foot and, most importantly for blunt configurations, a normal shock density ratio from 4 to 12." Part of this wide range of simulation parameters is due to the use of gases other than air as the test gas.

Due to the dissociation of the air in the shock layer at flight conditions, values of the density ratio (ϵ) of 0.05 to 0.067 are encountered. Obviously, the density ratio (ϵ) in this paragraph is the inverse of the normal-shock density ratio used in the quote from Miller presented in the previous paragraph. For ground tests at hypersonic Mach numbers where no dissociation occurs, the density ratio across the shock wave is determined by the specific-heat ratio of the gas. Thus, for wind tunnels using air or nitrogen as the test gas and operating at perfect-gas conditions, the value of epsilon is (1/6), or greater. By using tetrafluoromethane (CF_4) as the test gas in a Mach 6 tunnel, values of epsilon as low as 0.08 can be obtained at total temperatures less than 800 K, with no dissociation.[45] Using hexafluoroethane (Freon-116, C_2F_6) as a test gas, values of epsilon as low as 0.0667 have been obtained for Mach 5.4 flow, with the total temperature below 600 K. The values of epsilon and the Reynolds number simulation of CF_4 and C_2F_6 facilities are compared in Fig. 4.25 with the values of those parameters for several facilities at the Langley Research Center. Also presented in Fig. 4.25 are two entry trajectories typical of very blunt, high-drag configurations.

Hunt et al.[45] cite investigations that have shown that, although the aerodynamic characteristics of blunt configurations are essentially independent of Mach number, they are strongly dependent on real-gas effects, such as the excitation of vibrational energy levels or dissociation. These real-gas effects have been shown to correlate as a function of the density ratio across the strong bow shock wave for blunt configurations.

When extremely high freestream Mach numbers are required, helium may be used as a test gas. Anfimov[13] notes that using helium as

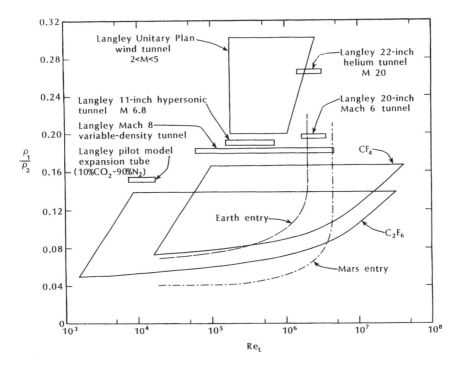

Fig. 4.25   Capability of obtaining density ratio with C_2F_6 and CF_4, as taken from Ref. 45.

a test gas allows Mach numbers exceeding 35 in the TSNIIMASH experimental facilities. Two high Mach number facilities at the Langley Research Center's HFC use helium as the test gas. Because helium can be expanded to temperatures as low as 3 to 4 °R without liquefying at the low pressures that exist in the wind-tunnel test section, Mach numbers of 28 can be obtained without heating the helium. As noted by Miller,[44] "These helium tunnels are the only high Mach number facilities for which the freestream flow and the flow within the shock layer of the model are thermally and calorically perfect." However, the ratio of the specific heats (γ) is 5/3 for helium, which is much higher than that typical of hypersonic flows in air (whether it behaves as a perfect gas or exhibits high-temperature effects). On the positive side, because the test gas is unheated, relatively inexpensive wood and plastic models can be used.

Nitrogen finds use as a test gas for several reasons. Nitrogen has roughly the same molecular weight and has the same perfect-gas value of γ as air. However, nitrogen can go to higher temperatures without dissociating and can go to lower temperatures before liquefaction occurs. Thus, one can achieve higher Mach numbers using

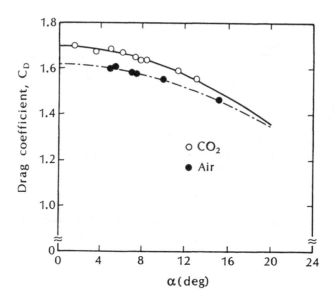

Fig. 4.26   Effect of gas composition on the drag coefficient of the Viking probe, U_∞ = 3.4 km/s, Re_D = 0.8 × 10⁶, as taken from Ref. 46.

nitrogen as the test gas in a "conventional" wind tunnel. For instance, in the Hypersonic Nitrogen Tunnel at the Langley Research Center, the nitrogen expands from the reservoir where the total pressure ranges from 2000 to 5500 psia and the total temperature ranges from 2800 to 3500 °R to a freestream Mach number of 17 with a test time of 3600 s. According to Miller,[44] this facility produces the highest value of the hypersonic viscous-interaction parameter of any NASA facility. Furthermore, when chemical reactions do occur, NO is not formed, so the test gas is environmentally cleaner. On the other hand, because there is no oxygen, real-gas effects typical of hypersonic flight in air cannot be simulated.

For many flowfields, e.g., hypersonic flow past a slender body with control surfaces, γ varies throughout the flowfield. Thus, the use of a test gas with a constant γ, even though approximately correct for the shock layer, still introduces questions regarding simulation.

Kirk et al.[46] note that gas properties can significantly affect the aerodynamics of probe configurations. This is shown in the comparison of the Viking drag measurements obtained for tests in CO_2 and those obtained for tests in air at one Reynolds number/velocity condition which is reproduced in Fig. 4.26. The data representing the use of different test gases were obtained in a ballistic free-flight range. These gases represent the atmosphere of other planets, such as Mars.

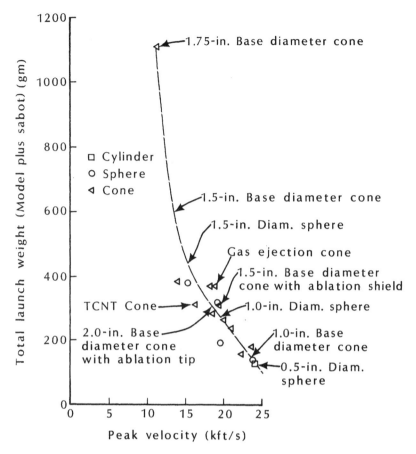

Fig. 4.27 Launching capabilities for typical models for Range G (AEDC), as taken from Ref. 14.

4.3.4 Ballistic Free-Flight Ranges

Range facilities offer a free-flight capability wherein a subscale model is launched to hypersonic velocities through a long-range tank having a controlled environment. Present high-performance ballistic ranges have a two-stage light-gas gun, which is used to launch the model, a tank in which the sabot is separated from the model and the gun gases are trapped, and a test section where the major portion of the instrumentation is located. The sabot is used to encase the model during the high acceleration loads during launch. After launch, the sabot pieces are separated from the model, usually by aerodynamic forces, and the model is free to fly through the test section.

The launch velocities that can be obtained in Range G at AEDC[14] are reproduced in Fig. 4.27. The higher velocities are obtained using light-weight, small-diameter models.

Increased capabilities can be obtained by coupling a free-flight range to a counter flow facility, e.g., a shock tunnel. For the Hypersonic Free-Flight Aerodynamic Facility (HFFAF) at the Ames Research Center (NASA), a shock tunnel produces a Mach 7 flow in the test section counter to the ballistic flight of the model producing an increase in the relative Mach number. As a result, a wide range of Mach numbers (0.2 to 22, 29 with counterflow) and Reynolds numbers (250 to 3×10^7) are possible.[47] The test gas can be readily changed to any nontoxic gas. Thus, one can readily simulate flight in planetary atmospheres in a ballistic free-flight range.[46]

Trimmer et al.[3] state:

> Disadvantages inherent to aeroballistic ranges include: 1) large accelerations generate large stresses in the model structure and generally restrict the model to be small and of uncomplicated shape at the higher velocities; 2) experimental measurements are difficult to obtain; and 3) small model scale gives rise to nonequilibrium effects at high velocities in realistic atmospheres.

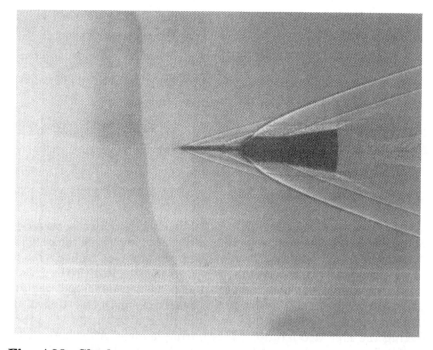

Fig. 4.28 Shadowgraph showing flowfield structure for the Apollo launch configuration with Launch Escape System, courtesy Marshall Space Flight Center (NASA).

Strawa et al.[47] concur, noting:

> Despite the wide capability of the ballistic range, data acquisition can be a problem. With present guns, the models used are relatively small and simple. Complex models can be flown, but they require gentle launches and, in most cases, they must not have lift at trim conditions. At the present time, all high speed data are acquired remotely.

Thus, most of the data are obtained through optical techniques, e.g., shadowgraph pictures (see Fig. 4.28), radiometer measurements of emissions from the gas cap, and flight-trajectory measurements including model position, orientation, and time of flight. However, advances in instrumentation and in telemetry can provide measurements not available to earlier experimenters.

A variety of materials-related tests have been conducted in Range G (AEDC). These tests include[14] ablation, erosion, nosetip transition, heat transfer, and transpiration cooling.

By evacuating the test section, one can measure the drag coefficients at extremely low Reynolds numbers typical of flight at noncontinuum conditions. Data from the AEDC ballistic range that were reproduced by Bailey and Hiatt[48] are presented in Fig. 4.29. The importance of obtaining drag data at these low Reynolds numbers is that the drag coefficient increases markedly as the slip-flow and free-molecular-flow regimes are approached.

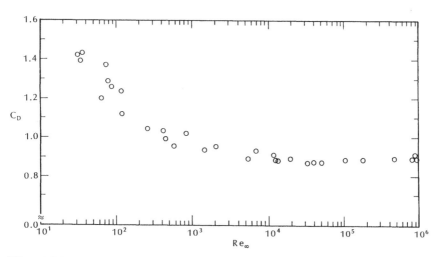

Fig. 4.29 Variation of sphere drag coefficient with Reynolds number at hypersonic speeds for $8.3 \leq M_\infty \leq 10.0$, using ballistic range data reproduced in Ref. 48.

Strawa et al.[47] note that the requirement to launch larger models at high speeds dictates an improvement of the present launching capabilities. They discuss several new technologies that are under development.

4.4 EXPERIMENTAL DATA AND MODEL DESIGN CONSIDERATIONS

Every one of the reasons that were put forth at the outset of this chapter for conducting ground-based test programs involves obtaining data. Therefore, specifying the appropriate data to be obtained is one of the most important factors in planning a test program, if we are to achieve the objectives.

Because reliance on CFD as a design tool continues to increase, an ever-increasing number of test programs are conducted to validate and to calibrate the numerical codes. CFD *code calibration* implies comparison of CFD code results with the experimental data over a range of configuration geometries and of flow conditions in order to provide a measure of the code's ability to predict specific parameters, such as the drag or the heat transfer to the surface. Since the data used in a code calibration exercise are either of the global type, e.g., drag and pitching moment, or surface measurements, e.g., pressure and heat transfer, agreement between a measurement and the corresponding computed value does not necessarily imply that all the features of the flow are correctly modeled. CFD *code validation* implies detailed surface and flowfield measurements compared with computed parameters to verify the code's ability to accurately model the critical physics and the chemistry of the flow. Thus, in addition to global measurements and surface measurements, code validation exercises require data defining the detailed structure of the flowfield, e.g., pitot probe data and electron beam data.

4.4.1 Heat-Transfer Data

Although an extended discussion of instrumentation is beyond the scope of this text, we will review techniques to measure heat transfer. The measurement of heat transfer to the surface bounding a flow is of interest to the fluid physicist, because it can provide information on the thermodynamic, chemical, or mechanical state of the fluid itself, e.g., the stagnation enthalpy, the degree of dissociation, whether the boundary layer is laminar or turbulent, etc. Furthermore, if one seeks to calibrate a computational code by comparing a computed flowfield parameter with the corresponding experimental value, matching the heat-transfer distribution is more challenging than matching the pressure distribution.

Thompson[49] notes that there are three conceptual methods of measuring heat flux.

1. The heat flux may set up a temperature gradient in a thin material layer. The heat flux for these "sandwich" gages can be related to the temperature gradient and to the material properties.

2. The incident heat flux may be caught in a thermal mass which acts as a calorimeter whose transient temperature change can be related to the heat flux.

3. The heat input may be balanced against a calibrated heat removal in a steady-state process.

Neumann[50] cautions that, although these categories represent fundamentally different methods for determining aerodynamic heating, there is not a unique relationship between the methods stated and their physical embodiment in a gage. The same physical gage, as for instance a wafer of material with thermocouples attached both to the heated surface and to the backface surface, can be used to determine the heat flux by any of the three stated methods. The differences among the methods have to do with the interaction between the thermal diffusion time and the test time.

Note that heat transfer is not a measured quantity. Successful determination of the heat transfer requires the measurement of specific temperatures located in a "sensor" that can be accurately modeled and that does not disturb the thermal environment.

Because of the wide variety of test parameters, such as model design, run time, and thermal environment, a large variety of test techniques have been developed to measure heat transfer. Let us categorize the techniques in one of two broad classes:

1. Temperature-sensitive surface coatings which include:

    a. Coatings that change phase at a specific temperature,[51] and

    b. Coatings of thermographic phosphor that produce nonuniform ultraviolet light distribution,[52] and

2. Point-discrete sensors, such as:

    a. Thermocouples used in the thin-skin technique,[50, 52]

    b. Thin-film resistance thermometers on a pyrex or quartz substrate,[50, 53]

    c. Schmidt-Boelter gages in which the temperature difference across a slab that is backed by a sink is related to the incident heat flux at the surface,[1] and

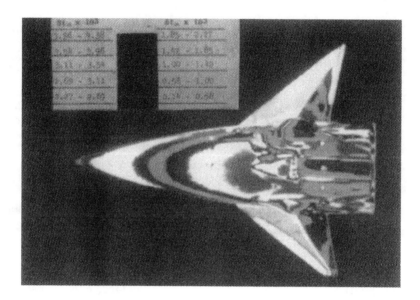

Fig. 4.30   Heat-transfer distribution obtained using thermo-graphic phosphor.

     d. Gardon gages in which the heat flows through a constan-tan disk to a copper heat sink (which is also the main structural component) establishing a radial temperature gradient which is used to define the incident heat flux.[52]

For the techniques employing temperature-sensitive surface coat-ings, a "massless" temperature indicator is painted on the surface of a model made of an insulative material. These techniques assume that the model wall temperature response is that of a semi-infinite slab subjected to an instantaneous and constant heat-transfer coefficient. Thus, one obtains the heat-transfer distribution for the entire model, rather than only at the specific locations where the sensors are lo-cated. See Fig. 4.30. Obtaining the heat-transfer distribution for the entire model is a principal advantage of this technique. However, the most severe heating rates are often missed, because they trigger the coating's response too quickly to be properly timed and/or recorded. In fact, extremely severe heating can trigger the coating's response while the model is being injected into the test stream. Thin sections of the model limit the application of the semi-infinite slab thermal model. Thus, one cannot use temperature-sensitive surface coatings to determine the heating rates near the wing leading edge, where the heating may be critical. Application of temperature-sensitive coat-ings are slow and labor intensive. Furthermore, not only are the data reduction techniques slow and labor intensive, the data are of limited accuracy.

The discrete measurement techniques offer the advantages of electrical outputs from sensors at well-defined locations. Thus, the resultant data can be handled using automated data reduction and plotting routines. The advantages and the disadvantages of various test techniques as summarized by Matthews et al.[1] are reproduced in Table 4.2. It should be noted that other experimentalists might have different opinions of the advantages and of the disadvantages of a particular technique.

The effect of the flowfield on the heat transfer can be characterized by the heat-transfer coefficient (h) which appears in Newton's law of cooling:

$$h = \frac{\dot{q}}{T_{aw} - T_w} \qquad (4\text{-}7)$$

Assuming that the heat-transfer coefficient is primarily determined by the flowfield and, therefore, constant for a given flow condition, the heat flux will vary as the wall temperature changes. Note that, as the wall temperature approaches the adiabatic wall temperature, the heat-transfer rate goes to zero. If the heating occurs over an extended period of time, the boundary-layer flow may change significantly, producing a significant change in h. Thus, to produce useful heat-transfer data, a sensor must provide both the surface heating rate and the wall temperature. Although an accurate determination of the adiabatic wall temperature is significant, it is difficult to measure. For hypersonic flows, numerical approximations or an accepted convention are generally used in its place.

Neumann[50] warns that "If a gage is not thermally matched to the model wall, then it not only disturbs the flowfield but induces lateral conduction of heat between the gage and the surrounding model structure. In this case, the gage is now measuring the sum of the aerodynamic heating caused by flowfield deceleration and that of conduction." Increased test duration leads to increases in the model temperature, probably creating nonuniform surface temperatures. Since this may lead to unanticipated and unmeasured thermal paths, increased test duration is not always desirable.

4.4.2 Flow Visualization Techniques

A variety of flow visualization techniques can be used to develop a considerable understanding of the flowfield. Large density gradients in the flowfield, such as those associated with shock waves, boundary layers, etc., can be made visible using various optical methods. An interferometer provides a direct measure of the changes in density. The schlieren technique provides a measure of the density gradients, whereas the shadowgraph technique depends on the second derivative

Table 4.2　Advantages and disadvantages of test techniques, as taken from Ref. 1

	ADVANTAGES	DISADVANTAGES
Discrete Measurements		
• Thin-skin	High quality data, dense spacing	Expensive model fab., conduction effects
• Coax gage	Easy to install, contourable, durable	Low output, short test times
• Schmidt-Boelter gage	High output, slightly contourable, very durable	Limited experience to date, but promising
• Gardon gages (high temp., low temp.)	Years of experience, fast response	Gage attrition rate, not contourable
• Thin-film	Dense spacing, fast response can be used on small radii	Limited experience to date, relatively difficult installation
Thermal Mapping		
• Phase-change paint	Vivid illustration of hot spots High spatial resolution	Slow run rate, data presentation can be confusing
• IR scanning camera	Computer-generated plots and color maps, and nonintrusive	Spatial resolution
• Thermographic phosphor	High run rate, good spatial resolution	Difficulties in data reduction and presentation
Materials Testing		
• Screening test	High run rate, many samples run in short time	Only relative ranking between samples
• Materials characterization	Characterization can allow material evaluation at selected flight conditions	Requires testing on material over a range of several parameters, $\dot{q}$, T_w, shear
• Component survivability	Provides "yes" or "no" answer for simulated flight condition	No quantitative data for "other" flight conditions
• Component Thermal response	Can provide temperatures for a variety of simulated conditions	Data valid only for specific component tested

of the density. Thus, as was illustrated in Figs. 4.20 and 4.28, the variations in the density allow us to see shock waves, boundary layers, and trip-induced flowfield perturbations.

Which optical technical technique is used depends not only on the level of detail required, i.e., first or second derivative of the density, but on the density itself. Different flow visualization techniques have been used to study the interaction of a Reaction Control System (RCS) jet with the hypersonic flowfield over a slender cone. A 7 deg. half-angle cone at an angle-of-attack of 10 deg. was subjected to the Mach 14 airstream of the 20-Inch Hypersonic Wind Tunnel at the Flight Dynamics Directorate of the Wright Laboratory. Since the total pressure in the reservoir was 1000 psia and the total temperature was 1900 °R, the freestream density in the test section was approximately $0.02\rho_{SL}$. As can be seen in Fig. 4.31a, the schlieren technique provided virtually no insight into the interaction between the RCS jet and the cone flowfield for this low-density flow. Clearly, alternative techniques are needed for low-density flows. The electron-beam fluorescence (EBF) technique was developed in the 1960s as a spectroscopic method for determining gas density, velocity, and temperature at low density. As noted by Cattolica et al.,[54] "The density distribution across the boundary layer can be measured using electron-beam excitation of the nitrogen molecule. By measuring the fluorescence signal from the N_2^+ ion, number densities up to 10^{16} cm^{-3} have been measured, with a linear dependence of signal on density." Improved resolution of the interaction flowfield is evident in a black-and-white electron-beam photograph, which is presented in Fig. 4.31b. The best resolution is obtained with the color electron-beam photograph, which is presented in Fig. 4.31c.

Another flow visualization technique records the motion of a thin film of oil as it responds to flow of air around the model. The resultant oil film on the surface can provide insights into surface shear and separation. An oil-flow pattern for the leeward nose region of the Shuttle Orbiter at an angle-of-attack of 30 deg. in Tunnel B at AEDC is presented in Fig. 4.32. The oil-flow pattern portrays a complex flowfield, dominated by viscous/inviscid interactions. A free-vortex shear-layer separation takes place as the air flows around the nose. Although cross-flow separation takes place, the leeward flow maintains a strong axial component (as indicated by the streamwise streak near the leeward plane-of-symmetry). This flow reattaches when it encounters the windshield. The high shear forces remove the oil from the windshield. Another pair of separation vortices form downstream of the canopy, reattaching to the leeward surface of the payload bay (downstream of the canopy) and leaving a feather pattern near the leeward plane-of-symmetry. The oil-flow pattern also provides qualitative insights into the heat transfer, since there is a correlation between heat transfer and skin friction. Refer to Fig. 2.11.

(a) Black-and-white schlieren

(b) Black-and-white electron beam

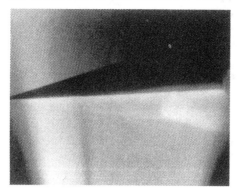

(c) Color electron beam

Fig. 4.31 Photographs of sonic RCS jet exhausting into the leeside-cone flowfield, $M_1 = 14$, $\alpha = +10$ deg., $\theta_c = 7$ deg., provided by Flight Dynamics Directorate, Wright Laboratory.

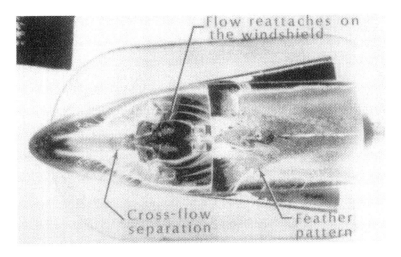

Fig. 4.32   Oil-flow pattern over the leeward surface of the nose region of the Space Shuttle Orbiter at α = 30 deg.

These two flow-visualization techniques have been combined in Fig. 4.33, where the oil-flow pattern is superimposed on a schlieren photograph for Mach 8 flow over an indented nose cone. Recall that a schlieren photograph captures the planar traces of three-dimensional, curved shock waves. Thus, the shock wave does not actually intersect the concave surface in the leeward plane-of-symmetry. Evident in the photograph are the complex viscous/inviscid shock/shock interaction on the windward side, a free-shear layer originating at the inflection of the bow shock wave on the leeward side near the nose, and the cross-flow streamlines in the oil-flow pattern. The inclination of the bow shock wave is such that the downstream flow is supersonic. However, the inclination of the outboard shock wave is such that the downstream flow is subsonic, as indicated in Fig. 4.33. Thus, a complex viscous/inviscid interaction containing a shear layer and a jet can be seen originating at the intersection of the two shock waves. As will be discussed in Chap. 9, the heating in the region where the shear-layer/jet combination strikes the surface can be an order of magnitude greater than the undisturbed heating. Thus, we also need surface-pressure measurements and heat-transfer measurements to provide a more complete picture of the flowfield.

Since the shock stand-off distance and, therefore, the shock interaction are affected by the gas chemistry, the designer must account for real-gas effects. If this is to be done using computational fluid dynamics, the flow model must accurately represent the phenomena depicted in Figs. 4.32 and 4.33. Thus, the development of CFD codes

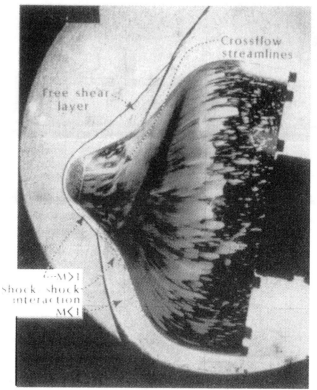

Fig. 4.33   Flow over an indented nose cone in a Mach 8 stream.

and the validation of the models used therein must always make use of quality experimental data.

4.4.3 Model Design Considerations

For a hypersonic wind-tunnel program, the cost of the model and the facility operation may approach one million dollars. Thus, the model design and its instrumentation will have an important impact in meeting the test objectives within available funding. For instance, consider the test objective of determining the perturbations to the heat transfer and surface pressure in the shock/shock interaction region associated with the deflection of a control surface. One could use a model of Stycast (a plastic) coated with a temperature-sensitive paint which would yield a map of the entire heat-transfer distribution. However, the thermal mapping technique leads to reduced precision and tends to complicate data reduction and analysis procedures. Furthermore, it will be necessary to have a second model in order to obtain the surface-pressure data. Conversely, one could use a metal model instrumented with point-discrete sensors, e.g., Schmidt-Boelter gages and static-pressure orifices. However, the more severe heating rates and surface pressures may not be measured, because the sensors were not located in the (relatively small) critical areas. Another consideration for the designer of this model is whether or not to use motor-driven control surfaces. Although the motor-driven control surfaces complicate the model design slightly, the additional cost is a small percentage of the total model cost. Furthermore, the savings in test time are significant. Thus, the motor-driven control-surface technique can be very cost-effective.

In most hypersonic wind-tunnel tests, the model is held in place by a support sting. Ericsson and Reding[55] note that "All support systems cause interference of one kind or another," and that "Support interference is much more of a problem in dynamic than in static tests." Sting interference becomes a problem when communication is opened up between the aft body and the base recirculation region. Thus, tests conducted at hypersonic, low-density flow conditions and tests conducted at a Reynolds number such that boundary-layer transition occurs on the aft body near the base may be affected by sting interference problems. In hypersonic, low-density flow, a thick boundary layer opens up the communication between the aft body and the wake. The presence of the support affects the lip shock via the wake recompression pressure, propagated upstream through the near-wake compression region. The resulting change of shoulder pressure is felt upstream of the base to an extent roughly proportional to the boundary-layer thickness.

Although there may be no sting size small enough to eliminate all interference, it is, in many cases, possible to minimize the interference to an acceptable value by using a very slender sting support. For a model with a flat base, sting support interference is usually not a problem for hypersonic flows.

For slender configurations, such as the NASP or waveriders, precise knowledge of the angle-of-attack and of the sideslip angle is required, since small misalignments could have a significant effect on the measurements. Thus, the model support system is an important consideration for such configurations.

However, the model design process offers unique challenges and risks. A wedge model whose plate surface was 0.6096 m (2.000 ft) by 0.4318 m (1.417 ft) was designed for tests in Tunnel B (AEDC). The test procedure is such that, once the tunnel flow is established, the operating conditions are held continuously. Thus, operations on the model, such as model changes, cooling the model between tests, etc., are accomplished in a sealed chamber beneath the test section. To take data, a door in the tunnel floor is opened, the model is injected into the test section, the model is then pitched and/or rolled on the sting until the proper orientation is obtained, and the data are taken.

The unsteady loads to which the wedge model was subjected as it passed through the wall boundary layer, crossed the test section, and hit the stop, ripped the model from the sting. This expensive failure occurred in the threads that transmitted the load from the model to the sting.

Another interesting problem encountered by the author is associated with model size. Ordinarily, the model is as large as can be accommodated in the test section. With larger models, one can better represent configuration details and can have room for more point-discrete sensors. Thus, in an attempt to assess the effect of surface protuberances, e.g., shear pads and antennae, and cavities, e.g., windows, on the convective heat transfer to the Apollo Command Module, a blockage study was run to determine the largest model that could be tested in Tunnel C (AEDC). Inexpensive, uninstrumented, featureless models of different scales were placed in the Mach 10 stream of Tunnel C in December. It was found that a 14.0-inch diameter model could be tested without choking the flow. Thus, a 13.86-inch diameter model (i.e., a 0.090 scale Apollo Command Module) with a variety of protuberances and cavities was built and instrumented. During the actual test program, which was conducted in June, tunnel blockage occurred at the low Reynolds number conditions. The cause of this problem has never been identified. However, possible causes include: (1) differences in the tunnel wall boundary layer (and, therefore, a reduction in the effective test-section area) due to differences in the temperature of the cooling water from December to June, (2) flowfield differences produced by the protuberances on the instrumented model as compared with the smooth blockage model, (3) reduction in the pressure capability of the power plant is reduced during the summer, and (4) problems caused by the impingement of the bow shock wave on the window cavities. "Blockage model" tests are often used prior to actual test program, so that the appropriate model scale can be defined.

The uniformity of the flow in the test section can be an important factor in determining the model size. Variations of the flow properties in the radial direction or in the axial direction can restrict the model dimensions.

It is not the intent of the comments in this section to cause needless fears about ground-based testing. Instead, since ground-based testing has a fundamental role in the development of a hypersonic vehicle, these stories should stress the importance of planning in the conduct of a successful test program.

4.5 FLIGHT TESTS

Flight tests are relatively expensive. They take a long time to plan and to execute successfully. Furthermore, it is difficult to obtain quality data at well-defined test conditions. Flight tests will never replace ground-based tests nor CFD in the design process. Nevertheless, flight tests are critical to our understanding of the hypersonic aerothermodynamic environment, since they provide data which cannot be obtained elsewhere.

4.5.1 Flight-Test Objectives

Neumann[56] suggests a variety of reasons for conducting flight tests. To the four reasons suggested by Draper et al.,[57] which are:

1. To demonstrate interactive technologies and to identify unanticipated problems;

2. To form a catalyst (or a focus) for technology;

3. To gain knowledge not only from the flights but also from the process of development; and

4. To demonstrate technology in flight so that it is credible for larger-scale applications.

Neumann added three reasons of his own:

5. To verify ground-test data and/or to understand the bridge between ground-test simulations and actual flight;

6. To validate the overall performance of the system; and

7. To generate information not available on the ground.

There are two types of flight-test programs: (1) Research and Development (R and D) programs and (2) flights of prototype or operational vehicles. R and D programs are focused on technical issues, which drive the design of the vehicle and its flight operations. For example, Project Fire provided calorimeter heating measurements on a large-scale blunt body entering the earth's atmosphere at an initial velocity of 11.35 km/s. The various phases of the trajectory and the sequence of major events prior to re-entry are depicted in Fig. 4.34, which is taken from Ref. 58. As shown in Fig. 4.35, the forebody of the "Apollo-like" re-entry capsule was constructed of three beryllium calorimeter shields, which were alternated with phenolic-asbestos heat shields. The phenolic-asbestos layers were used to protect the succeeding beryllium layers until the desired exposure times just prior to and after peak heating. This multiple-layer arrangement provided three distinct experimental or data-gathering periods during re-entry, as shown in Fig. 4.36. Selected data from the Project Fire program will be presented in Chap. 5, "Stagnation-Region Flowfield."

Flight-test data have been obtained on prototype vehicles, e.g., the unmanned Apollo Command Module, 017 and 020, as discussed

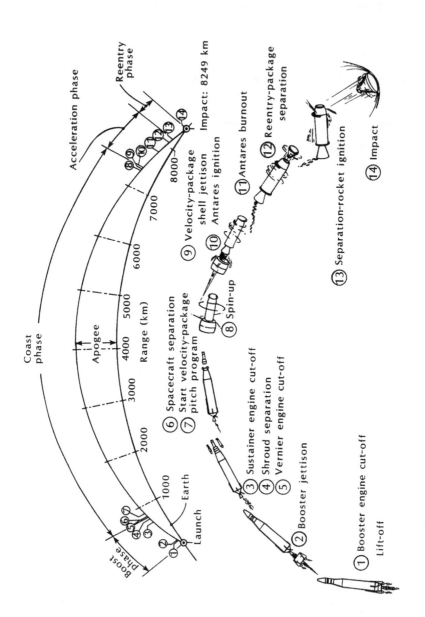

Fig. 4.34 Sequence of major events for a Project Fire flight, as taken from Ref. 58.

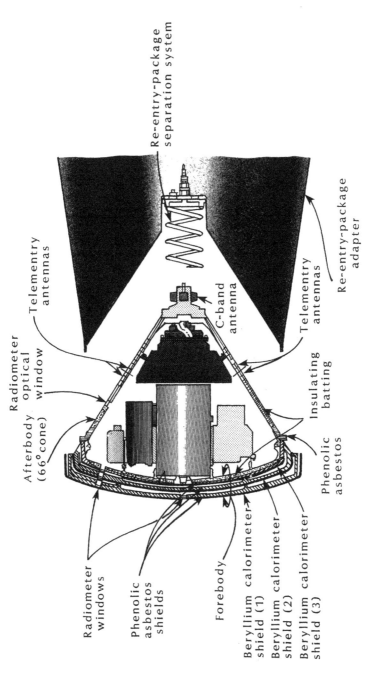

Fig. 4.35   Sketch of Project Fire re-entry package and adapter illustrating layered arrangement of forebody heat shielding, as taken from Ref. 58.

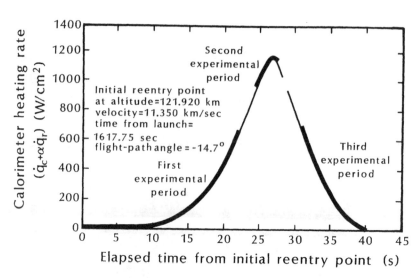

Fig. 4.36 Typical forebody heating rate and experimental data periods during Project Fire II re-entry, as taken from Ref. 58.

in Ref. 25, and on operational vehicles, e.g., the Space Shuttle. Aerothermodynamic parameters based on flight-test data obtained from thermocouples embedded in the Space Shuttle Thermal Protection System (TPS) were used to expand the flight envelope for the Orbiter. As noted by Hodge and Audley,[59] "Requirements for the technique include an analytical model for the simulation of the heat transfer to a point on the TPS, flight test maneuvers which cause thermocouples imbedded near the surface of the TPS to respond sufficiently above the noise levels, and a parameter estimation program to reduce flight thermocouple data. ... The data reduction program correlates heating with variables such as angle of attack, sideslip, control surface deflection, and Reynolds number. This technique could also be used for wind tunnel data reduction."

Since R and D flight-tests are designed to provide data relating to specific issues, they are most likely to provide information about known unknowns, i.e., problems that have already been identified during the design process and that are subjected to further study. One is most likely to discover unknown unknowns, i.e., to identify unanticipated problems, during flight tests of prototype vehicles or of operational vehicles. Such was the case for the pitching moment of the Space Shuttle Orbiter, which will be discussed in Sec. 4.6, and for the shock/shock interaction that caused severe heating to the X-15 pylon that supported the Hypersonic Research Engine (see Chap. 9).

4.5.2 Flight-Test Data

Williamson[60] divides flight-test measurements into three groups. These include atmospheric properties measurements, offboard vehicle related sensor measurements, and onboard vehicle related measurements. Atmospheric measurements determine wind conditions, atmospheric density, temperature, and pressure profiles as a function of altitude. These are extremely important parameters. Recall the discussion presented in Sec. 1.3, where the existence of large deviations from the Standard Atmosphere's density were discovered during the Shuttle flight tests. Typical ground-based support equipment includes radar and optical sensors both at the launch and at the re-entry area.

Onboard instrumentation for flight-test vehicles cannot be viewed as "off-the-shelf" items to be selected from vendor's catalogs. The discussion in this section will be confined to considerations related to heat-transfer data.

Neumann[56] notes:

> The Rockwell Space Shuttle was a fortuitous case where the selection of a thermal protection system from a weight consideration also was an ideal choice from a heat transfer measurement standpoint. It was a classical semi-infinite slab thermal model. The platinum wire thermocouples placed just below the surface were ideal for the sensed temperature that is the basis of heat transfer inference. They were extremely fine wire having little mass and they were placed along the surface as "isothermal" thermocouples such that conduction away from the thermocouple junction was minimized.

Serious problems can occur when the heat sensor is not properly integrated into the flight structure such that minimal thermal distortion is produced. Hearne et al.[61] examined the response of a slug calorimeter mounted in an ablative heat shield. The calorimeter is assumed to consist of a circular sensor surrounded by concentric guard rings, as shown in Fig. 4.37. The surface conditions are characterized by (1) two separate discontinuous changes in the wall temperature followed by an isothermal sensing element, (2) a step change in surface blowing, (3) a discontinuity in surface reactivity, and (4) protrusion of the sensor resulting in a pressure hump. Thus, Hearne et al.[61] conclude:

> Except at early times the net heat transfer to the calorimeter will bear little resemblance to the unaffected heat flux. Reradiation and the distortion of the convection coefficient created by the disturbances in the surface conditions

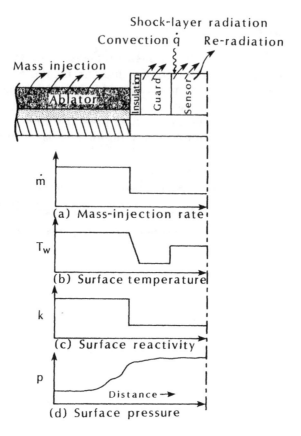

Fig. 4.37 Interactions between a calorimeter and its ablative environment, as taken from Ref. 61.

will be significant in altering the heat flux. Evaluation of the reradiation is straightforward since the data pertinent to its evaluation (sensor surface temperature and surface emissivity) will be known. However, determination of the convective heating requires that the complex influence of streamwise variations in surface conditions on the boundary layer characteristics be accounted for.

Throckmorton[62] analyzed the effect of solar radiation and, for the wing, cross radiation from the relatively hot Orbiter fuselage on the re-entry heating rates to the leeside of the Shuttle Orbiter. Because the convective heat transfer to the leeward surface is low, the radiant heat transfer had a significant effect on the measurements. For heat-transfer rates determined from calorimeter measurements, solar radiation is much more significant than was observed for thermocouple measurements. This results because the solar absorptivity

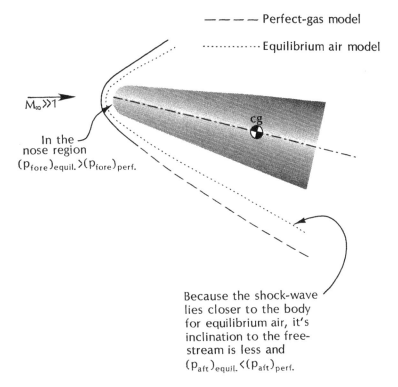

Fig. 4.38 Sketch of hypersonic flow past an inclined spherically-blunted cone comparing perfect-gas and equilibrium-air pressures.

of the calorimeter surface was high. For heating rates measured by calorimeters, radiation (solar plus cross) generally accounted for 15–30 percent of the total energy input.

The comments made in this section emphasize that the instrumentation must be carefully chosen and modeled in order to obtain valid data.

4.6 THE IMPORTANCE OF INTERRELATING CFD, GROUND-TEST DATA, AND FLIGHT-TEST DATA

Woods et al.[39] note that pre-flight predictions based on the aerodynamics in the Aerodynamics Design Data Book (ADDB) indicated that a 7.5 deg. deflection of the body flap would be required to trim the Space Shuttle Orbiter for the center-of-gravity and for the vehicle configuration of STS-1. In reality, the body flap had to deflect to much larger values ($\delta_{BF} \sim 16$ deg.) to maintain trim at the proper angle of attack ($\alpha = 40$ deg.). The deflection of 16 deg. was close

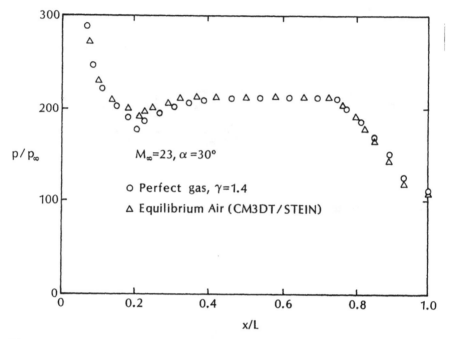

Fig. 4.39 A comparison of perfect-gas and of equilibrium-air computations of the windward pitch-plane pressure distribution for the Space Shuttle Orbiter at $\alpha = 30$ deg., as taken from Ref. 26.

to the limit of possible deflections. Comparisons of equilibrium-air calculations and perfect-gas calculations indicate that at least part of this so-called hypersonic anomaly is due to real-gas effects at very high Mach numbers. At Mach 8, the flight data and the ADDB values agreed.[39]

Consider the flow depicted in the sketch of Fig. 4.38. For perfect air ($\gamma = 1.4$), $\rho_2 = 6\rho_1$ across the normal portion of the shock wave; whereas $\rho_2 = 15\rho_1$ for air in thermodynamic equilibrium ($\gamma = 1.14$). Thus, for the equilibrium-air model, the shock layer is thinner and the inclination of the bow shock wave relative to the freestream is less than that for the perfect-air model. Tangent-cone theory applied to the afterbody region shows a decrease in pressure with decreasing gamma. Using this simplistic flow model, one would expect the equilibrium-air pressures on the aft end of the vehicle to be less than those for perfect air. Computations of the flowfield for simplified Orbiter geometries as reported by Woods et al.[39] and by Maus et al.[26] indicate that this is the case. The calculations of Maus et al. are reproduced in Fig. 4.39.

Maus et al.[26] note further that the stagnation pressure increases with decreasing gamma. Thus, as presented in Fig. 1.18, the

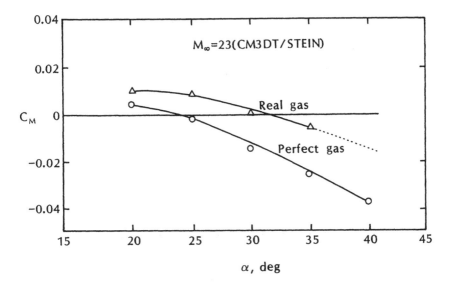

Fig. 4.40 A comparison of perfect-gas and of equilibrium-air computations of the pitching moment for the Space Shuttle Orbiter at M_∞ = 23, as taken from Ref. 26.

equilibrium-air value for the stagnation pressure is greater than that for perfect air. This too is reflected in the nose region pressures presented in the more rigorous solutions of Woods et al. and Maus et al.

The differences between the equilibrium-air pressure distribution and the perfect-air pressure distribution may appear to be relatively small. Indeed, there is little difference in the normal force coefficients for the equilibrium-air model and for the perfect-air model. However, because the equilibrium-air (real-gas) values are higher at the nose and lower at the tail, the real-gas effects tend to drive C_M more positive. The pitching moments for the Space Shuttle Orbiter at M_∞ = 23 presented by Maus et al.[26] are reproduced in Fig. 4.40.

Thus, detailed studies incorporating wind-tunnel data, flight-test data, and CFD solutions, as reported in Refs. 26 and 39, provide insights into a sophisticated aerodynamic problem.

REFERENCES

[1] Matthews, R. K., Nutt, K. W., Wannenwetsch, G. D., Kidd, C. T., and Boudreau, A. H., "Developments in Aerothermal Test Techniques at the AEDC Supersonic-Hypersonic Wind Tunnels," AIAA Paper 84-1803, Snowmass, CO, June 1984.

[2] Martellucci, A., "The Challenging Process of Validating CFD Codes," AIAA Paper 90-1402, Seattle, WA, June 1990.

[3] Trimmer, L. L., Cary, Jr., A. and Voisinet, R. L., "The Optimum Hypersonic Wind Tunnel," AIAA Paper 86-0739CP, West Palm Beach, FL, Mar. 1986.

[4] Neumann, R. D., "Experimental Methods for Hypersonics: Capabilities and Limitations," presented at the Second Joint Europe/U.S. Short Course in Hypersonics, U.S. Air Force Academy, Jan. 1989.

[5] Miller, C. G., "Experimental Investigation of Gamma Effects on Heat Transfer to a 0.006 Scale Shuttle Orbiter at Mach 6," AIAA Paper 82-0826, St. Louis, MO, June 1986.

[6] Romere, P.O., and Young, J. C., "Space Shuttle Entry Longitudinal Aerodynamic Comparisons of Flight 2 With Preflight Predictions," *Journal of Spacecraft and Rockets,* Vol. 20, No. 6, Nov.–Dec. 1983, pp. 518–523.

[7] Whitnah, A. M., and Hillje, E. R., "Space Shuttle Wind Tunnel Testing Program Summary," NASA RP-1125, 1984.

[8] Neumann, R. D., "Defining the Aerothermodynamic Methodology," J. J. Bertin, R. Glowinski, and J. Periaux (eds.), *Hypersonics, Volume I: Defining the Hypersonic Environment,* Birkhäuser Boston, MA, Boston, 1989.

[9] Sharma, S. P., Park, C., and Dannenberg, R. E., "Operating Characteristics of a 60 cm and a 10 cm Electric Arc-Driven Shock Wave," AIAA Paper 88-0142, Reno, NV, Jan. 1988.

[10] Rose, P.H., and Stark, W. I., "Stagnation Point Heat-Transfer Measurements in Dissociated Air," *Journal of the Aerospace Sciences,* Vol. 25, No. 1, Jan. 1958, pp. 86–97.

[11] Moeckel, W. E., and Weston, K. C., "Composition and Thermodynamic Properties of Air in Chemical Equilibrium," NACA TN-4265, Apr. 1958.

[12] Hertzberg, A., "The Application of the Shock Tube to the Study of the Problems of Hypersonic Flight," *Jet Propulsion,* Vol. 26, No. 7, July 1956, pp. 549–554, 568.

[13] Anfimov, N., "TSNIIMASH Capabilities for Aerogasdynamical and Thermal Testing of Hypersonic Vehicles," AIAA Paper 92-3962, Nashville, TN, July 1992.

[14] Trimble, M. H., Smith, R. T., and Matthews, R. K., "AEDC High-Temperature Testing Capabilities," Arnold Engineering Development Center, AEDC TR-78-3, Apr. 1978.

[15] Laster, M. L., private transmittal, Mar. 1992.

[16] Laganelli, A. L., and Martellucci, A., "Experimental Surface and Boundary-Layer Measurements in a Hypersonic Boundary Layer with Nonuniform Blowing," Yovanovich (ed.), *Heat Transfer with Thermal Control Applications,* Vol. 39 of Progress in Astronautics and Aeronautics, AIAA, New York, 1975, pp. 395–414.

[17] Bertin, J. J., Nipper, M. J., and Conine, W. D., "Effect of Gas Composition on the Ablation Performance of Phenolic Nylon," *Journal of Spacecraft and Rockets,* Vol. 7, No. 10, Oct. 1970, pp. 1253–1256.

[17A] Daum, F. L., and Gyarmathy, G., "Condensation of Air and Nitrogen in Hypersonic Wind Tunnels," *AIAA Journal,* Vol. 6, No. 3, Mar. 1968, pp. 458–465.

[18] Pate, S. R., "Effects of Wind Tunnel Disturbances on Boundary-Layer Transition with Emphasis on Radiated Noise: A Review," AIAA Paper 80-0431, Colorado Springs, CO, Mar. 1980.

[19] Boudreau, A. H., "Characterization of Hypersonic Wind Tunnel Flowfields for Improved Data Accuracy," AGARD Paper, Arnold Engineering Development Center (AEDC), Sep. 1987.

[19A] Bouslog, S. A., An, M. Y., and Derry, S. M., "Orbiter Windward Surface Boundary-Layer Transition Flight Data," presented at the Shuttle OEX Aerothermodynamics Symposium, Langley Research Center (NASA), Apr. 1993.

[20] Cooper, G. C., and Eaves, R. H., "Increased Reynolds Number Capability at the AEDC Hypersonic Tunnel C," presented at the 68th STA Meeting, Oct. 1987, Columbus, OH.

[21] Ames Research Staff, "Equations, Tables, and Charts for Compressible Flow," NACA Rept. 1135, 1953.

[22] Anonymous, "Hypersonic Shock Tunnel: Description and Capabilities," Calspan Corporation, Buffalo, NY, Mar. 1973.

[23] Bertin, J. J., "Wind-Tunnel Heating Rates for the Apollo Spacecraft," NASA TM X-1033, Jan. 1965.

[24] Lee, D. B., Bertin, J. J., and Goodrich, W. D., "Heat-Transfer Rates and Pressure Measurements Obtained During Apollo Orbital Entries," NASA TN D-6028, Oct. 1970.

[25] Lee, D. B., and Goodrich, W. D., "The Aerothermodynamic Environment of the Apollo Command Module During Superorbital Entry," NASA TN D-6792, Apr. 1972.

[26] Maus, J. R., Griffith, B. J., Szema, K. Y., and Best, J. T., "Hypersonic Mach Number and Real-Gas Effects on Space Shuttle Aerodynamics," *Journal of Spacecraft and Rockets,* Vol. 21, No. 2, Mar.–Apr. 1984, pp. 136–141.

[27] Griffith, B. J., Maus, J. R., Majors, B. M., and Best, J. T., "Addressing the Hypersonic Simulation Problem," *Journal of Spacecraft and Rockets,* Vol. 24, No. 4, July–Aug. 1987, pp. 334–341.

[28] Morrisette, E. L., Stone, D. R. and Whitehead, Jr., A. H., "Boundary-Layer Tripping with Emphasis on Hypersonic Flows," C. S. Wells (ed.), *Viscous Drag Reduction,* Plenum Press, New York, 1969.

[29] Nestler, D. E., and McCauley, W. D. "A Study of a Boundary-Layer Trip Concept at Hypersonic Speeds," AIAA Paper 81-1086, Palo Alto, CA, June 1981.

[30] Braslow, A. L., "Review of the Effect of Distributed Surface Roughness on Boundary-Layer Transition," NATO AGARD Rept. 254, Apr. 1960.

[31] Sterrett, J. R., Morrisette, E. L., Whitehead, Jr., A. H., and Nicks, R. M., "Transition Fixing for Hypersonic Flow," NASA TN D-4129, Oct. 1967.

[32] Bertin, J. J., Tedeschi, W. J., Kelly, D. P., Bustamante, A. C., and Reece, E. W., "Analysis of the Expansion-Fan Flowfield for Holes in a Hypersonic Configuration," *AIAA Journal*, Vol. 27, No. 9, Sep. 1989, pp. 1241-1248.

[33] Ericsson, L. E., and Reding, J. P., "Reynolds Number Criticality in Dynamic Tests," AIAA Paper 78-166, Huntsville, AL, Jan. 1978.

[34] Liepmann, H. W., and Roshko, A., *Elements of Gas Dynamics*, John Wiley & Sons, New York, 1957.

[35] Potter, J. W., "Transitional, Hypervelocity Aerodynamic Simulation and Scaling," J. N. Moss and C. D. Scott (eds.), *Thermophysical Aspects of Re-entry Flows*, Vol. 103 of Progress in Astronautics and Aeronautics, AIAA, New York, pp. 79-96, 1985.

[36] Hayes, W. D., and Probstein, R. F., *Hypersonic Flow Theory*, Academic Press, New York, 1959.

[37] Holden, M. S., "Theoretical and Experimental Studies of the Shock Wave-Boundary Interaction on Curved Compression Surfaces," *Proceedings of the 1969 Symposium, Viscous Interaction Phenomena in Supersonic and Hypersonic Flow*, University of Dayton Press, Dayton, OH, 1970.

[38] Love, E. S. "Advanced Technology and the Space Shuttle," *Aeronautics and Astronautics*, Vol. 11, No. 2, Feb. 1973, pp. 30-66.

[39] Woods, W. C., Arrington, J. P., and Hamilton II, H. H., "A Review of Preflight Estimates of Real-Gas Effects on Space Shuttle Aerodynamic Characteristics," *Shuttle Performance: Lessons Learned*, NASA CP-2283, Part 1, Mar. 1983.

[40] Wilhite, A. W., Arrington, J. P., and McCandless, R. S., "Performance Aerodynamics of Aeroassisted Orbital Transfer Vehicles," H. F. Nelson (ed.), *Thermal Design of Aeroassisted Orbital Transfer Vehicles*, Vol. 96 of Progress in Astronautics and Aeronautics, AIAA, New York, 1985, pp. 165-185.

[41] Bertin, J. J., and Goodrich, W. D., "Effects of Surface Temperature and Reynolds Number on Leeward Shuttle Heating," *Journal of Spacecraft and Rockets*, Vol. 13, No. 8, Aug. 1976, pp. 473-480.

[42] Trimmer, L. L., and Matthews, R. K., "Supersonic Aerothermal Testing—A New Requirement," AIAA Paper 78-773, San Diego, CA, Apr. 1978.

[43] Compiled by Peneranda, F. E., and Freda, M. S., "Aeronautical Facilities Catalogue: Volume 1: Wind Tunnels," NASA RP-1132, Jan. 1985.

[44] Miller, C. G., "Langley Hypersonic Aerodynamic/ Aerothermodynamic Testing Capabilities – Present and Future," AIAA Paper 90-1376, Seattle, WA, June 1990.

[45] Hunt, J. L., Jones, R. A., and Smith, K. A., "Use of Hexafluoroethane to Simulate the Inviscid Real-Gas Effects on Blunt Entry Vehicles," NASA TN D-7701, Oct. 1974.

[46] Kirk, D. B., Intrieri, P. F., and Seiff, A., "Aerodynamic Behavior of the Viking Entry Vehicle: Ground Test and Flight Results," *Journal of Spacecraft and Rockets,* Vol. 13, No. 4, July–Aug. 1978, pp. 208–212.

[47] Strawa, A. W., Chapman, G. T., Canning, T. N., and Arnold, J. O., "The Ballistic Range and Aerothermodynamic Testing," AIAA Paper 88-2015, San Diego, CA, May 1988.

[48] Bailey, A. B., and Hiatt, J., "Free-Flight Measurements of Sphere Drag at Subsonic, Transonic, Supersonic, and Hypersonic Speeds for Continuum, Transition, and Near-Free- Molecular Flow Conditions," Arnold Engineering Development Center, AEDC TR-70-291, Mar. 1971.

[49] Thompson, W. P., "Heat Transfer Gages," in *Fluid Dynamics,* Vol. 18, Part B, pp. 663–685, Academic Press, New York, 1981.

[50] Neumann, R. D., "Aerothermodynamic Instrumentation," *Special Course on Aerothermodynamics of Hypersonic Vehicles,* reprint from AGARD Rept. No. 761, 1990.

[51] Jones, R. A., and Hunt, J. L., "Use of Fusible Temperature Indicators for Obtaining Quantitative Aerodynamic Heat-Transfer Data," NASA TR R-230, Feb. 1966.

[52] Trimmer, L. L., Matthews, R. K., and Buchanan, T. O., "Measurement of Aerodynamic Heat Rates at the AEDC von Karman Facility," Reprint from ICIASF (International Congress on Instrumentation in Aerospace Simulation Facilities), 1973 Record, Sep. 1973.

[53] Wannenwetsch, G. D., Ticatch, L. A., Kidd, C. T., and Arterbury, R. L., "Measurement of Wing-Leading-Edge Heating Rates on Wind Tunnel Models Using the Thin Film Technique," AIAA Paper 85-0972, Williamsburg, VA, June 1985.

[54] Cattolica, R. J., Schmidt, R. L., and Palmer, R. E., "Feasibility of Non-Intrusive Optical Diagnostic Measurements in Hypersonic Boundary Layers for Flight Experiments," AIAA Paper 90-0627, Reno, NV, Jan. 1990.

[55] Ericsson, L. E., and Reding, J. P., "Review of Support Interference in Dynamic Tests," *AIAA Journal,* Vol. 21, No. 12, Dec. 1983, pp. 1652–1666.

[56] Neumann, R. D., "Designing a Flight Test Program," notes from a Hypersonic Short Course at The University of Texas at Austin, Nov. 1986.

[57] Draper, A. C., Buck, M. L., and Selegan, D. R., "Aerospace Technology Demonstrators/Research and Operational Options," AIAA Paper 83-1054, Dayton, OH, Apr. 1983.

[58] Cornette, E. S., "Forebody Temperatures and Calorimeter Heating Rates Measured During Project Fire II Reentry at 11.35 Kilometers Per Second," NASA TMX-1035, Nov. 1966.

[59] Hodge, J. K., and Audley, D. R., "Aerothermodynamic Parameter Estimation from Space Shuttle Thermocouple Data During Transient Flight Test Maneuvers," AIAA Paper 83-0482, Reno, NV, Jan. 1983.

[60] Williamson, W. E., Jr., "Hypersonic Flight Testing," J. J. Bertin, R. Glowinski, and J. Periaux (eds.), *Hypersonics, Volume II: Computation and Measurement of Hypersonic Flows,* Birkhäuser Boston, Boston, MA, 1989.

[61] Hearne, L. F., Chin, J. H., and Woodruff, L. W., "Study of Aerothermodynamic Phenomena Associated with Reentry of Manned Spacecraft," Lockheed Missiles and Space Company Rept. Y-78-66-1, May 1966.

[62] Throckmorton, D. A., "Influence of Radiant Energy Exchange on the Determination of Convective Heat Transfer Rates to Orbiter Leeside Surfaces During Entry," AIAA Paper 82-0824, St. Louis, MO, June 1982.

PROBLEMS

4.1 Tunnel B at the Arnold Engineering Development Center (AEDC) in Tennessee is one of the classic (continuous-flow) hypersonic wind tunnels in the United States. The nominal test-section Mach number (M_1) is 8. By varying the stagnation pressure, the test-section Reynolds number can be varied by over an order of magnitude. At the low end of the Reynolds number range, the total pressure (p_{t1}) in the stagnation chamber (or reservoir) of the tunnel is 50 psia and the total temperature (T_{t1}) is 890 °F.

(a) What is the static temperature in the test section (T_1)? To what altitude of the 1976 U.S. Standard Atmosphere does this temperature correspond; i.e., what temperature altitude is simulated by this wind-tunnel condition?

(b) What is the (freestream) static pressure in the test section (p_1)? To what altitude of the 1976 U.S. Standard Atmosphere does this pressure correspond; i.e., what is the pressure altitude simulated by this wind-tunnel condition?

(c) A spherical model, 1 ft in diameter, is placed in Tunnel B. What is the Reynolds number in the test section,

$$Re_D = \frac{\rho_1 U_1 D}{\mu_1} ?$$

Use Sutherland's equation, Eq. (2-4b), to calculate the test-section viscosity (μ_1).

(d) If the drag coefficient is 0.93, i.e.,

$$C_D = 0.93$$

what is the total force acting on the model?

(e) What are the pressure and temperature at the stagnation point downstream of the normal portion of the bow shock wave; i.e., what are p_{t2} and T_{t2}?

4.2   Consider again a flow in Tunnel B at AEDC. For this problem, we are interested in the high end of the Reynolds number range. Thus, the nominal test-section Mach number (M_1) is 8. In the stagnation chamber (or reservoir) of the tunnel, the total pressure (p_{t1}) is 850 psia and the total temperature (T_{t1}) is 1350 °R.

(a) What is the static temperature in the test section (T_1)? To what altitude of the 1976 U.S. Standard Atmosphere does this temperature correspond; i.e., what temperature altitude is simulated by this wind-tunnel condition?

(b) What is the (freestream) static pressure in the test section (p_1)? To what altitude of the 1976 U.S. Standard Atmosphere does this pressure correspond; i.e., what is the pressure altitude simulated by this wind-tunnel condition?

(c) A spherical model, 1 ft in diameter, is placed in Tunnel B. What is the Reynolds number in the test section,

$$Re_d = \frac{\rho_1 U_1 D}{\mu_1} ?$$

Use Sutherland's equation, Eq. (2-4b), to calculate the test-section viscosity (μ_1).

(d) If the drag coefficient is 0.93, i.e.,

$$C_D = 0.93$$

what is the total force acting on the model?

(e) What are the pressure and temperature at the stagnation point downstream of the normal portion of the bow shock wave; i.e., what are p_{t2} and T_{t2}?

General Information for Problems 4.3 and 4.4

A wide range of Mach number/Reynolds number combinations can be obtained in the Hypersonic Shock Tunnel at Calspan in Buffalo, New York. The handbook for this facility states, "For air supply temperatures above 5800 °R, at which some dissociation takes place, care must be taken to include nonequilibrium effects in the nozzle expansion."

Use the perfect-gas relations for the calculations of Problems 4.3 and 4.4. Although a rigorous analysis would include real-gas effects for these conditions, the perfect-gas relations will give reasonable estimates of the flow properties for the purpose of these two problems.

4.3 The stagnation flow properties in the reservoir of the Hypersonic Shock Tunnel are a total pressure (p_{t1}) of 4400 psia and a total temperature (T_{t1}) of 3750 °R. The freestream Mach number in the test section (M_1) is 16.

(a) What is the freestream static temperature in the test section (T_1)? Based on the 1976 U.S. Standard Atmosphere, what is the temperature altitude simulated by this test condition?

(b) What is the freestream static pressure in the test section (p_1)? Based on the 1976 U.S. Standard Atmosphere, what is the temperature altitude simulated by this test condition?

(c) A Space Shuttle model 1.5 ft in length is to be tested in the wind tunnel at these flow conditions. What is the freestream Reynolds number, based on model length, i.e.,

$$Re_{\infty,L} = \frac{\rho_1 U_1 L}{\mu_1}$$

(d) Assuming that the Chapman-Rubesin coefficient is 1, use Eq. (4-3) to calculate the viscous-interaction parameter $\overline{V}_\infty$, and use Eq. (4-5) to calculate the induced perturbation parameter χ_{orig}.

(e) What are the pressure and temperature at the stagnation point downstream of the normal portion of the bow shock wave; i.e., what are p_{t2} and T_{t2}? To give you an idea of the error introduced by the perfect-gas assumption, measured values of p_{t2} were approximately 1.2 psia.

4.4   The stagnation flow properties in the reservoir of the Hypersonic Shock Tunnel are a total pressure (p_{t1}) of 18,400 psia and a total temperature (T_{t1}) of 4000 °R. The freestream Mach number in the test section (M_1) is 8.1.

(a) What is the freestream static temperature in the test section (T_1)? Based on the 1976 U.S. Standard Atmosphere, what is the temperature altitude simulated by this test condition?

(b) What is the freestream static pressure in the test section (p_1)? Based on the 1976 U.S. Standard Atmosphere, what is the temperature altitude simulated by this test condition?

(c) A Space Shuttle model 1.5 ft in length is to be tested in the wind tunnel at these flow conditions. What is the freestream Reynolds number, based on model length, i.e.,

$$Re_{\infty,L} = \frac{\rho_1 U_1 L}{\mu_1} \; ?$$

(d) Assuming that the Chapman-Rubesin coefficient is 1, use Eq. (4-3) to calculate the viscous-interaction parameter $\overline{V}_\infty$, and use Eq. (4-5) to calculate the induced perturbation parameter χ_{orig}.

(e) What are the pressure and temperature at the stagnation point downstream of the normal portion of the bow shock wave; i.e., what are p_{t2} and T_{t2}? To give you an idea of the error introduced by the perfect-gas assumption, measured values of p_{t2} were approximately 130 psia.

4.5   To obtain increased Mach number capability, one might use helium as a test gas. The test-section Mach number (M_1) is 22. The total temperature (T_{t1}) is 860 °R. The total pressure in the nozzle stagnation chamber (p_{t1}) is 2500 psia. Use the perfect-gas relations. For helium:

$$\gamma = \frac{5}{3}; \quad R = 386 \, \frac{\text{ft lbf}}{\text{lbm °R}}$$

$$\mu = 2.36 \times 10^{-8} \, \frac{T^{1.5}}{T + 176} \, \frac{\text{lbf s}}{\text{ft}^2}$$

(a) What is the static temperature in the test section (T_1)?

(b) A spherical model, 4 inches in diameter, is placed in this helium tunnel. What is the corresponding Reynolds number in the test section, i.e.,

$$Re_D = \frac{\rho_1 U_1 D}{\mu_1} = ?$$

(c) What is the density ratio across the bow shock wave?

STAGNATION-REGION FLOWFIELD

5.1 INTRODUCTION

As the hypersonic flow approaching the body decelerates, a severe thermal environment can be created. To determine the stagnation-region flowfield, one must know the velocity/altitude conditions for flight or the Mach-number/freestream conditions for ground-based tests, as well as the leading-edge geometry. Blunting the leading edge reduces the convective heat-transfer rate at the stagnation point but increases the pressure drag acting on the vehicle. The increase in drag associated with blunting the leading edge may be critical to the force balance for a vehicle with an airbreathing propulsion system. However, because there have been significant improvements in high-temperature materials, one can use smaller nose-radius designs if required. The heat transfer can also be reduced by flying at higher altitudes where the freestream density is relatively low. However, the lower the density, the more the flow around the vehicle departs from equilibrium at higher flight speeds. The decisions on how to manage the thermal environment, e.g., whether the nose radius should be large or small and what trajectory should be flown, will depend on a variety of factors, including the propulsion system (if any), the thermal protection system, the time of flight, and costs.

In this chapter, we will analyze the flow along the stagnation streamline and the heating in the stagnation region.

5.2 THE STAGNATING STREAMLINE

The nomenclature for the stagnating streamline is illustrated in Fig. 5.1 (and is consistent with that presented in Fig. 1.13). The freestream flow is designated by the subscript 1, that downstream of the shock wave is designated by the subscript 2. For the sketch of Fig. 5.1, the shock wave is infinitesimally thin, i.e., the property changes are assumed to occur instantaneously from the freestream conditions. The assumption that the shock wave is negligibly thin in comparison with the shock-layer thickness will be used throughout the text unless otherwise stated.

Immediately after crossing the shock wave, the air particles begin to equilibrate their energies with the local thermodynamic environment. However, the property changes occur over a finite distance, since numerous collisions between the molecules and the atoms are

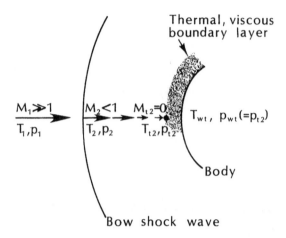

Fig. 5.1 Stagnation streamline.

required to establish the post-shock properties. At the higher altitudes, where the freestream density is relatively low, the required collisions may take a large fraction of the shock layer. For these flows, the nonequilibrium character of the flow must be modeled. See Fig. 2.3.

Along the stagnation streamline, the flow decelerates from the conditions at point 2 to the stagnation conditions at point $t2$. For a perfect-gas flow, the deceleration is isentropic and both the temperature and the pressure increase in going from point 2 to point $t2$. However, as indicated in Figs. 5.2 and 5.3, nonequilibrium chemistry

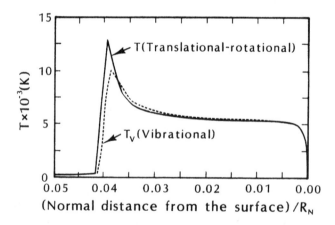

Fig. 5.2 The stagnation-line distributions of T_v and of T for a Martian aerobraking trajectory: R_N = 2.3 m; altitude = 35.9 km; velocity = 6.1 km/s, as taken from Ref. 2.

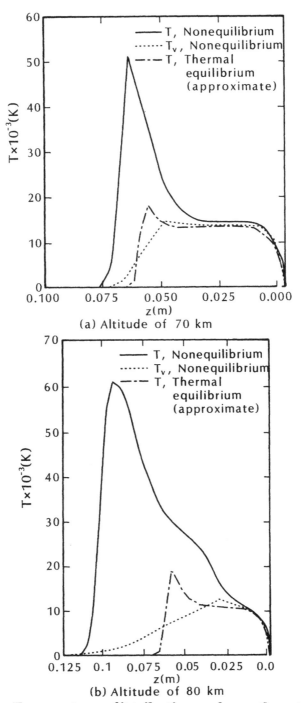

Fig. 5.3 **Temperature distributions along the stagnation streamline for** R_N **= 2.0 m; velocity = 12 km/s, as taken from Ref. 3.**

can have a significant effect on the temperature distribution along the stagnation streamline.

5.2.1 Pressure

Referring to the discussions in Chap. 1, the momentum equation for a one-dimensional, steady, inviscid flow is:

$$p_1 + \rho_1 U_1^2 = p_2 + \rho_2 U_2^2 \tag{5-1}$$

For hypersonic flows:

$$p_1 \ll \rho_1 U_1^2$$
$$p_2 \gg \rho_2 U_2^2$$

Thus, the pressure downstream of a normal-shock wave is of the order of the freestream momentum $(\rho_1 U_1^2)$. Therefore, the dissociation phenomenon does not *appreciably* affect the pressure. Note the use of *appreciably* and that these comments are made for the stagnating streamline. Since dissociation affects the density of the flow downstream of the shock, it will affect the shock stand-off distance. The altered shock shape will produce changes in the local pressure which, although small in magnitude, may significantly affect the pitching moment. As discussed earlier, discrepancies between preflight predictions and the actual pitching moment of the Space Shuttle Orbiter could be attributed to chemistry effects and/or to low-density effects.

5.2.2 The Temperature and the Density

For a one-dimensional, steady, inviscid, adiabatic flow:

$$h_1 + \frac{U_1^2}{2} = h_2 + \frac{U_2^2}{2} = H_t = e + \frac{p}{\rho} + \frac{U^2}{2} \tag{5-2}$$

As discussed in Chap. 2, the specific internal energy is the sum of the energy in heavy particle translation, the energy in rotation, the energy in vibration, and the latent chemical energy of the species.

When a relatively low supersonic flow crosses a normal-shock wave, the deceleration of the air molecules leads to an increase in the internal energy, represented by the translational temperature. When the flow velocity is increased so that the post-shock temperature is raised above 800 K (1440 °R), vibrational excitation of the oxygen molecules begins. Since the vibration phenomenon absorbs energy, the temperature does not increase to its perfect-gas value. Thus, when vibrational excitation occurs, the density rise across the shock wave (ρ_2/ρ_1) is greater than for a perfect-gas flow at the same Mach number. At these conditions, the air is thermally perfect but not calorically perfect.

When the shock Mach number is increased further, the number of vibrationally excited molecules becomes significant. If these vibrationally excited molecules collide with another molecule, they may dissociate. The dissociation process removes a molecule (diatomic species), producing two atoms. Furthermore, since the dissociation process absorbs energy, the temperature increase is reduced. The dissociation of O_2 begins at approximately 2500 K (4500 °R) and is complete at approximately 4222 K (7600 °R). The dissociation of N_2 begins above these temperatures.

Exercise 5.1:

Consider a blunt vehicle flying at 6.1 km/s at an altitude of 35.9 km. What is the temperature downstream of a normal-shock wave, assuming the air behaves as a perfect gas? What is the temperature downstream of a normal-shock wave, using the relations for air in thermochemical equilibrium?

Solution:

The freestream properties at 35.9 km, as determined using Table 1.1, the 1976 U.S. Standard Atmosphere, are:

$$p_1 = 508.8 \text{ N/m}^2 \qquad\qquad T_1 = 239.0 \text{ K}$$
$$\rho_1 = 0.00742 \text{ kg/m}^3 \qquad\qquad a_1 = 309.92 \text{ m/s}$$

Thus, the freestream Mach number is 19.68.

For the perfect-gas model for air, let us use Eq. (1-16).

$$\frac{T_2}{T_1} = \frac{(7M_1^2 - 1)(M_1^2 + 5)}{36M_1^2} = 76.25$$

Thus,
$$T_2 = 18,225 \text{ K}$$

For the equilibrium-air assumption, let us follow the procedure used in Exercise 1.1.

$$\rho_2 U_2 = \rho_1 U_1$$
$$p_2 + \rho_2 U_2^2 = p_1 + \rho_1 U_1^2$$
$$h_2 + 0.5U_2^2 = h_1 + 0.5U_1^2$$

The approximate value of the temperature downstream of a normal-shock wave is:

$$T_2 = 6750 \text{ K}$$

Using the tabular values for normal-shock wave parameters in equilibrium air, as presented in Ref. 1,

$$\frac{T_2}{T_1} = 26.57$$

Thus,

$$T_2 = 6350 \text{ K}$$

The difference between these two equilibrium-air values of T_2 reflect differences in the models for the thermodynamic properties.

The distributions of T and of T_v along the stagnation streamline are presented in Fig. 5.2 for an Aerobrake whose nose radius is 2.3 m flying through the Martian atmosphere at 6.1 km/s at an altitude of 35.9 km. The stand-off distance is 4.2 percent of the nose radius. The two temperatures are nearly equilibrated throughout the flowfield. For about half of the flowfield, the temperatures are nearly constant, indicating thermochemical equilibrium.

Stagnation streamline solutions have been computed by Mitcheltree and Gnoffo[3] for a 2 m nose radius vehicle flying at 12 km/s, both at 70 km and at 80 km. Solutions were obtained for two different models for the thermochemical environment: (1) nonequilibrium flow using a two-temperature model (T and T_v) which can allow thermal nonequilibrium and/or chemical nonequilibrium, and (2) (forced) equilibrium flow. The temperature distributions are presented in Figs. 5.3a and 5.3b. Based on the correlations of Fig. 2.3, one would expect significant nonequilibrium effects at these flight conditions. The temperature distributions of Fig. 5.3 clearly support this expectation. Immediately downstream of the shock wave, the translational temperature (T) increases rapidly. However, since thousands of collisions are needed to excite molecules vibrationally from the ground state to the upper states (from which dissociation occurs preferentially), the vibrational temperature (T_v) increases much more slowly. Since this process is density dependent, the differences between T and T_v are much greater at 80 km (see Fig. 5.3b), as one would expect. For the thermochemical-equilibrium assumption, there is only one temperature. Near the surface, the temperature computed assuming thermochemical equilibrium approaches the two temperatures computed using the nonequilibrium-flow model. This occurs because a fully-catalytic, constant-temperature (1500 K) wall-boundary condition is used in all of the computations. However, in general, the gas state would approach equilibrium when moving along a ray toward the body.

The temperature profiles indicate that the shock stand-off distance is greater when one accounts for the nonequilibrium-flow effects than when the flow is assumed to be in thermochemical equilibrium. This is consistent with the sketch of Fig. 2.2. The difference between the shock stand-off distances, as computed for the two flow models, is approximately 40 percent at 80 km.

5.2.3 The Chemistry

The thermochemical state of the air at the stagnation point was incorporated into the descriptions of the flowfield in the previous section. Flight conditions were described which correspond to perfect-gas flows, to flows that are in global equilibrium (i.e., chemical and thermal equilibrium), to flows that are in thermal equilibrium but in chemical nonequilibrium, and to flows that are in chemical and thermal nonequilibrium. The thermochemical state of air at the stagnation point of a 0.3 m radius sphere, as defined by Gupta et al.,[4] is reproduced in Fig. 5.4.

Roman numerals are used to designate regions of the altitude/velocity space in which the air is composed of a certain number of species, e.g., two, five, seven, or eleven. These boundaries are similar to those defined by Hansen and Heims,[5] which were presented in Fig. 1.14. Capital letters are used to designate regions of the altitude/velocity space relating to chemical and thermal nonequilibrium.

To determine whether or not thermochemical nonequilibrium would exist for a different-sized vehicle, we can use the principle of binary scaling.[6,7] Binary scaling requires that one match the product of density and linear scale $(\rho_\infty L)$ to simulate those chemical rate processes that involve two collision partners, i.e., flows for which three-body recombination is relatively unimportant. Thus, for experiments employing reduced-scale models to obey binary scaling, it is necessary to perform them at a higher density in the same ratio as the size is reduced. Although this binary scaling provides some insight into the trends which should be expected, a full flowfield analysis is required to establish the thermochemical state for the entire flowfield.

5.2.4 Shock Stand-Off Distance

The thermochemical phenomena that occur at high temperatures absorb energy, causing the temperature downstream of the shock wave to be lower than the value in a perfect gas flow where $\gamma = 1.4$.

Regions with Chemical Thermal Nonequilibrium	
Region	Aerothermal Equilibrium
A	Chemical and thermal equilibrium
B	Chemical nonequilibrium with thermal equilibrium
C	Chemical and thermal nonequilibrium

Chemical Species in High Temperature Air		
Region	Air Chemical Model	Species Present
I	2 species	O_2, N_2
II	5 species	O_2, N_2, O, N, NO
III	7 species	O_2, N_2, O, N, NO, NO^+, e^-
IV	11 species	O_2, N_2, O, N, NO, O_2^+, N_2^+, O^+, N^+, NO^+, e^-

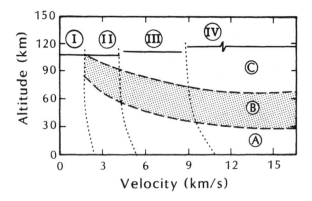

Fig. 5.4. Thermochemical state of the air at the stagnation point of a 0.3 m radius sphere, as taken from Ref. 4.

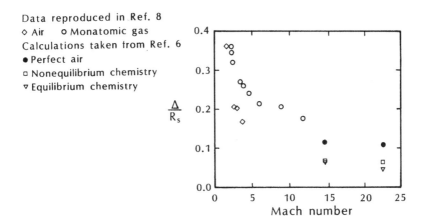

Fig. 5.5. The shock stand-off distances over a sphere, as taken from Refs. 6 and 8.

The resultant increase in density causes the thickness of the shock layer to be less than the perfect-gas value.

Experimental values of the shock stand-off distance (Δ) divided by the radius of curvature of the bow shock (R_s) for spheres in air and in a monatomic gas ($\gamma = 5/3$), as reproduced in Ref. 8, are presented in Fig. 5.5. The shock stand-off distances decrease with Mach number both for air and for the monatomic gas. Calculations by Hall et al.[6] of the shock stand-off distances for a 1.0-ft radius sphere at velocities of 15,000 ft/s and of 23,000 ft/s at an altitude of 200,000 ft are also presented in Fig. 5.5. The shock stand-off distances are greatest for the perfect-air model and least for the equilibrium-chemistry model. These calculations are qualitatively consistent with experimentally-determined shock stand-off distances.

It is possible to represent a blunt-body flowfield in which real-gas effects are represented by using a gas with the appropriate γ. The effective isentropic exponent for stagnation-point conditions varies from 1.08 to 1.22 for velocities from 3 km/s to 14 km/s and for altitudes from 30 km to 100 km.[9] Thus, as was discussed in Chap. 4, when tetrafluoromethane is used as the test gas in a wind tunnel, a density ratio (ρ_1/ρ_2) of 0.08 can be obtained at Mach 6. At these conditions, tetrafluoromethane behaves as a perfect gas. Similarly, one can approximate the shock stand-off distance using a gas with the appropriate γ. The shock stand-off distance is presented in Fig. 5.6 as a function of the freestream Mach number for different γ (perfect) gases.[9] However, it would not be the same γ for all Mach numbers nor would it necessarily yield correct values of other parameters, e.g., T_2/T_1.

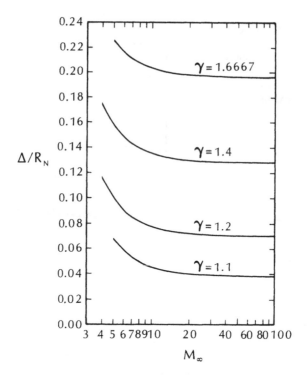

Fig. 5.6. Shock stand-off distances for a sphere as a function of γ for equilibrium flows, as taken from Ref. 9.

5.3   STAGNATION-POINT CONVECTIVE HEAT TRANSFER

The aerodynamic heat transfer to a body in a hypersonic stream is affected by two phenomena that are not normally present at low velocities. The first phenomenon which may affect the heat transfer is possible dissociation and ionization of air due to the higher static temperatures in the shock layer. Because dissociation and ionization (and their reverse process, recombination) proceed at finite rates, thermochemical equilibrium may not be achieved throughout the flowfield. Furthermore, if the atoms and the ions which diffuse toward the surface recombine at the wall, there will be a high specific energy release, significantly increasing the heat transferred by normal molecular conduction. Since a considerable fraction of the heat may be transported by atomic diffusion toward the wall followed by recombination at the surface, it would be possible to eliminate this fraction of the heat transfer by using a noncatalytic surface. However, the use of a noncatalytic surface is effective only if the atoms do not first recombine in the gas before reaching the wall. If the wall is catalytic, the atom concentration will be reduced to its equilibrium value at the wall temperature. Whether the atoms recombine in the boundary layer or at the wall will have no great effect on the convective heat transfer.

5.3.1 Heat-Transfer Relations

The magnitude of the heat transfer from a compressible boundary layer composed of dissociating gases can be approximated by:

$$|\dot{q}_w| = \left(k\frac{\partial T}{\partial y}\right)_w + \left(\sum \rho D_i h_i \frac{\partial C_i}{\partial y}\right)_w \tag{5-3}$$

where the heat is transported by conduction and by diffusion.[10] The molecular diffusion coefficient of the ith-component of the mixture is represented by the symbol D_i.

Some analysts model dissociating air as a binary mixture, i.e., a mixture of molecules (O_2 and N_2) and of atoms (O and N). One binary diffusion coefficient D_{12} will suffice to calculate the mass flux of the atoms through the molecules (and vice versa), since:

$$D_{12} = D_{21}$$

If C_A is the mass fraction of atoms, $(1 - C_A)$ is the mass fraction of molecules. Thus, Eq. (5-3) can be written:

$$|\dot{q}_w| = \left(k\frac{\partial T}{\partial y}\right)_w + \left(\rho D_{12}(h_A - h_M)\frac{\partial C_A}{\partial y}\right)_w \tag{5-4}$$

For atoms:

$$h_A = \int_0^T c_{pA}\ dT + h_A^o$$

where h_A^o is the dissociation energy per unit mass of atomic products. For molecules:

$$h_M = \int_0^T c_{pM}\ dT$$

The heat of formation (h_M^o) is taken as zero for the molecules and is equal to minus the heat of dissociation for the atoms.

Subtracting the expressions for the enthalpy of the molecules from that for the enthalpy of atoms and noting that:

$$h_A^o \geq \int_0^T (c_{pA} - c_{pM})dT$$

one obtains:

$$h_A - h_M = h_A^o + \int_0^T (c_{pA} - c_{pM})\ dT \approx h_A^o \tag{5-5}$$

Combining Eqs. (5-4) and (5-5), we obtain:

$$|\dot{q}_w| = \left(k\frac{\partial T}{\partial y}\right)_w + \left(\rho D_{12} h_A^o \frac{\partial C_A}{\partial y}\right)_w \qquad (5\text{-}6)$$

Introducing the definition for the Lewis-Semenov number (Le):

$$Le = \frac{\rho D_i c_p}{k} \qquad (5\text{-}7)$$

Eq. (5-6) becomes:

$$|\dot{q}_w| = \left(\frac{k}{c_p}\frac{\partial h}{\partial y}\right)_w + \left(\frac{k}{c_p} Le \; h_A^o \frac{\partial C_A}{\partial y}\right)_w \qquad (5\text{-}8)$$

When the Lewis number is approximately unity,

$$|\dot{q}_w| = \left[\frac{k}{c_p}\frac{\partial}{\partial y}(h + h_A^o C_A)\right]_w \qquad (5\text{-}9)$$

Thus, the heat flux is determined by the chemical enthalpy $(h + C_A h_A^o)$ difference between the freestream and the wall. As a result, whether the atoms recombine in the boundary layer or at the wall makes relatively little difference when the Lewis number is approximately 1. Despite the importance of the processes of dissociation and of recombination in determining the thermodynamic state of the air throughout the flowfield, their effect on the stagnation-point heat transfer is secondary.

5.3.2 Equations of Motion

In this chapter, we are considering the stagnation region of a sphere or a cylinder of a finite radius of curvature immersed in a hypersonic stream. Under the conditions of interest to this chapter, a distinct thermal boundary layer will exist independent of the detached bow shock wave. Since the boundary-layer thickness varies as the inverse square root of the Reynolds number and the shock stand-off distance is independent of the Reynolds number, the shock wave and the boundary layer merge at the very low Reynolds numbers, i.e., high altitudes. The altitudes above which these merged viscous flows occurred are depicted in Fig. 1.8.

For the flows of interest in this chapter, the boundary-layer thickness is small compared with the nose radius of curvature (and with the curvature of the detached shock wave). Only the gas which has passed through the normal portion of the shock wave close to the axis- (or plane-) of-symmetry will enter the boundary layer. The conditions at the edge of the boundary layer are those at the stagnation point downstream of the normal-shock wave (subscript $t2$ in Fig. 5.1). For the blunt-body stagnation-region flow, the entropy at the edge of the boundary layer is essentially constant, and the flow is Mach-number independent. As a result, ground-based tests provide a good simulation of these flows for TPS applications. Following the work of Fay and Riddell[11] to calculate the stagnation-point heat-transfer rate, it will be assumed that the flow has reached thermochemical equilibrium by the time it reaches the stagnation point.

Since the boundary-layer thickness is small compared to the body radius of curvature, the cross-section radius from the axis-of-symmetry to a point in the boundary layer (r) is essentially equal to r_o, the cross-section radius of the body. Furthermore, r_o is a unique function of x. Thus, the overall continuity equation is:

$$\frac{\partial(\rho u r_o^k)}{\partial x} + \frac{\partial(\rho v r_o^k)}{\partial y} = 0 \qquad (5\text{-}10)$$

where $k = 0$ for a two-dimensional flow and $k = 1$ for an axisymmetric flow. As indicated in Fig. 5.7, the x-coordinate is measured from the stagnation point (or from the leading edge) and the y-coordinate is measured along the outward normal from the body surface.

The x-momentum equation, which is unchanged by the fact that the air may be partially dissociated, is written in the usual form for boundary-layer applications:

$$\rho u \frac{\partial u}{\partial x} + \rho v \frac{\partial u}{\partial y} = -\frac{\partial p}{\partial x} + \frac{\partial}{\partial y}\left(\mu \frac{\partial u}{\partial y}\right) \qquad (5\text{-}11)$$

However, the fact that the gas is partially dissociated will affect the values of the viscosity coefficient μ.

The usual thin boundary-layer approximations lead to the conclusion that the pressure variation across the boundary layer is negligible, i.e.,

$$\frac{\partial p}{\partial y} \approx 0 \qquad (5\text{-}12)$$

Thus, the pressure at a point in the boundary layer is equal to the value at the edge of the boundary layer (p_e), which comes from the inviscid solution. Since the inviscid value is a function of x only,

$$\frac{\partial p}{\partial x} = \frac{dp_e}{dx} = -\rho_e u_e \frac{du_e}{dx} \qquad (5\text{-}13)$$

can be used in Eq. (5-11).

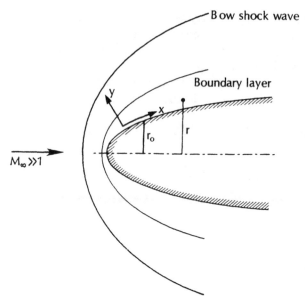

Fig. 5.7 Coordinate system used in analyzing the boundary-layer flow over an object in a hypersonic stream.

The energy equation is given by:

$$\rho u \frac{\partial H}{\partial x} + \rho v \frac{\partial H}{\partial y} = \frac{\partial}{\partial y}\left(\frac{\mu}{Pr}\frac{\partial H}{\partial y}\right) + \frac{\partial}{\partial y}\left[\mu\left(1 - \frac{1}{Pr}\right)\frac{\partial}{\partial y}\left(\frac{u^2}{2}\right)\right]$$

$$+ \frac{\partial}{\partial y}\left[\rho D_{12}\left(1 - \frac{1}{Le}\right)\sum(h_i - h_i^o)\frac{\partial C_i}{\partial y}\right] \tag{5-14}$$

where H is the stagnation enthalpy,

$$H \approx h + \frac{u^2}{2} \tag{5-15a}$$

Pr is the Prandtl number,

$$Pr = \frac{\mu c_p}{k} \tag{5-15b}$$

and Le is the Lewis-Semenov number,

$$Le = \frac{\rho D_{12} c_p}{k} \tag{5-15c}$$

5.3.3 Similar Solutions for the Stagnation-Point Heat Transfer

There are many situations where a coordinate transformation can be used to reduce the governing partial differential equations for a laminar boundary layer to ordinary differential equations. Fay and Riddell[11] state:

> As is usual in boundary-layer problems, one first seeks solutions of restricted form which permit reducing exactly the partial differential equations to ordinary differential form. An easily recognizable case is that of the stagnation point flow, where, because of symmetry, all the dependent variables are chosen to be functions of y alone, except u which must be taken proportional to x times a function of y. This also appears to be the only case for which the exact ordinary differential equations may be obtained regardless of the recombination rate.

5.3.3.1 Similarity transformations. To simplify the problem of solving the system of partial differential equations, i.e., Eqs. (5-10) through (5-14), we seek a transformation that will produce a set of ordinary differential equations. The transformation of the equations of motion from the physical coordinates (x, y) to similarity coordinates (S, η) is discussed at length by Dorrance[10] and Hayes and Probstein.[12] Depending on the specific application, the evolving technology of similarity transformations includes contributions by Blasius, Mangler, Howarth, Illingworth, Levy, Dorodnitsyn, and Lees.

Using the similarity transformations suggested by Fay and Riddell[11]:

$$\eta(x_1, y_1) = \frac{\rho_w u_e r_o^k}{\sqrt{2S}} \int_0^{y_1} \frac{\rho}{\rho_w} \mathrm{d}y \qquad (5\text{-}16a)$$

and

$$S(x_1) = \int_0^{x_1} \rho_w \mu_w u_e r_o^{2k} \, \mathrm{d}x \qquad (5\text{-}16b)$$

where r_o is the cross-section radius of the body-of-revolution and k denotes whether the flow is axisymmetric ($k = 1$), e.g., a sphere, or two-dimensional ($k = 0$), e.g., a cylinder whose axis is perpendicular to the freestream.

5.3.3.2 Transformed equations. For a steady flow which can be described in two coordinates, i.e., either two-dimensional or axisymmetric, so that Eq. (5-10) defines continuity, one can introduce a stream function ψ that automatically satisfies this equation. Thus,

$$\rho u r_o^k = \frac{\partial \psi}{\partial y} \tag{5-17a}$$

and

$$\rho v r_o^k = -\frac{\partial \psi}{\partial x} \tag{5-17b}$$

Using the similarity transformations defined above, we can rewrite the stream function as:

$$\frac{\partial \psi}{\partial y} = \frac{\partial \psi}{\partial \eta}\frac{\partial \eta}{\partial y} = \frac{\partial \psi}{\partial \eta}\frac{\rho u_e r_o^k}{\sqrt{2S}} \tag{5-18}$$

If we define the dependent variable involving u using the dimensionless parameter,

$$\frac{\partial f}{\partial \eta} = \frac{u}{u_e} \tag{5-19a}$$

Combining Eqs. (5-17a), (5-18), and (5-19a),

$$\rho u r_o^k = \rho u_e r_o^k \frac{\partial f}{\partial \eta} = \rho u_e r_o^k \frac{\partial \psi/\partial \eta}{\sqrt{2S}} \tag{5-20}$$

Thus, f is a transformed stream function and the continuity equation is automatically satisfied. Therefore, the continuity equation will not explicitly appear in the following.

Introducing additional dimensionless dependent variables:

$$g = \frac{H}{H_e} \tag{5-19b}$$

$$\text{and} \quad z_i = \frac{C_i}{C_{1e}} \tag{5-19c}$$

The corresponding momentum equation is:

$$\frac{\partial}{\partial \eta}\left[C\frac{\partial^2 f}{\partial \eta^2}\right] + f\frac{\partial^2 f}{\partial \eta^2} + \frac{2S}{u_e}\frac{du_e}{dS}\left[\frac{\rho_e}{\rho} - \left(\frac{\partial f}{\partial \eta}\right)^2\right]$$
$$= 2S\left[\frac{\partial f}{\partial \eta}\frac{\partial^2 f}{\partial \eta \partial S} - \frac{\partial f}{\partial S}\frac{\partial^2 f}{\partial \eta^2}\right] \tag{5-21a}$$

while the energy equation is:

$$\frac{\partial}{\partial \eta}\left[\frac{C}{Pr}\frac{\partial g}{\partial \eta}\right] + f\frac{\partial g}{\partial \eta} + \frac{u_e^2}{H_e}\frac{\partial}{\partial \eta}\left[C\left(1 - \frac{1}{Pr}\right)\frac{\partial f}{\partial \eta}\frac{\partial^2 f}{\partial \eta^2}\right]$$

$$+ \frac{\partial}{\partial \eta}\left[\frac{C(Le - 1)}{Pr}\sum C_{1e}\frac{(h_i - h_i^o)}{H_e}\frac{\partial z_i}{\partial \eta}\right]$$

$$= 2S\left[\frac{\partial f}{\partial \eta}\frac{\partial g}{\partial S} - \frac{\partial f}{\partial S}\frac{\partial g}{\partial \eta}\right] \tag{5-21b}$$

where

$$C = \frac{\rho\mu}{\rho_w\mu_w} \tag{5-22}$$

is the Chapman-Rubesin factor.

For "similar" solutions, the dependent variables f, g, and z_i are functions of η alone. For this to be rigorously correct, the right-hand sides of Eqs. (5-21a) and (5-21b) must be zero. For the dependent variables to be a function of η alone, it is first necessary that the thermodynamic state variables be unchanging both at the edge of the boundary layer and along the wall as S increases. This condition is satisfied in the stagnation region. It is also satisfied for supersonic flow past a sharp cone or past a wedge. $(2S/u_e) \times (du_e/dS)$ is a constant for the stagnation point and is zero for the sharp cone or for the wedge. It is also necessary that the source term, i.e., the term involving $\dot{w}_i$, depend on η alone. This is satisfied for frozen flow $(\dot{w}_i = 0)$ and for a flow in thermodynamic equilibrium.

5.3.3.3 The Fay-Riddell correlation for the stagnation-point heat transfer.
Rearranging Eq. (5-3),

$$|\dot{q}_w| = \left[\frac{k}{c_p}\left(\frac{\partial h}{\partial y} + \sum \frac{\rho D_i c_p}{k}h_i\frac{\partial C_i}{\partial y}\right)\right]_w \tag{5-23}$$

Assume that the flow is such that the Lewis number is approximately equal to unity, the gas near the wall is in chemical equilibrium, and the surface temperature is much less than the external stream temperature. For these assumptions, the magnitude of the heat transferred to the wall is:

$$|\dot{q}_w| = \left(\frac{k}{c_p}\frac{\partial h}{\partial y}\right)_w \tag{5-24}$$

Since u equals zero at the wall,

$$\left.\frac{\partial h}{\partial y}\right|_w = \left.\frac{\partial H}{\partial y}\right|_w = \frac{\rho_w u_e r_o^k}{\sqrt{2S}} H_e \left(\frac{\partial g}{\partial \eta}\right)_w \qquad (5\text{-}25)$$

where $(\partial g/\partial \eta)_w$ is the total-enthalpy gradient at the wall. Combining Eqs. (5-24) and (5-25), the heat transfer from the gas to the wall is:

$$\dot{q}_w = \frac{\rho_w \mu_w}{Pr_w} \frac{u_e r_o^k H_e}{\sqrt{2S}} \left(\frac{\partial g}{\partial \eta}\right)_w \qquad (5\text{-}26)$$

Let us introduce the following approximations for the flow in the vicinity of the stagnation point. Referring to Fig. 5.8,

$$r_o \approx x \qquad (5\text{-}27a)$$

The velocity is approximately a linear function of x (an approximation which has been verified through detailed computations and

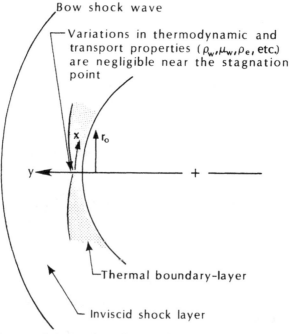

Fig. 5.8 Nomenclature for the stagnation-point heat-transfer formulation.

measurements of the stagnation-region flowfield), i.e.,

$$u_e \approx \left(\frac{du_e}{dx}\right)_{t2} x \tag{5-27b}$$

The air properties are essentially constant near the stagnation point, e.g.,

$$\rho_{w,t} \approx \text{constant}; \quad \mu_{w,t} \approx \text{constant} \tag{5-27c}$$

Thus, at the stagnation point,

$$S = \rho_{w,t}\mu_{w,t}\left(\frac{du_e}{dx}\right)_{t2}\int x x^{2k}\,dx = \rho_{w,t}\mu_{w,t}\left(\frac{du_e}{dx}\right)_{t2}\frac{x^{2(k+1)}}{2(k+1)} \tag{5-28}$$

and

$$\eta = \frac{\left(\frac{du_e}{dx}\right)_{t2} x x^k}{\sqrt{2S}}\int \rho\,dy = \left[\frac{\left(\frac{du_e}{dx}\right)_{t2}(k+1)}{\rho_{w,t}\mu_{w,t}}\right]^{0.5}\int \rho\,dy \tag{5-29}$$

Note that using Eq. (5-29) as the expression for the transformed y-coordinate, together with the definition for the heat transfer as given by Eq. (5-26), we find that, given the same flow conditions with the same wall temperature:

$$(\dot{q}_{t,\text{ref}})_{\text{axisym}} = \sqrt{2}(\dot{q}_{t,\text{ref}})_{2-\text{dim}} \tag{5-30}$$

i.e., the stagnation-point heat-transfer rate for a sphere (an axisymmetric configuration) is the $\sqrt{2}$ times that for a cylinder (a two-dimensional configuration).

Let us introduce expressions for the local Nusselt number (Nu_x) and for the local Reynolds number (Re_x).

$$Nu_x = \frac{hx}{k_{w,t}} = \frac{|\dot{q}_{t,\text{ref}}|c_{pw,t}x}{k_{w,t}(h_r - h_w)} \tag{5-31}$$

In Eq. (5-31), (unsubscripted) h is the local heat-transfer coefficient, h_r is the recovery enthalpy, and h_w is the enthalpy at the wall. Note that, at the stagnation point, $h_r = H_e = h_e = H_{t2}$.

$$Re_x = \frac{\rho_{w,t}u_e x}{\mu_{w,t}} = \frac{\rho_{w,t}}{\mu_{w,t}}\left(\frac{du_e}{dx}\right)_{t2}x^2 \tag{5-32}$$

Note, in this unique formulation for the Reynolds number, that the density and the viscosity are evaluated at the wall to be consistent with the coordinate transformations defined in Eqs. (5-16). We have also introduced the linear velocity approximation, i.e., Eq. (5-27b).

Combining Eqs. (5-26), (5-28), (5-31), and (5-32), we find that:

$$\frac{Nu_x}{\sqrt{Re_x}} = \sqrt{k+1}\left(\frac{\partial g}{\partial \eta}\right)_w \left(\frac{H_e}{h_r - h_w}\right) \qquad (5\text{-}33)$$

To determine the heat-transfer rate, we need to develop correlations from solutions of Eqs. (5-21a) and (5-21b). The heat-transfer parameter $(Nu_x/\sqrt{Re_x})$ determined from each integration depends not only upon the parameters and the boundary values of the dependent variables involved, but also upon the assumed variation of viscosity with temperature and with the composition of the gas.

Fay and Riddell[11] found that all their individual heat-transfer calculations could be correlated by:

$$\frac{Nu_x}{\sqrt{Re_x}} = 0.763(Pr_{w,t})^{0.4}\left[\frac{\rho_{t2}\mu_{t2}}{\rho_{w,t}\mu_{w,t}}\right]^{0.4}\left[1+(Le^{0.52}-1)\frac{h_d}{H_{t2}}\right] \qquad (5\text{-}34)$$

for velocities between 5,800 ft/s (1,768 m/s) and 22,800 ft/s (6,949 m/s) and at altitudes between 25,000 ft (7,620 m) and 120,000 ft (36,576 m). In Eq. (5-34), h_d is the average atomic dissociation energy multiplied by the atom mass fraction at the edge of the boundary layer.

Combining Eqs. (5-24), (5-29), and (5-33), the stagnation-point heat-transfer rate for a laminar boundary layer of a spherical cap may be written as:

$$\dot{q}_{t,\text{ref}} = \left(\frac{Nu_x}{\sqrt{Re_x}}\right)\sqrt{\rho_{w,t}\mu_{w,t}\left(\frac{du_e}{dx}\right)_{t2}}\left[\frac{H_{t2}-h_{w,t}}{Pr_{w,t}}\right] \qquad (5\text{-}35)$$

Combining Eqs. (5-34) and (5-35), we obtain the stagnation-point heat-transfer rate for an equilibrium boundary layer of Fay and Riddell[11]:

$$\dot{q}_{t,\text{ref}} = \frac{0.763}{(Pr_{w,t})^{0.6}}(\rho_{t2}\mu_{t2})^{0.4}(\rho_{w,t}\mu_{w,t})^{0.1}$$

$$\times\ (H_{t2}-h_{w,t})\left[1+(Le^{0.52}-1)\frac{h_d}{H_{t2}}\right]\left[\left(\frac{du_e}{dx}\right)_{t2}\right]^{0.5} \qquad (5\text{-}36)$$

5.3.3.4 Stagnation-point velocity gradient. In order to obtain numerical values for the stagnation-point heat-transfer rate using Eq. (5-36), we need to develop an expression for the inviscid flow velocity gradient at the stagnation point.

Let us use Euler's equation to evaluate the velocity gradient at the stagnation point:

$$\left(\frac{dp_e}{dx}\right)_{t2} = -\rho_e u_e \left(\frac{du_e}{dx}\right)_{t2} \approx -\rho_{t2} \left(\frac{du_e}{dx}\right)_{t2}^2 x \qquad (5\text{-}37)$$

Using the modified Newtonian flow pressure distribution [see Eq. (1-5)] to evaluate the pressure gradient at the stagnation point:

$$\left(\frac{dp_e}{dx}\right)_{t2} = -2p_{t2}\cos\phi\sin\phi\frac{d\phi}{dx} + 2p_\infty\cos\phi\sin\phi\frac{d\phi}{dx} \qquad (5\text{-}38)$$

Since $\sin\phi \approx \phi = \frac{x}{R_N}$, $\cos\phi \approx 1$; and $\frac{d\phi}{dx} = \frac{1}{R_N}$; we find that

$$\left(\frac{du_e}{dx}\right)_{t2} = \frac{1}{R_N}\sqrt{\frac{2(p_{t2} - p_\infty)}{\rho_{t2}}} \qquad (5\text{-}39)$$

Exercise 5.2:

Consider a sphere, 1.0 ft in radius, flying at 24,000 ft/s at an altitude of 240,000 ft. Assume that the wall temperature is 2500 °R and calculate the stagnation-point heat-transfer rate.

1. Use the perfect-gas relations to evaluate the properties of air.

2. Assume that the air behaves as a gas in thermochemical equilibrium to evaluate the properties of air.

Solution:

Referring to Table 1.1, the 1976 U.S. Standard Atmosphere, to obtain the freestream properties at 240,000 ft:

$$p_1 = 3.1608 \times 10^{-5}\ p_{SL} = 0.06689\ \frac{\text{lbf}}{\text{ft}^2}$$

$$T_1 = 381.62\ °\text{R}; \quad a_1 = 957.64\ \frac{\text{ft}}{\text{s}}$$

Thus,

$$M_1 = \frac{U_1}{a_1} = \frac{24,000}{957.64} = 25.0616$$

1. For the perfect-air model.

$$T_{t2} = T_{t1} = T_1(1 + 0.2M_1^2) = 48{,}319.5 \ {}^\circ\text{R}$$

$$\frac{p_{t2}}{p_1} = [1.2M_1^2]^{3.5} \left[\frac{2.4}{2.8M_1^2 - 0.4}\right]^{2.5} = 809.16$$

Thus,

$$p_{t2} = 54.124 \ \frac{\text{lbf}}{\text{ft}^2} = 0.02557 \ \text{atm}$$

Note that, even though the velocity is 24,000 ft/s, the stagnation pressure downstream of the normal portion of the shock wave is one-fortieth of the sea-level atmospheric pressure. Continuing with the perfect-gas relations:

$$\rho_{t2} = \frac{p_{t2}}{RT_{t2}} = \frac{54.124 \ \frac{\text{lbf}}{\text{ft}^2}}{\left(1716.16 \ \frac{\text{ft}^2}{\text{s}^2 \ ^\circ\text{R}}\right)(48{,}319.5 \ {}^\circ\text{R})} = 6.53 \times 10^{-7} \ \frac{\text{lbf s}^2}{\text{ft}^4}$$

$$\mu_{t2} = 2.27 \times 10^{-8} \frac{T_{t2}^{1.5}}{T_{t2} + 198.6} = 4.97 \times 10^{-6} \ \frac{\text{lbf s}}{\text{ft}^2}$$

Since the static pressure is constant across a thin boundary layer, $p_{t2} = p_{w,t}$. Thus,

$$\rho_{w,t} = \frac{p_{t2}}{RT_w} = \frac{54.124 \ \frac{\text{lbf}}{\text{ft}^2}}{\left(1716.16 \ \frac{\text{ft}^2}{\text{s}^2 \ ^\circ\text{R}}\right)(2500 \ ^\circ\text{R})} = 1.262 \times 10^{-5} \ \frac{\text{lbf s}^2}{\text{ft}^4}$$

$$\mu_{w,t} = 2.27 \times 10^{-8} \frac{T_w^{1.5}}{T_w + 198.6} = 1.05 \times 10^{-6} \ \frac{\text{lbf s}}{\text{ft}^2}$$

For perfect air, we will assume that the Prandtl number is 0.7 (throughout the flowfield). This is a reasonable assumption.

$$H_{t2} = c_p T_{t2} = \left(0.2404 \frac{\text{Btu}}{\text{lbm} \ ^\circ\text{R}}\right)(48{,}319.5 \ ^\circ\text{R}) = 11{,}616.0 \ \frac{\text{Btu}}{\text{lbm}}$$

$$h_{w,t} = c_p T_w = \left(0.2404 \ \frac{\text{Btu}}{\text{lbm} \ ^\circ\text{R}}\right)(2500 \ ^\circ\text{R}) = 601.0 \ \frac{\text{Btu}}{\text{lbm}}$$

The velocity gradient is:

$$\left(\frac{du_e}{dx}\right)_{t2} = \frac{1}{R_N}\left[\frac{2(p_{t2} - p_1)}{\rho_{t2}}\right]^{0.5} = 12{,}867.2 \ \text{s}^{-1}$$

Thus, assuming that the term containing the Lewis number is one, the stagnation-point heat-transfer rate is:

$$\dot{q}_{t,\text{ref}} = 2.4486 \, \frac{\text{Btu}}{\text{lbm}} \frac{\text{lbf s}}{\text{ft}^3}$$

But these are not the usual units for the heat-transfer rate.

$$\dot{q}_{t,\text{ref}} = \left(2.4486 \, \frac{\text{Btu}}{\text{lbm}} \frac{\text{lbf s}}{\text{ft}^3}\right) \left(32.174 \, \frac{\text{ft lbm}}{\text{lbf s}^2}\right)$$

$$\dot{q}_{t,\text{ref}} = 78.782 \, \frac{\text{Btu}}{\text{ft}^2 \text{s}}$$

2. To calculate the flow properties at the stagnation point downstream of the normal portion of the shock wave for air in thermochemical equilibrium, we will follow the procedure of Exercise 1.1.

$$\rho_1 = 4.2959 \times 10^{-5} \rho_{SL} = 1.021 \times 10^{-7} \, \frac{\text{lbf s}^2}{\text{ft}^4}$$

$$h_1 = \left(0.2404 \, \frac{\text{Btu}}{\text{lbm}\,^\circ\text{R}}\right) \left(381.62\,^\circ\text{R}\right)$$

$$= \left(91.74 \, \frac{\text{Btu}}{\text{lbm}}\right) \left(32.174 \, \frac{\text{lbm ft}}{\text{lbf s}^2}\right) \left(778.2 \, \frac{\text{ft lbf}}{\text{Btu}}\right)$$

$$= 2.297 \times 10^6 \, \frac{\text{ft}^2}{\text{s}^2}$$

Thus,

$$p_2 + \rho_2 U_2^2 = p_1 + \rho_1 U_1^2 = 58.876 \, \frac{\text{lbf}}{\text{ft}^2}$$

$$h_2 + \frac{1}{2}U_2^2 = h_1 + \frac{1}{2}U_1^2 = 290.297 \times 10^6 \, \frac{\text{ft}^2}{\text{s}^2}$$

$$\rho_2 U_2 = \rho_1 U_1 = 0.002451 \, \frac{\text{lbf s}}{\text{ft}^3}$$

Following the procedure of Exercise 1.1, we soon find that:

$$h_2 = 11,561.5 \, \frac{\text{Btu}}{\text{lbm}}$$

$$p_2 = 55.741 \, \frac{\text{lbf}}{\text{ft}^2} = 0.0263 \, \text{atm}$$

$$T_2 = 10,600 \, °R$$

$$z_2 = 1.668$$

Following the line of constant entropy (since $s_2 = s_{t2}$) from the conditions above to the total enthalpy of

$$H_{t2} = 290.297 \times 10^6 \frac{ft^2}{s^2} = 11,594 \frac{Btu}{lbm}$$

Using Fig. 1.17d, we find that:

$$T_{t2} = 10,650 \, °R$$

$$p_{t2} = 0.027 \, atm = 57.138 \, \frac{lbf}{ft^2}$$

$$z_{t2} = 1.670$$

We can compare this value of z_{t2} (1.670) as taken from the charts of Moeckel and Weston,[13] with that of Hansen,[14] as reproduced in Fig. 1.15. For $p_{t2} = 0.027 \, atm$ and $T_{t2} = 5917 \, K$, $z_{t2} \approx 1.7$ in Fig. 1.15, as we are well into Nitrogen dissociation at these conditions. We can now calculate the density:

$$\rho_{t2} = \frac{p_{t2}}{z_{t2} R T_{t2}} = \frac{57.138 \, \frac{lbf}{ft^2}}{(1.670)(1716.16 \, \frac{ft^2}{s^2 \, °R})(10,650 \, °R)}$$

$$\rho_{t2} = 1.87 \times 10^{-6} \frac{lbf \, s^2}{ft^4}$$

Using Fig. 2.4 and Eq. (2-4b) to calculate the viscosity,

$$\mu_{t2} = 1.1 \mu_{Eq.(2-4)} = 1.1 \left[2.27 \times 10^{-8} \frac{T_{t2}^{1.5}}{T_{t2} + 198.6} \right]$$

$$= 2.53 \times 10^{-6} \frac{lbf \, s}{ft^2}$$

For $p_{t2} = p_{w,t} = 57.138 \, lbf/ft^2$ and $T_w = 2500 \, °R$, $z_{w,t} = 1.00$. Thus, the density at the wall is:

$$\rho_{w,t} = \frac{p_{t2}}{z_{w,t} R T_w} = 1.332 \times 10^{-5} \frac{lbf \, s^2}{ft^4}$$

$$\mu_{w,t} = 2.27 \times 10^{-8} \frac{T_w}{T_w + 198.6} = 1.05 \times 10^{-6} \frac{lbf \, s}{ft^2}$$

Consider the assumption that the term containing the Lewis number is approximately 1. Fay and Riddell[11] note that the Lewis number does not change appreciably with temperature (below 9000 K) and, although the exact value is uncertain, is estimated to be about 1.4. With this assumption and the values of the properties calculated above, the stagnation-point heat-transfer rate is:

$$q_{t,\text{ref}} = 70.91 \frac{\text{Btu}}{\text{ft}^2 \text{s}}$$

The external flow properties are much more important than the wall values in determining the heat-transfer rate. The uncertainty in the heat transfer is about 40 percent of the uncertainty in the external viscosity. The physical reason for the importance of the external viscosity is that the growth of the boundary layer and, hence, the heat transfer to the wall depend mostly upon the external properties.

5.3.4 Additional Correlations for the Stagnation-Point Heat Transfer

Note that the perfect-air value of the stagnation-point heat-transfer rate is within 12 percent of the equilibrium-air value, as calculated using Eq. (5-36) for the flow conditions of Exercise 5.2. This relative agreement suggests that there are some basic parameters that govern the stagnation-point heat-transfer rate and that are independent of the air-chemistry model.

Exercise 5.3:

Starting with Eqs. (5-36) and (5-39), identify the basic parameters that can be used in correlations for the stagnation-point heat-transfer rate. Neglecting the Lewis number term,

$$\dot{q}_{t,\text{ref}} = \frac{0.763}{(\text{Pr}_{w,t})^{0.6}} (\rho_{t2} \mu_{t2})^{0.4} (\rho_{w,t} \mu_{w,t})^{0.1}$$

$$\times (H_{t2} - h_{w,t}) \left[\frac{1}{R_N} \left(\frac{2(p_{t2} - p_1)}{\rho_{t2}} \right)^{0.5} \right]^{0.5}$$

Solution:

$$\rho_{t2} \propto \frac{p_{t2}}{z_{t2} T_{t2}} \quad ; \quad \rho_{w,t} \propto \frac{p_{t2}}{z_{w,t} T_w}$$

$$\mu_{t2} \propto (T_{t2})^E \quad ; \quad \mu_{w,t} \propto (T_w)^E$$

where E is an exponent (approaching 0.5, based on Sutherland's equation, for high-temperature flows). Note also that $p_{t2} \gg p_1$. Thus,

$$\dot{q}_{t,ref} = K \left(\frac{p_{t2}}{R_N} \right)^{0.5} (H_{t2} - h_{w,t})\,(\text{factor}) \tag{5-40}$$

where K is a constant and the factor is

$$\text{Factor} = \frac{(T_{t2})^{0.4E}(T_w)^{0.1E}}{(z_{t2}T_{t2})^{0.15}(z_{w,t}T_w)^{0.1}}$$

For hypersonic flows, Eq. (1-11) would lead to the approximation that

$$H_{t2} - h_{w,t} \approx H_{t2} \approx 0.5U_1^2$$

Furthermore, based on Eq. (1-10),

$$p_{t2} \approx \rho_1 U_1^2$$

Thus, Eq. (5-40) could be written in terms of the freestream density and the freestream velocity,

$$\dot{q}_{t,ref} = \frac{K}{\sqrt{R_N}} (\rho_1)^{0.5}(U_1)^{3.0}(\text{factor}) \tag{5-41}$$

Using coordinate transformations similar to those described by Eqs. (5-16a) and (5-16b) and a stagnation pressure gradient given by Eq. (5-39) with $p_{t2} \gg p_\infty$, Sutton and Graves[15] investigated the convective heat transfer to the stagnation point of an axisymmetric blunt body for a mixture of arbitrary gases in chemical equilibrium. The objective of their study was to develop a general relation for $\dot{q}_{t,ref}$ for gas mixtures typical of those for planetary entry, e.g., the Martian atmosphere and the Jovian atmosphere, as well as the earth's atmosphere. The equation for heat transfer is

$$\dot{q}_{t,ref} = K\sqrt{\frac{p_{t2}}{R_N}} (H_{t2} - h_{w,t}) \tag{5-42}$$

where K, a heat-transfer factor, is a function of the molecular weight, the mass fraction, and a transport parameter of the base gas or of the base gases of the mixture.

Another engineering correlation for the calculation of the convective heat-transfer rate to the stagnation point of a sphere is reported by Scott et al.[16]

$$\dot{q}_{t,ref} = \frac{18,300(\rho_\infty)^{0.5}}{\sqrt{R_N}} \left[\frac{U_\infty}{10^4} \right]^{3.05} \tag{5-43}$$

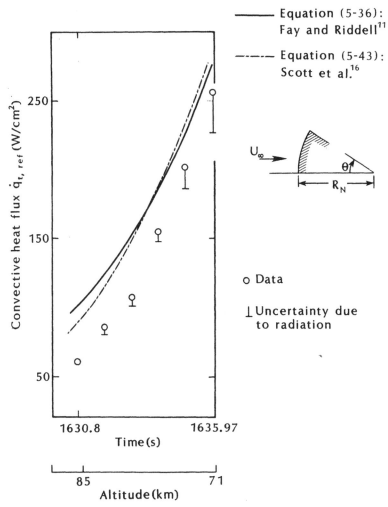

Figure 5.9   Stagnation-point convective heat flux during early data period of Fire II flight experiment re-entry, as taken from Ref. 16.

where the freestream density is expressed in kg/m^3, the freestream velocity in m/s, the nose radius in m, and the heat-transfer rate in W/cm^2. The comparisons of the stagnation-point heat-transfer rates as presented by Scott et al. are reproduced in Fig. 5.9. An effective nose radius (to account for the fact that the configuration is a truncated spherical cap) was used when applying the correlations, e.g., Eq. (5-36) due to Fay and Riddell[11] and Eq. (5-43) due to Scott et al.[16] The correlation formula gives the correct trend but produces heating rates that are consistently higher than the measured values.

Detra, et al.[17] have developed a correlation for flight applications

$$\dot{q}_{t,\text{ref}} = \frac{11{,}030}{(R_N)^{0.5}} \left(\frac{\rho_\infty}{\rho_{SL}}\right)^{0.5} \left(\frac{U_\infty}{U_{CO}}\right)^{3.15} \tag{5-44a}$$

In Eq. (5-44a), R_N is the nose radius of the sphere in m, ρ_{SL} is the density at sea level in kg/m^3, and U_{CO} is the circular orbit velocity in m/s. For the purposes of this text, the circular orbit velocity will be 7,950 m/s. The heat-transfer rate is in W/cm^2. The equivalent equation for English units is:

$$\dot{q}_{t,\text{ref}} = \frac{17{,}600}{(R_N)^{0.5}} \left(\frac{\rho_\infty}{\rho_{SL}}\right)^{0.5} \left(\frac{U_\infty}{U_{CO}}\right)^{3.15} \tag{5-44b}$$

where R_N is in feet, U_{CO} is 26,082 ft/s, and the heat-transfer rate is in Btu/ft^2s.

The heat-transfer rate at the stagnation point of a 1-foot (0.3048 m) sphere flying at 150,000 ft (45,721 m) has been calculated using Eq. (5-36) for both the perfect-air model and for the equilibrium-air model and using Eq. (5-44). The computed values are presented in Fig. 5.10 as a function of M_∞.

Exercise 5.4:

Use Eq. (5-43) and Eq. (5-44a) to calculate the convective heat-transfer rate to a sphere whose radius is 0.3 m that is flying at 4000 m/s at an altitude of 45 km.

Solution:

Using the 1976 U.S. Standard Atmosphere, Table 1.1,

$$\frac{\rho}{\rho_{SL}} = 1.6051 \times 10^{-3} ; \quad \rho_\infty = 0.001966 \frac{kg}{m^3}$$

Using Eq. (5-43),

$$\begin{aligned}
\dot{q}_{t,\text{ref}} &= \frac{18{,}300(\rho_\infty)^{0.5}}{\sqrt{R_N}} \left(\frac{U_\infty}{10^4}\right)^{3.05} \\
&= \frac{18{,}300(0.001966)^{0.5}}{(0.3)^{0.5}} \left(\frac{4{,}000}{10{,}000}\right)^{3.05} \\
&= 90.57 \frac{W}{cm^2}
\end{aligned}$$

Using Eq. (5-44a)

$$\dot{q}_{t,\text{ref}} = \frac{11,030}{\sqrt{R_N}} \left(\frac{\rho_\infty}{\rho_{SL}}\right)^{0.5} \left(\frac{U_\infty}{U_{CO}}\right)^{3.15}$$

$$= \frac{11,030}{(0.3)^{0.5}} (1.6051 \times 10^{-3})^{0.5} \left(\frac{4,000}{7,950}\right)^{3.15}$$

$$= 92.70 \frac{\text{W}}{\text{cm}^2}$$

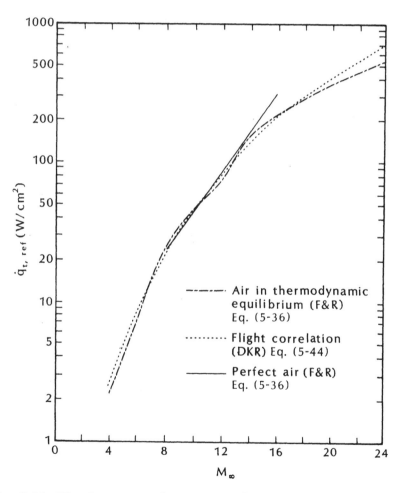

Fig. 5.10 The heat-transfer rate to the stagnation point of a sphere (R_N = 1 ft) at an altitude of 150,000 ft.

5.3.5 The Effect of Surface Catalycity on Convective Heat Transfer

Heat-transfer measurements made during the early flights of the Space Shuttle exhibited a significant surface catalycity effect. Since a considerable fraction of the heat may be transferred by atomic diffusion toward the wall followed by recombination on the surface, it would be possible to eliminate this fraction of the heat transfer by using a noncatalytic surface. The effect of surface catalycity on the heat-transfer parameter, $Nu_x/\sqrt{Re_x}$, as computed by Fay and Riddell,[11] is presented in Fig. 5.11. The heat-transfer parameter for one flight condition and wall temperature is presented for a catalytic wall and for a noncatalytic wall over a range of C_1, the recombination rate parameter. For very large values of C_1 (i.e., flows near equilibrium), there is essentially no difference between the catalytic wall value and the noncatalytic wall value, since few atoms reach the wall (with the wall temperature at 300 K, the equilibrium composition of air is molecules). For very low values of C_1 (i.e., essentially frozen flow), the heat released by the recombination of the atoms at the catalytic surface causes a high value for the heat-transfer parameter, $Nu_x/\sqrt{Re_x}$. Note that, for a catalytic wall, the value of the heat-transfer parameter varies only slightly with the value of the recombination-rate parameter. This reflects the fact that whether the atoms recombine within the boundary layer or at the wall makes no great difference since the energy is conducted about as readily by normal conduction as by diffusion when the Lewis number is approximately 1.

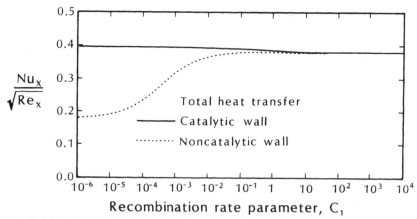

Fig. 5.11 The effect of surface catalycity on the heat-transfer rate parameter, $Nu_x/\sqrt{Re_x}$, as a function of the recombination-rate parameter C_1, $g_w = 0.0123$, $C_{A,e} = 0.536$, $Le = 1.4$, $Pr = 0.71$, $T_w = 300$ K, as taken from Ref. 11.

The correlations presented in Fig. 5.11 were based on boundary-layer solutions using state-of-the-art techniques of the late 1950s. A complete hypersonic flowfield analysis should include finite-rate chemistry, since the frozen-flow model and the equilibrium flow model represent limiting conditions. A viscous shock-layer (VSL) computer code (incorporating a subset of the Navier-Stokes equations) has been used to compute flowfields for slender, blunted cones over a range of nose radii, body half-angles, and altitude/velocity conditions.[18] The stagnation-point heat-transfer rates computed by Zoby et al. for a Mach 25 flow are presented in Fig. 5.12 as a function of the nose radius for an equilibrium flow, for a fully catalytic wall, and for a non-catalytic wall. Note that, for these calculations too, the heat transfer is essentially the same when the flow is assumed to be in equilibrium as when the finite-rate chemistry flow is bounded by a fully catalytic wall. However, the ratio of the heating for a noncatalytic wall to that for a fully catalytic wall is not independent of the nose radius. Note that, as the nose radius increases, the relation between the characteristic time for air particles to move through the stagnation region relative to the relaxation time for chemical reactions changes. Thus, for increasing nose radius, the heat-transfer rates computed using the finite-rate chemistry model (termed nonequilibrium flows) approach the values computed assuming that the flow is in equilibrium.

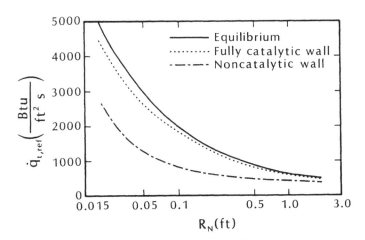

Fig. 5.12 The effect of wall catalycity on the stagnation-point heat-transfer rate as a function of nose radius, altitude 175,000 ft (53,340 m), M_∞ = 25, T_w = 2260 °R (1256 K), as taken from Ref. 18.

5.3.6 A Word of Caution About Models

In Chap. 4, we saw that data obtained in ground-based test facilities reflect the limitations associated with flow simulations. Furthermore, no matter how powerful the computer hardware, computed flowfield solutions reflect the limitations associated with the flow models and the numerical algorithms employed. The limitations both of ground-based tests and of computed flowfields have been a recurring theme of this text. Thus, the reader should appreciate the need for comparing data with computed results and vice versa.

In the early 1960s, researchers from two large organizations developed analytical techniques/experimental facilities for determining the stagnation-point heat transfer at superorbital velocities, i.e., velocities in excess of 7,950 m/s (26,082 ft/s). Buck et al.[19] noted that there was a controversy at that time regarding the magnitude and contributing factors causing the increased heating at superorbital velocities. The two dramatically different correlations that were presented in Ref. 19 are reproduced in Fig. 5.13. Since the two groups that developed these correlations contained many talented, competent people, the individual researchers and their organizations are not identified. Their identity is not important. What is important

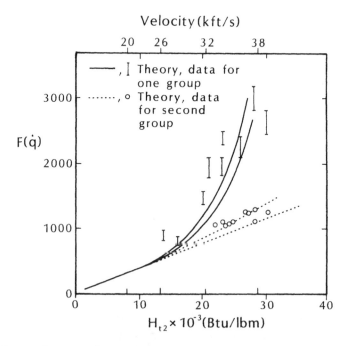

Fig. 5.13 The stagnation-point heat-transfer correlations at super-orbital speeds, as taken from Ref. 19.

is that each team of researchers presented both data and theoretical results that were consistent with their position. Eventually, the correlation giving the lower heat-transfer rates was found to be correct.

The objective of the discussion in this section is to point out to the reader that the analysis of hypersonic flows involving complex phenomena should be conducted with considerable thoroughness as well as competence. One must carefully validate models used to represent physical processes in CFD codes, in addition to evaluating the test conditions and instrumentation of experimental programs.

5.3.7 Non-Newtonian Stagnation-Point Velocity Gradient

The stagnation-point velocity gradient, as given by Eq. (5-39), is based on the pressure distribution for the modified Newtonian flow model. The pressure distribution over a considerable portion of a true hemisphere is reasonably well predicted by the modified Newtonian approximation. However, as will be discussed in Chap. 6, if the nose is truncated before $\phi = 45$ deg., the sonic point will move to the corner (as shown in Fig. 5.14b). The changes in the inviscid flowfield propagate throughout the subsonic region. As a result, the pressure decreases more rapidly with distance from the stagnation point. The resultant pressure gradient (or, equivalently, the velocity

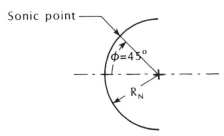

(a) Modified Newtonian flow model yields a reasonable pressure distribution

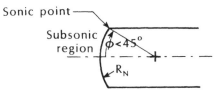

(b) Pressure distribution affected in the subsonic region

Fig. 5.14 Non-Newtonian stagnation-point velocity gradient.

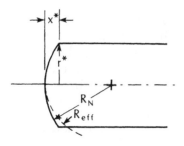

Fig. 5.15   Nomenclature for the velocity-gradient correlation.

gradient) can be calculated by determining the effective radius of a spherical cap (R_{eff}). Refer to Fig. 5.15. The correlation for the effective nose radius based on data obtained by Boison and Curtiss[20] for $M_\infty = 4.76$ is reproduced in Fig. 5.16. The bow shock wave shape and the stagnation-point velocity gradient are Mach number dependent. However, the values for Mach 4.76 provide a reasonable correlation for hypersonic flow.

Exercise 5.5:

The Apollo Command Module is an example of a configuration with a truncated spherical nose. Using the nomenclature of Fig. 5.15, approximate values for the Apollo Command Module are:

$$r^* = 1.956 \text{ m} \quad (6.417 \text{ ft})$$
$$R_N = 4.694 \text{ m} \quad (15.400 \text{ ft})$$

The term *approximate* is used, since (as will be discussed in Chap. 6) the Apollo Command Module has rounded corners. What is the actual stagnation-point velocity gradient relative to that based on the nose radius R_N?

Solution:

To calculate the actual stagnation-point velocity gradient, we will use Fig. 5.16 to determine the effective nose radius. Let us first determine the angle of the corner (ϕ_c):

$$\phi_c = \sin^{-1} \frac{r^*}{R_N} = 24.62 \text{ deg.}$$

Thus,

$$x^* = R_N(1 - \cos \phi_c) = 0.09091 \, R_N$$
$$r^* = 0.4167 \, R_N$$

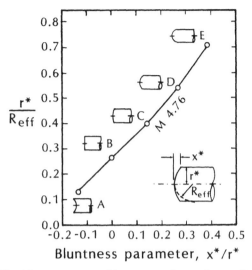

Fig. 5.16 Effective nose radius as a function of the bluntness parameters, using data for M_∞ = 4.76, as taken from Ref. 20.

So that

$$\frac{x^*}{r^*} = 0.218$$

Using Fig. 5.16,

$$\frac{r^*}{R_{\text{eff}}} = 0.475$$

Thus,

$$R_{\text{eff}} = 4.118 \text{ m } (13.510 \text{ ft})$$

Using Eq. (5.39),

$$\frac{\left(\dfrac{\mathrm{d}u_e}{\mathrm{d}x}\right)_{\text{eff}}}{\left(\dfrac{\mathrm{d}u_e}{\mathrm{d}x}\right)_{R_N}} = \frac{\dfrac{1}{\sqrt{R_{\text{eff}}}}}{\dfrac{1}{\sqrt{R_N}}} = \sqrt{\frac{R_N}{R_{\text{eff}}}} = 1.068$$

Using the correlation of Stoney,[21] which represents higher Mach number data, the ratio would be 1.055. Thus, the correlation of Fig. 5.16 provides a reasonable approximation for calculating the stagnation-point velocity gradient of spherical segments in a hypersonic stream.

5.3.8 Stagnation-Point Velocity Gradient for Asymmetric Flows

The stagnation-point velocity gradient calculated using Eq. (5-39) applies to modified-Newtonian flow over a spherical cap (axisymmetric flow) or over a cylinder whose axis is perpendicular to the freestream (two-dimensional flow). A correlation for the effective nose radius for those flows where truncation causes changes that propagate through the subsonic region was presented in Sec. 5.3.7. However, using an effective radius of curvature to characterize the stagnation-point velocity gradient may not be sufficient for many applications. For a three-dimensional stagnation-point flow, the heating rates are influenced rather significantly by the principal velocity gradients in the streamwise and the crosswise directions, which reflect the three-dimensionality of the stagnation region. Using K to define the ratio of the crosswise velocity gradient to the streamwise velocity gradient,

$$
K = \frac{\left(\dfrac{\partial w_e}{\partial z} \right)_{t2}}{\left(\dfrac{\partial u_e}{\partial x} \right)_{t2}} = \frac{R_x}{R_z}
$$

which is equal to the ratio of the two principal radii of curvature at the stagnation point. The heat transfer for a three-dimensional stagnation-point flow may be approximated as [22]:

$$
(\dot{q}_{t,\text{ref}})_{3D} = \sqrt{\frac{1 + K}{2}} \, (\dot{q}_{t,\text{ref}})_{\text{axisym}} \qquad (5\text{-}45)
$$

However, as noted by Goodrich et al.,[23] if the stagnation point is in a relatively large subsonic region, the local geometric radii of curvature may not properly define the local velocity gradients, when the local radii of curvature change significantly. In such cases, the velocity gradients should be based on more exact flowfield solutions. Since the subsonic zone size will change with the angle-of-attack and with gas chemistry, the effective radius is influenced by both of these factors. The influence of gas chemistry and of the model used to calculate the velocity gradient on the stagnation-point heating as computed for the Shuttle Orbiter by Goodrich et al.[23] is reproduced in Fig. 5.17. Since the Orbiter was at 31.8 deg. to 41.4 deg. angle-of-attack, the subsonic region is large and three-dimensional. The predictions were made using both conventional modified Newtonian-flow velocity gradients and the three-dimensional velocity gradients obtained from Euler flowfield solutions. The stagnation-point heating is significantly lower when the more rigorous flowfield models are used to compute the three-dimensional velocity gradients in the stagnation region.

5.3.9 Perturbations to the Convective Heat Transfer at the Stagnation Point

In some cases, the stagnation-point convective-heat-transfer rate measured during ground-test programs is significantly different than the value predicted using accepted theoretical correlations, e.g., Eq. (5-36). One such program with which the author was personally involved was the experimental program conducted to define the aerothermodynamic environment of the Apollo Command Module. There were numerous individual runs in which the experimentally-determined convective-heat-transfer rate was between 20 percent to 80 percent greater than the theoretical value. It should be noted that it is the author's experience that, for runs where the differences between experiment and theory exceeded 20 percent, the measurements were always high.

Weeks[24] presented an analytical treatment of the stagnation-region heating accounting for vorticity amplification and viscous dissipation. It was found that freestream turbulence might cause (1) increased stagnation-region heat transfer (in some cases over a 100 percent increase), (2) incorrect mechanical ablation, and (3) early boundary-layer transition. The effect of freestream turbulence on the onset of boundary-layer transition will be discussed in Chap. 7. Weeks[24] noted that "Convection of vorticity entropy and chemical modes and propagation of acoustic waves through the stagnation point boundary layer will alter the surface heat transfer rate and skin friction calculated by assuming laminar flow and laminar transport properties (including viscosity, conductivity, and species diffusion coefficients)."

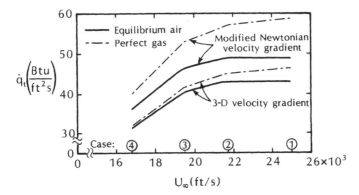

Fig. 5.17 The influence of gas chemistry and of velocity-gradient model on the stagnation-point heating for the Space Shuttle Orbiter, as taken from Ref. 23.

Holden[25] conducted a series of experimental studies in an attempt to identify other potential fluid mechanical mechanisms that might cause enhanced heating in the stagnation region of blunt bodies in hypersonic flow at high Reynolds numbers. The four mechanisms investigated were: (1) boundary-layer transition close to the stagnation region, (2) surface roughness, (3) surface blowing in the stagnation region, and (4) particle/shock-layer interactions. Holden[25] stated, "It is rationalized that, since the flow in the stagnation region is subsonic, pressure disturbances propagating forward from transition can promote increased heating in this region." Holden's correlation indicated that once transition had moved to within one-tenth of a body diameter from the stagnation point, there was a signficiant increase in the ratio of the measured to the theoretical heating rate. The reader should note that it is possible that the fact that transition occurred so near the stagnation point (as observed by Holden) may be due to freestream turbulence.

5.4 RADIATIVE HEAT FLUX

When a vehicle flies at very high velocity, the temperatures in the shock layer become sufficiently high to cause dissociation and ionization. With a knowledge of the thermodynamic and chemical properties of the gas in the shock layer and fundamental data on the radiation from the gas as a function of these properties, it is possible to determine the intensity of the radiation emitted per unit volume of the gas. The units for the radiation intensity (E) are W/cm^3. Page and Arnold[26] note, "It should be remarked that the species causing the predominant radiation varies with temperature. For example, at temperatures of 5,000 to 7,000 K, the NO-beta and NO-gamma molecular bands have large contributions; at temperatures of 7,000 to 9,000 K, the N_2^+-first negative molecular band is important; whereas increasing ionization at temperatures above 10,000 K causes free-free and free-bound radiation from the natural and ionized atoms of N and O to have large contributions to the total radiation."

5.4.1 The Radiation Intensity

Using a two-temperature model, Park[27] computed the stagnation streamline flowfield for the Fire II vehicle (roughly a 0.2-scale model of the Apollo Command Module). The temperature T, which represents the heavy-particle translational and molecular rotational energies, and T_v, which characterizes the molecular vibrational, electronic translational, and electronic excitation energies, as computed by Park[27] for an altitude of 73.72 km (i.e., a freestream density of 5.98×10^{-5} kg/m^3) are reproduced in Fig. 5.18a. The translational temperature (T) is 62,000 K just downstream of the shock wave and

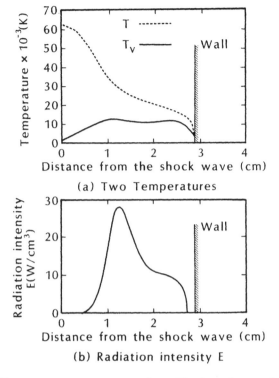

Fig. 5.18 **The temperatures and radiation intensity as computed along the stagnation streamline in the shock layer of Fire II at 73.72 km, as taken from Ref. 27.**

decreases monotonically. The vibrational temperature (T_v) peaks at 13,000 K. The computed radiation intensity is reproduced in Fig. 5.18b. Note that the peak intensity occurs during the broad peak in T_v, thereby suggesting that it is mostly T_v, which represents the electronic excitation temperature and radiation intensity. Furthermore, the peak radiation intensity occurs well away from the shock wave.

Once the distribution of the radiation intensity $(E$ in $W/cm^3)$ is known, the radiation falling on a unit area $(I$ in $W/cm^2)$ can be determined by integrating along the path (in this case the stagnation streamline). Radiation data are often obtained in shock tubes and in free-flight facilities. It should be emphasized that the total intensity of the radiation as measured from the one-dimensional flow pattern behind a normal-shock wave in shock-tube experiments and from the three-dimensional flow pattern along the stagnation streamline in the shock layer of a blunt body should by no means be expected to agree perfectly.

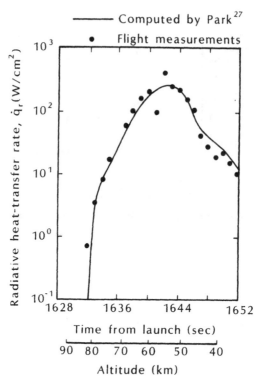

Fig. 5.19 Comparison between the calculated and the measured stagnation-point heat-transfer rates for Fire II, as taken from Ref. 27.

The shock-layer thickness has a significant effect on the radiation intensity I. Thus, since a large-nose-radius configuration produces a thicker shock layer, the radiative heat transfer to the body would be greater for a larger nose radius. Recall that the shock-layer thickness is greater for nonequilibrium flow than for equilibrium flow.

5.4.2 Stagnation-Point Radiative Heat-Transfer Rates

The stagnation-point radiative heat-transfer rates, as computed by Park,[27] are compared with data from the Fire II flight in Fig. 5.19. Park notes:

> The present calculation agrees closely with the measurements at altitudes below 81 km. At an altitude of 81 km, the calculation severely underestimates the heat flux. The discrepancy is most likely caused by the finite thickness of the shock wave which is neglected in the present continuum theory. The present theory assumes the shock wave

to be infinitesimally thin, and therefore that there is no chemical-kinetic process within the shock wave. At an altitude of 81 km, the true thickness of the shock wave becomes comparable to the thickness of the shock layer.

The analysis developed by Martin[28] indicates that the gas-to-surface radiation for a re-entry vehicle may be estimated as:

$$\dot{q}_{r,t} = 100 R_N \left(\frac{U_\infty}{10^4}\right)^{8.5} \left(\frac{\rho_\infty}{\rho_{SL}}\right)^{1.6} \qquad (5\text{-}46)$$

Martin notes "that up to satellite velocity one may treat surface heat transfer as arising exclusively in the aerodynamic boundary layer to the accuracy of the engineering approximations describing the heat transfer." According to Martin, radiation and convective heating, i.e., that from the aerodynamic boundary layer, become comparable for a 1-ft (0.3048 m) radius sphere at $U_\infty = 40$ kft/s (12.2 km/s).

Sutton[29] noted that, because ASTVs enter the atmosphere at relatively high velocities and have a large frontal area in order to generate the desired large drag forces, there is renewed interest in radiative heat-transfer technology. Assuming that the gas was in chemical equilibrium, Sutton generated radiatively coupled solutions of the inviscid, stagnation-region flowfields for a variety of environments. Presented in Fig. 5.20 are the calculations of the re-entry heating environment and the total heating-rate data for Fire II, which

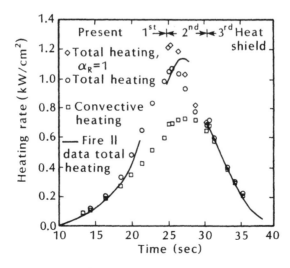

Fig. 5.20 Comparisons of the entry heating measurements for the flight of Fire II, as taken from Ref. 29.

was an "Apollo-like" configuration with a layered heat shield. The initial entry velocity for Fire II was in excess of 11 km/s (36,000 ft/s). Heat shields 1 and 2 were ejected sequentially during re-entry to expose a clean surface for the next data period. At peak heating, the calculations of Sutton indicated that 35 percent of the total heating was due to absorbed radiation.

Measurements from the stagnation region during the re-entry of Apollo Spacecraft 017[30] indicated a peak radiative heating rate of 100 Btu/ft^{2}s (114W/cm^2). This is roughly one-fourth of the maximum heating rate.

Park[31] presented calorimeter and radiometer data for the Apollo 4 flight. It was noted that the calorimeters were designed to measure the sum of the convective and radiative heat-transfer rates, while the radiometers measured only the radiative components. In the high altitude range (where the velocity is the highest), the calorimeter and the radiometer measured approximately the same heat fluxes. This can be true only if the convective heat-transfer rate is zero. In turn, it can be true only if the gases injected into the boundary layer by the ablation process, "blow off" the boundary layer, reducing the convective heat transfer to zero.

5.5 CONCLUDING REMARKS

The deceleration of the hypersonic flow creates high temperatures in the shock layer, causing a severe thermal environment. The stagnation-point convective heat transfer is roughly proportional to $(\rho_\infty)^{0.5}(U_\infty)^3/(R_N)^{0.5}$. Thus, the blunter the body, the lower the convective heat-transfer rate of the stagnation point. For nonequilibrium flows that occur at high altitudes, the convective heat transfer is affected by the surface catalycity.

For velocities greater than orbital velocities, the dissociation and possible ionization of the gas in the shock layer may produce significant radiation intensity. The stagnation-point radiative heat transfer is roughly proportional to $(\rho_\infty)^{1.6}(U_\infty)^{8.5}R_N$. Note that the blunter the body, the higher the stagnation-point radiative heat-transfer rate. For vehicles re-entering the atmosphere at velocities well in excess of orbital speeds, the radiative heat transfer may be of the same order as the convective heat transfer. For such cases, the designer must consider conflicting requirements regarding the nose radius (R_N).

Addition of CO_2 to the gas mixture (such as would occur in the Martian atmosphere) would increase the radiative heat-transfer rates because of the pressure of the strongly radiating molecule CN in the shock layer.

REFERENCES

[1] Wittliff, C. E., and Curtis, J. T., "Normal Shock Wave Parameters in Equilibrium Air," Cornell Aeronautical Lab. Rept. No. CAL-111, Nov. 1961.

[2] Candler, G., "Computation of Thermo-Chemical Nonequilibrium Martian Atmospheric Entry Flows," AIAA Paper 90-1695, Seattle, WA, June 1990.

[3] Mitcheltree, R., and Gnoffo, P., "Thermochemical Nonequilibrium Issues for Earth Reentry of Mars Mission Vehicles," AIAA Paper 90-1698, Seattle, WA, June 1990.

[4] Gupta, R. N., Yos, J. M., and Thompson, R. A., "A Review of Reaction Rates and Thermodynamic and Transport Properties for the 11-Species Air Model for Chemical and Thermal Nonequilibrium Calculations to 30,000 K," NASA TM-101528, 1989.

[5] Hansen, C. F., and Heims, S. P., "A Review of Thermodynamic, Transport, and Chemical Reaction Rate Properties of High Temperature Air," NACA TN-4359, July 1958.

[6] Hall, J. G., Eschenroeder, A. Q., and Marrone, P. V., "Blunt-Nose Inviscid Airflows with Coupled Nonequilibrium Processes," *Journal of the Aerospace Sciences,* Vol. 29, No. 9, pp. 1038–1051, Sep. 1962.

[7] Hornung, H. G., "Experimental Real-Gas Hypersonics," The 28th Lanchester Memorial Lecture, The Royal Aeronautical Society, Paper No. 1643, May 1988.

[8] Li, T.-Y., and Geiger, R. E., "Stagnation Point of a Blunt Body in Hypersonic Flow," *Journal of the Aeronautical Sciences,* Vol. 24., No. 1, pp. 25–32, Jan. 1957.

[9] Lomax, H., and Inouye, M., "Numerical Analysis of Flow Properties About Blunt Bodies Moving at Supersonic Speeds in an Equilibrium Gas," NASA TR R-204, July 1964.

[10] Dorrance, W. H., *Viscous Hypersonic Flow,* McGraw-Hill, New York, 1962.

[11] Fay, J. A., and Riddell, F. R., "Theory of Stagnation Point Heat Transfer in Dissociated Air," *Journal of the Aeronautical Sciences,* Vol. 25, No. 2, pp. 73–85, 121, Feb. 1958.

[12] Hayes, W. D., and Probstein, R. F., *Hypersonic Flow Theory,* Academic Press, New York, 1959.

[13] Moeckel, W. E., and Weston, K. C., "Composition and Thermodynamic Properties of Air in Chemical Equilibrium," NACA TN-4265, Aug. 1958.

[14] Hansen, C. F., "Approximations for the Thermodynamic and Transport Properties of High-Temperature Air," NACA TR R-50, Nov. 1957.

[15] Sutton, K., and Graves, R. A. Jr., "A General Stagnation-Point Convective Heating Equation for Arbitrary Gas Mixtures," NASA TR R-376, Nov. 1971.

[16] Scott, C. D., Ried, R. C., Maraia, R. J., Li, C. P., and Derry, S. M., "An AOTV Aeroheating and Thermal Protection Study," H. F. Nelson (ed.), *Thermal Design of Aeroassisted Orbital Transfer Vehicles,* Vol. 96 of Progress in Astronautics and Aeronautics, AIAA, New York, 1985, pp. 198–229.

[17] Detra, R. W., Kemp, N. H., and Riddell, F. R., "Addendum to Heat Transfer to Satellite Vehicles Reentering the Atmosphere," *Jet Propulsion,* Vol. 27, No. 12, pp. 1256–1257, Dec. 1957.

[18] Zoby, E. V., Lee, K. P., Gupta, R. N., and Thompson, R. A., "Nonequilibrium Viscous Shock Layers Solutions for Hypersonic Flow Over Slender Bodies," Paper No. 71, Eighth National Aero-Space Plane Technology Symposium, Monterey, CA, Mar. 1990.

[19] Buck, M. L., Benson, B. R., Sieron, T. R., and Neumann, R. D., "Aerodynamic and Performance Analyses of a Superorbital Re-Entry Vehicle," S. M. Scala, A. C. Harrison, and M. Rogers (eds.), *Dynamics of Manned Lifting Planetary Entry,* John Wiley & Sons, New York, 1963.

[20] Boison, J. C., and Curtiss, H. A., "An Experimental Investigation of Blunt Body Stagnation Point Velocity Gradient," *ARS Journal,* Vol. 29, No. 2, pp. 130–135, Feb. 1959.

[21] Stoney, W. E., Jr., "Aerodynamic Heating on Blunt-Nose Shapes at Mach Numbers Up to 14," NACA RML 58E05a, 1958.

[22] DeJarnette, F. R., Hamilton, H. H, Weilmuenster, K. J., and Cheatwood, F. M., "A Review of Some Approximate Methods Used in Aerodynamic Heating Analyses," *Journal of Thermophysics,* Vol. 1, No. 1, pp. 5–12, Jan. 1987.

[23] Goodrich, W. D., Li, C. P., Houston, C. K., Chiu, P. B., and Olmedo, L., "Numerical Computations of Orbiter Flowfields and Laminar Heating Rates," *Journal of Spacecraft and Rockets,* Vol. 14, No. 5, pp. 257–264, May 1977.

[24] Weeks, T. M., "Influence of Free-Stream Turbulence on Hypersonic Stagnation Zone Heating," AIAA Paper 69-167, New York, NY, Jan. 1969.

[25] Holden, M. S., "Studies of Potential Fluid-Mechanical Mechanisms for Enhanced Stagnation-Region Heating," AIAA Paper 85-1002, Williamsburg, VA, June 1985.

[26] Page, W. A., and Arnold, J. O., "Shock-Layer Radiation of Blunt Bodies at Reentry Velocities," NASA TR R-193, Apr. 1964.

[27] Park, C., "Assessment of Two-Temperature Kinetic Model for Ionizing Air," AIAA Paper 87-1574, Honolulu, HI, June 1987.

[28] Martin, J. J., *Atmospheric Re-entry, An Introduction to Its Science and Engineering,* Prentice-Hall, Englewood Cliffs, 1966.

[29] Sutton, K., "Air Radiation Revisited," H. F. Nelson (ed.), *Thermal Design of Aeroassociated Orbital Transfer Vehicles,* Vol. 96 of Progress in Astronautics and Aeronautics, AIAA, New York, 1985, pp. 419–441.

[30] Lee, D. B., and Goodrich, W. D., "The Aerothermodynamic Environment of the Apollo Command Module During Suborbital Entry," NASA TND-6792, Apr. 1972.

[31] Park, C., "Radiation Enhancement by Nonequilibrium in Earth's Atmosphere," *Journal of Spacecraft and Rockets,* Vol. 22, No. 1, pp. 27–36, Jan.–Feb. 1985.

PROBLEMS

5.1 In order to develop the required correlation for the solution of Eqs. (5-21a) and (5-21b), it is necessary to develop an expression for the velocity-gradient (or, equivalently, the pressure-gradient) parameter:

$$\beta = \frac{2S}{u_e} \frac{\mathrm{d}u_e}{\mathrm{d}S}$$

which appears in Eq. (5-21a). Using limiting approximations for the stagnation-region flow, similar to those presented in Eqs. (5-27) and (5-28), develop an expression β both for a two-dimensional flow (such as the stagnation point of a cylinder) and for an axisymmetric flow (such as the stagnation point of a sphere).

5.2 Consider a spherical vehicle, 1 m in radius, flying at 2.5 km/s at an altitude of 50 km. Using metric units, calculate the stagnation-point heat-transfer rate using the following three techniques. Assume that the wall temperature is 500 K.

(a) Use Eq. (5-36), using the perfect-gas relations to evaluate the properties of air.

(b) Use Eq. (5-36), assuming that the air behaves as a gas in thermochemical equilibrium to evaluate the properties of air.

(c) Use Eq. (5-44a). Assume that the circular-orbit velocity (U_{co}) is 7950 m/s.

5.3 Consider a sphere, 1 ft in radius, flying at 24,000 ft/s at an altitude of 240,000 ft. Using English units, calculate the stagnation-point heat-transfer rate using the three following techniques. Assume that the wall temperature is 2500 °R.

(a) Use Eq. (5-36), using the perfect-gas relations to evaluate the properties of air.

(b) Use Eq. (5-36), assuming that the air behaves as a gas in thermo-chemical equilibrium to evaluate the properties of air.

(c) Use Eq. (5-44b). Assume that the circular-orbit velocity (U_{CO}) is 26,082 ft/s.

5.4   Consider a vehicle that flies a trajectory containing the following points (taken from the Apollo design entry trajectory).

(a) Velocity = 36,000 ft/s; Altitude = 240,000 ft
(b) Velocity = 32,000 ft/s; Altitude = 230,000 ft
(c) Velocity = 27,000 ft/s; Altitude = 250,000 ft
(d) Velocity = 20,000 ft/s; Altitude = 220,000 ft
(e) Velocity = 16,000 ft/s; Altitude = 200,000 ft
(f) Velocity = 8,000 ft/s; Altitude = 150,000 ft

Assume that the nose radius is 2 ft and that the circular orbit velocity (U_{CO}) is 26,082 ft/s. Use Eq. (5-44b) to calculate the convective heating and use Eq. (5-46) to calculate the radiative heat flux at the stagnation point. Note the division between convective and radiative heating.

5.5   Repeat Problem 5.4 for a vehicle whose nose radius is 20 ft.

6

THE PRESSURE
DISTRIBUTION

6.1 INTRODUCTION

As noted in the earlier chapters, for high Reynolds number flows
where the boundary layer is attached and relatively thin (with the
streamlines relatively straight), the pressure gradient normal to the
wall is negligible, i.e., $(\partial p/\partial y) \approx 0$. Significant normal pressure gradients may occur: (1) for turbulent boundary layers at very high
local Mach numbers, (2) when there is significant curvature of the
streamlines, or (3) when there is a significant gas injection at the
wall, such as occurs with ablation. When $(\partial p/\partial y) \approx 0$, the pressure distribution of the inviscid flow at the edge of the boundary
layer also acts at the surface. Thus, as discussed in Chap. 3, a solution to the Euler equations, which are obtained by neglecting the
diffusive terms in the Navier-Stokes equations, yields the pressure
distribution acting near the surface. With this information, one can
generate the outer boundary conditions for the boundary-layer solutions (in a two-layer flow-model analysis), as well as estimates of
the normal force and of the moments acting on the configuration. Of
course, a single-layer flow-model analysis, e.g., the full Navier-Stokes
equations, would simultaneously provide the viscous forces and the
pressure forces.

However, the objective of this chapter is to examine the effects
of configuration (including vehicle geometry and angle-of-attack), of
viscous/inviscid interactions, and of gas chemistry on the pressure
distribution. Since the emphasis is on developing an understanding
of the relationship between the flowfield characteristics and the pressure distribution, much of the discussion will utilize simple, albeit
approximate flow models. Included in the discussion will be limitations to the validity of the approximate flow models.

6.2 NEWTONIAN FLOW MODELS

For the windward surface of relatively simple shapes, one can
assume that the speed and the direction of the gas particles in the
freestream remain unchanged until they strike the solid surface exposed to the flow. For this flow model (termed *Newtonian flow* since
it is similar in character to one described by Newton in the seventeenth century), the normal component of momentum of the imping-

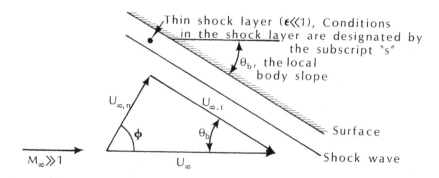

Fig. 6.1   Nomenclature for the Newtonian flow model.

ing fluid particle is wiped out, while the tangential component of momentum is conserved.

Thus, using the nomenclature of Fig. 6.1 and writing the integral form of the momentum equation for a constant-area streamtube normal to the surface,

$$p_\infty + \rho_\infty \left[U_{\infty,n} \right]^2 = p_\infty + \rho_\infty \left[U_\infty \sin \theta_b \right]^2 = p_s \qquad (6\text{-}1)$$

Rearranging so that the local pressure is written in terms of the pressure coefficient, one obtains:

$$C_p = \frac{p_s - p_\infty}{\frac{1}{2}\rho_\infty U_\infty^2} = 2 \sin^2 \theta_b = 2 \cos^2 \phi \qquad (6\text{-}2)$$

This equation for the pressure coefficient, Eq. (6-2), is based on the Newtonian flow model, where the 2 represents the pressure coefficient at the stagnation point (which is designated $C_{p,t2}$), since $\theta_b = 90$ deg. at the Newtonian stagnation point.

6.2.1 Modified Newtonian Flow

All of the computed values for the stagnation-point pressure coefficient that were presented in Chap. 1 were less than 2, the value for the Newtonian flow model. In fact, even when $M_1 = 24$, $C_{p,t2}$ is 1.838 for perfect air and is 1.932 for air in thermodynamic equilibrium. Thus, as noted by Lees,[1] it would be more appropriate to compare the ratio $C_p/C_{p,\max}$ with $\sin^2 \theta_b$ (or, equivalently, $\cos^2 \phi$). Such a comparison is presented in Fig. 6.2 using data for hemispherically capped cylinders (as taken from Ref. 2). Even though the data of Fig. 6.2 are for freestream Mach numbers from 1.97 to 4.76, the $\sin^2 \theta_b$ relation represents the data quite adequately for this blunt configuration. Therefore, an alternative representation of the pressure coefficient for hypersonic flow is

$$C_p = C_{p,t2} \sin^2 \theta_b = C_{p,t2} \cos^2 \phi \tag{6-3}$$

which will be termed *modified Newtonian flow*.

The modified Newtonian flow model can be used to obtain a quick, engineering estimate of the pressure distribution. Consider the axisymmetric configuration shown in Fig. 6.3, where $\hat{n}$ is a unit vector that is normal to the surface element dS and is positive in the inward direction, θ is the local surface inclination, and β is the angular position of a point on the surface of the body. The angle η, the angle between the velocity vector $\boldsymbol{V}_\infty$ and the inward normal $\hat{n}$, is given by:

$$\cos \eta = \frac{\boldsymbol{V}_\infty \cdot \hat{n}}{|\boldsymbol{V}_\infty||\hat{n}|} \tag{6-4}$$

where

$$\boldsymbol{V}_\infty = U_\infty \cos \alpha \hat{i} - U_\infty \sin \alpha \hat{j} \tag{6-5}$$

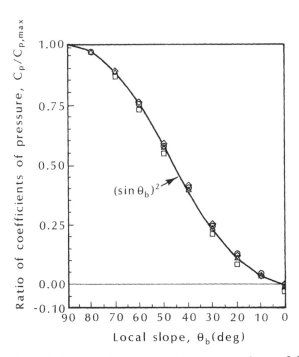

Fig. 6.2 Correlation between $C_p/C_{p,\mathrm{max}}$ ratio and local body slope, as taken from Ref. 2.

and

$$\hat{n} = \hat{i}\sin\theta - \hat{j}\cos\theta\cos\beta - \hat{k}\cos\theta\sin\beta \qquad (6\text{-}6)$$

Thus,

$$\cos\eta = \cos\alpha\sin\theta + \sin\alpha\cos\theta\cos\beta \qquad (6\text{-}7)$$

so that the pressure coefficient is:

$$C_p = C_{p,t2}\cos^2\eta \qquad (6\text{-}8)$$

In the (modified) Newtonian flow model, the freestream flow does not impinge on those portions of the body surface which are inclined away from the freestream direction and which may, therefore, be thought of as lying in the "shadow of the freestream." This is illustrated in Fig. 6.4.

Modified Newtonian flow is attractive because it requires only the angle between the freestream velocity vector and the inward normal to the local surface, which can often be calculated analytically, and provides reasonable estimates of the local pressure (as indicated in Fig. 6.2). This facilitates the calculation of streamlines. Furthermore, one can easily integrate the pressure distributions to obtain adequate estimates of the forces and moments. However, there are several flow phenomena which give rise to significant differences between the actual pressures and those predicted using modified Newtonian flow. Examples of where such differences occur are:

1. In the subsonic region of the shock layer of a "truncated" blunt body (such as the Apollo Command Module),

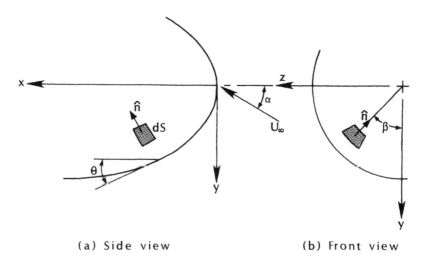

(a) Side view   (b) Front view

Fig. 6.3 Coordinate system nomenclature for axisymmetric configurations.

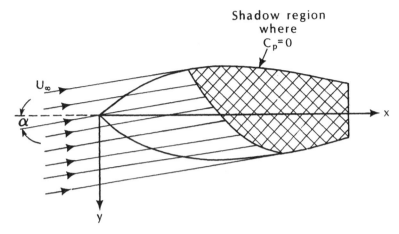

Fig. 6.4 Sketch illustrating region of the body which lies in the "shadow of the freestream."

2. In the rapid overexpansion and recompression of the flow near the nose region of a blunted sphere/cone, and

3. On the control surfaces (such as a body flap) where the turning of the local flow produces additional shock waves within the shock layer.

These flows will be discussed in Sec. 6.3.

There are other approximations which are based on the Newtonian flow model but which employ slightly different formulations. For instance, a term to account for the change in pressure due to streamline curvature is introduced in the Newton-Busemann model.

$$C_p = 2(\sin^2 \theta_b + \sin \theta_b \frac{d\theta_b}{dA} \int_O^A \cos \theta_b dA)$$

where A is the cross-sectional area of the body normal to the freestream direction. Another modification to the Newtonian flow model assumes that:

$$C_p = 2(\cos \eta)^N$$

where the exponent N is determined from experimental data. Based on the author's experience, these alternative approximations do not provide significant improvements to the pressure distributions based on the (modified) Newtonian flow model. If the modified Newtonian flow model does not provide pressures of suitable accuracy (as would be the case for flows 1–3 noted above), the author recommends the use of more rigorous formulations, e.g., Euler solutions.

6.2.2 Thin-Shock-Layer Requirements

The (modified) Newtonian flow model and the various theories for thin-shock-layer flows are based on the assumption that:

$$\rho_2 \gg \rho_\infty$$

For the hypervelocity flight of a blunted vehicle, the vibration, the dissociation, and the ionization energy modes are excited as the air passes through the flow shock wave. As a result, an appreciable amount of energy is absorbed by the excitation of the molecules, and the temperature does not increase as much as it would in the case of no excitation. As discussed in Chap. 5, the excitation process does not appreciably affect the pressure downstream of a normal-shock wave. However, since the excitation and the subsequent dissociation of the molecules absorb energy, the temperature in the shock layer does not achieve the perfect-gas (or frozen-flow) level. Thus, as the dissociation process is driven toward completion, the density ratio across the normal portion of the bow shock wave is two to three times the value obtained in conventional air (or nitrogen) hypersonic wind tunnels. For hypersonic flow over blunt bodies, the primary factor governing the shock stand-off distance and the inviscid forebody flow is the normal-shock density ratio. Therefore, *certain aspects* of a real gas in thermochemical equilibrium can be simulated in a wind tunnel by the selection of a test gas that has a low ratio of specific heat (γ), such as tetrafluoromethane (CF_4) as described in Chap. 4. Since $\gamma = 1.12$ for tetrafluoromethane, large values of ρ_2/ρ_∞ (e.g., 12.0) can be obtained[3] even when the stagnation temperature is 1530 °R. The simulation of the density ratio is not to imply, however, that the real-gas chemistry is simulated.

Jones and Hunt[4] studied flowfields for a family of blunted and of sharp large half-angle cones in hypersonic flows of helium, air, and tetrafluoromethane. The effective isentropic exponents for these gases are 1.67, 1.40, and 1.12, respectively. The pressure distributions obtained[4] on a sharp cone ($\theta_c = 50$ deg.) are presented as a function of the wetted distance along the conical surface and are reproduced in Fig. 6.5. The values of the surface-pressure ratio are very strongly dependent on the value of γ and, thus, the normal-shock density ratio. The level decreases with an increasing density ratio, the CF_4 data approaching the value predicted by Newtonian theory. The nondimensionalized pressure ratio is 20 percent lower when CF_4 is the test gas than when helium is.

The pressure distribution on the model in the helium stream falls off for s/s_{max} greater than 0.6. A similar decrease occurs for the model in the airstream when s/s_{max} exceeds 0.8. However, the pressures on the model in the CF_4 stream are essentially constant all the

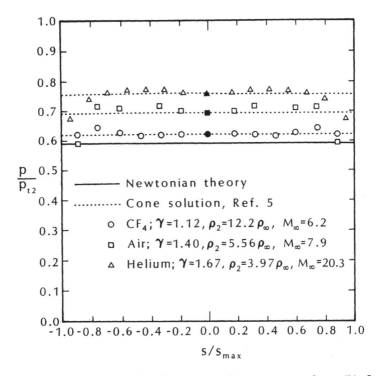

Fig. 6.5 Pressure distribution on a sharp cone, θ_c = 50 deg., α = 0 deg., as taken from Ref. 4.

way to the corner. Jones and Hunt[4] note that "The reason for the 'fall off' is because the subsonic flow in the shock layer must accelerate to sonic speeds at the sharp corner and since the CF$_4$ forebody flow is already very near sonic, it requires less acceleration at the corner than do the more subsonic helium and air forebody flows." This corresponds to the flow 1 listed in Sec. 6.2.1. The method of South and Klunker[5] provides reasonable estimates of the "constant" pressures in all these test gases. The method of Ref. 5 employs a direct method in that the body shape is given as one of the bounding coordinate surfaces. The shock wave is another bounding coordinate surface, and the governing differential equations are solved by integrating inward from the shock.

Using data obtained in wind tunnels where the test gas was either air or CF$_4$, Micol[3] and Miller and Wells[6] studied the effects of the normal-shock density ratio on the aerothermodynamic characteristics of the Aeroassist Flight Experiment (AFE) vehicle. Sketches of the configuration are shown in Fig. 6.6. The forebody configuration is derived from a blunted elliptic cone that is raked off at 73 deg. to the centerline to produce a circular raked plane. The blunt nose is

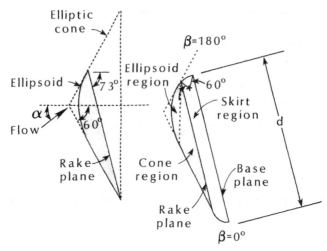

Fig. 6.6 Development of AFE configuration from original elliptic cone (symmetry plane shown), as taken from Ref. 6.

an ellipsoid with an ellipticity of 2. A skirt having an arc radius equal to one-tenth of the rake-plane diameter has been attached to the rake plane. The ellipsoid nose and the skirt are tangent to the elliptic cone surface at their respective intersections.

Schlieren photographs indicate that the shock-detachment distance at the stagnation point for CF_4 tests is less than one half the shock-detachment distance for the air tests, when $\alpha = 0$ deg. Furthermore, near the ellipsoid-cone juncture, an inflection was observed[6] in the shock wave for the CF_4 tests, indicating a flow overexpansion process. As evident in the pressure distributions presented in Fig. 6.7, the overexpansion becomes more pronounced as the angle-of-attack is decreased (refer to Fig. 6.6 for the sign convention for α). Referring to the pressure distributions for $\alpha = 0$ deg., which are presented in Fig. 6.7b, there is an overexpansion of the flow from the nose onto the conical surface when the test gas is CF_4 but not when it is air. At the juncture of the nose/conical surface, i.e., $(s/L) = 0.22$, the pressure coefficient ratio for CF_4 is 15 percent lower than the corresponding value for air. Miller and Wells[6] note, "Thus, typical of real-gas effects, the magnitude of the surface pressure in regions of compression such as the nose is relatively unaffected by an increase in density ratio; however, in regions of expansion, such as occur as the flow moves off the nose onto the conical section, the pressure decreases due to an increase in density ratio or decrease in the ratio of specific heat."

When the angle-of-attack is +10 deg., the configuration appears "relatively blunt" to the oncoming flow, with much of the forebody

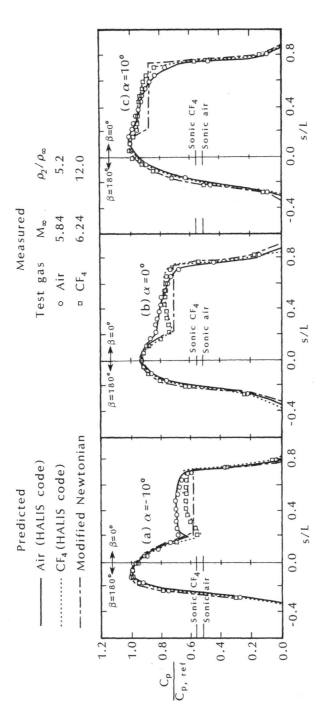

Fig. 6.7 Effects of normal-shock density ratio on the pressure distributions for the AFE, as taken from Ref. 6.

surface being almost perpendicular to the freestream. As a result, the pressure remains relatively constant over the face of the AFE vehicle. Furthermore, the normal-shock density ratio has a relatively small effect on the pressures, which are presented in Fig. 6.7c. However, when the angle-of-attack is decreased to −10 deg., the body appears "less blunt" to the oncoming flow. The stagnation point moves to $s \simeq -0.1L$ (on the $\beta = 180$ deg. plane) for both gases. For positive s, a significant fraction of the inviscid forebody flow achieves near-sonic conditions for the CF_4 test. The pressure distributions in the regions of expansion are very sensitive to the test gas.

The data are compared with the pressure distributions given by modified Newtonian theory and with those computed using the HALIS code.[7] The modified Newtonian model does not accurately represent the expansion process that occurs on the conical section nor does it represent the pressure dependence on the test gas. Since the shock layer is thinner for the CF_4 flows, i.e., the shock shape more closely follows the shape of the forebody, the modified Newtonian theory predicts the trend of the CF_4 data better than that of the air data. The pressures computed using the HALIS code were within 3 to 4 percent of the experimental values both for air and for CF_4.

The small-density-ratio requirement ($\epsilon \ll 1$) for Newtonian theory also places implicit restrictions on the body shape in order that the shock layer be thin. The range of applicability for Newtonian theory, as defined by Marconi et al.,[8] is reproduced in Fig. 6.8. Small-

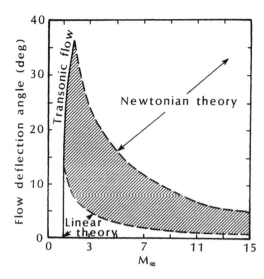

Fig. 6.8 Regions of applicability of inviscid flow theories for the surface pressure on a sharp cone, as taken from Ref. 8.

perturbation theory yields accurate results only for the flow over slender bodies at small angles-of-attack in a low supersonic Mach number stream. However, Newtonian theory provides useful results when the Mach number is large and/or the flow deflection angle is large. This is equivalent to strong shock assumption

$$M_\infty \sin \theta_b \gg 1$$

Exercise 6.1:

A sharp cone ($\theta_c = 15$ deg.) was exposed to a hypersonic stream of helium in the Ames 20-Inch Hypersonic Wind Tunnel. The freestream Mach number was 14.9 and the Reynolds number based on the radius of the base was 0.86×10^6. Surface-pressure measurements as a function of the circumferential angle β (refer to Fig. 6.3), as reported by Cleary and Duller,[9] are:

β (°)	$\alpha = 10°$ C_p	$\alpha = 20°$ C_p
0	0.359	0.663
30	0.326	0.550
60	0.234	0.327
90	0.112	0.093
120	0.060	0.028
150	0.037	0.009
180	0.039	0.014

Compare these measurements with the circumferential distributions predicted using the Newtonian flow model and using the modified Newtonian flow model.

Solution:

Equation (6-8) can be used to define the pressure coefficient both for the Newtonian flow model and for the modified Newtonian flow model. For a sharp cone, the local surface inclination is constant ($\theta = \theta_c$). Thus,

$$C_p = C_{p,t2} \left[\cos \alpha \sin \theta_c + \sin \alpha \cos \theta_c \cos \beta \right]^2$$

where $C_{p,t2} = 2$ for Newtonian flow and

$$C_{p,t2} = \left[\frac{p_{t2}}{p_1} - 1 \right] \frac{2}{\gamma M_1^2}$$

for modified Newtonian flow. For this hypersonic stream of helium ($M_1 = 14.9$ and $\gamma = 5/3$),

$$\frac{p_{t2}}{p_1} = \left[\frac{(\gamma + 1)M_1^2}{2}\right]^{\frac{\gamma}{(\gamma-1)}} \left[\frac{\gamma + 1}{2\gamma M_1^2 - (\gamma - 1)}\right]^{\frac{1}{(\gamma-1)}} \quad (1\text{-}17)$$

$$\frac{p_{t2}}{p_1} = 326.54$$

Thus, for modified Newtonian flow,

$$C_{p,t2} = 1.7596$$

For the larger angle-of-attack flow, i.e., $\alpha = 20$ deg., the angle-of-attack exceeds the inclination angle of the conical generator. Therefore, a portion of the conical surface will be in a shadow region (refer to Fig. 6.4), where $C_p = 0$ based on the Newtonian flow models. To locate the upper limits for β, i.e., (β_u), for this region,

$$\cos \alpha \sin \theta_c + \sin \alpha \cos \theta_c \cos \beta_u = 0$$

Thus,

$$\beta_u = \cos^{-1}\left[-\frac{\tan \theta_c}{\tan \alpha}\right] = 137.4° \quad \text{or} \quad 222.6°$$

Oil-flow patterns presented by Cleary and Duller[9] indicated flow separation at this angle-of-attack but affecting a somewhat smaller region.

The pressure distributions are presented in Fig. 6.9. For these relatively slender configurations, the pressure coefficients calculated using the Newtonian flow model are in better agreement with the data on the windward side. The differences between the measured pressures and the analytical values on the leeside ($\beta \geq 90$ deg.) are sufficient to present difficulties in computing the boundary layer using a two-layer technique.

6.3 DEPARTURES FROM THE NEWTONIAN FLOW MODEL

Three examples of flow phenomena which cause the pressure distributions to be significantly different from those predicted by the (modified) Newtonian flow model were identified at the end of Sec. 6.2.1. A discussion of these phenomena and how they affect the surface-pressure distributions is presented in this section.

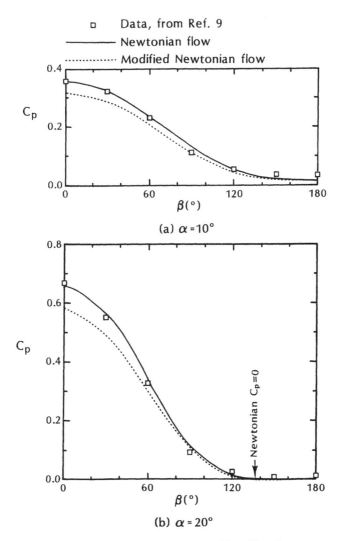

Fig. 6.9 Circumferential pressure distributions on a sharp cone (θ_c = 15 deg.,) for Exercise 6.1.

6.3.1 Truncated Blunt-Body Flows

As noted in the discussion of stagnation-region flowfields in Chap. 5 and in the surface-pressure distributions on a sharp, large half-angle cone in Fig. 6.5, the sonic point moves to the corner of blunt bodies. Since disturbances can propagate upstream in a subsonic flow, altering the locations of the sonic points affects the pressure distribution throughout the subsonic region.

Let us apply the modified Newtonian flow model to the flow over the spherical configuration, shown in Fig. 6.10. The expression for the modified Newtonian flow pressure coefficient, Eq. (6-3), can be rearranged to give:

$$\frac{p_s}{p_{t2}} = \sin^2 \theta_b + \frac{p_\infty}{p_{t2}} \cos^2 \theta_b = \cos^2 \phi + \frac{p_\infty}{p_{t2}} \sin^2 \phi \qquad (6\text{-}9)$$

As the boundary layer grows in the streamwise direction, air is entrained from the inviscid portion of the shock layer. Thus, when determining the fluid properties at the edge of the boundary layer, one must determine the entropy of the streamline at the boundary-layer edge. Note that the streamlines near the axis-of-symmetry of a blunt-body flow, such as those depicted in Fig. 6.10, have passed through that portion of the bow shock wave which is nearly perpendicular to the freestream flow. As a result, it can be assumed that all of the air particles at the edge of the boundary layer have essentially the same entropy. Therefore, the entropy at the edge of the boundary layer and, as a result, p_{t2} are the same at the streamwise stations.

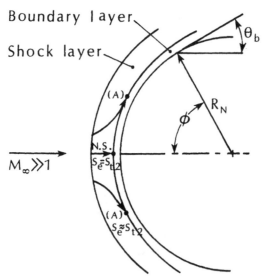

(A) The air at the edge of the boundary layer at point A has passed through the (nearly) normal part of the shock wave. Thus, the entropy of the air particles at the edge of the boundary layer (S_e) is essentially equal to S_{t2} at all stations. In essence, the air has passed through the normal part of the shock wave (NS) and has undergone an isentropic expansion (IE) to the local pressures.

Fig. 6.10 Sketch of the normal-shock/isentropic-expansion flow model for hypersonic flow over a blunt body.

The local flow properties are the same as if the air had passed through a normal-shock wave and had undergone an isentropic expansion to the local pressure (designated an NS/IE process). For such an isentropic expansion (IE), the ratio of p_s/p_{t2} can then be used to define the remaining flow conditions (for an equilibrium flow). Note that, if the flow expands isentropically to a point where the local static pressure (p_s) is approximately $0.5p_{t2}$, the flow is essentially sonic for all values of γ. Solving Eq. (6-9), we find that $p_s \approx 0.5p_{t2}$ when $\theta_b \approx 45$ deg. (i.e., the sonic points occur when the local body slope is 45 deg.). As indicated in relation to Fig. 5.14, if the nose is truncated before $\phi = 45$ deg., the sonic point moves to the corner, changing the flow over the entire spherical cap.

As shown in the sketch of Fig. 6.11, the Apollo Command Module is an example of a truncated spherical configuration. Because the windward heat shield of the Apollo Command Module is a truncated spherical cap, the actual sonic point, which occurs near the tangency point of the spherical heat shield and the toroidal surface, i.e., $\phi \approx 23$ deg., are inboard of the locations that they would occupy for a full spherical cap. As a result, the entire flowfield in the subsonic portion of the shock layer is modified, and the streamwise velocity gradients are relatively large in order to produce sonic flow at the "corners" of the Command Module.

The modified Newtonian pressure distribution for the zero angle-of-attack Apollo Command Module is compared in Fig. 6.12 with data obtained in Tunnel C at AEDC.[10] The experimental pressures measured in Tunnel C at a nominal freestream Mach number of 10 and at a Reynolds number ($Re_{\infty D}$) of 1.1×10^6 have been divided by the calculated value of the stagnation pressure behind a normal-

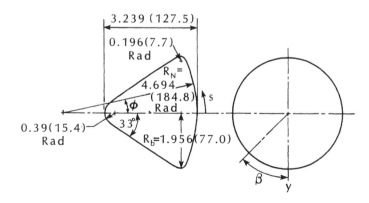

Fig. 6.11 Sketch of clean (no protuberances) Apollo Command Module. Dimensions in meters (inches).

shock wave (p_{t2}). For reference, an s/R_b ratio of 0.965 defines the tangency point of the spherical heat shield and the toroidal surface, while an s/R_b ratio of 1.082 corresponds to the maximum body radius. Thus, ϕ for the tangency point is 23 deg. and the local body slope changes from 67 deg. to 0 deg. in the region $s = 0.965R_b$ to $1.082R_b$. Note that significant differences exist between the modified Newtonian pressures and the measured values as one approaches the edge of the spherical heat shield. Because the velocity gradient at the stagnation point of a hemispherical segment is increased above the value for a full hemisphere, the stagnation-point heating rate will also be increased. As discussed in Chap. 5, investigations of stagnation point velocity gradients as a function of R_b/R_N have been reported by Stoney[11] and Boison and Curtiss,[12] among others.

Asymmetry due to angle-of-attack introduces an additional degree of freedom into the analysis of a blunt-body flowfield. For asymmetric flowfields, the stagnating streamline is not straight. As a result, the shape and the location of the stagnation streamline (which also serves as the dividing streamline) is unknown in advance. Hayes and Probstein[13] note that the stagnation point lies on the side of the

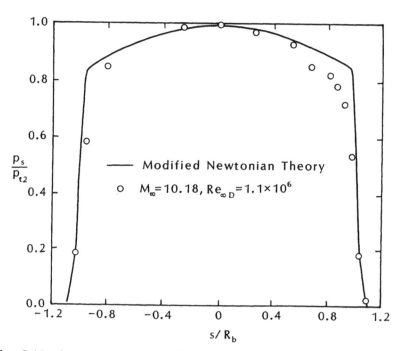

Fig. 6.12 A comparison of the modified Newtonian pressures and the experimental pressures for the Apollo Command Module at $\alpha = 0$ deg., as taken from Ref. 10.

normal point on the bow shock wave for which the body or shock curvature is greater and that the streamline of maximum entropy on the shock turns in the direction of decreasing body or shock curvature. This is illustrated in Fig. 6.13. The displacement between the point where the stagnation streamline crosses the bow shock wave and the point where the bow shock wave is normal to the freestream flow (i.e., the location of the maximum entropy streamline) is small when $\epsilon^{0.5}$ is sufficiently small. For the Apollo Command Module at nonzero angles-of-attack, the experimentally determined location of the stagnation point is nearer the axis of symmetry than is the modified Newtonian stagnation point (see Fig. 6.14). Based on the Apollo wind-tunnel data (surface pressures and shadowgraphs), the stagnation streamline lies nearer the axis-of-symmetry, which corresponds to the left-hand sketch of Fig. 6.13. Experimental pressures for the pitch plane of the Apollo Command Module at an angle-of-attack of 33 deg. are compared with the modified Newtonian pressures in Fig. 6.15. Because there is a large difference between the predicted and the actual stagnation point locations and because the rapid change in body slope on the toroid of the Command Module produces large pressure gradients, there are very important differences between the actual pressures and those predicted using the modified Newtonian flow model for $0.2R_b \leq s \leq 1.0R_b$. Furthermore, the heat-transfer rates computed using the experimental pressures will be dramatically different from those computed using the modified Newtonian pressures for this region. Away from the stagnation region ($s < 0$), there is good agreement between the measured pressures from the pitch plane and the modified Newtonian values.

From the pressure data presented thus far, it should be clear that, when the configuration is a truncated blunt body, such as the Apollo

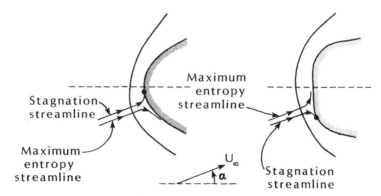

Fig. 6.13 Sketch of the stagnation streamline and the maximum entropy streamline for asymmetric blunt-body flows, as taken from Ref. 13.

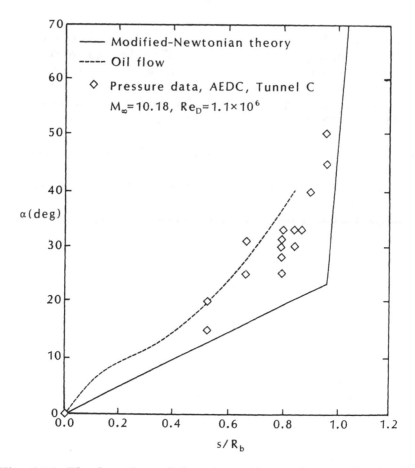

Fig. 6.14 The location of the stagnation point on the Apollo Command Module as a function of the angle-of-attack, as taken from Ref. 10.

Command Module, the entire subsonic flowfield is affected by the fact that the locations of the sonic regions are significantly different from their Newtonian values. Scott et al.[14] have investigated various design methodologies in determining the aerothermodynamic environment of an ellipsoidally-blunted raked-off elliptic cone, such as shown in Fig. 6.16. It was found that "the influence of configuration on the heat flux is unquestionably one of the most important considerations." Pressure distributions computed using the Navier-Stokes solutions for an ideal gas flow over this asymmetric configuration are reproduced in Fig. 6.17. The computations indicate that, by truncating the body, one obtains lower pressure over the remaining surface of the body.

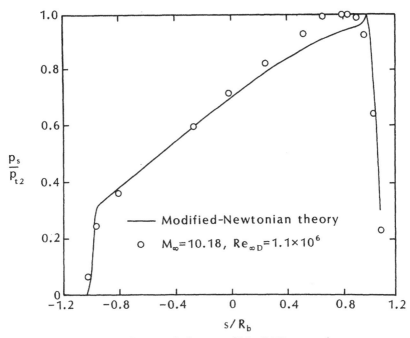

Fig. 6.15 A comparison of the modified Newtonian pressures for the Apollo Command Module and the experimental pressures at α = 33 deg., as taken from Ref. 10.

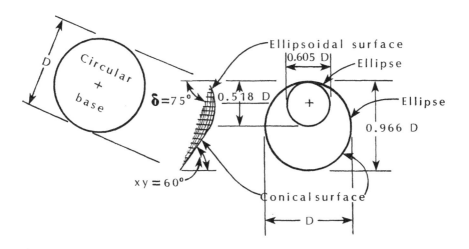

Fig. 6.16 The ellipsoidally-blunt raked-off elliptic cone used in the AOTV aeroheating study of Ref. 14.

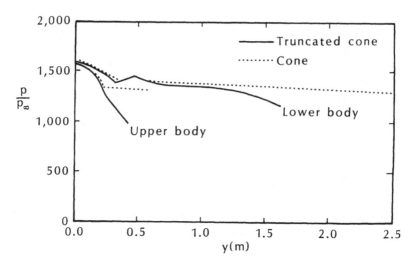

Fig. 6.17 The pressure distributions on ellipsoidally-blunt raked-off elliptic cone, as presented in Ref. 14 (θ_c = 60 deg.).

6.3.2 Nose Region of a Blunted Sphere/Cone

There are a variety of applications that can be satisfied only by ve-hicles that are slender, low-drag configurations. Blunting the nose of these slender bodies not only reduces the peak convective heat trans-fer but produces a variable entropy (and a variable stagnation pres-sure) inviscid shock-layer flow that can persist for more than $100R_N$. As discussed in Chap. 3, the entropy variations affect the properties at the edge of the boundary layer, the location where boundary-layer transition occurs, and (in fact) the ability to locate the edge of the boundary layer. However, blunting the nose also affects the shock shape and the pressure acting on the vehicle. These effects will be discussed using flowfields for spherically blunted cones, such as that shown in Fig. 6.18.

When the (modified) Newtonian flow model is used, the theo-retical values of the local pressure depend only on the local surface inclination relative to the freestream. Thus, the Newtonian pressure on the spherical cap would decrease with distance from the stagna-tion point on the spherical cap, becoming constant on the conical generator. However, in reality, the pressure on the surface of the blunted cone reflects the presence and interaction of compression and of expansion waves which originate from body curvature, three-dimensionality, reflection from the bow shock wave, and reflection from the sliplines due to the rotationality introduced by the curved bow shock. The surface-pressure distribution may be underexpanded

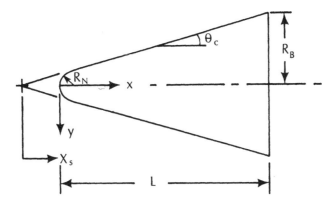

Fig. 6.18 Sketch of nomenclature for the blunted cone.

or overexpanded depending on the distance along the surface, the cone semi-vertex angle, and the freestream Mach number. When the ratio of p_j (the pressure at the tangency point between the spherical cap and the conical surface) to p_c (the asymptotic cone pressure) is much larger than unity, it is likely that underexpansion occurs everywhere, with no pressure less than the asymptotic pressure. For very slender cones, the surface pressure in the expanding flow may decrease so slowly that the asymptotic (sharp cone) value is not reached on the conical surface, resulting in an underexpanded flow. For larger cone half angles, the reflecting expansion waves encounter the conical surface, resulting in overexpanded pressures. Further downstream, the incident compression waves cause the surface pressure to increase, asymptotically approaching the sharp-cone value. For surfaces whose inclination to the freestream in increased, the minimum pressure point moves toward the nose, while the opposite is true if the surface inclination is decreased. If the incidence of the conical generator is increased further but the flow remains supersonic, the minimum pressure is followed by a compression that significantly overshoots the final asymptotic level. The interaction of the expansion waves and the bow shock wave can produce an inflection in the shock, which is characteristic of the flow over blunt cones. As noted by Traugott,[15] "Clearly, no shock inflection point would be expected if the surface pressure is of the underexpanded type. It is concluded that bow shock inflection points occur in the regions of overexpanded pressures." Thus, the pressure distribution and the shock shape depend on the Mach number and the conical half angle.

The interrelationship between the shock shape, the surface-pressure distribution, and the total-pressure profile is illustrated in the sketches of Fig. 6.19, which are taken from Ref. 16. The overexpansion in the static surface-pressure distribution, the inflection in

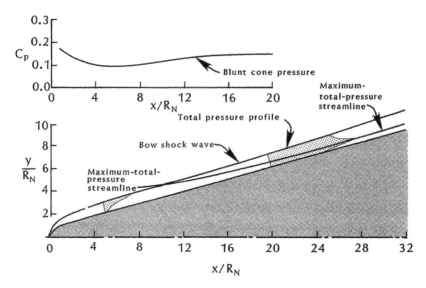

Fig. 6.19   Sketch of flowfield for blunt-nosed cones in a hypersonic stream, $M_\infty = 10$, $\alpha = 0$ deg., as taken from Ref. 16.

the shock wave, and the movement of the location of the maximum stagnation-pressure toward the surface are due to the axisymmetric character of the flow, i.e., the spreading of the streamlines. The inflection point in the bow shock wave reflects the "complex" flowfield which contains both expansion waves and compression waves. The inflection point corresponds to a local minimum in the shock inclination angle of the bow shock wave. As a result, there is a local maximum of the total pressure within the shock layer.

The total-pressure profiles computed by Cleary[16] for Mach 10 flow of air past a blunted cone are reproduced in Fig. 6.20. For $x = 60R_N$, the total pressures in the shock layer are essentially equal to the sharp-cone value except near the wall, where the peak stagnation pressure associated with the shock inflection persists. Further downstream, the boundary-layer growth will swallow this streamline and the viscous effects will cause the stagnation pressure to decrease.

Hecht and Nestler[17] have developed a correlation for the overexpansion/recompression pressure distribution for hypersonic flow past blunted cones. While the forebody pressure distribution may not be particularly sensitive to Mach number, this is not so for the pressure induced on the afterbody by the blunted region. The overexpansion and the subsequent recompression for the zero angle-of-attack measurements for a sphere/cone[16] are correlated in Fig. 6.21 using the parameters of Ref. 17. The measurements closely follow the correlation developed in Ref. 17 using hypersonic data on slender cones.

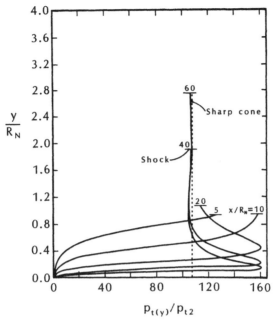

Fig. 6.20 Total-pressure distribution across the shock-layer for a spherically-blunted cone (θ_c = 15 deg.) M_∞ = 10, perfect air, as taken from Ref. 16.

The pressures computed using the modified Newtonian flow model and those computed using the pressure correlation of Ref. 17 (as reproduced in Fig. 6.21) are compared with experimental pressures in Fig. 6.22 and 6.23. The comparisons were made for different Mach numbers, for different angles-of-attack, and for different cone half angles (semi-vertex angles). Over the range of conditions considered, the pressures computed using the correlation presented in Ref. 17 are in better agreement with the measured pressures. The effect of the overexpansion/recompression becomes more pronounced as the cone half-angle is increased and as the angle-of-attack is increased. Thus, the differences between the experimental pressure distributions and those computed using the modified Newtonian flow model increase with θ_c and with α.

Exercise 6.2:

The following pressure measurements were reported by Cleary[16] for a spherically blunted cone ($\theta_c = 30$ deg.) in a Mach 10.6 stream of air.

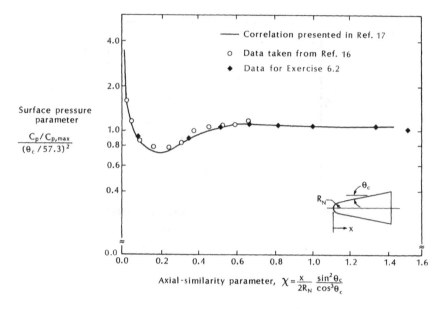

Fig. 6.21 Surface pressure correlation for sphere/cones, as taken from Ref. 17.

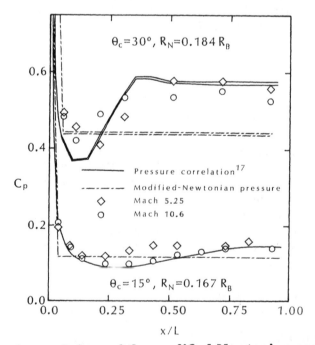

Fig. 6.22 A comparison of the modified Newtonian pressures, the pressures based on the correlation of Ref. 17, and the measured pressures from Ref. 16 for two different sphere/cones at $\alpha = 0$ deg.

x/R_N	C_p
0.476	0.476
1.841	0.452
2.667	0.530
3.492	0.545
4.429	0.546
5.238	0.555
6.984	0.562
7.794	0.541
8.810	0.516

Reduce these data in terms of the surface-pressure parameter and of the axial-similarity parameter of Hecht and Nestler[17] (see Fig. 6.21).

Solution:

To calculate the surface-pressure parameter, the value of $C_{p,max}(= C_{p,t2})$ is needed.

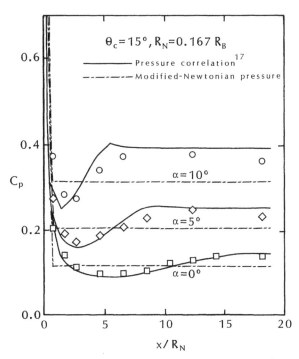

Fig. 6.23 A comparison of the modified Newtonian pressures, the pressures based on the correlation of Ref. 17, and measured pressures from Ref. 16 for different angles of attack, $M_\infty = 10.6$.

$$\frac{p_{t2}}{p_1} = \left[\frac{(\gamma+1)M_1^2}{2}\right]^{\frac{\gamma}{(\gamma-1)}} \left[\frac{\gamma+1}{2\gamma M_1^2 - (\gamma-1)}\right]^{\frac{1}{(\gamma-1)}}$$

$$\frac{p_{t2}}{p_1} = 145.131$$

$$C_{p,\text{max}} = \left[\frac{p_{t2}}{p_1} - 1\right] \frac{2}{\gamma M_1^2} = 1.833$$

Thus, the surface-pressure parameter is:

$$\frac{C_p/C_{p,\text{max}}}{(\theta_c/57.3)^2} = 1.9902C_p$$

The axial-similarity parameter is:

$$\chi = \frac{x}{2R_N} \frac{\sin^2 \theta_c}{\cos^3 \theta_c} = 0.1925 \frac{x}{R_N}$$

Thus,

χ	$\dfrac{C_p/C_{p,\text{max}}}{(\theta_c/57.3)^2}$
0.0916	0.947
0.3543	0.900
0.5133	1.055
0.6720	1.085
0.8524	1.087
1.0081	1.105
1.3441	1.119
1.5000	1.077
1.6955	1.027

6.3.3 Flow Turned Through Multiple Shock Waves

Compressive surfaces located on a hypersonic vehicle serve either to decelerate the flow or to increase the static pressure. The compressive surface shown in Fig. 6.24a serves to decelerate the flow prior to reaching the engine inlet and, therefore, to produce acceptable flow velocities in the combustion zone. Although this deceleration can be accomplished by a single, normal-shock wave standing just upstream of a pitot-type intake, "the loss in total pressure makes this simple solution unacceptable for flight Mach numbers in excess of $M_\infty = 1.5$. In reality, the intake designers use a number of (mostly oblique) shock waves to compress and decelerate the incoming flow."[18] The increased pressure on deflected control surfaces, such as the body flap and the rudder shown in Fig. 6.24b, provide the forces and the moments that are used to maintain or to change the angle between the vehicle axes and the freestream flow.

The compression surfaces depicted in Fig. 6.24 are located downstream of the bow shock wave and, therefore, "see" flow conditions markedly different from the true freestream conditions. Since the focus of this section, i.e., "Departures from the Newtonian Flow Model," are inviscid-flow phenomena which cause the pressure distributions to be significantly different from those predicted by the (modified) Newtonian flow model, the discussion of shock-wave/boundary-layer (viscous) interactions associated with compression surfaces will be deferred to the next section.

Consider the situation where the flow is to be turned (compressively) 20 deg. from the freestream direction. If the turning is accomplished gradually, through a series of infinitesimal turns, the deceleration process is isentropic. If the compressive turning of the flow is isentropic, the resultant static pressure is much higher than if the 20 deg. turn is accomplished by passing the air through a single oblique

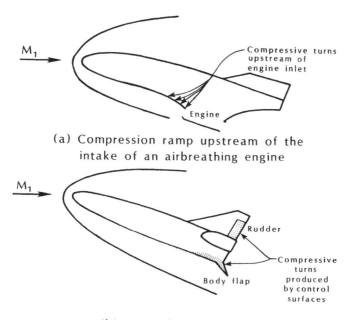

(a) Compression ramp upstream of the intake of an airbreathing engine

(b) Control surfaces

Fig. 6.24 Sketches illustrating surfaces where compressive turns in the flow occur within the shock layer flow.

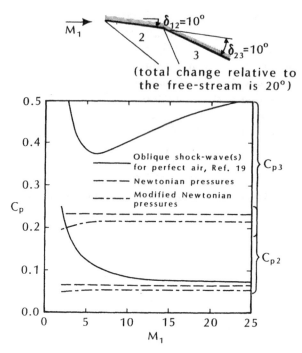

Fig. 6.25 A comparison of the Newtonian pressures and those for a pair of oblique shock waves generated when an inviscid supersonic flow encounters a double wedge configuration.

shock wave. Thus, the local pressure is a function of the freestream Mach number, the local flow direction, the gas-property model, and the process through which the flow was turned.

Consider a steady, inviscid, supersonic flow approaching a double-wedge configuration, as shown in the sketch of Fig. 6.25. Each wedge turns the flow 10 deg., so that the flow in region 3 has been turned 20 deg. from the freestream. The pressure coefficients for regions 2 and 3 have been computed using the Newtonian flow model, using the modified Newtonian flow model, and using the (two-dimensional) oblique shock relations for perfect air[19] for the flow from 1 to 2 and for the flow from 2 to 3. The pressure coefficients computed using these three approximations are presented as a function of Mach number in Fig. 6.25. As one would expect, the results for region 2 are similar to those presented in Fig. 1.4. For $M_1 < 10$, the Newtonian flow pressures are much lower than the corresponding wedge value for this relatively small deflection angle. However, as $M_1 \to \infty$, the pressure coefficients in region 2 become independent of the Mach number (the Mach number independence principle). The Newtonian flow pressures do not provide suitable estimates of the pressure in

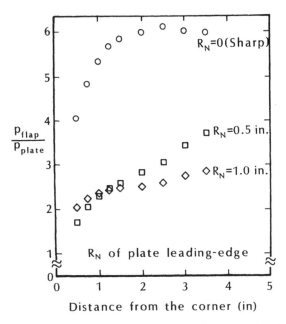

Fig. 6.26 Pressure recovery on a flap, δ_{plate} = 12 deg., δ_{flap} = 15 deg., M_∞ = 10.19, as taken from Ref. 20.

region 3 even as $M_1 \to \infty$. The Newtonian flow model does not account for the increased pressure recovery that occurs when the flow deflection is accomplished through incremental steps as opposed to a single turn.

6.4 SHOCK-WAVE/BOUNDARY-LAYER (VISCOUS) INTERACTION FOR TWO-DIMENSIONAL COMPRESSION RAMPS

When one neglects the effects of viscosity, the flow turns from region 2 to region 3 through a linear shock wave that extends to the corner; see Fig. 6.25. The surface pressure is constant in region 2 and is constant in region 3, with $p_3 > p_2$. In actuality, the body flaps are embedded in the flowfield of the carrier vehicle and are dominated by that flowfield. Thus, they do not "see" the freestream air. The subsonic portion of the boundary layer on the surface approaching a compression ramp provides a path for the disturbances such as the increased pressure produced by the shock wave to influence the flow upstream of the corner. The upstream influence may be relatively small, e.g., a slight thickening of the upstream boundary layer, or it may be dramatic, e.g., a large separation bubble at the corner with a complex, lambda-shaped shock wave.

6.4.1 The Effect of Nose Blunting on the Interaction

Both the inviscid and the viscous portions of the approach flow in which the control surface is embedded are important when characterizing the local flowfield. Wind-tunnel pressure measurements presented by Neumann[20] indicate that the entropy layer associated with a blunted nose affects the pressure recovery on a flap. Presented in Fig. 6.26 are pressure measurements made on a flap which is located 25.0 in. (63.5 cm) from the leading edge of a plate deflected 12 deg. to the Mach 10.19 freestream. Pressure measurements sensed at orifices located on the 15 deg. deflected flap are presented for three levels of nose bluntness for the plate: $R_N = 0$, i.e., a sharp leading edge, $R_N = 0.5$ in., and $R_N = 1.0$ in. Measurements of the Mach number profiles approaching the flap indicated a persistent downstream influence of the entropy variation due to nose blunting. The pressure measurements presented in Fig. 6.26 indicate that the pressure recovery on the deflected flap was greatest for the sharp leading-edge configuration. Thus, when making calculations to determine the control-surface effectiveness, it is important to properly model the approach flow, including the rotational flow external to the boundary layer. Recall the discussion associated with Fig. 3.30 and Exercise 3.4.

6.4.2 Parameters Which Influence the Shock-Wave/Boundary-Layer Interaction

The extent of the upstream influence of the shock-wave/boundary-layer interaction depends on the size of the subsonic portion of the approach boundary layer and the strength of the shock wave produced by the turning of the flow. Thus, the parameters which are likely to influence the extent of an interaction are:

1. Whether the approach boundary layer is laminar or turbulent,

2. The Mach number of the approach flow,

3. The Reynolds number of the approach flow,

4. The surface temperature,

5. The deflection angle of the ramp, and

6. The chemical state of the gas.

When the conditions are such that the ramp-induced pressure rise feeds forward and separates the approach boundary layer, the flow pattern contains the phenomena shown in the sketch of Fig. 6.27. The adverse pressure gradient causes the approach boundary layer to thicken, producing a series of compression waves. Starting

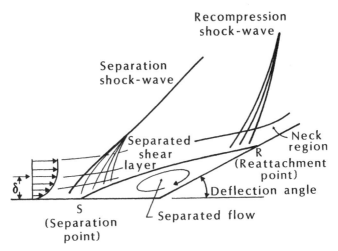

Fig. 6.27 Sketch of a separated-flow pattern for a shock-wave/boundary-layer interaction at a ramp.

near the wall, the shock is first seen to be curved, the curvature being due to its propagation through a rotational layer in which the Mach number changes from one streamline to another. Outside of the boundary layer, the shock wave is linear since the incoming flow is uniform. (The rotational flow downstream of the blunted nose described in the previous section would not be uniform.) A separated free-shear layer impinges on the ramp, reattaching and turning through a recompression shock-wave system. Separating the flow coming toward the flap surface is a highly disturbing process which may cause transition of the free-shear layer. The location of the separation and the reattachment points are indicated on the pressure distribution obtained by Roshko and Thomke[21] for a supersonic, turbulent boundary layer approaching a 25 deg. ramp, which is reproduced in Fig. 6.28. Delery[22] notes that, "Once separation has occurred, there is a large disymmetry between the pressure rises at separation and reattachment, the latter being much more important. As pressure rise to separation does not depend on downstream conditions, an increase in the overall pressure rise necessarily entails a higher pressure rise at reattachment."

The subsonic region of a laminar boundary layer is larger than that for a turbulent boundary layer. Thus, the pressure rise required to separate a laminar boundary layer is much lower than that required to separate a turbulent boundary layer. Neumann[20] notes that, for adiabatic flows $(T_w \simeq T_r)$, initially laminar boundary layers can separate at as small a deflection angle as 5 deg. Korkegi[23] states that a 10 deg. deflection is sufficient to separate a laminar boundary

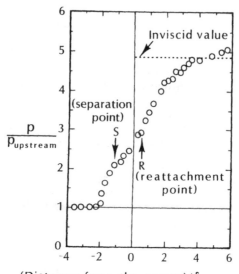

Fig. 6.28   Pressure distribution in the vicinity of a flap corner, $M_\infty = 2.95$, $Re_{\delta_{\text{upstream}}} = 3.17 \times 10^6$ ($\delta_{\text{upstream}} = 4.21$ in.), $\delta_{\text{plate}} = 0$ deg., $\delta_{\text{flap}} = 25$ deg., as taken from Ref. 21.

layer. Because cooling the wall significantly reduces the thickness of a laminar boundary layer (reducing the size of the subsonic region), separation is strongly dependent on the surface temperature. Korkegi cites data showing that, for highly cooled laminar boundary layers ($T_w \approx 0.1 T_t$), compression angles as high as 25 deg. for a wedge and 35 deg. for a cylindrical/flare model did not produce separation. Since the temperature of the surface will probably change dramatically during the course of the vehicle's flight, it is important to use data for the shock-wave/boundary-layer interaction that simulate this parameter (T_w/T_t), as well as other parameters.

Turbulent boundary layers are not only able to withstand larger compressions without separating, but the shock/turbulent-boundary-layer interactions are less sensitive to Reynolds number and to the wall temperature. However, their dependency on the Mach number is significant. Turbulent regions of separation are relatively short and generally exhibit a small "knee" in the pressure distribution (refer to the pressure distribution for a turbulent boundary layer that was presented in Fig. 6.28). This pressure distribution is in contrast to the pressure distribution for a laminar separation, which has a long, constant-pressure plateau. Korkegi[23] cites data showing that the plateau and peak pressures exhibit relatively little sensitivity to Reynolds number, when the approach boundary layer is turbulent.

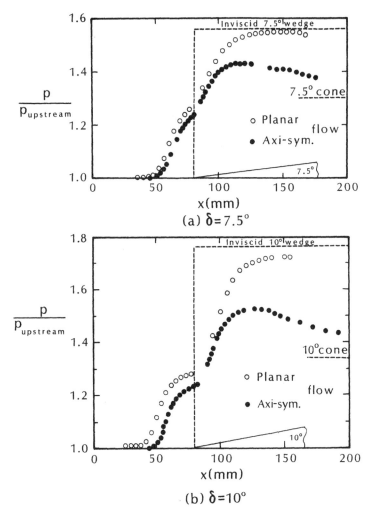

Fig. 6.29 Comparison between planar and axisymmetric flows, as taken from Ref. 26.

Dolling and Murphy[24] note that wall-pressure-fluctuation measurements have shown that the unsteady nature of the separation shock wave in a shock/boundary-layer interaction can generate an intermittent wall-pressure signal. As a result, shock/boundary-layer interactions can produce severe acoustic loads, e.g., 160 dB upstream of the inlet and 185 dB in the vicinity of control surfaces for an air-breathing hypersonic vehicle.[25] The simultaneous coupling of such severe fluctuating-pressure loading with high heating rates can present a severe challenge to the designer of the vehicle structure.

6.4.3 Interaction for Nonplanar Flows

Ginoux[26] compared static-pressure distributions that were measured near the corner of a two-dimensional flat-plate/flap model and of a hollow-cylinder/flare model in a Mach 2.25 stream at the von Karman Institute. Pressure distributions for deflection angles of 7.5 deg. and of 10.0 deg. are presented in Fig. 6.29. For both values of δ, a flare (i.e., the axisymmetric flow) causes a smaller overall pressure rise and a shorter separation length than a flap (i.e., the planar flow) of the same deflection angle.

6.5   TANGENT-CONE AND TANGENT-WEDGE APPROXIMATIONS

In Chap. 3, it was noted that empirical correlations complemented by analytical techniques were used in the design of the Space Shuttle. As shown in Fig. 3.2, some portions of the vehicles were represented by cones and other portions by wedges. When using the tangent-cone method or the tangent-wedge method, it is assumed that the local pressure depends solely on the surface inclination. Thus, for the tangent-cone method, the local static pressure acting on a surface inclined an angle δ to the freestream flow is equal to the pressure acting on a cone whose conical half-angle (θ_c) is equal to δ. Similarly, for the tangent-wedge method, the local static pressure is assumed to be that for a wedge whose deflection angle is equal to the local slope.

The assumption that the surface pressure on a conical generator is (to a first order) dependent on the local surface inclination is supported by data reported by Cleary,[16] which are reproduced in Fig. 6.30. The pressure distributions were obtained on blunted spheres/cones, one with a semi-vertex angle (θ_c) of 15 deg., the other with a semi-vertex angle of 30 deg. For both models, $R_N = 1.00$ in. (2.54 cm). Note that the pressure measurements for rays lying in the vertical plane-of-symmetry are essentially the same for a given inclination angle for the four inclination angles tested. Areas of relatively poor agreement are the greater overshoot data of the 15 deg. model windward ray and the slightly greater pressures of the 30 deg. model leeward ray when at angle-of-attack. Although the pressure on a conical generator depends, to a first order, on the local surface inclination, such is not the case for the convective heat transfer along a conical ray. As will be discussed in Chap. 7, because of crossflow, the heat transfer to the windward ray of a 15 deg. cone at an angle-of-attack of 15 deg. will be different from the heat transfer to a conical generator of a 30 deg. cone at zero angle-of-attack, even though the flow is deflected 30 deg. for both cases.

Experimental pressure distributions from the windward pitch-plane of the Shuttle Orbiter are compared with the tangent-cone

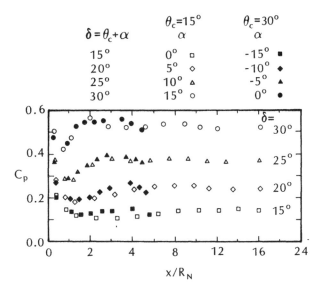

Fig. 6.30 A comparison of experimental pressure distributions for the same inclinations of the blunted cone surfaces, R_N = 1.00 in., γ = 1.4, M_∞ = 10.6, Re/ft = 1.0 × 10 ^{6}, as taken from Ref. 16.

and the tangent-wedge models in Fig. 6.31, as taken from Ref. 27. Because the spreading of the axisymmetric streamlines for supersonic flow past a cone produces a thinner shock layer than for the wedge, the shock inclination and, therefore, the surface pressure is lower for the tangent-cone model.

Cox and Crabtree[28] have developed expressions for the pressure coefficients on slender shapes as $M_\infty \to \infty$. For a slender wedge:

$$C_{p_w} = (\gamma + 1)\theta_w^2 \tag{6-10}$$

and for a slender cone:

$$C_{p_c} = \frac{2(\gamma + 1)(\gamma + 7)}{(\gamma + 3)^2}\theta_c^2 \tag{6-11}$$

Note that both expressions reduce to the Newtonian value:

$$C_p = 2\theta^2 \tag{6-12}$$

as $\gamma \to 1$.

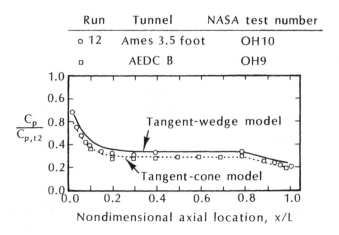

Run	Tunnel	NASA test number
o 12	Ames 3.5 foot	OH10
□	AEDC B	OH9

Fig. 6.31 Pressure-coefficient comparisons from the lower-surface center line of the Space Shuttle Orbiter. Tests were conducted in Tunnel B (AEDC) and in the 3.5-Foot Hypersonic Wind Tunnel (Ames Research Center), as taken from Ref. 27.

6.6 THE NEED FOR MORE SOPHISTICATED FLOW MODELS

The flow models that have been used to calculate the pressure distributions discussed previously in this chapter employ relatively major simplifications. Even so, for many cases, these relatively simple flow approximations provide a reasonable approximation of the pressure distribution. As will be discussed in Chap. 8, integration of these approximate pressure distributions yield engineering estimates of the normal force coefficients and the pitching moment coefficients. Thus, they serve as valuable tools for the conceptual design process. However, the discussions of the problems encountered with the body-flap deflection of the Space Shuttle Orbiter that were presented in Chaps. 3 and 4 underscore how much slight differences in the pressure distribution can affect the pitching moment for the basic configuration. Furthermore, for the deflected flap, one must properly model the physics of the viscous/inviscid interaction (recall the discussion of the shock/boundary-layer interaction in Sec. 6.4), in order to properly estimate the incremental change in the force and the moment coefficients due to elevon and body-flap deflections.

6.6.1 Spherically Blunted Conic Configurations

Viscous effects can have a significant impact on the pressure distribution for simple shapes. Miller and Gnoffo[29] reported the existence of embedded shock waves on the leeward surface of blunted conic con-

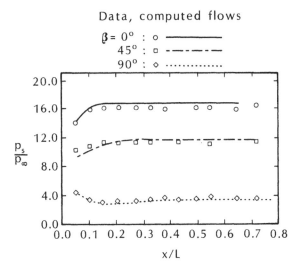

Data, computed flows

$\beta = 0°$: ○ ——————
      45° : □ —·—·—·—
      90° : ◇ ············

Fig. 6.32 A comparison of the pressure distributions for a blunt cone at $\alpha = 21$ deg., $M_\infty = 6$, as taken from Ref. 30.

figurations which occurred when the fore-cone angle-of-attack (α_f) exceeded the fore-cone semi-vertex angle $(\theta_{c,f})$. They concluded that the flow expands around the spherical nose to a supersonic condition and recompresses because of the pressure of the cone section. The recompression starts as a continuous compression on the surface just downstream of the sphere/cone junction, and the Mach lines in the compression region converge so that an embedded shock wave is formed.

An engineering code developed by Zoby and Simmonds[30] computes the inviscid flowfield over hyperboloids, ellipsoids, paraboloids, and sphere cones at 0 deg. angle-of-attack. The method is based on the Maslen technique.[31] Maslen applies the von Mises transformation to the governing equations so that the independent variables are the distance along the shock wave and the stream function. The effect of the normal velocity component on the pressure distribution is incorporated using an improvement suggested by Maslen.[32] The local flow properties and the geometry are computed along rays normal to the shock wave until the body (the zero streamline) is reached. The process is repeated until the computed body matches the actual body geometry within some specified tolerance. The pressure distributions, thus computed, are in good agreement with experimental pressures as illustrated in Fig. 6.32. Presented are values from Ref. 30 for a 12.84 deg. blunt cone at an angle-of-attack of 21 deg. in a Mach 6 stream. Values are presented only for $0° \leq \beta \leq 90°$, since a pressure minimum was observed in the experimental data at

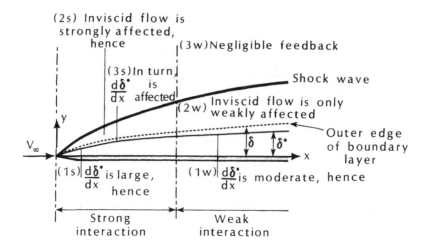

Fig. 6.33 Flowfield sketch illustrating the characteristics of the strong-interaction region and the weak-interaction region, as taken from Ref. 33.

$\beta = 135$ deg. This occurrence of a pressure minimum was attributed to a flow separation. Once flow separation occurs, viscous/inviscid interactions have a significant effect on the flowfield. As a result, two-layer flow models may fail to represent important phenomena or may simply not be able to provide flow solutions. In such cases, the entire flowfield (viscous and inviscid regions) must be incorporated into a single model.

6.6.2 Flat-Plate and Wedge Configurations

Correlations for the viscous/inviscid-interaction-induced pressure perturbations were presented in Sec. 3.2. The similarity parameter that governs the pressure perturbations for laminar flows, χ,

$$\chi = \frac{M_\infty^3 \sqrt{C_\infty}}{\sqrt{Re_{\infty,x}}} \qquad (6\text{-}13)$$

Thus, the pressure perturbations due to the viscous/inviscid interactions are greatest when the shock layer is the thinnest (i.e., the Mach number is relatively high) and the boundary layer is thickest (i.e., the Reynolds number is relatively low).

Anderson[33] divides the hypersonic viscous interaction over a flat plate into two regions: the strong-interaction region immediately downstream of the leading edge and the weak-interaction region further downstream. The flow is depicted in the sketch of Fig. 6.33, which is taken from Ref. 33. The strong-interaction region is characterized by the following phenomena:

1s. In the leading-edge region, the rate of growth of the boundary-layer displacement thickness is large, i.e., $d\delta^*/dx$ is large. Thus, the oncoming flow is turned compressively.

2s. The compressive turning of the flow produces a shock wave at the leading edge of the plate. As a result, the inviscid flow is strongly affected.

3s. The changes in the inviscid flow, of course, affect the boundary layer, so that $d\delta^*/dx$ is not the flat-plate value but a perturbed-flow value.

In the strong-interaction region, the boundary layer is a large fraction of the shock layer, even merging with the shock wave at some conditions.

The y-coordinate of the shock wave increases more rapidly with x than does the y-coordinate of the displacement thickness. Thus, as one proceeds downstream, the boundary layer becomes an increasingly smaller fraction of the shock layer (refer to Fig. 6.33). As a result, the pressure perturbations are less in the weak-interaction region. As shown in the sketch, the weak-interaction region is characterized by the following phenomena:

1w. The boundary layer is a relatively small fraction of the shock layer and the rate of growth of δ^* is moderate, i.e., $d\delta^*/dx$ is relatively small.

2w. As a result, the inviscid flow is only weakly affected by the presence of the boundary layer.

3w. Since the interaction-induced changes in the inviscid flow are small, the boundary layer is essentially the same as that of the unperturbed flow.

The pressure distribution for a flat-plate flow due to the hypersonic viscous interaction ($V = 0.22$), as presented by Hirschel,[34] is reproduced in Fig. 6.34. Near the leading edge (in the strong-interaction region) the pressure is 10 times the flat-plate (or the freestream) value. The pressure decreases rapidly with distance from the leading edge. For strong interactions, the more rigorous flow models, e.g., full Navier-Stokes or Direct Simulation Monte Carlo, must be used if one desires to compute flowfields of suitable accuracy.

6.7 PRESSURE DISTRIBUTIONS FOR A REACTING GAS

The large amount of kinetic energy present in a hypersonic freestream is converted by flow work into increased pressure and by

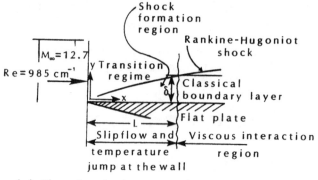

(a) Sketch of viscous-induced interaction

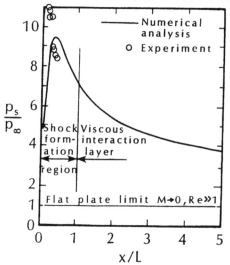

(b) Pressure distribution

Fig. 6.34 Pressure distribution due to the hypersonic viscous interaction for a flat plate, as taken from Ref. 34.

subsequent molecular collisions into high thermal energy surrounding the body. All of this gives rise to chemical reactions affecting the composition of the gas. Since the time scale associated with the chemical reactions differs from the time scale associated with the fluid motion, nonequilibrium effects may be important. There are a variety of models that can be used to represent the gas chemistry. The effects of the gas-chemistry model on the surface-pressure distribution will be discussed in this section, first for the stagnation region of a blunt body and then for wedges and for cones.

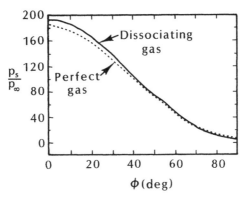

Fig. 6.35 Surface-pressure distributions over a blunt-ogive forebody for $M_\infty = 12, Re_{R_N} = 54{,}294$, as taken from Ref. 36.

6.7.1 Pressures in the Stagnation Region

Calculations presented in Sec. 1.3 indicate that the stagnation-region pressures downstream of a normal-shock wave as computed equilibrium-air properties are only slightly greater than those calculated using the perfect-air model. It was noted in Sec. 1.3 that "the stagnation point pressure for hypersonic flow is (to first order) independent of flow chemistry."

Candler and MacCormack[35] have computed the flowfield over a blunted sphere/cone ($R_N = 0.1524$ m, $\theta_c = 9$ deg., $L = 1.295$ m) at flight conditions ranging from near-thermochemical equilibrium to strong nonequilibrium. Pressure distributions were computed for $M_\infty = 25.9$ at 71 km using a seven-species (N_2, O_2, NO, NO^+, N, O, and e^-) reacting-air chemistry model and a perfect-air (frozen chemistry) model. Candler and MacCormack[35] note "The surface pressure is almost identical for each case. This result is expected because the pressure distribution is essentially a result of change in normal momentum which is not affected by the reaction of the gas." Actually, the stagnation-region pressure as computed by Candler and MacCormack for the "real-gas" flow model is slightly higher than that for the perfect-air model.

Shang and Josyula[36] also computed a nonequilibrium, hypersonic flowfield over a blunted body. They noted:

> if one neglects the presence of a trace amount of nitric oxide and limits the electronic excitation to dissociation, the chemically reacting system can be reasonably represented by a four species gaseous mixture. This simplified gaseous mixture has even more attractive features, in that the molecular/atomic weights of oxygen and nitrogen

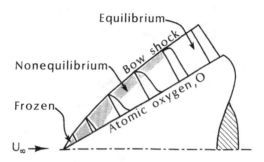

Fig. 6.36 Sketch of profiles of atomic oxygen in the shock layer and the bow shock wave for hypersonic flow past a sharp cone, as taken from Ref. 37.

are relatively close and the characteristic temperatures for dissociation are also sufficiently different between species. Thus, the assumption of a binary heterogeneous mixture for transport properties of air can be applied with plausible physical justification.

Using a single-temperature formulation, the dissociation/ recombination reactions were computed using a rate equation compatible with the Lighthill dissociating gas approximation. The surface-pressure distributions over a blunt-ogive body, as computed by Shang and Josyula[36] are reproduced in Fig. 6.35. The dissociating-gas calculations yield a 4 percent higher value than the perfect-gas result at the stagnation point. The differences between the two solutions eventually disappear as the flow proceeds downstream. (The oscillatory pulse in the pressure distribution is attributed to a sensitivity to the change of grid-point clustering.)

6.7.2 Pressures for Wedges and for Cones

While the stagnation-region pressure is to a first order independent of the gas chemistry, such is not the case for hypersonic flow past a flat plate inclined to the freestream, i.e., a wedge, or past a circular cone. Computations by Rakich et al.[37] of a nonequilibrium flow over a pointed cone at zero incidence indicated that even though the body is conical, the flow has a scale that depends on the time constant for pertinent species reactions. Sketches of the oxygen profiles and of the bow shock shape, as taken from Ref. 37, are reproduced in Fig. 6.36. Near the apex of the cone, there exists a region where there is not sufficient time for reactions to occur. In this region, the flow is conical, and the species concentrations are equal to the freestream

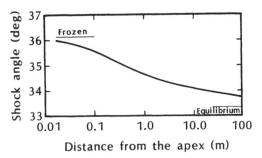

Fig. 6.37 Shock angle for nonequilibrium flow over a 30 deg. wedge, as taken from Ref. 37.

values upstream of the shock wave, i.e., the flow is frozen. The shock angle is equal to that for a perfect-gas flow with a constant ratio of specific heats. Note that the concentration of atomic oxygen is low near the apex (as evident in the first oxygen profile in Fig. 6.36). Proceeding downstream from the apex, the species begin to equilibrate but the flow has not reached equilibrium. Far downstream of the apex, the reactions go to completion, which is the equilibrium region. As indicated in the atomic oxygen profiles, there is an increase in the atomic concentration, which is relatively constant across the shock layer. Because energy is absorbed in the dissociation process, the temperature is lower and the density higher in the equilibrium region.

Because of the varying chemistry (frozen to nonequilibrium to equilibrium), the bow shock is curved. The shock inclination is greatest near the apex (the frozen region) and the least far downstream (the equilibrium region). The variation of shock angle with distance for hypersonic flow over a 30 deg. wedge, as computed by Rakich et al.,[37] is reproduced in Fig. 6.37. Rakich et al. conclude that a weak entropy layer forms near the body because the streamline at the apex passes through a stronger shock than the outboard streamlines. Thus, even for a pointed cone at zero incidence, a weak entropy layer occurs for nonequilibrium flow.

A comparison of the experimental and of the theoretical shock-layer "thickness," as presented by Spurk,[38] is reproduced in Fig. 6.38. The data were obtained in an expansion tube with a hypersonic stream of oxygen flowing past a sharp cone whose semi-vertex angle is 45 deg. For the freestream conditions of the experiments, the flow around the conical bodies is purely supersonic. The data are compared with computations for the equilibrium assumption ($\tau_{vib} = \tau_{diss} = 0$) and with computations for the frozen-flow assumption ($\tau_{vib} = \tau_{diss} = \infty$).

In Ref. 39, Park compared the data of Spurk[38] with computations made by Candler using (a) frozen-flow chemistry, (b) equilibrium-

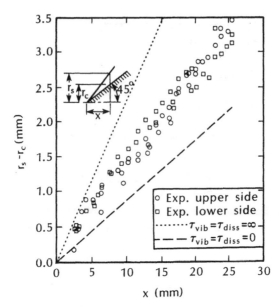

Fig. 6.38   Comparison of the experimental and the theoretical shock-layer "thickness" for $U_\infty = 6350$ m/s, $\theta_c = 45$ deg., as taken from Ref. 38.

flow chemistry, (c) finite-rate chemistry using a one-temperature model, and (d) finite-rate chemistry using a multi-temperature model. Park[39] concluded that, "only the multi-temperature model agrees with the experimental data." Further, "Near the apex, the shock angle is large and is nearly equal to that of a frozen (that is, perfect-gas) flow value. At large distances from the apex, the shock angle approaches that of an equilibrium flow. The wall pressure is dictated mostly by the local shock angle; it decreases with the distance. This decrease in wall pressure causes a positive (nose-up) pitching moment to a lifting body. As a result, the trim angle-of-attack and the slope of the pitching moment curve are both affected by this phenomenon."

The curvature of the bow-shock wave due to nonequilibrium effects causes a reduction in the post-shock pressure. Using the shock-wave angles presented in Fig. 6.37, Park[40] has calculated the post-shock pressure variation with distance from the apex. The pressures, thus calculated, are reproduced in Fig. 6.39. These results are consistent with the simplistic arguments used to explain a nose-up pitching moment for the spherically blunted cone, as discussed in Chap. 4.

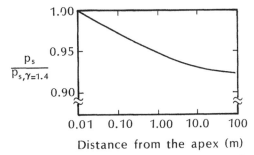

Fig. 6.39 Post-shock pressure as calculated using the shock wave angles presented in Fig. 6.37, as taken from Ref. 40.

6.8 PRESSURES IN SEPARATED REGIONS

The low pressures in a separated-flow region can have a significant effect on the forces and moments acting on a configuration. In order for a viscous boundary layer to separate from the wall, it is necessary that the pressure increases in the streamwise direction, i.e., there is an adverse pressure gradient along the flow path. For a two-dimensional boundary-layer separation, skin friction vanishes at the separation line. However, for three-dimensional flows, it is not necessary that the two components of skin friction vanish for separation to occur. There are two types of three-dimensional separation:

1. A bubble type of separation, and

2. A free-vortex-layer type of separation.

The bubble type of separation is characterized by zero skin-friction at the separation line with subsequent reversal of the flow in the streamwise direction. The free-vortex-layer separation is depicted in the sketch of Fig. 3.25. For the free-vortex-layer separation, the circumferential component of the velocity component close to the body is reversed, but the direction of the meridional velocity remains unchanged. Oil-flow patterns on the surface of a blunted cone at angles-of-attack in excess of θ_c indicated that the flow near the leeward plane-of-symmetry continued from the attached region, through the region of zero lateral skin friction, and into the vortex region. Such an oil-flow pattern indicates that the longitudinal component of skin friction was always finite.

At angles-of-attack greater than those necessary to equalize the cone base pressure and the leeward pressure, the leeward flow expands to pressure levels below the base pressure and a secondary separation occurs on the rear portion of the cone's leeward surface. As discussed by Stetson and Friberg,[41] the secondary-separation region contains subsonic, reverse flow. Thus, at these relatively high angles-

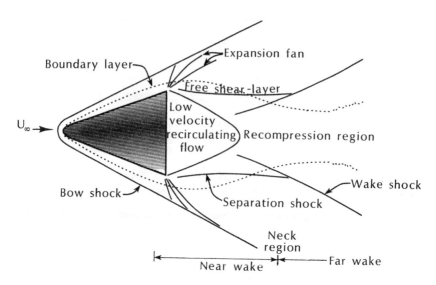

Fig. 6.40   Sketch of near-wake flowfield.

of-attack, there are two distinct separation patterns, the vortex-layer separation followed by a bubble type.

Because of its complexity, the separated flow in the base region or to the leeward side of a lifting entry configuration is a function of many variables. The variables to be considered include:

1. The Reynolds number,

2. The surface temperature,

3. The angle-of-attack,

4. The heat-shield material and mass addition,

5. The Mach number,

6. The configuration (both the forebody geometry and the after-body geometry), and

7. The gas composition.

A sketch of the hypersonic flow around an entry configuration is presented in Fig. 6.40. These features exist (in a general sense) either for a two-dimensional flow or an axisymmetric flow and both for blunt and for slender configurations. Under the influence of viscous and of pressure forces, the boundary layer separates from the body to form a viscous mixing layer. The adverse pressure gradient which causes

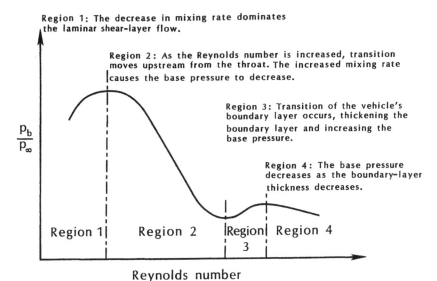

Fig. 6.41   The effect of Reynolds number on the base pressure, as taken from Ref. 42.

the viscous boundary layer to separate "at the corner" is produced by a separation shock. To conserve mass, all of the streamlines below a dividing streamline are turned back into the base region by the pressure rise (adverse pressure gradient) which occurs at the neck. Thus, there is a low-velocity, recirculating flow between the neck and the base. The gas particles outside the dividing streamline pass through the neck and continue downstream. In the neighborhood of the neck, the existence of the trailing "wake shock" produces a pressure rise as the flow turns downstream.

6.8.1 The Effect of the Reynolds Number

The complex process through which the vehicle's boundary layer separates, creating a free-shear layer, part of which is turned back into the near wake when it encounters the adverse pressure gradient at the neck, clearly must be Reynolds number-dependent. A correlation between the base pressure and the Reynolds number has been developed by Crocco and Lees.[42] Four distinct regimes, as illustrated in Fig. 6.41, were identified and their characteristic flow phenomena discussed.

1. At sufficiently low Reynolds number, both the boundary layer of the vehicle and that part of the viscous wake in which the major portion of the recompression occurs are laminar. Within this region, as the Reynolds number increases, the laminar mixing

rate decreases and the base pressure increases. The decrease in the mixing rate is apparently more than enough to offset the decrease in the thickness of the boundary layer at separation.

2. After the base-pressure ratio reaches a local maximum (which occurs when the mixing rate has its smallest value), it decreases rapidly as the Reynolds number increases. As the Reynolds number is increased, transition moves upstream from the throat (with a corresponding order-of-magnitude increase in the local mixing rate). The increased mixing rate apparently is a much more important factor than the accompanying increase in the thickness of the mixing layer, which, by itself, would cause the base-pressure ratio to increase. Once the transition in the shear layer has moved close to the aft end of the vehicle, the base-pressure ratio continues to decrease as the Reynolds number increases. The decrease is associated with the decrease in boundary-layer thickness at the aft end of the vehicle and continues as long as the boundary layer on the vehicle remains laminar.

3. At a sufficiently high Reynolds number, transition of the vehicle's boundary layer will occur near the aft end. There is a consequent thickening of the boundary layer. Since the mixing rate in the viscous wake is relatively unaffected, the base-pressure ratio increases with Reynolds number.

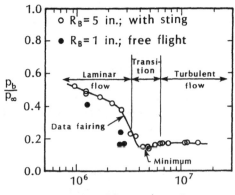

Fig. 6.42 The effect of Reynolds number on base-pressure ratio in laminar and turbulent flow showing minimum at transition for sharp cone, θ_c = 10 deg., M_∞ = 4, α = 0 deg., as taken from Ref. 43.

4. Finally, a high Reynolds number regime is reached for which the transition point has moved far upstream on the vehicle and the fact that the boundary-layer thickness at the trailing edge decreases with Reynolds number becomes a dominant factor. The base-pressure ratio, therefore, decreases slowly (but noticeably) with increasing Reynolds number.

Base-pressure data for a sharp cone ($\theta_c = 10$ deg.) at zero angle-of-attack presented by Cassanto[43] are reproduced in Fig. 6.42. Comparing the Reynolds number dependence of Cassanto's data with the correlation of Fig. 6.41, these measurements correspond to regions 2, 3, and 4.

Base-pressure measurements obtained during the flights of two slender, slightly blunted RVs having an approximately flat base-geometry have been reported by Bulmer in Ref. 44. Histories of the base pressure obtained using a transducer with a 0 to 1 psia range are reproduced in Fig. 6.43. Note that there is a sudden jump in each pressure history. This jump occurred simultaneously with the onset of boundary-layer transition as determined using accelerometer data and using heat-shield temperature data. Thus, the onset of transition is believed responsible for the sudden increase in base pressure. A sharp increase in the base-pressure ratio at the onset of transition was also evident in the flight-test data of Cassanto and Hoyt.[45]

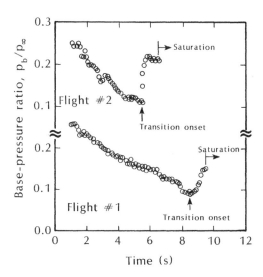

Fig. 6.43 Base-pressure ratio measurements from flight tests of two slender, slightly blunted RVs, as taken from Ref. 44.

6.8.2 The Effect of Mass Addition

The data of Cassanto and Hoyt[45] also indicate the effect of the injection of the gases from the ablative thermal-protection system (TPS). The base pressure was consistently higher for flights where there was medium mass injection from the ablator than for flights where there was relatively low mass injection from the ablating TPS. Cassanto and Hoyt noted, "The thicker boundary layer, which includes the mass flow from the ablation process, is swept into the wake region and tends to enlarge the neck of the wake and force it downstream, thus changing (decreasing) the wake expansion angle." Cassanto and Hoyt continue, "The expansion angle is similar to what would be expected in laminar flow and results in a higher base pressure ratio."

6.8.3 The Effect of Mach Number

A correlation of the effect of Mach number on the base pressure has been developed by Cassanto[43] using measurements from flight tests and from ground tests. The correlation, which is reproduced in Fig. 6.44, presents the ratio of the base pressure to the local cone pressure, i.e., that on the conical generator prior to separation, as a function of the local Mach number.

6.8.4 The Effect of Configuration

The base pressure is a function of the forebody and the afterbody geometry. Cassanto[43] noted, "Low drag bodies tend to produce low

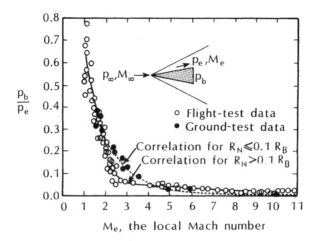

Fig. 6.44 The base-pressure ratio as a function of the local Mach number for a turbulent boundary layer, as taken from Ref. 43.

base pressure ratios while high drag bodies (typical of planetary entry configurations) tend to produce base pressure ratio levels of approximately 2 to 5 higher." Afterbody geometry also affects the base-pressure ratio. Cassanto et al.[46] studied the effect of a bulbous-, or dome-shaped, base geometry. It was found that an increase in the dome radius of the model caused a decrease in base pressure, with the full-dome model having a 20 percent lower base pressure than the flat-based model. The reason for the base-pressure variation is attributed to variations in the local flow conditions at the separation point and the flow turning angle.

Exercise 6.3:

A sharp cone ($\theta_c = 10°$, $L = 0.20$ m) flies at an altitude of 10 km at a velocity of 1.20 km/s. Compare the base pressure as calculated using (a) the modified Newtonian flow model, (b) the correlation of Fig. 6.42, and (c) the correlation of Fig. 6.44.

Solution:

Using Table 1.1, the freestream conditions at 10 km are:

$$p_\infty = (2.6153 \times 10^{-1})1.01325 \times 10^5 = 2.650 \times 10^4 \text{ N/m}^2$$
$$\rho_\infty = (3.3756 \times 10^{-1})1.2250 = 0.4135 \text{ kg/m}^3$$
$$\mu_\infty = (0.81461)1.7894 \times 10^{-5} = 1.458 \times 10^{-5} \text{ kg/s} \cdot \text{m}$$
$$a_\infty = 299.53 \text{ m/s}$$

Thus,
$$M_\infty = \frac{1200}{299.53} = 4.006$$

$$Re_{\infty,L} = \frac{(0.4135)(1200)(0.20)}{1.458 \times 10^{-5}} = 6.808 \times 10^6$$

Furthermore, from the dimensions of the cone,

$$R_B = 0.0353 \text{ m} = 1.39 \text{ in.}$$

The Mach number, the Reynolds number, the base radius, and the angle-of-attack all correspond to the test conditions of Figs. 6.42 and 6.44.

(a) For the modified Newtonian flow model,

$$C_p = 0 \quad \text{and} \quad p_b = p_\infty = 2.650 \times 10^4 \text{ N/m}^2$$

(b) Using Fig. 6.42 with $Re_{\infty,L} = 6.808 \times 10^6$ (the boundary layer is turbulent),

$$\frac{p_b}{p_\infty} = 0.17$$

Thus, $p_b = 0.451 \times 10^4$ N/m^2.

(c) Using the charts in NACA Report 1135 (Ref. 19)

$$C_{p_c} = 0.079 \quad ; \quad M_c = 3.51$$

For the present application, the conditions at the edge of the boundary layer (designated by e) are the same as those for a sharp-cone flow (designated by c)

$$\frac{p_c}{p_\infty} = 1 + \frac{\gamma}{2} M_\infty^2 C_{p_c} = 1.89$$

For $M_c = M_e = 3.51$, the correlation of Fig. 6.44 (which is for a turbulent boundary layer) yields

$$\frac{p_b}{p_c} = 0.0917$$

$$\frac{p_b}{p_\infty} = \left(\frac{p_b}{p_c}\right)\left(\frac{p_c}{p_\infty}\right) = 0.173$$

Thus,

$$p_b = 0.459 \times 10^4 \text{ N/m}^2$$

Note that the values of the base pressure using the two graphical correlations are essentially the same and are considerably lower than the modified Newtonian value.

6.8.5 A Comment

Note that even for relatively simple shapes such as sharp cones or slightly blunted cones with a relatively flat base, there will be a radial pressure gradient across the base. Furthermore, for sting-supported models in a wind tunnel, the presence of the sting will affect the recompression process and the wake flowfield. Measurements sensed by orifices at different locations but at the same test conditions may differ in magnitude from the data presented in Figs. 6.42 through 6.44. Thus, these data are intended mainly to illustrate trends and the effect of various parameters.

6.9 CONCLUDING REMARKS

There are a variety of relatively simple techniques that yield pressure distributions suitable for conceptual design studies and for evaluations of the effect of various configuration parameters. These techniques rely primarily on a knowledge of the turning angles of the flow and the freestream conditions. However, there are many flow phenomena that restrict their usefulness. Furthermore, even for conditions where these correlations provide a reasonable estimate of the pressure levels, serious deficiencies can exist between the pitching moments calculated using them and the actual pitching moments.

For better estimates of the pressure field, it is necessary to use CFD codes. For a thin (attached) boundary layer, whether it is laminar or turbulent, the pressure variation across the boundary layer is negligible. Thus, the surface pressure is given by the inviscid flow and Euler solutions can be used. The surface pressures are essentially independent of the Reynolds number.

When flow separation and viscous/inviscid interactions affect the pressure distributions, the statements of the previous paragraph are no longer true. The CFD codes used to compute the flowfields must model the viscous characteristics of the flow.

REFERENCES

[1] Lees, L., "Hypersonic Flow," *Proceedings of the Fifth International Aeronautical Conference,* Los Angeles, Institute of the Aeronautical Sciences, 1955, pp. 241–275.

[2] Isaacson, L. K., and Jones, J. W., "Prediction Techniques and Heat-Transfer Distributions over Bodies of Revolution in High Subsonic to Low Supersonic Flight," Naval Weapons Center, NWC TP-4570, Nov. 1968.

[3] Micol, J. R., "Simulation of Real-Gas Effects on Pressure Distributions for a Proposed Aeroassist Flight Experiment Vehicle and Comparison to Prediction," AIAA Paper 87-2368, Monterey, CA, Aug. 1987.

[4] Jones, R.A., and Hunt, J. L., "Measured Pressure Distributions on Large-Angle Cones in Hypersonic Flows of Tetrafluoromethane, Air, and Helium," NASA TND-7429, Dec. 1973.

[5] South, J. C., Jr., and Klunker, E. B., "Methods for Calculating Nonlinear Conical Flows," in *Analytic Methods in Aircraft Aerodynamics,* Oct. 1969, NASA SP-228, 1970.

[6] Miller, C. G., III, and Wells, W. L., "Wind-Tunnel Based Definition of the AFE Aerothermodynamic Environment," J. J. Bertin, J. Periaux, and J. Ballmann (eds.), *Advances in Hypersonics, Vol-*

ume 1: Defining the Hypersonic Environment, Birkhäuser Boston, Boston, MA, 1992.

[7] Weilmuenster, K J., and Hamilton, H. H., II, "Calculations of Inviscid Flow Over Shuttle-Like Vehicles at High Angles of Attack and Comparisons with Experimental Data," NASA TP-2103, May 1983.

[8] Marconi, F., Salas, M., and Yeager, L., "Development of a Computer Code for Calculating the Steady Super/Hypersonic Inviscid Flow Around Real Configurations, Volume I – Computational Technique," NASA CR-2675, Apr. 1976.

[9] Cleary, J. W., and Duller, C. E., "Effects of Angle of Attack and Bluntness on the Hypersonic Flow Over a 15° Semiapex Cone in Helium," NASA TND-5903, Aug. 1970.

[10] Bertin, J. J., "The Effect of Protuberances, Cavities, and Angle of Attack on the Wind Tunnel Pressure and Heat-Transfer Distribution for the Apollo Command Module," NASA TMX-1243, Oct. 1966.

[11] Stoney, W. E., Jr., "Aerodynamic Heating on Blunt-Nose Shapes at Mach Numbers Up to 14," NACA RML-58E05a, 1958.

[12] Boison, J. C., and Curtiss, H. A., "An Experimental Investigation of Blunt Body Stagnation Point Velocity Gradient," *ARS Journal,* Vol. 29, No. 2, Feb. 1959, pp. 130–135.

[13] Hayes, W. D., and Probstein, R. F., *Hypersonic Flow Theory, Volume 1: Inviscid Flows,* Academic Press, New York, 1966.

[14] Scott, C. D., Reid, R. C., Maraia, R. J., Li, C. P., and Derry, S. M., "An AOTV Aeroheating and Thermal Protection Study," H. F. Nelson (ed.), *Thermal Design of Aeronautics Orbital Transfer Vehicles,* Vol. 86 of Progress in Astronautics and Aeronautics, AIAA, New York, 1985, pp. 198–229.

[15] Traugott, S. C., "Some Features of Supersonic and Hypersonic Flow About Blunted Cones," *Journal of the Aerospace Sciences,* Vol. 29, No. 4, Apr. 1962, pp. 389–399.

[16] Cleary, J. W., "An Experimental and Theoretical Investigation of the Pressure Distribution and Flow Fields of Blunted Cones at Hypersonic Mach Numbers," NASA TND-2969, Aug. 1965.

[17] Hecht, A. M., and Nestler, D. E., "A Three-Dimensional Boundary-Layer Computer Program for Sphere-Cone Type Re-entry Vehicles, Volume I – Engineering Analysis and Description," Air Force Flight Dynamics Laboratory AFFDL-TR-78-67, June 1978.

[18] Stollery, J. L., "Some Aspects of Shock-Wave Boundary Layer Interaction Relevant to Intake Flows," Paper 17 in AGARD Conference Proceedings No. 428, *Aerodynamics of Hypersonic Lifting Vehicles,* Nov. 1987.

[19] Ames Research Staff, "Equations, Tables, and Charts for Compressible Flow," NACA Rept. 1135, 1953.

[20] Neumann, R. D., "Defining the Aerothermodynamic Methodology," J. J. Bertin, R. Glowinski, and J. Periaux (eds.), *Hypersonics, Volume I: Defining the Hypersonic Environment*, Birkhäuser Boston, Boston, MA, 1989.

[21] Roshko, A., and Thomke, G. J., "Supersonic, Turbulent Boundary-Layer Interaction with a Compression Corner at Very High Reynolds Number," Proceedings of the 1969 Symposium, *Viscous Interaction Phenomena in Supersonic and Hypersonic Flow*, University of Dayton Press, Dayton, OH, 1970.

[22] Delery, J. M., "Shock Interference Phenomena in Hypersonic Flows," presented at the Third Joint Europe/U.S. Short Course in Hypersonics, RWTH Aachen, Oct. 1990.

[23] Korkegi, R. H., "Viscous Interactions and Flight at High Mach Numbers," AIAA Paper 70-781, Los Angeles, CA, June 1970.

[24] Dolling, D. S., and Murphy, M. T., "Unsteadiness of the Separation Shock Wave Structure in a Supersonic Compression Ramp Flowfield," *AIAA Journal*, Vol. 21, No. 12, Dec. 1983, pp. 1628–1634.

[25] Venneri, S. L., Presentation to the U.S.A.F. Scientific Advisory Board, June 1991.

[26] Ginoux, J. J., "High Speed Flows Over Wedges and Flares with Emphasis on a Method of Detecting Transition," Proceedings of the 1969 Symposium, *Viscous Interaction Phenomena in Supersonic and Hypersonic Flow*, University of Dayton Press, Dayton, OH, 1970.

[27] Edwards, C. L. W., and Cole, S. R., "Predictions of Entry Heating for Lower Surface of Shuttle Orbiter," NASA TM-84624, July 1983.

[28] Cox, R. N., and Crabtree, L. F., *Elements of Hypersonic Aerodynamics*, Academic Press, New York, 1965.

[29] Miller, C. G., III, and Gnoffo, P. A., "Pressure Distributions and Shock Shapes for 12.84°/7° On-Axis and Blunt-Nose Biconics in Air at Mach 6," NASA TM-83222, Nov. 1981.

[30] Zoby, E. V., and Simmonds, A. L., "Engineering Flowfield Method with Angle-of-Attack Applications," AIAA Paper 84-0303, Reno, NV, Jan. 1984.

[31] Malsen, S. H., "Inviscid Hypersonic Flow Past Smooth Symmetric Bodies," *AIAA Journal*, Vol. 2, No. 6, June 1984, pp. 1055–1061.

[32] Maslen, S. H., "Axisymmetric Hypersonic Flow," NASA CR-2123, Sep. 1972.

[33] Anderson, J. D., Jr., *Hypersonic and High Temperature Gas Dynamics*, McGraw-Hill, New York, 1989.

[34] Hirschel, E. H., "The Influence of the Accommodation Coefficients on the Flow Variables in the Viscous Interaction Region of an Hypersonic Slip-Flow Boundary-Layer," *Zeitschrift für Flugwissenschaften,* Vol. 20, No. 12, Dec. 1972, pp. 470–475.

[35] Candler, G., and MacCormack, R., "The Computation of Hypersonic Ionized Flows in Chemical and Thermal Nonequilibrium," AIAA Paper 88-0511, Reno, NV, Jan. 1988.

[36] Shang, J. S., and Josyula, E., "Numerical Simulations of Non-Equilibrium Hypersonic Flow Past Blunt Bodies," AIAA Paper 88-0512, Reno, NV, Jan. 1988.

[37] Rakich, J. V., Bailey, H. E., and Park, C., "Computation of Nonequilibrium, Supersonic Three-Dimensional Inviscid Flow over Blunt-Nosed Bodies," *AIAA Journal,* June 1983, Vol. 21, No. 6, pp. 834–841.

[38] Spurk, J. H., "Experimental and Nonequilibrium Flow Studies," *AIAA Journal,* Vol. 8, No. 6, June 1970, pp. 1039–1045.

[39] Park, C., "Modeling of Hypersonic Reacting Flows," J. J. Bertin, R. Glowinski, and J. Periaux (eds.), *Advances in Hypersonic Flows, Volume 2: Modeling Hypersonic Flows,* Birkhäuser Boston, Boston, MA, 1992.

[40] Park, C., *Nonequilibrium Hypersonic Aerothermodynamics,* John Wiley, New York, 1990.

[41] Stetson, K. F., and Friberg, E. G., "Communication Between Base and Leeward Region of a Cone at Angle of Attack in Hypersonic Flow," ARL 69-0115, July 1969.

[42] Crocco, L., and Lees, L., "A Mixing Theory for the Interaction Between Dissipative Flows and Nearly Isentropic Streams," *Journal of the Aeronautical Sciences,* Vol. 19, No. 10, Oct. 1952, pp. 649–676.

[43] Cassanto, J. M., "A Base Pressure Experiment for Determining the Atmospheric Pressure Profile of Planets," *Journal of Spacecraft and Rockets,* Vol. 10, No. 4, Apr. 1973, pp. 253–261.

[44] Bulmer, B. M., "Detection of Flight Vehicle Transition from Base Measurements," *Journal of Spacecraft and Rockets,* Vol. 10, No. 4, Apr. 1973, pp. 280–281.

[45] Cassanto, J. M., and Hoyt, T. L., "Flight Results Showing the Effect of Mass Addition on Base Pressure," *AIAA Journal,* Vol. 8, No. 9, Sep. 1970, pp. 1705–1707.

[46] Cassanto, J. M., Schiff, J., and Softley, E. J., "Base Pressure Measurements on Slender Cones with Domed Afterbodies," *AIAA Journal,* Vol. 7, No. 8, Aug. 1969, pp. 1607–1608.

PROBLEMS

6.1   Consider the spherically blunted biconic shown in the sketch of Fig. 6.45. If $R_N = 0.15$ m, $\theta_{c1} = 30$ deg., and $\theta_{c2} = 5$ deg., determine the Newtonian pressure distribution as a function of s/R_N, where s is the wetted distance measured along the surface from the stagnation point. Assume that the angle-of-attack is 0 deg.

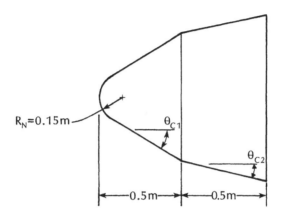

RN=0.15m

θ_{C1}

θ_{C2}

—0.5m— —0.5m—

Fig. 6.45   Configuration for Problems 6.1 and 6.2.

6.2   Consider the spherically blunted biconic shown in the sketch of Fig. 6.45. If $R_N = 0.15$ m, $\theta_{c1} = 30$ deg., and $\theta_{c2} = 5$ deg., determine the modified Newtonian pressure distribution as a function of s/R_N, where s is the wetted distance measured along the surface from the stagnation point. Assume a perfect-air stream at Mach 10, with the model at an angle-of-attack of 20 deg. Calculate the pressure distributions for $\beta = 0$ deg., 90 deg., and 180 deg.

6.3   Assuming a normal-shock/isentropic expansion, where does the sonic point occur for Problem 6.1? Assuming a normal-shock/isentropic expansion, where does the sonic point occur for $\beta = 0$ deg. and for $\beta = 180$ deg. for Problem 6.2?

6.4   Consider a sharp cone whose semi-vertex angle ($\theta_c = 9$ deg.). The cone is in a Mach 8 stream of perfect air at angle-of-attack of 5 deg. Calculate the pressure in the windward pitch plane and the leeward pitch plane for the following models.

(a) Newtonian flow
(b) Modified Newtonian flow

(c) Tangent cone
(d) Equation (6-11)

6.5   Consider a "flat-plate" wing inclined 15 deg. to the Mach 6 stream of air, with a body flap deflected 5 deg., as shown in the sketch of Fig. 6.46. Calculate the pressures in regions 2 and 3 using the modified Newtonian flow model and using the oblique shock relations presented in the charts of Ref. 19.

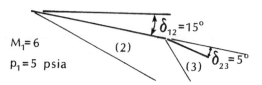

Fig. 6.46   Sketch for Problem 6.5.

6.6   A sharp cone with a semi-vertex angle of 10 deg., $L = 0.40$ m, flies at zero angle-of-attack at an altitude of 14 km at 1180 m/s. Calculate the pressure on the conical surface and the base pressure for the following two models:

(a) Modified Newtonian flow for both pressures
(b) The charts of Ref. 19 for the conical pressure and Fig. 6.42 for the base pressure

6.7   Neglecting the viscous effects, what is the drag coefficient for the cone of Problem 6.6 for each of the two models?

$$C_D = \frac{D}{\frac{1}{2}\rho_\infty U_\infty^2 \, \pi R_B^2}$$

THE BOUNDARY LAYER
AND CONVECTIVE HEAT TRANSFER

7.1 INTRODUCTION

The strong shock wave that forms ahead of a blunt body, such as the Apollo Command Module, traveling at hypersonic speeds, converts the kinetic energy of the freestream air molecules into "thermal" energy. This thermal energy downstream of the shock wave is partitioned into increasing the translational kinetic energy of the air molecules (increasing the temperature) and, once the temperature is sufficiently high, into exciting other molecular energy states such as vibration, dissociation, and ionization. Most of the energy remains with the air particles, as they flow around the vehicle. As a result, the convective heat transfer to the surface is relatively low. (As discussed in Chap. 5, radiative heat transfer becomes significant when a blunt vehicle flies at superorbital velocities.)

For a slender vehicle, such as an airbreathing cruiser, traveling at hypersonic speeds, the inclination of the bow shock wave relative to the freestream flow is small. Thus, the velocity of the air particles does not change dramatically as the flow crosses the shock wave. Except in the boundary layer near the surface of the vehicle, the kinetic energy of the air particles remains high. However, the deceleration of the air particles due to the effects of viscosity near the surface, coupled with the requirement to maintain the surface temperature below material failure limits, produces large increases in the temperature in the boundary layer and, therefore, large temperature gradients normal to the surface. Refer to the discussion related to Fig. 3.30. These large temperature gradients produce very high convective heat-transfer rates to the vehicle surface.

In order to determine the skin friction and the convective heat-transfer rates, one must be able to obtain solutions for the thin boundary layer near the surface. The calculation of the heat-transfer rates with suitable accuracy is much more difficult than the calculation of the surface pressure to the same accuracy. Because the flow near the surface is subsonic, the boundary layer is a mixed subsonic/supersonic flow. In order to accurately represent the large velocity gradients and temperature gradients normal to the surface, it is necessary to develop suitable (innovative) computational grid schemes. Gaps in the understanding of physical phenomena, such as boundary-layer transition and turbulence, lead to the use of semi-empirical and empirical correlations in the flow models. These correlations are often developed using a limited data base and should not be extrapolated to conditions outside of that data base.

The challenges to the numerical analyst that were described briefly in the previous paragraph exist even for the boundary layers of the perfect-gas flows characteristic of conventional hypersonic wind tunnels. For higher-enthalpy hypersonic flows, real-gas phenomena, such as dissociation, present additional challenges to the analyst. When the gas chemistry is not in equilibrium, the convective heat transfer depends on the surface catalycity. For still higher enthalpy flows, the vehicle is covered by an ablative thermal-protection system (TPS). "Cooking" the TPS material absorbs a fraction of the incident heat and releases gases that percolate through the charred ablator into the boundary layer. The injection of the pyrolysis products from the charring ablator into the boundary layer adds further complexity to the numerical-flow model. Since $v_w \neq 0$, the wall is no longer a streamline. Furthermore, additional mass conservation equations are required to account for the products of ablation.

As discussed in Chap. 3, values for the skin friction and for the convective heat transfer can be determined using analytical techniques of varying rigor. The available tools include relatively simple analytical correlations, boundary-layer codes (as part of a two-layer flow model), viscous-shock-layer (VSL) or parabolized Navier-Stokes (PNS) codes, and full Navier-Stokes codes. By now the reader should realize that there is no universally preferred tool. The relatively simple analytical correlations find use in "Conceptual Design Codes." The additional effort required to develop more rigorous flow models and numerical algorithms for full Navier-Stokes codes is justified when one must account for phenomena which are not modeled in the simpler codes or when benchmark solutions are needed in order to establish procedures for extrapolating correlations developed at wind-tunnel conditions to flight conditions.

Li[1] notes that very accurate numerical solutions can be obtained using boundary-layer formulations with relatively few points by incorporating second-order effects.[2] Stetson[3] notes that "not only are there differences between PNS code and boundary-layer code results, but these differences are very sensitive to the grid density utilized in the PNS code." The PNS calculations were repeated using 60 grid points, using 120 grid points, and finally using 120 grid points but with increased density of grid points in the boundary layer until solutions of suitable accuracy were obtained. Stetson[3] states that "for boundary-layer profiles on relatively simple configurations, perhaps the old boundary-layer codes have been too quickly abandoned." Clearly the choice of what is the most appropriate numerical tool depends on the application and on the required accuracy.

The objective of the discussion in this chapter is to gain insight into the phenomena which affect the skin friction and the convective heat-transfer rates.

7.2 BOUNDARY CONDITIONS

For high-altitude flight, where the Reynolds number is low and the Knudsen number is high, the viscous region adjacent to the wall takes up a large fraction of the shock layer. As discussed in Sec. 1.3.1, when the flow becomes more rarefied, the spatial region that influences the state of the gas adjacent to the surface increases. As a result, there are significant velocity-slip and temperature-jump effects. Such flows will not be discussed further in this chapter.

Of interest to this chapter are continuum flows. The boundary conditions downstream of the bow shock wave are defined by the Rankine-Hugoniot relations. At the wall, the no-slip and the no-temperature-jump boundary conditions apply. When the Reynolds number is high, the boundary layer is thin and occupies only a small fraction of the shock layer. As reported by McWherter et al.,[4] the inviscid/boundary-layer method results are in good agreement with data and with PNS computations, when the boundary layer is thin. However, the results obtained with the inviscid/boundary-layer method deviated significantly from the experimental data for cases of high viscous interaction.

7.2.1 The Conditions at the Edge of the Boundary Layer

The conditions at the edge of the boundary layer, i.e., at the interface between the viscous boundary layer and the inviscid portion of the shock layer, were discussed in Sec. 3.4.2. When the bow shock wave is curved, the flow in the inviscid portion of the flowfield is rotational. As a result, neither the velocity gradient nor the temperature gradient is zero at the edge of the boundary layer. One way to locate the edge of the boundary layer is to determine the location where the total-enthalpy gradient goes to zero.[5] An alternate way of defining the edge of the boundary layer is to use the point where $H(y)$ first equals H_∞ within some increment, such as was done in Exercise 3.4. Gupta et al.[6] use 0.995 for the value of $H(y)/H_\infty$ at the edge of the boundary layer. Recall, however, that the total enthalpy within the boundary layer may locally exceed H_∞ for some flows, making it more difficult to locate the edge of the boundary layer in these cases.

Even when the numerical technique treats the shock layer in a unified fashion, a definition for the edge of the boundary layer may be useful, since, for some schemes, the grid spacing is tied to the expected gradients. The significance of entropy variations in the shock layer on the boundary-layer-edge location and on the edge properties was discussed in Sec. 3.4. Thus, the discussion will not be repeated here.

7.2.2 The Conditions at the Wall

When the vehicle flies at relatively high hypersonic velocities, it may be necessary to use active cooling or an ablative coating in order to protect the exposed surface from the extreme heating rates. Gases are released during the ablation process and are injected into the boundary layer. The ablation products transpiring into the boundary layer change the boundary conditions at the wall, reduce the convective heat transfer, and promote boundary-layer transition, i.e., cause transition to occur early. Similar effects can be obtained by the controlled injection of gas through a porous (nonablating) wall. The gas injection may be restricted to a limited surface area, such as from a slot in the surface, (known as film cooling) or may occur over a wide surface area (known as transpiration cooling). The "price paid" for the benefits gained through the use of film cooling or of transpiration cooling include additional cost in developmental testing and analysis and in hardware complexity. For the rest of this section, we will discuss the boundary conditions for a continuum flow past a solid surface.

7.2.2.1 Velocity at the wall.
Consider a continuum flow where the x-coordinate is aligned with the surface and the y-coordinate is perpendicular to the surface, as shown in Fig. 7.1. The assumption of "no slip at the wall" associated with continuum flows implies that

$$u_w = 0 \qquad\qquad (7\text{-}1a)$$

Since there can be no flow through a solid wall,

$$v_w = 0 \qquad\qquad (7\text{-}1b)$$

Equivalently, the wall is a streamline (for which the stream function can be set equal to zero).

7.2.2.2 Temperature at the wall.
In order to determine the temperature-related boundary condition at a solid wall, consider the energy balance for the element shown in Fig. 7.1. Let us neglect conduction through the (end) surfaces perpendicular to the exposed surface. This one-dimensional assumption implies that the changes in the flowfield and in the wall temperature in directions tangent to the vehicle surface are relatively small. Since the heat flux is one dimensional (in the negative y-direction), the flux per unit area can

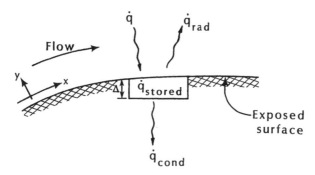

Fig. 7.1 Energy balance for a surface element of the thermal-protection system (TPS).

be used. For the element at the surface of the thermal-protection system (TPS),

$$\dot{q} = \dot{q}_{\text{stored}} + \dot{q}_{\text{cond}} + \dot{q}_{\text{rad}} \tag{7-2}$$

In Eq. (7-2), the incident convective heat flux is

$$\dot{q} = h(T_r - T_w) \tag{7-3a}$$

The rate at which energy is stored in the element is

$$\dot{q}_{\text{stored}} = \rho_w c_w \Delta \frac{dT_w}{dt} \tag{7-3b}$$

This storage term is zero for steady-state calculations. The rate at which energy is conducted through the backface of the element is

$$\dot{q}_{\text{cond}} = k_w \frac{dT}{dy} \tag{7-3c}$$

The rate at which energy is radiated from the exposed surface is

$$\dot{q}_{\text{rad}} = \varepsilon \sigma T_w^4 \tag{7-3d}$$

The Stefan-Boltzmann constant (σ) is 5.67×10^{-8} W/(m^2 K^4) or 4.76×10^{-13} Btu/(ft^2 s $^\circ$R^4).

As noted by Hirschel et al.,[7] there are three types of thermal boundary conditions at the wall. Assume that the energy stored in the element is negligible.

1. The wall temperature is prescribed. With T_w given, the incident convective heat flux, Eq. (7-3a), and the rate at which heat is

radiated from the surface, Eq. (7-3d), are fixed. One can then use Eq. (7-2) to determine the rate at which heat is conducted through the backface of the element into the adjacent material.

2. The wall is assumed to be adiabatic, i.e., $\dot{q}_{cond} = 0$. If $\dot{q}_{rad}$ is significant, the wall temperature is determined from the energy flux balance. If $\dot{q}_{rad}$ is negligible, the energy balance requires that

$$T_r = T_w$$

i.e., the wall temperature is equal to the recovery temperature (which is, therefore, also known as the adiabatic wall temperature). The recovery temperature is given by

$$r = \frac{T_r - T_e}{T_{te} - T_e} \tag{7-4}$$

where r, the recovery factor, is given by

$$r = \sqrt{Pr}$$

for a laminar boundary layer and is given by

$$r = \sqrt[3]{Pr}$$

for a turbulent boundary layer. For perfect gas,

$$\frac{T_r}{T_{te}} = r + \frac{1 - r}{\left(1 + \dfrac{\gamma - 1}{2} M_e^2\right)} \tag{7-5}$$

3. The incident convective heat flux is prescribed. As a result, the wall temperature is determined from the energy-flux balance.

Thus, for hypersonic flight, the thermal design of the vehicle structure and the material selection depends on the convective heating rate. Hopkins and Inouye[8] noted that, "At supersonic speeds ($M \simeq 3$), the surface temperature is essentially the adiabatic wall temperature. At hypersonic speeds ($M \simeq 7$), however, the external surface temperatures generally will be $0.3 - 0.5$ of the adiabatic wall temperatures as a result of considerable radiative cooling and internal heat transfer."

Instrumentation placed on the Shuttle Orbiter provided valuable aerothermodynamic data during the re-entry phases of the first five flights of the Space Transportation System (STS), i.e., the Space Shuttle. Thermocouples, located approximately 0.38 mm beneath the surface coating of the tiles, provided temperature histories from which heating-rate histories were inferred. The incident convective heat flux was assumed to be equal to

$$\dot{q} = 1.06\varepsilon\sigma T_w^4 \qquad (7\text{-}6)$$

The factor 1.06 was obtained using an inverse thermal math model[9] to account for conduction into the tile, Eq. (7-3c), and for the fact that the thermocouple was not on the external surface. Scott[10] stated that "Over the range of time in the trajectory and temperatures considered in this paper, a correction factor of 1.06 is accurate to within 2 or 3 percent."

Exercise 7.1:

Determine the surface temperature at the stagnation point of the Shuttle Orbiter at the following conditions:

(a) $U_\infty = 7.20$ km/s; altitude= 75 km; and

(b) $U_\infty = 2.96$ km/s; altitude= 48 km

According to Ref. 11, the effective nose radius for the Shuttle Orbiter is 28.33 in. (0.7196 m). Although the effective nose radius for computing the stagnation-region flowfield would be a function of the angle-of-attack [$\alpha = 40.0$ deg. for (a) and 34.8 deg. for (b)], this variation will be neglected for this exercise.

Solution:

Let us use the energy balance represented by Eq. (7-2) to determine the surface temperature. Neglecting the energy stored in the element and incorporating the heat conducted out of the element into a factor such as was done in Eq. (7-6), the right-hand side of the equation becomes

$$\dot{q} \simeq 5.67 \times 10^{-8} T_w^4 \frac{W}{m^2} \left(10^{-4} \frac{m^2}{cm^2}\right)$$

Note that the product of the factor times the surface emissivity has been assumed equal to unity, which is a reasonable approximation for a surface whose emissivity is near unity.

For the stagnation-point heat-transfer rate, we will use Eq. (5-44a) modified to account for a finite wall temperature.

$$\dot{q} = \frac{11,030}{R_N^{0.5}} \left(\frac{\rho_\infty}{\rho_{SL}}\right)^{0.5} \left(\frac{U_\infty}{U_{CO}}\right)^{3.15} \left(1 - \frac{h_w}{H_{te}}\right) \qquad (7\text{-}7)$$

The heating rate in Eq. (7-7) has the units of W/cm^2.

(a) Use Table 1.1a to determine the freestream properties at 75 km:

$$\rho_\infty = 3.2589 \times 10^{-5} \rho_{SL} = 3.992 \times 10^{-5} \ \frac{\text{kg}}{\text{m}^3}$$

$$a_\infty = 289.40 \ \frac{\text{m}}{\text{s}}$$

Although it is not needed for this exercise, let us calculate the freestream Mach number:

$$M_\infty = \frac{U_\infty}{a_\infty} = \frac{7200}{289.40} = 24.88$$

To calculate the factor $(1 - \frac{h_w}{H_{te}})$, let us introduce the approximations:

$$H_{te} \simeq = 0.5 U_\infty^2 = 0.5(7200)^2 = 25.92 \times 10^6 \ \frac{\text{m}^2}{\text{s}^2}$$

$$= 25.92 \times 10^6 \ \frac{\text{J}}{\text{kg}} = 11,163.5 \ \frac{\text{Btu}}{\text{lbm}}$$

$$p_{t2} \simeq \rho_\infty U_\infty^2 = (3.992 \times 10^{-5})(7200)^2$$

$$= 2069.53 \ \frac{\text{N}}{\text{m}^2} = 0.0204 \ \text{atm}$$

The energy balance for these conditions is:

$$9.581 \times 10^{12} \left(1 - \frac{h_w}{H_{te}}\right) = T_w^4$$

For the first iteration, neglect the correction for the wall enthalpy and solve for T_w.

$$T_w = 1759 \ \text{K}$$

With $T_w = 1759$ K and $p_{t2} = 0.02$ atm,

$$\frac{h_w}{H_{te}} = 0.076$$

Having obtained a value of $h_w = 0.076 H_{te}$, for the second iteration we will use $(1 - \frac{h_w}{H_{te}}) - 0.92$ and solve for T_w.

$$T_w = 1723 \ \text{K} = 3100 \ {}^\circ\text{R}$$

which is only slightly different than the value ($T_w = 1759$ K) obtained in the first iteration. For this value of T_w, the value of the wall enthalpy term is 0.92 (as assumed) and the iterative process is complete.

(b) Use Table 1.1a to determine the freestream properties at 48 km:

$$\rho_\infty = 1.0749 \times 10^{-3} \rho_{SL} = 1.317 \times 10^{-3} \ \frac{kg}{m^3}$$

$$a_\infty = 329.80 \ \frac{m}{s}$$

Thus,

$$M_\infty = \frac{2960}{329.80} = 8.98$$

Using the approximations for the stagnation conditions downstream of the normal-shock wave,

$$H_{te} \simeq = 0.5 U_\infty^2 = 4.38 \times 10^6 \ \frac{m^2}{s^2} = 4.38 \times 10^6 \ \frac{J}{kg}$$

$$= 1887 \ \frac{Btu}{lbm}$$

$$p_{t2} \simeq \rho_\infty U_\infty^2 = 11,536 \ \frac{N}{m^2} = 0.1139 \ atm$$

$$3.346 \times 10^{12} \left(1 - \frac{h_w}{H_{te}}\right) = T_w^4$$

For the first iteration, neglect the correction for wall enthalpy. Thus, $T_w \simeq 1352$ K. With $T_w = 1352$ K and $p_{t2} = 0.1139$ atm, $h_w/H_{te} = 0.356$.

For the second iteration, assume

$$1 - \frac{h_w}{H_{te}} = 1 - 0.30 = 0.70$$

As a result, $T_w = 1237$ K (2227 °R). With $T_w = 1237$ K and $p_{t2} = 0.1139$ atm, $h_w/H_{te} = 0.305$, which is very close to the value assumed at the start of the iteration. Thus, we have obtained the wall temperature.

Note that $(1 - h_w/H_{te})$ is 0.92 for the conditions of part (a) and is 0.70 for the conditions of part (b).

Exercise 7.2:

During the Apollo program, the "thin-skin thermocouple" technique was used to determine the convective heat-transfer rate to a

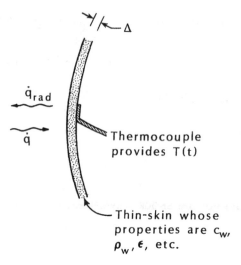

Fig. 7.2 Thin-skin thermocouple technique to determine the heating rate for Exercise 7.2.

wind-tunnel model in a hypersonic stream. As shown in the sketch of Fig. 7.2, a thermocouple was spot-welded to the back surface of the model skin to obtain the temperature history.

Tabulated (handbook) data for the stainless steel skin included:

$$\rho_w = 483.84 \ \frac{\text{lbm}}{\text{ft}^3}$$

$$c_w = 0.11 \ \frac{\text{Btu}}{\text{lbm} \ °\text{R}}$$

$$\varepsilon = 0.20$$

The skin thickness was 0.040 in. and the temperature history yielded:

$$\frac{dT}{dt} = 54 \ \frac{°\text{F}}{\text{s}} \quad \text{when} \quad T_w = 150 \ °\text{F}$$

Neglecting conduction losses, what is the heat-transfer rate?

Solution:

Using Eq. (7-2)

$$\dot{q} = \rho_w c_w \Delta \frac{dT}{dt} + \varepsilon \sigma T_w^4$$

$$\dot{q} = \left(483.84 \ \frac{lbm}{ft^3}\right) \left(0.11 \ \frac{Btu}{lbm \ ^\circ R}\right) \left(\frac{0.040}{12} \ ft\right) \left(54 \ \frac{^\circ R}{s}\right)$$

$$+ (0.20) \left(4.756 \times 10^{-13} \ \frac{Btu}{ft^2 \ s \ ^\circ R^4}\right) (610 \ ^\circ R)^4$$

Note, for the first term,

$$\frac{dT}{dt} = 54 \ \frac{^\circ F}{s} = 54 \ \frac{^\circ R}{s}$$

since a change in temperature in $^\circ$F is exactly equal to the change in $^\circ$R. For the second term, the wall temperature must be stated in absolute units,

$$T_w = 150 \ ^\circ F = 610 \ ^\circ R$$

Thus,

$$\dot{q} = 9.580 + 0.013 = 9.593 \ \frac{Btu}{ft^2 \ s}$$

Note that the radiative component is roughly 0.1 percent of the total heating and can be neglected.

During this test program, the measured heating rates were compared with theoretical values. The theoretical values were computed using the Fay and Riddell equation and the tunnel test conditions. For this test, it was found that the measured heating rate was much lower than the theoretical value. It required considerable effort before it was found that the specific heat value was in error. Measurements showed that

$$c_w = 0.143 \ \frac{Btu}{lbm \ ^\circ R}$$

and not the handbook value of 0.11 Btu/(lbm $^\circ$ R). Thus, the corrected "experimental" heating rate was

$$\dot{q} = 12.453 + 0.013 = 12.466 \ \frac{Btu}{ft^2 \ s}$$

This value agreed with the theoretical value.

There are two points to be made. First, data should always be compared with theory and vice versa. Second, when differences occur, one should not automatically assume that the problem lies with the experiment or with the computation. Careful, objective study is needed to identify the source of the discrepancy.

7.2.2.3 Reactions at the wall.   As noted in Chap. 2, a nonequilibrium gas state may result when the air particles pass through a strong shock wave or when they undergo a rapid expansion. In either case, the nonequilibrium state occurs because there have not been sufficient collisions to achieve equilibrium during the time characteristic of the fluid motion. However, near the wall, the temperature is usually below the temperatures at which oxygen and nitrogen recombine if flow is in equilibrium (roughly, 2200 K and 4200 K, respectively). Thus, if the boundary-layer flow is in equilibrium, the atoms would recombine near the "cool" surface. As the atoms recombine, they release energy to the gas in the boundary layer. (That is, recombination is an exothermic process.) This added heat release tends to increase the temperature in the boundary layer near the wall. The relatively large temperature gradient at the wall creates a correspondingly large heat flux. The equilibrium heat flux is often used as a reference value. When the gas flow in the boundary layer is in equilibrium, the rates of formation are already sufficiently fast that a catalytic surface will not further increase the formation of molecules.

However, at very low density, there are not sufficient collisions taking place as the atoms approach the cool wall to accommodate the recombination process. Thus, although the gas state is far from the equilibrium state, the reactions proceed so slowly that, for all practical purposes, they do not proceed at all. For nonequilibrium flows, the catalytic character of the surface is important. A catalyst may serve as a third body to carry away the energy released in the recombination process, may serve as a template to hold the atoms enhancing the recombination process, etc. If the wall is partially catalytic, some of the atoms striking the wall recombine; others do not. This is illustrated in the sketch of Fig. 7.3. When the wall is fully catalytic, all of the atoms that strike the surface recombine, releasing all of their energy of dissociation to the surface.

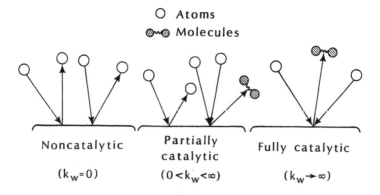

Fig. 7.3 The effect of surface catalycity on atom recombination at the surface.

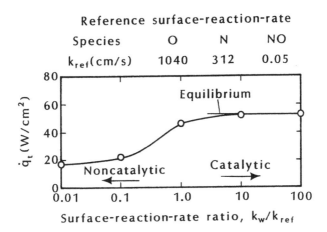

Fig. 7.4   The effect of catalytic recombination rate on the Shuttle stagnation-point heating, U_∞ = 6.61 km/s, alt = 68.88 km, α = 40.2 deg., as taken from Ref. 12.

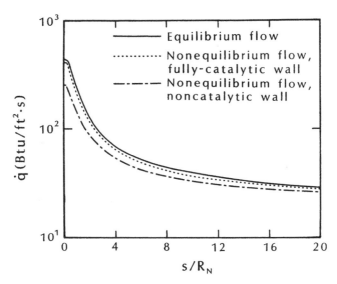

Fig. 7.5   The effect of flow chemistry on the heat-transfer distributions for a 20 deg. hyperbola, U_∞ = 21,330 ft/s, altitude = 211,000 ft, as taken from Ref. 13.

The effect of surface catalycity on the stagnation-point heating rate for the Space Shuttle Orbiter flying at 6.61 km/s at 68.88 km is indicated in the computations of Rakich and Lanfranco,[12] which are reproduced in Fig. 7.4. The catalytic efficiency is specified in the computations in terms of the surface reaction rate (k_w), which is also known as the catalytic recombination rate. For a noncatalytic surface, $k_w = 0$. For a fully catalytic surface, $k_w \to \infty$. Note that, for the computations of Rakich and Lanfranco, the heat-transfer rate to a fully catalytic surface $(k_w \to \text{large})$ approaches the equilibrium value. Note that a significant reduction in heating results for the nonequilibrium flow when the surface is noncatalytic.

The heat-transfer-rate distributions for a 20 deg. hyperbola flying at 21,330 ft/s at an altitude of 211,000 ft as computed by Zoby et al.[13] are reproduced in Fig. 7.5. (For the interested reader, the geometry of a hyperboloid can be seen in Fig. 3.21.) The flowfield computations were made using a viscous-shock-layer (VSL) code with different chemistry models: equilibrium chemistry and nonequilibrium chemistry (both for a noncatalytic and for a fully catalytic wall). The heating rates for a nonequilibrium flow bounded by a fully catalytic wall are close to those for an equilibrium flow. Near the nose region (i.e., for s small), the heating rates to a noncatalytic surface in a nonequilibrium flow are significantly below those for the other two models. However, at about 20 nose radii downstream, the differences in the heating rates based on the nonequilibrium chemistry model (for a noncatalytic and for a fully catalytic wall) and those based on the equilibrium chemistry assumption are relatively small. The computed heat-transfer rates indicate that the influence of nonequilibrium flow chemistry decreases with distance. This occurs when the flow distances approach the relaxation distance required for the chemical reactions.

Note that the heat flux to the wall is essentially the same whether the atoms in a (low-density) nonequilibrium flow recombine at a fully catalytic wall or whether the density is sufficiently high that the flow chemistry is in equilibrium and the atoms recombine within the boundary layer near the wall. Although the heat-transfer rates for a nonequilibrium flow past a fully catalytic wall and for an equilibrium flow are approximately equal, the two flows are distinctly different. The differences must be properly modeled in the numerical algorithm used to compute the flowfields.

7.3 THE METRIC, OR EQUIVALENT, CROSS-SECTION RADIUS

In the two-layer approach, the inviscid solution provides the flow conditions at the edge of the boundary layer, e.g., the static pressure and the entropy, as shown in the sketch of Fig. 7.6. Having defined the pressure and the entropy distribution, the remaining flow proper-

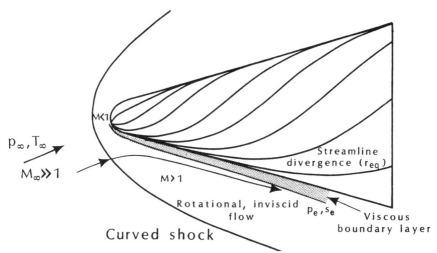

Fig. 7.6   Characterization of a flow around a re-entry vehicle at hypersonic speed.

ties can be calculated (if the flow is in thermochemical equilibrium) and the inviscid streamlines generated. In developing the similarity transformations for a laminar boundary layer in Chap. 5, it was clear that the streamline divergence had a significant effect on the convective heat transfer. For a given pressure distribution and for given edge properties, the stagnation-point heat-transfer rate to a sphere (an axisymmetric flow) was greater than that to a cylinder (a two-dimensional flow) by a factor of $\sqrt{2}$. See Eq. (5-30). The cross-section radius from the axis-of-symmetry (r) provided a "measure" of the divergence of the streamlines for the flows discussed in Chap. 5. Streamline divergence has a significant effect on the convective heating for general flows. As indicated in the sketch of Fig. 7.6, the spreading of adjacent streamlines can be used to define a scale factor or "equivalent cross-section radius" (r_{eq}), which is analogous to the cross-section radius (r) for an axisymmetric flow. The equivalent cross-section radius, determined from the divergence of the streamlines, allows a three-dimensional flow to be modeled by axisymmetric analogue, or small cross-flow assumption (Ref. 14).

Let us define a coordinate system x, y, and $\tilde{\beta}$ where x is the wetted distance along the streamline, y is perpendicular to the surface, and $\tilde{\beta}$ is constant along a streamline, as shown in Figs. 7.7 and 7.8. In this coordinate system, differentials of the wetted lengths for a coordinate displacement are given by:

$$dl_i = h_i dx_i \qquad (7\text{-}8)$$

where h_i is a scale factor, or metric,[15] giving the ratio of differ-

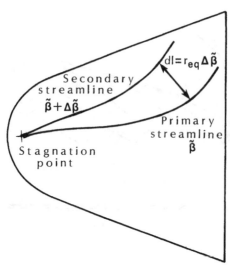

Fig. 7.7 Planform sketch illustrating the nomenclature for the two-streamline method.

ential distance (dl_i) to the differential of the coordinate parameter (dx_i). For the streamwise coordinate system, $\tilde{\beta}$ is constant along a streamline, as shown in Figs. 7.7 and 7.8. Thus, $h_{\tilde{\beta}}$ is a measure of the spreading of adjacent streamlines, analogous to the cross-section radius (r) for axisymmetric flow. For this reason, $h_{\tilde{\beta}}$ is called the equivalent cross-section radius. It will be represented by the symbol r_{eq} for the rest of this text.

The essence of the method lies in the application of the axisymmetric form of the boundary-layer equations along with the inviscid surface streamlines, with the axisymmetric radius (r) replaced by the radius of the equivalent body-of-revolution (r_{eq}). Thus, each surface inviscid streamline on a three-dimensional body corresponds to a different equivalent body-of-revolution at zero angle-of-attack.

A simplified method of approximating the streamline direction on the body surface assumes that the direction of a streamline is in the plane containing the normal to the surface and the direction of the freestream flow. Streamlines determined in this way denote the family of paths of steepest descent from the stagnation point with respect to the direction of the flow at infinity. The method is called the steepest descent[16] or simplified streamline method.

The equivalent cross-section radius (or metric) is computed using the relations that contain the first and second derivatives of the surface pressure with respect to the longitudinal and the circumferential coordinates. Fivel[17] notes that such methods work well when the body shape is relatively simple and the pressure field is well de-

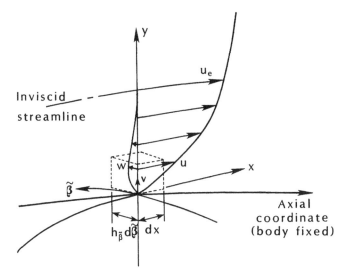

Fig. 7.8 Local (streamwise) boundary-layer coordinate system.

fined. However, the derivatives of the pressure cannot be determined accurately for complex geometries and/or irregular input data. As a result, the metrics computed for such cases are relatively poor. Hecht and Nestler[18] note that the calculation of r_{eq} can become unstable at high angle-of-attack. It was found that this was due to errors in the evaluation of pressure-gradient terms.

Thus far, we have spoken of "inviscid surface streamlines." Hecht and Nestler[18] note that, "the streamlines are calculated in an inviscid sense, and in interpreting these streamlines as boundary layer paths we assume them to be loci along which inviscid flow is entrained into the viscous boundary-layer flow." The distinction between *surface streamlines* and *inviscid streamlines* may be important when comparing the computed streamlines with those obtained experimentally using oil-flow techniques.

In discussing the comparison between experimental results and numerical solutions, Leigh and Ross[19] state, "It should be noted at the outset that there is considerable question as to whether an oil-streak curve represents the inviscid streamline direction at the edge of the boundary layer, the shear-stress direction at the actual body surface, or some streamline direction in the boundary-layer crossflow." Providing the oil-film is sufficiently thin relative to the boundary-layer height, the oil-film lines are believed to represent the shear-stress direction within the boundary layer. The reduced velocity of the fluid particles in the boundary layer makes the motion of these particles more sensitive to the transverse pressure gradients than that of the fluid particles at the edge of the boundary layer (see Fig. 7.8).

Brandon and DeJarnette[20] have computed inviscid streamlines on a blunt, tangent ogive using Euler's equation and a modified Newtonian pressure distribution. The edge properties were evaluated using either the normal-shock/isentropic-expansion (NS/IE) model, i.e., one in which the entropy remains constant at the value immediately downstream of a normal-shock wave, or the variable-entropy model. For the variable-entropy model, the entropy of the streamline entering at the edge of the boundary layer is determined by locating the position where that streamline has crossed the shock wave. That position is found by integrating the mass flow across the bow shock wave until it equals the mass flux within the boundary layer at the station of interest. A modified Maslen method is used to calculate the shock shape and the inviscid flowfield up to the inviscid surface streamline. The distance along a surface streamline is interpreted as the distance along an equivalent axisymmetric body, and the metric, which is a measure of the streamline divergence, is interpreted as the radius of the equivalent axisymmetric body.

The effect of entropy-layer swallowing on the divergence of an inviscid streamline at the edge of the boundary layer for a blunt, tangent ogive at $\alpha = 20$ deg. is shown in Fig. 7.9. Brandon and DeJarnette[20] note that:

> Entropy-layer swallowing starts to affect streamline divergence at about five nose radii downstream of the stagnation point. For the variable entropy solution the stream-

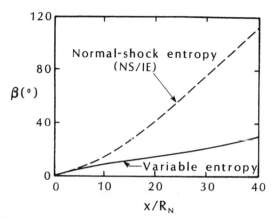

Fig. 7.9 Effect of entropy-swallowing on streamline divergence for a blunt, tangent ogive at $\alpha = 20$ deg., as taken from Ref. 20.

line does not diverge nearly as much as for the normal-shock entropy solution. Near the end of the model the streamline for the variable entropy solution has diverged to an angle of about 30 deg., while the streamline for the normal-shock solution has wrapped around the leeward side of the model to about $\beta = 110$ deg. This overprediction in the streamline spreading tends to increase the normal-shock entropy heating rates around the circumference, and compensates for the decrease in heating resulting from the normal-shock entropy assumption.

Riley et al.[21] stated:

After reviewing the procedure to calculate the streamlines and metrics, the use of boundary-layer edge density, velocity, and Mach number is determined to be incompatible with the calculation procedure. The use of these properties results in computed coordinates that are not the surface streamline coordinates but correspond to a series of different streamlines as the equations are integrated over the vehicle. The density, velocity, and Mach number in the surface streamline equations should be based on normal-shock entropy conditions for proper inviscid surface streamlines to be computed. However, variable-entropy edge conditions should be used in the heat-transfer calculations.

7.4 CONVECTIVE HEAT TRANSFER AND SKIN FRICTION

Solutions of the boundary layer, subject to the appropriate boundary conditions, will provide the temperature, the velocity, and the gas-components distributions adjacent to the surface from which one can determine the convective heat-transfer rate and the skin friction. Techniques of various rigor can be used to generate distributions for the convective heat transfer and for the skin friction. The techniques include relatively simple correlations, analytical solutions based on similarity transformations, boundary-layer solutions as part of a two-layer flowfield model, and the viscous region of a single-layer flowfield model. Regardless of which technique is used, one must model the physical phenomena characteristic of the boundary layer in its various states: laminar, transitional, or fully turbulent. A discussion of laminar, transitional, and fully turbulent boundary layers follows. (Note: Since radiative heat transfer is negligible for most of the flows discussed in this chapter, the terms *convective heat transfer* and *heat transfer* will be used interchangeably, unless otherwise stated.)

7.4.1 Eckert's Reference Temperature

In the 1950s and early 1960s, there were numerous attempts to develop correlations which would approximate the results from numerical solutions of the boundary layer. One of the most recognized of these correlations is the Eckert reference temperature (or reference enthalpy) approach.[22] In this approach, the heat-transfer rates are calculated using the relations developed for incompressible flows with the temperature-related parameters evaluated at Eckert's reference temperature (T^*). According to Ref. 22,

$$T^* = 0.5(T_e + T_w) + 0.22r(T_{te} - T_e) \qquad (7\text{-}9\text{a})$$

For the reference enthalpy method,

$$h^* = 0.5(h_e + h_w) + 0.22(h_r - h_e) \qquad (7\text{-}9\text{b})$$

To calculate the corresponding skin friction, Reynolds analogy is used to relate the skin friction to the convective heat-transfer rate.

The Eckert reference-enthalpy technique was used in the calculation of the heating environment for the design of the Space Shuttle Orbiter. Lee and Harthun[23] noted that, "Aft of the 20% body length station ($x/L = 0.2$), Eckert's flat plate reference enthalpy method was used to predict the laminar flow heating. Streamline divergence and cross-flow effects were taken into account by calibration factors which represent the ratio of wind-tunnel data to the theoretical calculations. Scaling to flight was accomplished by applying the calibration factors to the theoretical calculations adjusted for real-gas conditions."

The process described by Lee and Harthun is illustrated in Fig. 3.2. The curve referred to as "the analytical solution" represents calculations made using the Eckert flat-plate reference-enthalpy method (Eckert FPRE). By comparing the heat transfer measured in the wind tunnel with the heat transfer calculated using the Eckert FPRE method, one obtains the calibration factor. A major source of the difference between a measured value and the corresponding analytical solution is due to streamline divergence and cross-flow effects. (Refer to the discussion of streamline divergence and cross-flow effects presented in Sec. 7.3).

7.4.2 Laminar Boundary Layers

Solutions for the equations of motion for a laminar boundary layer in the stagnation region were discussed in Chap. 5. "Similarity" coordinate transformations were introduced, i.e., Eqs. (5-16a) and

(5-16b), to reduce the governing partial differential equations for a laminar boundary layer to ordinary differential equations.

7.4.2.1 Similarity solutions.

Hayes and Probstein[24] discuss the conditions for similar solutions (i.e., solutions in which the dependent variables are independent of the streamwise coordinate and are functions only of the coordinate normal to the surface). It is stated that:

> There are only two physical situations in which we may expect similarity to hold for a general fluid in equilibrium. One situation is that in which u_e and p_e are constant along the boundary layer, with the solutions termed "constant-pressure" solutions. Included in this category are the attached flow on a wedge or on a cone. The other situation is that in which there is a stagnation point for the inviscid flow and the boundary layer we are interested in lies on the body within the stagnation region. We term the solutions "stagnation point" solutions. In this case the requirement of equilibrium is not needed, but the similar solution breaks down outside the stagnation region.

As noted by Hayes and Probstein[24]:

> Another important physical case is the hypersonic case, in which M_e is very large and u_e^2/H_e is close to 2. In this hypersonic case, similarity may be obtained only if the fluid has certain self-similar properties. A nondissociating perfect gas is such a self-similar fluid.

The similarity transformations introduced by Fay and Riddell[25] for a stagnation-region, laminar boundary layer were discussed in Chap. 5. In seeking a natural coordinate system related to the Cartesian coordinate system through appropriate transformations in which the derivatives of the dependent variables become separable and ordinary differential equations result, Dorrance[26] obtained the following independent variable transformations:

$$\eta = \frac{\rho_e u_e r^k}{(2\xi)^{0.5}} \int \frac{\rho}{\rho_e} dy \qquad (7\text{-}10a)$$

and

$$\xi = \int \rho_e \mu_e u_e r^{2k} dx \qquad (7\text{-}10b)$$

In these equations, r is the radial distance from the axis-of-symmetry to a point in the boundary layer. See Fig. 7.10. However, for ap-

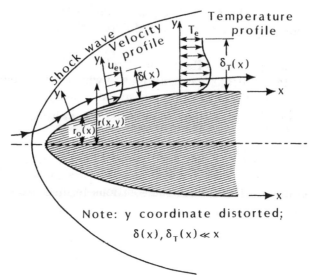

Fig. 7.10   Schematic drawing illustrating the nomenclature for the boundary-layer flow.

plications to thin boundary layers, one usually uses r_0, the radial distance from the axis-of-symmetry to a point on the surface, which is a function of x only.

The relations of Eq. (7-10a) and Eq. (7-10b) are the Lees-Dorodnitsyn transformations for steady, compressible, two-dimensional ($k = 0$) or axisymmetric ($k = 1$) laminar boundary layers.

Exercise 7.3:

Using the Lees-Dorodnitsyn transformations, compare the laminar heat-transfer rate at a point on a cone (which is a distance x along the surface of the cone from the apex) to that at a point on a wedge (which is also distance x from the leading edge). For both configurations, assume that the bow shock wave is weak and attached and that the flow properties at the edge of the boundary layer, i.e., u_e, p_e, ρ_e, are identical for the two cases. The wall temperature is the same for both cases and is independent of x.

Solution:

The heat-transfer rate to the surface is determined from the temperature gradient of the air at the wall:

$$\dot{q} = \left(k\frac{\partial T}{\partial y} \right)_{y=0} = \frac{k_w}{c_{pw}} \left(\frac{\partial h}{\partial y} \right)_w \qquad (7\text{-}11)$$

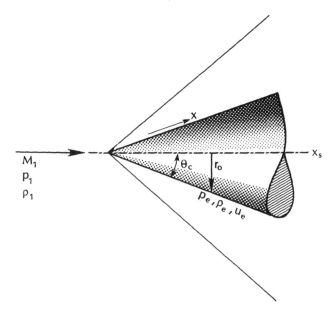

Fig. 7.11   Sketch for the cone flowfield for Exercises 7.3 and 7.4.

At the wall,

$$\frac{\partial h}{\partial y} = \frac{\partial H}{\partial y} - u\frac{\partial u}{\partial y} = \frac{\partial H}{\partial y}$$

Let

$$g = \frac{H}{H_e}$$

Thus,

$$\dot{q} = \frac{k_w H_e}{c_{pw}} \left(\frac{\partial \eta}{\partial y}\right)_w \left(\frac{\partial g}{\partial \eta}\right)_w \qquad (7\text{-}12)$$

To calculate $(\partial \eta/\partial y)_w$, refer to Eqs. (7-10a) and (7-10b). For a thin boundary layer on a sharp cone, $k = 1$ and

$$r \simeq r_0 = x \sin \theta_c$$

as shown in the sketch of Fig. 7.11. Thus,

$$\xi = \rho_e \mu_e u_e \left(\frac{x^3}{3} \sin^2 \theta_c\right)$$

As a result,

$$\dot{q}_{\text{cone}} = \frac{k_w H_e}{c_{pw}} \frac{u_e x \sin \theta_c \rho_w}{\left[2\rho_e \mu_e u_e \frac{x^3}{3} \sin^2 \theta_c\right]^{0.5}} g'(0)$$

where $g'(0)$ represents the derivative of g with respect to η evaluated at the wall. Simplifying:

$$\dot{q}_{cone} = H_e \left[\frac{\rho_w k_w}{c_{pw}}\right] \left[\frac{3 u_e}{2\rho_e \mu_e x}\right]^{0.5} g'(0)$$

For a wedge (or an inclined flat plate), $k = 0$ and

$$\xi = \rho_e \mu_e u_e x$$

As a result,

$$\dot{q}_{wedge} = \frac{k_w H_e}{c_{pw}} \frac{u_e \rho_w}{[2\rho_e \mu_e u_e x]^{0.5}} g'(0)$$

Simplifying:

$$\dot{q}_{wedge} = H_e \left[\frac{\rho_w k_w}{c_{pw}}\right] \left[\frac{u_e}{2\rho_e \mu_e x}\right]^{0.5} g'(0)$$

Since the properties at the inner and at the outer edges of the boundary layer are identical, $g'(0)$ is the same for both flows. Thus,

$$\dot{q}_{cone} = \sqrt{3}\, \dot{q}_{wedge} \tag{7-13}$$

Both a flat plate (or a wedge) and a cone at zero angle-of-attack provide a zero pressure-gradient boundary layer that can be described by two coordinates. With identical boundary conditions (at the wall and at the edge), the mean boundary-layer thickness and the velocity and the temperature profiles are identical for the two cases, when

$$x_c = 3 x_{wedge}$$

Exercise 7.4:

A sharp cone ($\theta_c = 15$ deg.) is exposed to a hypersonic stream of air where $M_1 = 10.6$, $p_{t1} = 1200$ psia, and $T_t = 2000\ °\text{R}$. Compare the following heat-transfer measurements[27] with those calculated using the Eckert's reference temperature technique.

$\dfrac{x_s}{L}$	0.207	0.250	0.293	0.380	0.552	0.638	0.724	0.811	0.897
$\dfrac{\dot{q}}{\dot{q}_0}$	0.122	0.0968	0.0775	0.0692	0.0664	0.0636	0.0581	0.0581	0.0554

The reference parameters are $L = 22.39$ in. and $\dot{q}_0 = 19.0$ Btu/ft^{2}s. For this problem, x_s is the distance from the apex along the axis-of-symmetry, L is the length of the cone along the axis-of-symmetry, and x is the distance from the apex along a conical generator. (See Fig. 7.11.) Assume that the air behaves as a perfect gas.

Solution:

First calculate the freestream static properties in the test section. For $M_1 = 10.6$,

$$\frac{p_1}{p_{t1}} = 0.1596 \times 10^{-4}; \quad \frac{T_1}{T_t} = 0.4260 \times 10^{-1}$$

Thus, $p_1 = 0.01915$ psia; $T_1 = 85.20$ °R. Although it is not required for this exercise, let us calculate the unit Reynolds number, since Cleary[27] states that it is 1.20×10^6/ft.

$$\rho_1 = \frac{p_1}{RT_1} = \frac{\left(0.01915 \frac{\text{lbf}}{\text{in}^2}\right)\left(144 \frac{\text{in}^2}{\text{ft}^2}\right)}{\left(1716.16 \frac{\text{ft}^2}{\text{s}^2 \, °\text{R}}\right)(85.20 \, °\text{R})} = 1.886 \times 10^{-5} \frac{\text{lbf s}}{\text{ft}^4}$$

$$U_1 = M_1(49.02\sqrt{T_1}) = 4796.22 \frac{\text{ft}}{\text{s}}$$

Using Sutherland's equation to calculate the viscosity:

$$\mu_1 = 2.27 \times 10^{-8} \frac{T_1^{1.5}}{T_1 + 198.6} = 6.29 \times 10^{-8} \frac{\text{lbf s}}{\text{ft}^2}$$

With these perfect-gas values, the unit Reynolds number is:

$$Re_1 = \frac{\rho_1 U_1}{\mu_1} = 1.438 \times 10^6 \text{ /ft}$$

which is considerably greater than the value reported by Cleary. The difference between the viscosity that is calculated using Sutherland's equation and the actual viscosity (see Exercise 4.2) would account for part, but not all of the difference.

Using the charts of Ref. 28 to calculate conditions at the edge of the boundary layer on the cone for $M_1 = 10.6$ and $\theta_c = 15$ deg.,

$$C_{pe} = \frac{p_e - p_1}{q_1} = 0.141$$

$$M_e = 5.85$$

For $M_e = 5.85$, $\frac{T_e}{T_t} = 0.1275$, and $T_e = 255$ °R

$$p_e = p_1 \left(1 + \frac{\gamma}{2} M_1^2 C_{p_e}\right) = 0.2315 \text{ psia}$$

Assume $T_w = 560$ °R. For a laminar boundary layer, $r = \sqrt{Pr} = 0.8367$.

$$T^* = 0.5(T_e + T_w) + 0.22r(T_{te} - T_e) = 728.71 \text{ °R}$$

$$\rho^* = \frac{p_e}{RT^*} = 2.666 \times 10^{-5} \frac{\text{lbf s}^2}{\text{ft}^4}$$

$$\mu^* = 4.815 \times 10^{-7} \frac{\text{lbf s}}{\text{ft}^2}$$

$$u_e = 5.85(49.02\sqrt{255}) = 4579.30 \frac{\text{ft}}{\text{s}}$$

Note that even though the Mach number dropped from 10.6 in the freestream to 5.85 at the edge of the boundary layer, the velocity only changed from 4796.22 ft/s to 4579.30 ft/s. A similar relation would exist for the flow approaching the inlet of the airbreathing engine of a hypersonic cruiser, i.e., while the Mach number would be considerably less than the freestream value, the local velocity would be relatively near the freestream velocity.

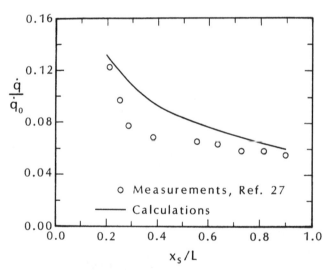

Fig. 7.12   Comparison of the calculated and the measured heat-transfer distribution for hypersonic flow past a sharp cone, $M_1 = 10.6$, $\theta_c = 15$ deg.

$$Re_x^* = \frac{\rho^* u_e x}{\mu^*} = 2.535 \times 10^5 x$$

Using the results from Exercise 7.3,

$$St = \frac{0.575}{(Pr^*)^{0.667} (Re_x^*)^{0.5}} = \frac{1.448 \times 10^{-3}}{\sqrt{x}}$$

So that the heat transfer is:

$$\dot{q} = \rho^* u_e c_p (T_r - T_w) St$$

To calculate the recovery temperature,

$$T_r = T_e + r(T_{te} - T_e) = 1714.97 \text{ }^\circ R$$

$$\dot{q} = \frac{1.5798}{\sqrt{x}} \frac{\text{Btu}}{\text{ft}^2 \text{ s}}$$

The heat-transfer rates, thus calculated, are compared with the measurements reported by Cleary[27] in Fig. 7.12.

When there is a streamwise pressure gradient in the external flow, the velocity and the temperature profiles for a laminar boundary layer are no longer similar, i.e., independent of ξ (or x) in the transformed η coordinate. However, if the external and the surface conditions vary sufficiently slowly with the ξ (or x) coordinate, a *local similarity* approximation can be made. For this technique, *local similarity solutions* are patched together so that the *history* of the gradients in the external flow is represented by the ξ-coordinate and its effect on η.

Lees[29] and Kemp et al.[30] used the local similarity approximation to calculate the laminar heat-transfer-rate distribution around highly cooled blunt bodies in a hypersonic stream. Combining the logic used in developing Eq. (7-12),

$$\dot{q} = \frac{k_w H_e}{c_{pw}} \left(\frac{\partial \eta}{\partial y}\right)_w \left(\frac{\partial g}{\partial \eta}\right)_w$$

with the similarity transformation for η, Eq. (7-10a),

$$\dot{q} = \frac{\rho_w \mu_w u_e r^k}{Pr_w \sqrt{2 \int \rho_e \mu_e u_e r^{2k} dx}} H_e[g'(0)] \qquad (7\text{-}14)$$

The data against which the local similarity calculations will be compared were presented in dimensionless fashion as the ratio of the local heating rate divided by a reference heating rate. (To obtain a dimensionless ratio that is independent of the flow conditions, one might use the corresponding ratio of heat-transfer coefficients.) The convective heat transfer to the stagnation point of the sphere whose radius is R_{ref} will be used as the reference heating rate, $\dot{q}_{t,ref}$. R_{ref} can be the actual nose radius of the spherical cap or an arbitrary reference radius, such as one-foot reduced-to-model scale (as is done for the Space Shuttle Orbiter). For the transformation described by Eq. (7-10a):

$$\dot{q}_{t,ref} = \frac{\rho_{w,t}\mu_{w,t}u_e r^k}{Pr_w\sqrt{2\xi}}\ H_{t2}[g'(0)]_t$$

Using the common approximations for the stagnation-point region that:

$$u_e \simeq \left(\frac{du_e}{dx}\right)_{t2} x; \quad r \simeq x$$

where x is the wetted surface distance from the stagnation point of the reference sphere. Using these approximations in Eq. (7-10b):

$$\dot{q}_{t,ref} = \frac{\rho_{w,t}\mu_{w,t}}{Pr_{w,t}}\ \left[\frac{2\left(\frac{du_e}{dx}\right)_{t2}}{\rho_{t2}\mu_{t2}}\right]^{0.5} H_{t2}\,[g'(0)]_t \qquad (7\text{-}15)$$

Dividing the expression for $\dot{q}$, i.e., Eq. (7-14), by the reference value, i.e., Eq. (7-15), and nondimensionalizing the local flow parameters, one obtains:

$$\frac{\dot{q}}{\dot{q}_{t,ref}} = \frac{F(\xi)}{\left[\left(\frac{d(u_e/U_1)}{dx}\right)_{t2}\right]^{0.5}}\ \frac{[g'(0)]}{[g'(0)]_t} \qquad (7\text{-}16)$$

where

$$F(\xi) = \frac{Pr_{w,t}}{Pr_w}\ \frac{\frac{\rho_w}{\rho_{w,t}}\frac{\mu_w}{\mu_{w,t}}\frac{u_e}{U_1}\left(\frac{r}{R_{ref}}\right)^k}{2\left[\int\frac{\rho_e}{\rho_{t2}}\frac{\mu_e}{\mu_{t2}}\frac{u_e}{U_1}\left(\frac{r}{R_{ref}}\right)^{2k}dx\right]^{0.5}} \qquad (7\text{-}17)$$

The required flow properties can be calculated when the pressure and the entropy distributions are known as well as the surface temperature.

From solutions of the transformed, laminar boundary-layer equations, Kemp et al.[30] developed a correlation for the ratio of enthalpy gradients at the wall:

$$\frac{[g'(0)]}{[g'(0)]_t} = \frac{1+0.096\sqrt{\beta p}}{1.06788}\ \frac{1-g(0)}{1-[g(0)]_t} \qquad (7\text{-}18)$$

where βp, the local pressure-gradient parameter in the similarity solution, is

$$\beta p = \frac{2\xi}{u_e} \frac{du_e}{d\xi} \qquad (7\text{-}19)$$

In reality, since βp varies with ξ, the concept of *local similarity* is employed. Using the values of enthalpy for an adiabatic wall as suggested by Goodrich et al.[11] instead of the inviscid value of the total enthalpy, one obtains:

$$\frac{[g'(0)]}{[g'(0)]_t} = \frac{1 + 0.096\sqrt{\beta p}}{1.06788} \frac{h_{aw} - h_w}{h_{aw,t} - h_{w,t}} \qquad (7\text{-}20)$$

Thus, the value for the nondimensionalized local heat-transfer rate is calculated using Eqs. (7-16), (7-17), and (7-20).

Using the equations presented in this section, the heat-transfer distribution was computed for the hypersonic flow of a perfect gas past a sphere. The modified Newtonian pressures, Eq. (6-3), with a normal-shock/isentropic-expansion flow model were used to define the local properties (Fig. 6.10). The nominal freestream conditions include a Mach number of 18, a static pressure of 0.04429 p_{SL} (which corresponds to a pressure altitude of 70,000 ft) and a static temperature of 181.93 °R. Included in Fig. 7.13 for comparison are the heat-

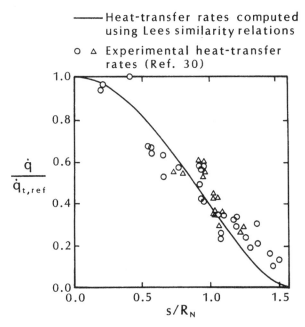

Fig. 7.13 **Heat-transfer distribution on hemisphere / cylinder.**

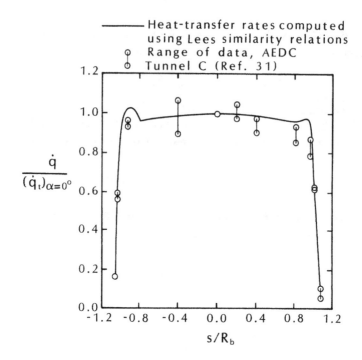

Fig. 7.14 The heating distribution for the windward pitch plane of the Apollo Command Module at $\alpha = 0$ deg., as taken from Ref. 31.

transfer data for the flow of dissociated air past a highly cooled body, as reported by Kemp, et al.[30] These data were obtained in a shock tube for M_s between 7.65 and 13.5. The computed heat-transfer distribution is in reasonable agreement with the experimental heating rates.

The heating-rate distribution for the windward pitch plane of the Apollo Command Module at zero angle-of-attack is presented in Fig. 7.14. Included for comparison are the range of experimental values from Tunnel C at AEDC[31] and the heating-rate distribution calculated using the equations of Lees[29] which employ the similarity transformation defined in Eqs. (7-10a) and (7-10b). Note that the theoretical heat-transfer distribution presented in Fig. 7.14 uses the experimental pressure distribution presented in Fig. 6.12 in the evaluation of the local flow properties. Furthermore, the correlations of Stoney[32] for a hemispherical segment were used to obtain the velocity gradient for the Apollo Command Module:

$$\sqrt{\left(\frac{du_e}{dx}\right)_{t2,\alpha=0°}} = 1.055 \left[\frac{1}{R_N}\sqrt{\frac{2(p_{t2} - p_1)}{\rho_{t2}}}\right]^{0.5}$$

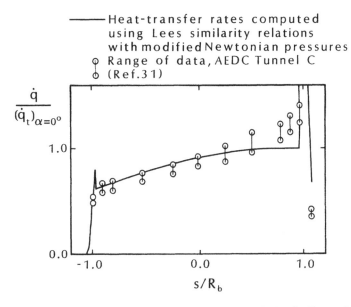

Fig. 7.15 The heat-transfer distribution in the windward pitch plane of the Apollo Command Module at α = 20 deg., as taken from Ref. 31.

in the calculation of the reference heating rate, that at the stagnation point of the Apollo Command Module at zero angle-of-attack.

The theoretical heat-transfer-rate distribution for the windward pitch plane of the Apollo Command Module at an angle-of-attack of 20 deg., as computed using Lees similarity relations, is presented in Fig. 7.15. Modified Newtonian pressures and the normal-shock/isentropic-expansion (NS/IE) assumption are used to calculate the local flow parameters at the edge of the boundary layer. The computation of the theoretical heat-transfer distribution incorporates a metric (i.e., the radius of the equivalent body-of-revolution, r_{eq}) based on the inviscid surface streamlines calculated using Euler's equation (in a variation of the method described in Ref. 33) to approximate the three-dimensional character of the boundary layer. Included for comparison are data from Tunnel C (AEDC), as reported in Ref. 31. Although there is significant scatter in the measured heating rates, the computed heat-transfer rates are in good agreement with the measured values over much of the spherical heat shield. However, there are significant differences near the stagnation point. The differences are attributed (in part, at least) to differences between the modified Newtonian pressures and the actual pressure distribution. Recall that the actual stagnation point is inboard of the modified Newtonian location (refer to Fig. 6.14) and that the actual pressure gradients (and, hence, also the actual velocity gradients) are

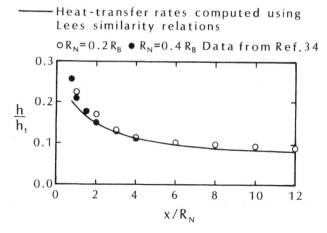

Fig. 7.16   Comparison of blunt-cone heat-transfer distribution with similarity theory; helium, M_1 = 14.9, α = 0 deg., as taken from Ref. 34.

not as severe as predicted by the modified Newtonian model. If we had used the experimental pressure distribution when evaluating the terms in Eq. (7-17), there would have been better agreement between the computed and the measured heating rates. Note also that the peak heating does not occur at the stagnation point, which is on the spherical cap (see Fig. 6.14), but on the torroidal segment, where the maximum values of the inviscid velocity gradient occur.

Although the local similarity approximations were developed by Lees[29] and by Kemp et al.[30] for blunt bodies, Cleary and Duller[34] used the technique to calculate the heat-transfer distribution on a blunted, slender cone in a Mach 14.9 stream of helium. The comparison between theory and experiment is reproduced in Fig. 7.16. The theoretical heat-transfer distributions were calculated using the measured pressure distributions. The agreement between theory and experiment is good for both bluntness ratios. However, Cleary and Duller[34] state, "but for higher angles of attack theory tends to underestimate the measured heating. Since theory does not account for increases in h due to the effects of entropy gradients or cross flow, the differences shown at high angles of attack may be due, in part, to these effects." The use of r_{eq}, the cross-section radius for the equivalent body-of-revolution, can be used in Eq. (7-17) to account for streamline divergence.

The experimental heat-transfer distribution for the windward plane-of-symmetry of the Space Shuttle Orbiter at an angle-of-attack of 30 deg. that was presented in Ref. 35 is reproduced in Fig. 7.17. The data are presented in dimensionless form as the experimental local heat-transfer coefficient divided by the computed stagnation-

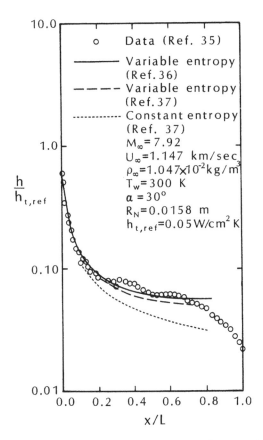

Fig. 7.17 Comparisons of the predicted and experimental heat-transfer distributions from the windward pitch-plane of the Shuttle Orbiter at α = 30 deg.

point heat-transfer coefficient for a scaled 0.3048-meter spherical nose radius. Included for comparison are the heat-transfer distributions as computed by Zoby[36] for a variable-entropy flow and by Bertin et al.[37] both for a variable-entropy flow and for a constant-entropy (NS/IE) flow. The laminar heat-transfer distribution presented by Zoby is computed using an incompressible Blasius relation (a similarity transformation) with compressibility effects accounted for by Eckert's reference-enthalpy method. The method is described in Ref. 38. The heat-transfer distributions of Bertin et al.[37] were computed using the nonsimilar boundary-layer code described in Ref. 39. Note that accounting for the entropy-layer swallowing, i.e., variable entropy, produces increased heating rates. This can be seen by comparing the two computations of Bertin et al. (which used the same metric, r_{eq}, distributions for both entropy models). It is also interesting to note that, despite the significant differences between the equations used by Zoby[36] and those used by Bertin et al.,[37] the heat-transfer distributions computed using the two variable-entropy models are in good agreement.

The relatively simple techniques described thus far in this chapter can be used to generate approximate theoretical values of the heat-transfer distributions. Since these approximations are easily computed, they can be compared with measurements to evaluate ground-based-test results and can be used in conceptual design codes. Indeed, as has been discussed, these relatively simple techniques were used throughout the design process for the Apollo Command Module and for the Space Shuttle Orbiter. The success of these techniques is due in part to the relative simplicity of the flowfield for these rather blunt vehicles. Furthermore, as noted in Sec. 3.2, the uncertainties inherent in the simple flow models force the introduction of conservation into the estimates of the heating environment prior to the first flight.

For many applications, the computer code must incorporate more rigorous flow models. Such applications include flows where gas chemistry is important, flows where, although the inviscid flow is supersonic in the streamwise direction and the subsonic flow in the viscous sublayer is always positive in the streamwise direction, cross-flow separations are permitted, and flows involving important viscous/inviscid interactions.

7.4.2.2 Detailed computations of the viscous flow. As noted in Chap. 3, there are a variety of techniques which can be used to generate detailed solutions of the viscous flow. These include two-layer models, i.e., dividing the shock-layer flow into an inviscid region and a viscous boundary layer, and single-layer models, e.g., the parabolized Navier-Stokes formulations and the full Navier-Stokes formulations.

The effect of chemical reactions on the temperature in a hypersonic boundary layer is indicated in the computations presented in Fig. 7.18. Presented are the computed temperature distributions across the boundary layer in the windward pitch-plane near the nose of the Space Shuttle Orbiter. The freestream conditions ($M_\infty = 29.86$; altitude $= 246,000$ ft; $\alpha = 41.4$ deg.) represent a time early in the re-entry phase of the flight. Boundary-layer solutions were computed using a finite-difference formulation[39] and an integral-matrix formulation.[40] Variable-spacing grid schemes automatically concentrated points near the wall where the gradients are the greatest. For the purposes of the computations presented in Fig. 7.18 (which included calibrating the code of Ref. 39), the air was assumed to be in thermochemical equilibrium. Although the freestream Mach number is 29.86, the local edge Mach number is near-sonic. As a result, there is no region of locally high tempera-

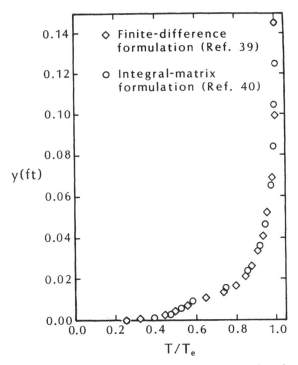

Fig. 7.18   Temperature distribution across a laminar boundary layer at $x = 0.107L$, Orbiter Flight Design Trajectory; $M_\infty = 29.86$; altitude = 246,000 ft; angle-of-attack = 41.4 deg.

tures near the wall due to viscous dissipation. However, both codes indicate an inflection point in the temperature profiles in the region where the temperature is on the order of 4500 °R to 5000 °R. In this temperature range, the oxygen molecules begin to dissociate at the static pressure of this flow (which is approximately 0.01 atm).

Blunted biconics find applications as a moderate lift-to-drag orbit-transfer vehicle, as a Mars-sample return vehicle, and as a volumetrically efficient re-entry vehicle. Detailed studies of the flowfield for a spherically blunted 12.84 deg./7 deg. biconic have been conducted at the NASA Langley Research Center. A sketch of the wind-tunnel model, as presented in Ref. 41, is reproduced in Fig. 7.19. Using leeward surface oil-flow patterns, Miller et al.[42] described the flow for a blunt cone at incidence, e.g., $\alpha = 12$ deg. in Fig. 7.19b.

A large, favorable pressure gradient exists on the leeward side just downstream of the nose causing the boundary layer to remain attached. In this region close to the nose, the lateral component of skin friction is in the direction of

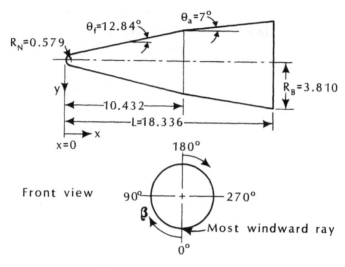

(a) Model configuration

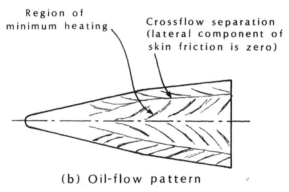

(b) Oil-flow pattern

Fig. 7.19 Sketches of a spherically blunted biconic model and of the leeside oil-flow pattern, as taken from Ref. 41; all dimensions in cm.

the most leeward ray; however, at larger distance downstream of the nose, the lateral component changes direction and is now away from the most leeward ray. When this change in direction occurs, two symmetrical separation lines develop downstream of the region where the lateral component of the skin friction passes through zero. The longitudinal component of skin friction in the vicinity of the most leeward ray is finite, as revealed by oil that flowed from the attached region just downstream of the nose, through the region of zero lateral skin friction, and into the vortex region.

Clearly, the assumptions inherent in the parabolized Navier-Stokes (PNS) formulation are consistent with the flowfield described in the previous paragraph. As described by Chaussee[43]:

> The parabolized approximation to the Navier-Stokes equations assumes that the flow is supersonic in the streamwise direction and that the subsonic flow in the viscous sublayer is always positive in the streamwise direction. Thus, flows with large streamwise separation and flow reversals are excluded from the foregoing assumptions. However, cross-flow separations are permitted. Under these assumptions, the Navier-Stokes equations become parabolized in the streamwise direction, enabling a marching solution procedure, which is computationally desirable and efficient.

Heating-rate distributions for laminar flow over a spherically blunted 12.84 deg./7 deg. biconic were reported by Miller et al.[41] The effect of angle-of-attack on the longitudinal heating rate for the leeward plane-of-symmetry ($\beta = 180$ deg.) is presented in Fig. 7.20 for $M_\infty = 9.86$. At a specific location, the heating decreases with increasing angle-of-attack up to $\alpha = 6$ deg. At $\alpha = 10$ deg., the heating to the aft-cone section increases with x/L. Miller et al. note that:

> The departure of the heating distribution at the higher angles of attack from those at the lower angles of attack is attributed to cross-flow separation. This flow separation results in the formation of longitudinal, counter-rotating, primary vortices that reattach along the most leeward ray forming a "stagnation line"; heating along the most leeward ray is thus augmented by flow reattachment. The distributions of Fig. 7.20 show that cross-flow separation occurs first on the rearward portion of the aft-cone section, and the separation region moves upstream with increasing angle of attack.

Included for comparison are the heating rates predicted with a parabolized Navier-Stokes code developed by Gnoffo. Although the PNS code qualitatively predicts the heating distributions, it underpredicts the heating levels by roughly 10 to 15 percent.

The effect of angle-of-attack on the circumferential heating distribution just upstream of the fore-cone/aft-cone junction ($x = 0.54L$) and on the aft-cone section ($x = 0.85L$) is illustrated by the measurements presented in Fig. 7.21 for $M_\infty = 9.86$. Heating on the midmeridian ray ($\beta = 90$ deg.) of the fore-cone section and of the aft-cone section is nearly independent of angle-of-attack (over the

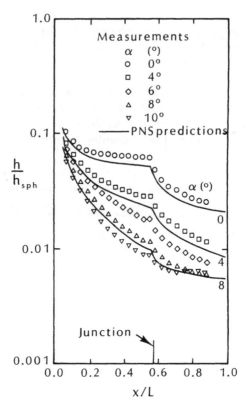

Fig. 7.20 Effect of angle-of-attack on longitudinal heating distribution along the most leeward ray (β = 180 deg.) M_∞ = 9.86, $Re_{\infty L}$ = 3.2 × 10^5, 0 deg. $\leq \alpha \leq$ 10 deg., as taken from Ref. 41.

range considered) "which is typical of cones."[41] At low angles-of-attack (i.e., $\alpha \leq 12$ deg. on the fore cone and $\alpha \leq 8$ deg. to 10 deg. on the aft cone), the heating decreases continuously in the circumferential direction. The first hint of the occurrence of cross-flow separation (a decrease, then an increase in the heating as the flow expands from the midmeridian ray to the most leeward ray) occurs at the angles-of-attack noted in the previous sentence. The circumferential extent of flow separation increases as the angle-of-attack is increased to $\alpha = 20$ deg. Again, there is qualitative agreement between the predictions of Gnoffo's PNS code and the experimental values with the predictions about 10 to 15 percent below the data.

Clearly, advanced CFD codes are needed to generate solutions for comparison with data and for use in generating flowfield solutions at flight conditions. However, the CFD codes employ approximations in the numerical algorithms and in the flow models. Thus, in developing computational tools, it is critical that one conducts systematic studies of the various computational parameters and their effect on the flowfield as compared to experiment. The computational parameters include grid spacing (both for the vehicle and for the flowfield), smoothing techniques, turbulence modeling, transport properties, chemistry models, etc. Note that it is much more difficult to compute the heat transfer than it is to compute the surface pressure. Thus, comparisons with heat transfer are a much better measure of the validity of the models used in the code.

7.4.3 Boundary-Layer Transition

Procedures for calculating the convective heat transfer when the boundary layer is laminar were discussed in the previous section. If the boundary layer were turbulent, the rate at which heat is

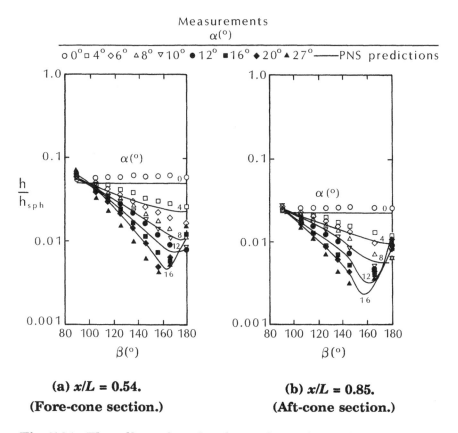

(a) $x/L = 0.54$.

(Fore-cone section.)

(b) $x/L = 0.85$.

(Aft-cone section.)

Fig. 7.21   The effect of angle-of-attack on circumferential heating distribution for a biconic, $M_\infty = 9.86$, $Re_{\infty L} = 3.2 \times 10^5$, as taken from Ref. 41.

Fig. 7.22   Spark shadowgraph of the boundary-layer transition process on a pencil model, M_∞ = 3.9, Re/in = 2.2 × 10⁶, as taken from Ref. 44.

transferred to the surface would be much greater. Furthermore, viscous/inviscid interactions, such as the interaction between a shock wave and the boundary layer, depend on the character of the boundary layer, i.e., whether it is laminar, transitional, or turbulent. Thus, in order to define accurately the aerothermodynamic environment of a vehicle in a hypersonic stream, the designer must develop criteria for the conditions defining when transition of the boundary layer from the laminar state to a turbulent state occurs. This is not an easy task, because the transition process is a complex one. Furthermore, when and if boundary-layer transition occurs depends on many, interrelated parameters.

7.4.3.1 General comments about the transition process.   It is generally believed that boundary-layer transition occurs when some disturbances to the boundary layer grow to a critical amplitude that produces a breakdown of the laminar boundary layer. Not all disturbances produce boundary-layer transition. In some instances, the disturbances attenuate and the boundary layer remains laminar. In other cases, the disturbances grow as they travel downstream in the boundary layer, until they finally produce a turbulent boundary layer. The initial boundary-layer disturbance amplitude depends upon the characteristics of the disturbances to which the boundary layer is exposed, the receptivity of the boundary layer to these disturbances, and the extent of the initially stable region.

During the initial stages of the transition process, the change in the strength of a particular disturbance is modeled by stability

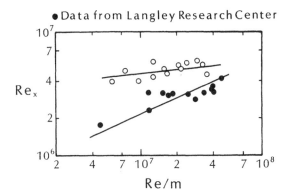

Fig. 7.23   Comparison of (previous) transition Reynolds number measurements for the beginning of transition as determined using thermocouple data. $M_\infty = 7.5$, $\theta_c = 5$ deg., $T_w = 0.4T_t$, **as taken from Ref. 45.**

theory. As the disturbances grow, three-dimensional unstable waves and hairpin eddies develop. Subsequently, turbulent spots (such as first observed by Emmons) occur. A turbulent spot, or burst, is evident in the spark shadowgraph of the pencil model in free flight at Mach 3.9, which is taken from Ref. 44 and is reproduced in Fig. 7.22. Note that the burst extends well above the laminar boundary layer into the inviscid flow. The shadowgraph indicates that the boundary layer is laminar downstream (as well as upstream) of the burst at the instant the photograph was taken. If there were sensitive heat-transfer gages in the region of the burst, they would sense locally high, fluctuating heating rates. Gages located downstream of the burst would sense heat-transfer rates at the laminar level. After the burst moves downstream, the heating will return to the laminar level. Thus, in the transitional region, the flow intermittently exhibits turbulent characteristics. Toward the end of the transition zone, the intermittently turbulent flow becomes dominant.

All boundary layers are subjected to disturbances. These disturbances may originate in the freestream, e.g., vorticity, entropy variations, and sound fluctuations, or they may originate at the surface, e.g., surface roughness and surface vibrations. Therefore, it is important to know what disturbances are likely to grow and what parameters influence the growth of these unstable disturbances.

Configuration differences influence the transition process. Note that much of the boundary-layer transition data presented are for sharp cones and for blunted cones. Stetson[3] notes that "disturbances

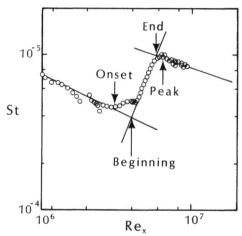

Fig. 7.24 Heat-transfer distribution from thermocouple measurements made in the Langley Research Center tunnel, as taken from Ref. 45.

begin to grow sooner on a plate (smaller Re_c), but they grow slower than in a cone boundary layer. This result would suggest that, for a quiet environment (thus a long distance of disturbance growth before transition), plate transition Reynolds number should be greater than cone transition Reynolds numbers." Measurements made in a quiet wind tunnel indicate cone-to-flat-plate transition Reynolds number ratios near 0.8 for low-noise freestream conditions to about 1.2 for higher-noise conditions.

Heating-rate distributions[45] were used to investigate transition on two similar sharp cones ($\theta_c = 5$ deg.), one tested in the Ames Research Center's 3.5-Foot Hypersonic Wind Tunnel and the second tested in the Langley Research Center's 18-Inch Variable Density Wind Tunnel. The Reynolds number at the beginning of transition, as determined using thermocouple data, are reproduced in Fig. 7.23. Although the models are similar and the test conditions are close, there are considerable differences between the transition Reynolds numbers from the two programs. To investigate these transition Reynolds number differences, additional measurements were made as reported in Ref. 45. "However, since the parameters affecting boundary-layer transition are numerous and complex, the same models and detection techniques (thermocouples and thin film gages) were used in both facilities. This was done in an attempt to alleviate any inconsistencies due to the effects of unknown model influence and difference between the various detection techniques."

The streamwise heat-transfer distribution, as measured using the thermocouple technique obtained in the Langley facility, is repro-

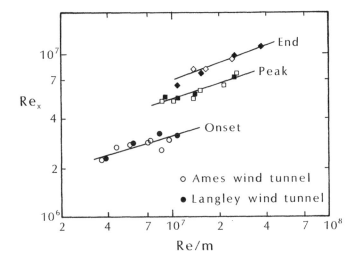

Fig. 7.25   Comparison of thin-film transition data. $M_\infty = 7.5$, $\theta_c = 5$ deg., $T_w = 0.4T_t$, **as taken from Ref. 45.**

duced in Fig. 7.24. Based on this heat-transfer distribution, the authors of Ref. 45 *identify* three *stages* of the transition process:

1. The *onset* of the transition, where the Stanton numbers first consistently exceed the laminar value,

2. The *beginning* of transition, obtained by fairing straight lines through the laminar and the transitional data, and

3. The *end* of transition, obtained by fairing straight lines through the turbulent and the transitional data.

Also included is the location of the peak Stanton number. It might be noted that the increase in the heating (i.e., Stanton number) between the "onset" and the "beginning" of the transition evident in the data of Fig. 7.24 is somewhat unusual. It may be a factor in the differences between the data from Ames and those from Langley.

The reader should note that a heat-transfer distribution determined using typical techniques, e.g., thermocouples and calorimeters, provides time-averaged information. In reality, by the time the heat transfer significantly exceeds the laminar values, the transition process is relatively advanced. Thus, sensitive instrumentation, capable of sensing time-dependent fluctuations, is required to gain insights into the transition process. Nevertheless, since a heat-transfer distribution, such as that presented in Fig. 7.24, presents graphic, relatively easily obtained data, many experimentalists use such data to determine the *onset* of transition, the *end* of transition, etc.

The Reynolds numbers for the onset of transition, for the end of transition, and for the peak Stanton number (as determined using

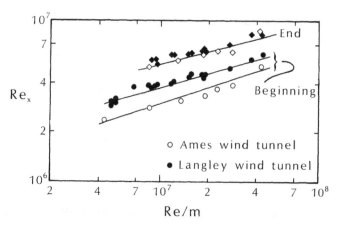

Fig. 7.26 Comparison of thermocouple transition data. M_∞ = 7.5, θ_c = 5 deg., T_w = 0.4T_t, as taken from Ref. 45.

thin-film gages) are presented in Fig. 7.25 as a function of the local unit Reynolds number. The agreement between the data from the two facilities is excellent. However, comparison of the thermocouple data obtained in the two wind tunnels still shows some differences when the conventional beginning and end of transition are used. The *beginning of transition* Reynolds numbers determined from the heat-transfer distributions determined using thermocouples are reproduced in Fig. 7.26. Owen et al.[45] conclude, "A comparison between the thermocouple heat transfer and thin-film gage fluctuation techniques has shown that anomalous heat transfer data can affect the thermocouple technique of transition detection which could lead in some cases to inconsistent results. But, when consistently defined onset and peak are used for comparison, there is excellent agreement between the two methods."

Using data from nine wind tunnels ($3 \leq M_\infty \leq 8$), Pate and Schueler[46] developed a correlation to predict wind-tunnel transition Reynolds numbers which was dependent only on the tunnel-wall turbulent boundary layer, aerodynamic noise parameters (displacement thickness and skin-friction coefficient), and tunnel test-section circumference. In a review article on the effects of wind-tunnel disturbances on boundary-layer transition, Pate[47] cited three distinct disturbance modes that could affect transition locations determined through wind-tunnel testing: vorticity (turbulence) fluctuations, entropy fluctuations (temperature spottiness), and pressure fluctuations (sound waves or radiated aerodynamic noise). Aerodynamic noise could result from sound radiation from the wall's turbulent boundary layer, from shimmering Mach waves from wall roughness or waviness, and from wall vibrations.

The discussion presented thus far indicates that the transition locations determined in ground-based tests may be influenced by dis-

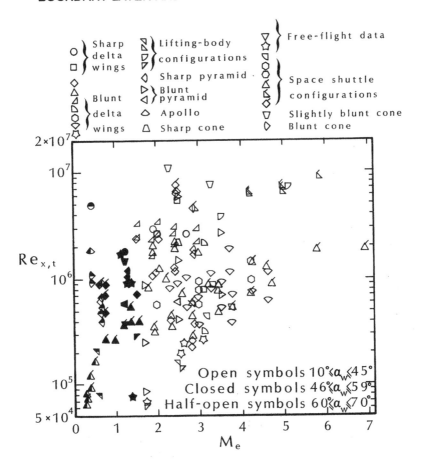

Fig. 7.27   Transition Reynolds number as a function of local Mach number, as taken from Ref. 48.

turbances in the freestream flow. Furthermore, the transition location depends on the technique used (e.g., heat-transfer distribution, flow visualization, etc.) and on the criteria (e.g., onset, beginning, end, etc.). In addition, numerous flowfield parameters affect transition. The following will be discussed in this chapter: the local Mach number, surface cooling (as characterized by the wall-temperature to total-temperature ratio), the unit Reynolds number, nose blunting (including the effect of entropy swallowing), cross-flow (or three-dimensional effects), surface roughness, and mass injection. Other parameters, such as gas chemistry, may influence transition.

Thus, the experimentally determined transition location depends on the nominal test conditions, on the quality of the flow in the facility in which the data are obtained, and on the detection technique that is used to determine transition. It should not be surprising then that significant scatter results when one presents transition locations representing diverse configurations tested in various facilities. One such collection of data, taken from Ref. 48, is reproduced in Fig. 7.27. In 1969, Morkovin[49] stated that "Time and again some of the implicit parameters are changed and the prediction bands may be rendered inapplicable to the new problem." In 1975, Reshotko[50] stated that "These efforts, however, have yielded neither a transition theory nor any even moderately reliable means of predicting transition Reynolds number." In 1992, Stetson[3] stated that "There is no transition theory. All transition prediction methods are empirical."

Clearly, the designer of a hypersonic vehicle who must develop a means of predicting boundary-layer transition faces a considerable challenge. Slender, high-beta re-entry vehicles that enter the Earth's atmosphere at a large negative flight-path angle do not slow significantly before quickly reaching low altitudes and, thus, fly at very high Reynolds numbers. As a result, the thermal-protection system is designed for a turbulent boundary layer. For blunt, low-beta vehicles that enter the earth's atmosphere at a small negative flight-path angle, such as the Hermes and the Space Shuttle Orbiter, significant deceleration occurs at high altitudes, i.e., low Reynolds numbers. Thus, over most of the trajectory, the boundary layer is laminar. The reader should note, however, that there was considerable concern before the first Shuttle flight that surface roughness due to tile misalignment (within the fabrication tolerances) would cause early transition. In fact, at the time this book goes to press, the effect of surface roughness (due to tile misalignment and to gap-filler material) is of continued concern to the operators of the Space Shuttle Orbiter. However, as will be discussed in Sec. 7.4.3.8, boundary-layer transition did not occur until the Mach number was less than 10. The problem is more serious for airbreathing vehicles which fly for long periods of time at intermediate altitudes where uncertainties in transition criteria are important. Dr. Tom Weeks of the Flight Dynamics Directorate at the Wright Laboratory envisions the day when "manned hypersonic vehicles may well carry boundary layer transition capability coupled with AI based, robust flight controls." What can be done to develop transition correlations to be used during the design process? One should make use of the extensive data base of transition measurements, review stability theory and stability experiments, and try to understand the various transition correlations and their limitations. The transition data base provides insights into the effects of an individual parameter. Stability theory and stability experiments (where disturbance growth is measured in the boundary layer prior to transition) can also be used to provide insights into the effects of parametric variations.

7.4.3.2 Linear stability theory. The idea that the origin of turbulence can be related to the instability of laminar flow dates back to the nineteenth century. Linear stability theory has been continuously developed by numerous investigators. Mack[51] states, "One major use of the stability theory in the past has been the calculation of the initial point of instability, or critical Reynolds number, as a function of some parameter of the mean flow." Mack[52] states further, "No criterion is known by which to judge where the processes that follow linear instability, such as nonlinear wave interactions or secondary instability, start in a hypersonic boundary layer. In a hypersonic boundary layer, the velocity fluctuations are small and the density fluctuations large." Stetson[3] notes that:

> Subsonic and low supersonic boundary layers contain relatively low frequency, vorticity disturbances called first mode disturbances (Tollmien-Schlichting waves) and hypersonic boundary layers contain the first mode and Mack mode disturbances... The second mode disturbances are expected to be the dominant instability in most hypersonic

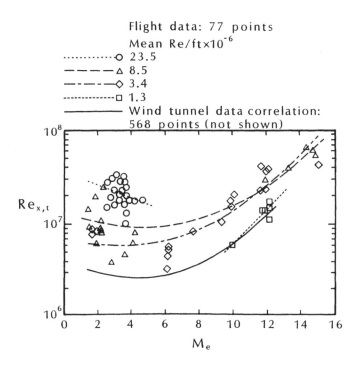

Fig. 7.28 Correlation of transition Reynolds number data on sharp cones in wind tunnels and in flight, as taken from Ref. 56.

boundary layers ... Mack's calculations indicated that the
first mode was strongly stabilized by cooling; however,
complete stabilization was not possible since more cooling was required to stabilize oblique disturbances than
two-dimensional disturbances and the higher modes were
destabilized by surface cooling. Thus, if second mode
disturbances are the major instabilities, then a cold surface would be expected to produce a smaller transition
Reynolds number than a hot surface. Mack has warned
that parameters such as pressure gradients and mass addition or removal may also affect second mode disturbances
in a different manner than first mode disturbances.

Thus, stability theory can contribute to an understanding of the
complex relationship between boundary-layer transition and surface
cooling and other parametric effects.

Researchers continue to explore promising new techniques for
computing the stability of compressible boundary layers. One such
technique, utilizing the parabolized stability equations (PSE), is described in Refs. 53, 54, and 55.

7.4.3.3 The effect of Mach number.

Wind-tunnel results at hypersonic
Mach numbers have consistently shown a large increase in the transition Reynolds number with increasing Mach number. This conclusion is supported by the stability calculations of Mack,[51] which
indicate that, at hypersonic Mach numbers, the maximum amplification rates decrease as the Mach number increases. A decrease in the
maximum amplification rate would be expected to result in larger
transition Reynolds numbers.

The variation of transition onset with local Mach number and
with local unit Reynolds number, as presented by Beckwith,[56] is
reproduced in Fig. 7.28. The flight transition Reynolds numbers are
generally higher than the correlation line for the wind-tunnel data,
presumably due to the freestream disturbances associated with wind-
tunnel flows. Note that a large increase in the transition Reynolds
number with unit Reynolds number is evident in the flight data.

7.4.3.4 The effect of surface cooling.

Mack[51] notes that:

With the Mach number and unit Reynolds number constant, the external flow is unchanged as the boundary layer
is cooled. As is well known this apparently simple experimental situation has led to an astonishing variety of
observations. A delay of transition by cooling, transition
reversal (decrease of Re_t with cooling after an initial increase), transition re-reversal (increase of Re_t after a previ-

ous decrease), unexpectedly early transition and no effect of cooling whatever are some of the phenomena which have been found in different types of experimental facilities, in different examples of the same kind of facility, and in the same facility under different flow conditions.

As noted in Sec. 7.4.3.2, stability theory can contribute to an understanding of the complex relationship between boundary-layer transition and surface cooling. Hypersonic boundary layers would be expected to have both first- and second-mode disturbances, with the second-mode disturbances as the dominant disturbances. Surface temperature is seen to have a potentially large effect on hypersonic boundary-layer transition, with wall cooling expected to be stabilizing for first-mode disturbances and destabilizing for second-mode disturbances.

Some of the difficulties in defining the effect of cooling may be due to the parameters used to correlate test conditions. "Cooling" is often represented by a temperature ratio, e.g., T_w/T_t. A given temperature ratio can be obtained in a wind-tunnel test either by changing T_w or by changing T_t. Mack[52] made stability calculations for the boundary layer on a sharp cone in a Mach 8 stream with total temperatures of 512 K (922 °R) and 728 K (1310 °R). Mack noted, "Increasing the stagnation temperature has a considerable stabilizing influencing *(sic)* at $M_e = 6.8$. The amplification rate is lowered at almost all frequencies, and unstable frequency band is narrowed by about 15%." Stetson[3] follows with the comment, "A reduction in the second mode amplification rates would be expected to increase the transition Reynolds number. If this is a consistent trend, then the larger stagnation temperatures in flight should produce larger transition Reynolds numbers than found in wind tunnels, independent of the environmental effects."

As noted earlier, the effect of surface cooling is often represented by the temperature ratio T_w/T_t. The reader should note that T_w/T_t can be reduced either by cooling the model or by increasing the total temperature of the wind tunnel. Cooling the wall to achieve a particular value of T_w/T_t may have a different impact on the transition process than heating the tunnel flow to achieve the same value of T_w/T_t. The addition of heat to the tunnel flow may result in increased disturbances to the freestream flow, e.g., vorticity fluctuations and temperature spottiness, thereby destabilizing the flow. Thus, decreasing T_w/T_t by increasing the stagnation temperature in a wind tunnel may produce lower transition Reynolds numbers. This statement is intended to remind the reader that boundary-layer transition is a complex process that depends on many interrelated parameters.

7.4.3.5 The effect of the unit Reynolds number. The data presented in Figs. 7.25 and 7.28 indicated that the transition Reynolds number increased with the unit Reynolds number. Because wind-tunnel-related disturbances affect the onset of boundary-layer transition, there has been uncertainty as to whether or not the unit Reynolds number effect would occur in flight. Because of the concern that the relation between the transition Reynolds number and the unit Reynolds number was related to disturbances peculiar to data obtained in ground-test facilities, Potter[57] examined the possible sources of disturbances in a transition study conducted in a free-flight range. Potter investigated: (1) finite angles-of-attack and oscillatory motion; (2) surface roughness under conditions of cold walls and large unit Reynolds number, i.e., thin boundary layer; (3) vibration of the model resulting from launch accelerations; and (4) nonuniform surface temperature owing to aerodynamic heating. Potter[57] noted, "The principal conclusion is that a unit Reynolds number effect was found in the free-flight range environment. None of the range-peculiar conditions investigated thus far appears to offer an explanation for this result." A siren was used to perturb the "quiet" range conditions and to investigate the effect of noise on transition measurements. While the siren produced an rms level of $\tilde{p}/p_\infty$ more than 200 times greater than the measured maximum in the "quiet" range, the ratio remained an order of magnitude less than would be expected in a typical supersonic wind tunnel. Note that the ratio $\tilde{p}/p_\infty$ is the fluctuating sound pressure amplitude divided by the freestream pressure. The data, which are reproduced in Fig. 7.29, indicate that the particular noise spectrum imposed by the siren did not produce a significant effect on boundary-layer transition.

Exercise 7.5:

A sharp cone ($\theta_c = 10$ deg.) with x_L (the total length of a conical generator) of 1.0 m flies at Mach 8 at an altitude of 30 km at zero angle-of-attack. Using the correlations of Fig. 7.28, what is x_t? Assume a perfect-gas model for the air.

Solution:

Using Table 1.1a, the freestream static pressure is:

$$p_1 = 0.011813p_{SL} = 1197 \text{ N/m}^2$$

and the freestream static temperature is:

$$T_1 = 226.509 \text{ K}$$

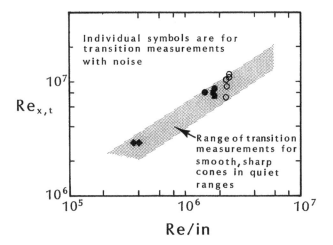

Fig. 7.29 Influence of elevated noise level on the transition Reynolds number, as taken from Ref. 57.

Although it is not required information, the velocity of the missile is:

$$U_1 = M_1 a_1 = (8.0)(301.71) = 2413.7 \text{ m/s}$$

Using the charts of Ref. 28 to obtain the pressure coefficient and the inviscid Mach number for the flow at the edge of the cone's boundary layer for $M_1 = 8$, $\theta_c = 10$ deg.:

$$C_{p_e} = 0.07$$
$$M_e = 6$$

Thus,

$$p_e = p_1(1 + \frac{\gamma}{2}M_1^2 C_{p_e}) = 4951 \text{ N/m}^2$$

To calculate the other properties at the edge of the boundary layer, we need T_e.

$$T_e = T_{te} \bigg/ \left(1 + \frac{\gamma - 1}{2}M_e^2\right)$$

Since we are assuming the perfect-gas model,

$$T_{te} = T_{t1} = T_1\left(1 + \frac{\gamma - 1}{2}M_1^2\right) = T_t$$

$$T_{te} = 226.509(13.8) = 3125.8 \text{ K}$$

Recall the statements from Chap. 2, that the viscosity of air is independent of the pressure for temperatures below 3000 K and that the dissociation of oxygen occurs at temperatures of the order of 3000 K (the specific value is pressure-dependent). Thus, the perfect-gas assumption should be acceptable for this flow. Thus,

$$T_e = 381.2 \text{ K}$$

$$\rho_e = \frac{p_e}{R T_e} = 0.04525 \frac{\text{kg}}{\text{m}^3}$$

$$\mu_e = 1.458 \times 10^{-6} \frac{T_e^{1.5}}{T_e + 110.4} = 2.207 \times 10^{-5} \frac{\text{kg}}{\text{s} \cdot \text{m}}$$

$$u_e = M_e a_e = (6.0)(391.40) = 2348.4 \text{ m/s}$$

To use Fig. 7.28, we need the unit Reynolds number,

$$Re/\text{m} = \frac{\rho_e u_e}{\mu_e} = \ = 4.814 \times 10^6/\text{m}$$

However, the unit Reynolds numbers presented in Fig. 7.28 are in English units. To convert,

$$Re/\text{ft} = (Re/\text{m})(\text{m/ft}) = (4.814 \times 10^6)(0.3048)$$
$$Re/\text{ft} = 1.467 \times 10^6$$

For the flight data, this unit Reynolds number is represented by the square symbols and the dashed-line correlation. Unfortunately, the correlation of the flight data does not extend below Mach 10. However, since the flight-data correlation for this unit Reynolds number follows the wind-tunnel correlation, we will use that correlation for $M_e = 6$. Thus,

$$Re_{x,t} = \frac{\rho_e u_e x_t}{\mu_e} \approx 3.0 \times 10^6$$

Solving for x_t,

$$x_t = \frac{Re_{x,t}}{Re/\text{m}} = \frac{3.0 \times 10^6}{4.814 \times 10^6} = 0.623 \text{ m}$$

However, because of the uncertainty in using the correlation of the wind tunnel data to extend the correlation of the flight data, additional information should be sought if the transition location has a significant impact on the design.

7.4.3.6 The effect of nose blunting.　Small nosetip bluntness has been found to stabilize the laminar boundary layer on the frustrum of a cone. Measurements of the fluctuation spectra in the boundary layer for a 7 deg. half-angle cone with a 0.15 in. nosetip radius ($R_N = 0.03R_B$) by Stetson[3] indicated that disturbances of all frequencies were damped and remained stable to $s = 121R_N$. This corresponds to a location on the conical frustrum where most of the entropy layer generated by the nosetip has been swallowed by the boundary layer. However, at the same local Reynolds number based on length (specifically, 5.1×10^6), the boundary-layer disturbances on the sharp cone had grown to sufficient amplitudes to initiate second-mode wave-breakdown (presumably, an early stage of the transition process). Thus, blunting the nose slightly (three percent for these measurements) completely stabilized the laminar boundary layer to local Reynolds numbers for which transition onset was observed on a sharp cone flow.

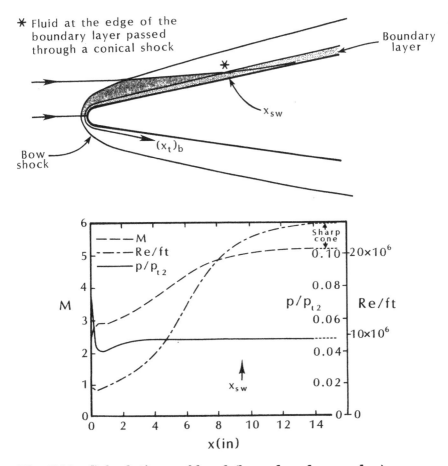

Fig. 7.30   Calculations of local (boundary-layer edge) properties for $M_\infty = 5.9$, $\theta_c = 8$ deg., $R_N = 0.02 R_B$, as taken from Ref. 58.

For a slender cone in a hypersonic stream, the variable-entropy flow produced by blunting the nose affects the properties at the edge of the boundary layer for a considerable distance. The local properties at the edge of the boundary layer for a slightly blunted $(R_N = 0.02R_B)$ cone $(\theta_c = 8$ deg.$)$ in a Mach 5.9 stream, as calculated by Stetson,[58] are reproduced in Fig. 7.30. Note that there are large reductions in the unit Reynolds number and in the Mach number relative to the sharp-cone values. Also shown is the surface distance to the entropy-layer-swallowing position (x_{sw}), as calculated for this flow by Stetson using the method of Rotta. (The parameters which affect the entropy-swallowing length are discussed in the text related to Fig. 3.19.)

Some indication of the effect of nosetip bluntness on boundary-layer transition is indicated in the data presented in Fig. 7.31, which are taken from Ref. 58. Presented in Fig. 7.31a is the ratio: $(x_t)_b/(x_t)_s$, i.e., the blunted-cone transition location divided by the corresponding sharp-cone transition location. Presented in Fig. 7.31b are the ratios of the transition Reynolds numbers and of the unit Reynolds numbers at the transition location. Thus, the dimensionless parameter represented by the ordinate of Fig. 7.31a is the product of the two parameters represented by the ordinates of Fig. 7.31b.

$$\frac{(x_t)_b}{(x_t)_s} = \frac{(Re_{x,t})_b}{(Re_{x,t})_s} \frac{(Re/ft)_s}{(Re/ft)_b} \qquad (7\text{-}21)$$

As noted by Stetson,[58] "a transition Reynolds number obtained from an experimentally determined transition location and the calculated boundary layer edge conditions at that location does not provide general information regarding the stability characteristics of the boundary layer." Furthermore, the transition Reynolds numbers, thus obtained, do not account for the history of the boundary-layer disturbances, which have been growing in a changing flowfield. Nevertheless, informative trends can be derived from these data.

Because many parameters influence boundary-layer transition, different phenomena may play dominant roles in the transition process at various regions of the entropy layer. Three regions are identified in Fig. 7.31. For region 1, the entropy layer is essentially swallowed, i.e., $(x_t)_b$ is approximately equal to x_{sw}. For region 1, nose blunting moves transition slightly aft, i.e., $(x_t)_b$ is slightly greater than $(x_t)_s$, with an increase in the transition Reynolds number, i.e., $(Re_{x,t})_b$ is greater than $(Re_{x,t})_s$.

For region 2, where $(x_t)_b$ is about $0.1x_{sw}$, the transition Reynolds number is reduced by a factor of 2 from the sharp-cone value, even though the blunted-cone transition length is much greater than the sharp-cone transition length, since the unit Reynolds number reduction dominated, e.g., $(Re/ft)_s$ is approximately 6.5 times $(Re/ft)_b$

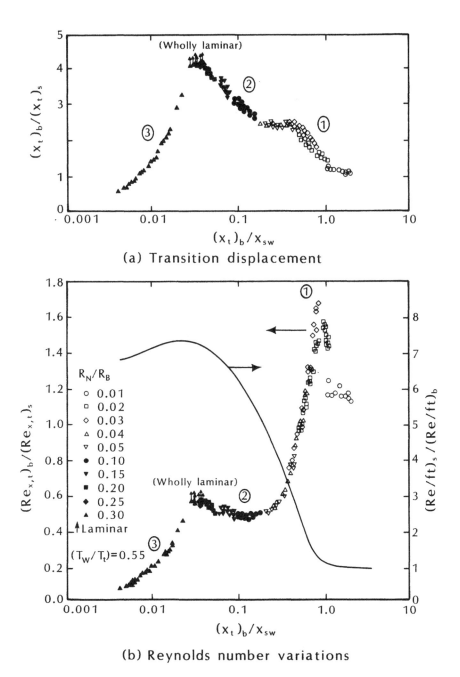

(a) Transition displacement

(b) Reynolds number variations

Fig. 7.31 The effect of nosetip bluntness on cone frustrum transition at $M_\infty = 5.9$, as taken from Ref. 58.

when $(x_t)_b = 0.1x_{sw}$. The left boundary of region 2 corresponds to the maximum rearward displacement of transition due to blunting. However, the maximum displacement is not quantified, since the boundary layer remained laminar along the entire length of the model (as indicated by the arrows and the notation "wholly laminar").

For region 3, where $(x_t)_b$ is less than $0.03x_{sw}$, the transition Reynolds numbers became smaller, even though the pressure gradients for the inviscid flow were favorable. Thus, the transition process appears to be dominated by nose region phenomena.

Stetson[3] notes:

> There is another nosetip consideration that should be included—the very low transition Reynolds numbers associated with transition *on* the nosetip and the region of the frustrum just downstream of the nosetip. Nosetip transition Reynolds numbers can be as much as two orders of magnitude less than cone frustrum transition Reynolds number. This situation requires that a separate transition criteria be applied to this portion of a configuration.

Stetson[3] also notes:

> Transition experiments have shown that there is a definite cut-off in the increased stability benefits to be derived from nosetip bluntness... It is speculated that, as the nosetip radius is increased, these nosetip region disturbances have a greater distance to grow in the vicinity of the nosetip and exceed some threshold amplitude which forces continued growth downstream.

7.4.3.7 The effect of cross-flow.   As noted by Stetson[58]:

> Intuition derived from boundary-layer transition results at zero angle of attack is not very helpful in predicting the transition trends on a sharp cone at angle of attack. The effect of angle of attack is to increase the local Reynolds number and decrease the local Mach number on the windward ray. One might logically assume that transition would then move forward on the windward ray with increases in angle of attack. On the leeward ray, the local Reynolds number decreases and the local Mach number increases. Based upon results obtained at zero angle of attack, it might be expected that transition would move rearward on the leeward ray with increases in angle of attack. In reality just the opposite of these trends occur. Transition experiments[59,60] with a sharp cone have con-

sistently found a rearward movement of transition on the
windward ray and a forward movement on the leeward
ray. These trends are also supported by boundary-layer
stability theory.

Data illustrating the effect of angle-of-attack on the transition location, as taken from Ref. 60, are reproduced in Fig. 7.32. In addition
to the trends noted above for sharp cones, data are presented for
blunted cones. For blunted cones, transition initially moved rearward
along the windward ray as the angle-of-attack increased. However,
there was a forward movement at higher angle-of-attack. Transition
does not continue to move rearward, since the effect of bluntness
diminishes with angle-of-attack.

7.4.3.8 The effect of surface roughness. Surface roughness obviously
perturbs the boundary layer. The size of the roughness elements
necessary to move the transition location significantly upstream becomes very large for supersonic Mach numbers. For roughness elements that promote boundary-layer transition, the growth of the
roughness-induced disturbances cannot be described by linear theory. In fact, the process by which roughness-induced transition occurs "bypasses" many of the steps in the classical boundary-layer
transition process. The designer of hypersonic vehicles encounters
two types of surface roughness:

1. Roughness elements intentionally placed on a wind-tunnel
   model to induce boundary-layer transition to occur at the location expected in flight, and

2. Roughness due to surface misalignments and gaps, such as occurred with the thermal protection tiles on the Space Shuttle
   Orbiter.

As a result of the combination of relatively small model scale and
of the test conditions in hypersonic wind tunnels, it may by necessary
to roughen the model surface to produce a turbulent boundary layer,
i.e., to trip the boundary layer. Potter and Whitfield[61] noted that
the ratio of the trip height k to the boundary-layer thickness at the
trip location δ_k, i.e., k/δ_k, "increases approximately as the first power
of M_e at hypersonic Mach numbers," when transition is fixed at the
tripping element. This means that, for a hypersonic flow, the trip is
usually as large as, or larger than, the undisturbed boundary layer.
Thus, trips that are large enough to trip the boundary layer
will perturb the external flow as well, producing shock waves,
shock/boundary-layer interactions, etc., such as shown in Fig. 4.20.
Sterrett et al.[62] note, "Previous work on boundary-layer transition

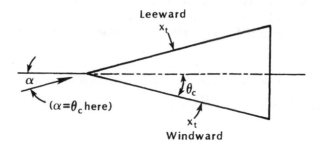

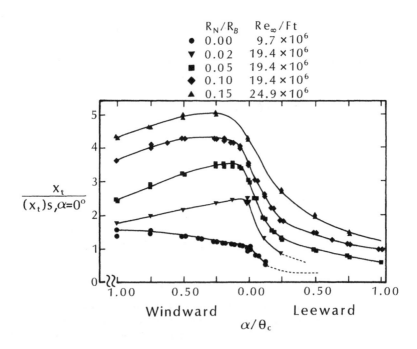

Fig. 7.32 The transition location as a function of angle-of-attack and bluntness. $M_\infty = 5.9$, $\theta_c = 8$ deg., as taken from Ref. 60.

has indicated the process by which trips (roughness) produce turbulent flow is for the trips to produce some type of vortex flow downstream of tripping element." The effects of the vortical flow can produce augmented heating, i.e., heating rates that exceed the theoretical turbulent values by as much as 20 percent, e.g., Ref. 63, and "defects" in the pitot-pressure profiles that persist for a considerable distance downstream, e.g., Ref. 64.

Just as the transition process itself depends on numerous parameters, the size, the shape, and the distribution of roughness elements required to trip a flow depends on the application. Nestler and McCauley[63] note:

> For each bluntness, the flow without trips was completely laminar for all Reynolds numbers. For the smallest nose bluntness (0.25 in.), significant roughness tripping is already occurring at $Re_\infty/\text{ft} = 1.5 \times 10^6$ while transition onset is just beginning for the larger bluntness. This is to be expected because as one approaches a sharp nose, smooth body transition occurs on the frustrum for these freestream Reynolds numbers. Thus transition onset is most easily caused for small bluntness, with increasing roughness being required to overcome the normal transition delay (for no trips) being caused by increasing bluntness.

Furthermore, relatively large trips that were located on the blunt spherical cap of the Apollo Command Module (see Fig. 6.11) tripped the boundary layer (in the locally subsonic flow). However, the heating rates indicated that the boundary layer quickly relaminarized in the presence of the strong, favorable pressure-gradient. Further, in comparing a variety of types and of configurations for boundary-layer trips, Boudreau[65] noted that "distributed roughness trips produce relatively small disturbances (shock waves) in the flowfield compared to equally effective spherical trips."

Although boundary-layer trips should be avoided where possible, there are some instances where the turbulent boundary layer must be simulated and it cannot be obtained unless trips are used. The object is to find the most effective trip that has the smallest effect on the parameters that are to be obtained during the test program, e.g., heat transfer, aerodynamic forces, etc.

Prior to the first Space Shuttle flight, there was concern as to what was the maximum surface roughness (in the form of tile misalignment) that could be tolerated before the onset of boundary-layer transition would be affected. A series of wind-tunnel test-programs were conducted using 0.0175-scale Space Shuttle Orbiter models. To study the effect of tile misalignment, selected tiles were precisely etched (or deposited, depending on the misalignment height) on the windward surface, so that they were slightly above the model surface. The misalignment tiles formed a herringbone pattern (symmetric about the plane-of-symmetry) covering the windward surface of the Orbiter model up to the tangent line of the chines from $x = 0.02L$ to $0.8L$. The raised tiles, selected randomly represented 25 percent of the tiles in the area of interest. The selected tiles were 0.267 cm (0.105 in.) square. The misalignment heights were 0.0025 cm (0.0010 in.), designated k_1, and 0.0051 cm (0.0020 in.), designated k_2. A photograph of the model showing the randomly misaligned tiles is included in Fig. 7.33.

The effects of the height of the misaligned tiles and the surface temperature on the heat-transfer distribution for the plane-of-symmetry when the Orbiter is at an angle-of-attack of 30 deg. are illustrated in the data presented in Fig. 7.33, which is taken from Ref. 66. Data are presented from tests where the freestream Mach number was 8.0, the freestream Reynolds number based on the model length was 7×10^6, and the surface temperature was either 300 K ($0.40 \, T_t$) or in the range of 96–127 K ($0.128T_t$–$0.171T_t$). The heat-transfer distributions are presented as the dimensionless ratio $h/h_{t,\text{ref}}$, which involves the experimental value of the local heat-transfer coefficient divided by the theoretical value of the heat-transfer rate to the stagnation point of a 0.00533 m (0.0175 ft) radius sphere, as calculated using the theory of Fay and Riddell.[25] For purposes of data presentation, the recovery factor has been set equal to unity.

Heat-transfer distributions for theoretical solutions of a nonsimilar laminar boundary layer using the small cross-flow axisymmetric analogue to represent the three-dimensional character of the boundary layer, which were calculated using the numerical code of Ref. 39, are included in Fig. 7.33. Required as input for the boundary-layer code were the static pressures, the entropy at the edge of the boundary layer, and the radius of the "equivalent" body-of-revolution. Because the bow shock wave is curved, the entropy varies throughout the shock layer. Thus, for this flow, which is designated as the *variable entropy* flow model, the local flow properties at the edge of the boundary layer were evaluated using the entropy and the surface-pressure distributions calculated using the numerical code of Ref. 11. The variation of the local entropy at the edge of the boundary layer, due to variation in the local boundary-layer thickness (which is Reynolds number dependent), was believed to be of second-order importance for the present application and was, therefore, neglected. The heating distributions calculated using the variable-entropy flow model are in good agreement with the laminar data, especially for $x \geq 0.2L$. Theoretical values for the nondimensionalized heat-transfer coefficient were slightly greater for the colder

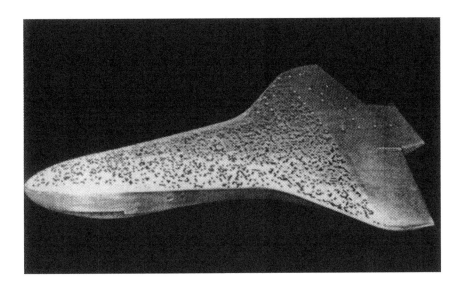

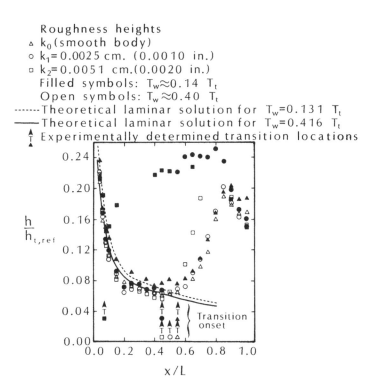

Roughness heights
△ k_0 (smooth body)
○ $k_1 = 0.0025$ cm. (0.0010 in.)
□ $k_2 = 0.0051$ cm. (0.0020 in.)
   Filled symbols: $T_w \approx 0.14\ T_t$
   Open symbols: $T_w \approx 0.40\ T_t$
------ Theoretical laminar solution for $T_w = 0.131\ T_t$
—— Theoretical laminar solution for $T_w = 0.416\ T_t$
△ Experimentally determined transition locations

Fig. 7.33   The effect of tile-misalignment height and of surface temperature on the heat-transfer distribution, $M_\infty = 8.0$, $Re_{\infty,L} = 7.0 \times 10^6$, $\alpha = 30$ deg., as taken from Ref. 66.

surface, i.e., $T_w = 0.131T_t$. The increased heating is attributed to the increased velocity gradients and viscous dissipation, which result when the boundary-layer thickness is decreased by surface cooling, i.e., as T_w/T_t decreases. Note that the dimensionless heat-transfer coefficients for the cooled wall $(T_w \approx 0.14T_t)$, as indicated by the filled symbols, are greater than those for the room-temperature surface $(T_w \approx 0.40T_t)$, as indicated by the open symbols. Thus, these data from the laminar region indicate that cooling the wall increased the dimensionless heat-transfer coefficient and are consistent with the theoretically determined effect of surface temperature. Study of additional data from these tests indicates that there was no clear correlation between the experimental values of the nondimensionalized local heating and the surface temperature. This lack of correlation is of little significance, since the theoretical solutions predict less than a 15 percent change in heat-transfer level, and the experiment was not designed to resolve changes of this magnitude.

The heat-transfer distributions were used to determine the "point" at which boundary-layer transition location occurred in the plane-of-symmetry. The experimentally determined transition location was that point at which the heat transfer deviated from the laminar distribution. The locations of transition onset for the selected tests are presented in Fig. 7.33. At the higher surface temperature, the misaligned tiles moved transition only slightly upstream relative to the smooth-body transition location (refer to the open symbols in Fig. 7.33). However, at the lower temperature, tiles which were misaligned 0.0025 cm (k_1) moved transition forward by $0.1L$, and increasing the tile misalignment height to 0.0051 cm (k_2) moved transition well forward. For lower surface temperature and for greater tile misalignment, transition occurred very near the nose, approaching a minimum transition length.

Using wind-tunnel data alone, several transition prediction methodologies were developed. These methods all had one thing in common: the transition parameter, e.g., x_t/L, $Re_{x,t}$, $(Re_\theta/M_e)_t$, used for making predictions of roughness-induced transition was correlated in terms of a departure from the same parameter correlated using smooth-body data alone. This departure was functionally related to a roughness parameter, e.g., δ^*/k or Re_k, using the wind tunnel data.

The correlation judged best suited for making pre-flight predictions was based on a departure of Re_θ/M_e from the smooth-body transition values as a function of Re_k at $x = 0.1L$, designated $Re_{k,x=0.1L}$. This was selected because the smooth-body prediction methodology was already based on Re_θ/M_e. This correlation, reproduced from Ref. 67, is shown in Fig. 7.34. The normalized, or relative, transition parameter ξ is the ratio of Re_θ/M_e at roughness-induced transition conditions to Re_θ/M_e at smooth surface conditions, i.e., $\xi = (Re_\theta/M_e)_{t,R}/(Re_\theta/M_e)_{t,S}$. $Re_{k,x=0.1L}$ is the reference roughness Reynolds number calculated at the 10 percent centerline station for the flow conditions at which boundary-layer transition occurred. Several factors were involved in selecting $x = 0.1L$ as the reference location for calculating the Re_k in the correlation of the transition parameter accounting for the effects of surface roughness and temperature. Initially, the Orbiter was expected to have relatively uniform surface roughness, which suggested that any point might be adequate. Furthermore, at the time the correlation was developed, it was relatively difficult to calculate Re_k over the complete vehicle, but Re_k could be calculated relatively easily and accurately at $x = 0.1L$. However, the most important factor was the observation that when surface roughness and temperature effects did cause transition to move forward, transition would move to this general area. In addition, this location was at the end of the favorable pressure gradient region coming from the nose. Therefore, this location, which had the largest values of Re_k for low pressure gradient areas, would better reflect the effects of surface roughness on transition for the rest of the vehicle, which also had very low pressure gradients.

Also noted on Fig. 7.34 are the values of $Re_{k,x=0.1L}$ which cause transition to move forward at various rates on the Orbiter. Surface roughness and cooling had no measurable effects on transition lo-

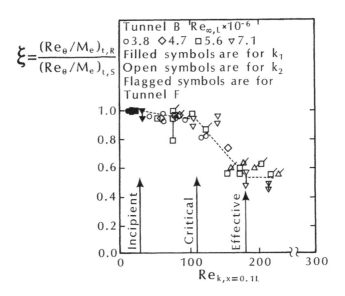

Fig. 7.34 Correlation of the relative transition parameter as a function of Re_k at $x = 0.1L$, as taken from Ref. 67.

cation for values of $Re_{k,x=0.1L}$ less than 30. This is defined as the incipient value of $Re_{k,x=0.1L}$, because it is the value that causes ξ to decrease just below unity. The effects of roughness and cooling on transition are minimal until $Re_{k,x=0.1L}$ reaches a value of 110, which is called the critical value. Above the critical value, the relative transition parameter decreases rapidly (transition moves rapidly upstream toward the nose) until $Re_{k,x=0.1L}$ reaches a value of 180. This is the minimum value of $Re_{k,x=0.1L}$ which will move transition furthest forward for the test conditions. Therefore, 180 is defined as the effective value of $Re_{k,x=0.1L}$, because the roughness elements are now serving as effective tripping devices.

The freestream entry conditions for the Orbiter Flight Test (OFT) program are depicted in Fig. 7.35 in terms of $Re_{\infty,L}$ and M_∞. The range of these parameters used for the wind-tunnel test program is also shown for reference. Identification of the conditions for transition onset and completion is also indicated for each trajectory. The time of transition onset is defined to be when transition reaches the $x = 0.99L$ station. The time of transition completion corresponds to transition reaching $x = 0.10L$, The important point is that transition travels the length of the centerline over a relatively narrow range of relatively low freestream Mach numbers. That is, transition onset occurs near Mach 10 and is essentially complete by Mach 7. These conditions occur well past peak heating and, therefore, contribute

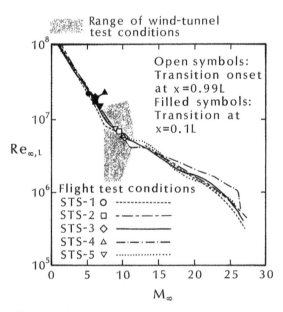

Fig. 7.35 Comparison of freestream flow parameters for flight and wind-tunnel test conditions, as taken from Ref. 68.

very little to the heat load experienced by the Orbiter during entry. Also note that, in terms of these parameters, the wind-tunnel conditions provided a reasonable coverage of the flight conditions.

Correlations of the measurements of the onset of transition along the Orbiter centerline are presented in Fig. 7.36 for the first five flights of the Space Transportation System (STS), i.e., the Space Shuttle Orbiter. Three different parameters are used to correlate transition data: $Re_{x,t}$, $(Re_\theta/M_e)_t$ and $Re_{k,t}$. The parameter $(Re_\theta/M_e)_t$ has found popular use in several design programs, e.g., the Space Shuttle Orbiter. The success of this parameter may be due (at least in part) to the Shuttle's blunt nosetip, high angle-of-attack, rough surface, and locally supersonic flow (with essentially a constant Mach number). Braslow[69] and others have shown that the roughness Reynolds number, i.e., Reynolds number based on the roughness height and the flow properties at the top of the roughness element,

$$Re_k = \frac{\rho_k u_k k}{\mu_k} \qquad (7\text{-}22)$$

successfully correlates the effect of roughness on boundary-layer transition over a range of local Mach numbers from 0 to 4. Included in Fig. 7.36 are indications of the corresponding transition data obtained in wind tunnels, as indicated by the "correlation" for smooth models and the range for rough models.

Note that, when transition occurs on the aft half of the configuration, i.e., $x \geq 0.5L$, the values for all three transition parameters based on flight data are considerably greater than the corresponding wind-tunnel values. Since the boundary layer is relatively thick on the aft half of the vehicle, the misaligned tiles are buried within the boundary layer. Roughness-induced perturbations probably do not dominate the transition process. Therefore, noise and other flow disturbances, which are unique to ground-based tests, contribute to the transition process in the wind tunnel, causing the tunnel-based parameters to be smaller than those derived from the flight data. However, when transition occurs on the first half of the model, i.e., $x < 0.5L$, the values of all three transition parameters based on flight-test data are only slightly greater than the corresponding wind-tunnel values. Since the boundary layer is relatively thin, the misaligned tiles induce a larger perturbation to the flow. Thus, if the roughness-induced perturbations dominate the transition process, differences due to tunnel noise would not be as significant.

7.4.3.9 The effect of mass injection. Gaseous injection into the boundary layer may occur due to the outgassing from an ablative heat shield. It also may be used as transpiration cooling in order to reduce the convective heat transfer to a vehicle. Transpiration cooling

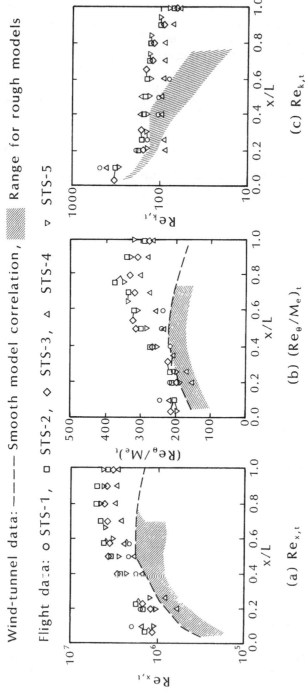

Fig. 7.36 Comparisons of local flow conditions for transition along the Orbiter windward pitch plane at flight and wind-tunnel conditions, as taken from Ref. 68.

results when a gas is injected perpendicular to the surface to reduce the heat transfer to that surface. In either case, gaseous injection into the boundary layer (including the flow rate, the distribution, and the molecular weight) has a significant effect on the boundary-layer transition. An experimental investigation of the effect of gaseous injection into the boundary layer of a sharp cone ($\theta_c = 12$ deg.) was conducted, as described in Ref. 70, in the Vought Hypervelocity Wind Tunnel (VHWT). Three gases were used as injectants: nitrogen, N_2 (MW=28), methane, CH_4 (MW=16), and freon, or monochlorodifluoromethane, $CHClF_2$ (MW=86). A sketch of the model is presented in Fig. 7.37.

Several trends have been observed in relation to transpiration cooling.

1. In the laminar region, the higher the injection rate, the lower the heating rate. That is, for a given injectant, at a given location, the heating rate decreases as the injection rate increases, if the flow remains laminar.

2. The lower the molecular weight, the more effective the gas is as a coolant. At a given location, at a given injection rate, the lighter the gas, the more reduction in heat transfer, providing the flow remains laminar. This effect is present because the effectiveness of an injectant in reducing the heating rate depends on its ability to move away from the wall. The lighter gases are able to move more quickly into these regions.

3. Lighter gases are more destabilizing to the boundary layer. As a result, premature transition may occur due to injection and turbulent heating rates might exist in a region that, without injection, would have had lower heating rates. A heavier molec-

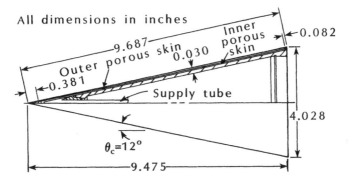

Fig. 7.37 Double-skin sharp-cone model used for mass-injection studies, as taken from Ref. 70.

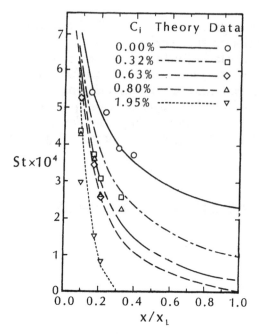

Fig. 7.38 The effect of the uniform injection of nitrogen on the heat transfer to a sharp cone, $M_\infty = 12$, $Re_\infty / \text{ft} = 6 \times 10^6$, nitrogen stream, $\theta_c = 12$ deg., as taken from Ref. 70.

ular weight gas will give a result similar to cooling the wall (i.e., increasing the density near the surface), and thus, have less effect on the transition of the boundary layer.

The effect of transpiration cooling on the heat transfer is indicated in the data and in the computations for a sharp 12 deg. cone located in a Mach 12 nitrogen stream, which are presented in Fig. 7.38. For these tests, nitrogen is injected uniformly through the porous conical surface. Thus, even though the inviscid flow meets the conditions for a similar boundary layer, the wall boundary conditions (uniform injection) do not. Heat-transfer measurements are presented only for those thermocouples in the laminar region. The fact that fewer thermocouples appear as C_i increases indicates that the onset of transition moves forward as the mass-injection rate increases. The dimensionless, total flow rate through the porous skin is:

$$C_i = \frac{\int \rho_w v_w \, dA_{\text{cone}}}{\rho_\infty U_\infty \pi R_B^2} \tag{7-23}$$

The fact that the theoretical heat transfer calculated for a laminar boundary layer for a $C_i = 0.0195$ goes to zero indicates that

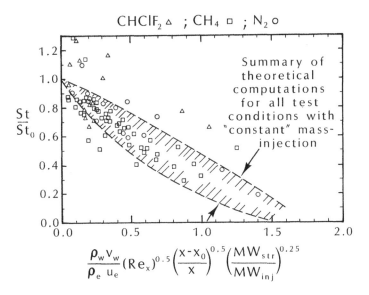

Fig. 7.39 The effect of uniform gas injection on the laminar Stanton number, as taken from Ref. 70.

boundary-layer "blow-off" would have occurred had it remained laminar. However, the injected gases promoted transition and the turbulent boundary layer remained attached. The agreement between the measured Stanton numbers and the calculated values is best for no injection and for the highest injection rate. A possible source of the discrepancy at the lower injection rates may be due to possible differences between the calibrated injection distribution and the actual injection distribution which existed during the test. (Only the total flow rate, not the exact distribution, was measured during a run.)

The effect of mass injection on the ratio of the Stanton number with injection to the Stanton number without injection, i.e., St/St_0, is presented for laminar boundary layers in Fig. 7.39 and for turbulent boundary layers in Fig. 7.40. The laminar data are presented as a function of:

$$\frac{\rho_w v_w}{\rho_e u_e}\,[Re_x]^{0.5}\left(\frac{x-x_0}{x}\right)^{0.5}\left(\frac{MW_{\text{str}}}{MW_{\text{inj}}}\right)^{0.25}$$

where x_0 is the length of the nonporous tip (which could not be avoided in the model construction) and x is distance from the apex along a conical generator. An increase in $(\rho_w v_w/\rho_e u_e)$ directly reflects an increase in the local injection. The factors $[Re_x]^{0.5}$ and

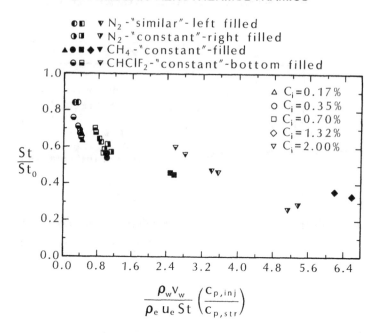

Fig. 7.40   The effect of gas injection on the turbulent Stanton number, as taken from Ref. 70.

$[(x - x_0)/x]^{0.5}$ increase as one proceeds downstream, i.e., as the boundary layer thickens for a particular test condition.

The reduction in heat transfer for turbulent boundary layers is presented as a function of:

$$\frac{\rho_w v_w}{\rho_e u_e St} \left(\frac{c_{p,\text{inj}}}{c_{p,\text{str}}} \right)$$

The use of the specific-heat ratio in the correlation parameter for turbulent flow was suggested by Hearne et al.[71] Data are presented both for uniform mass injection and for similar mass injection, i.e., the mass injection is proportional to $x^{-0.5}$.

The reduction in the transition Reynolds number for the tests with constant injection is presented in Fig. 7.41 as a function of:

$$F \left(\frac{MW_{\text{str}}}{MW_{\text{inj}}} \right)^{0.25}$$

where:

$$F = \frac{\int_{x_0}^{x_t} \rho_w v_w \, dA_{\text{cone}}}{\rho_\infty U_\infty A_{b,t}} \qquad (7\text{-}24)$$

where the numerator represents the gas injected through the surface of the cone up to the transition location. Since $A_{b,t}$ is the circular area of the cone at the *point* of transition, the denominator represents the freestream mass flow rate across an area equal to the cross section of the cone at the transition location. The parameter of Eq. (7-24) was suggested by Marvin and Akin.[72] Also included in Fig. 7.41 are the correlation line:

$$\frac{Re_{x,t}}{Re_{x,t,o}} = 1 - 0.25 \left(\frac{MW_{str}}{MW_{inj}} \right)^{0.25} F \qquad (7\text{-}25)$$

and the data of Ref. 72, which also were for a "constant" injection distribution.

7.4.4 Turbulent Boundary Layers

In this section, we will consider flows where transition has occurred and the boundary layer is fully turbulent. A turbulent boundary layer is one in which irregular fluctuations (mixing or eddying motions) are superimposed on the mean flow. In describing a turbulent flow, the local value of any dependent variable can be expressed as

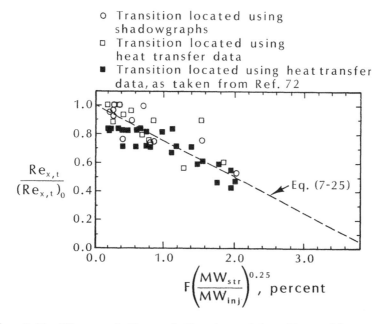

Fig. 7.41 The variation of the transition Reynolds number with the correlation parameter of Ref. 72 (for uniform mass-injection), as taken from Ref. 70.

the sum of a time-averaged, mean value plus an instantaneous fluctuating component; i.e.,

$$f = \bar{f} + f' \qquad (7\text{-}26)$$

where the time-averaged mean value of any dependent variable is defined as:

$$\bar{f} = \frac{1}{\Delta t} \int_{t_0}^{t_0 + \Delta t} f \, dt \qquad (7\text{-}27)$$

where the integration interval Δt is much longer than any significant period of fluctuation. If the mean value of $\bar{f}$ is itself independent of time, the flow is considered a "steady" turbulent flow.

The velocity fluctuations occur in the direction of the mean flow and at right angles to it, and they affect macroscopic lumps of fluid. Measurements of the fluctuating velocity components in a supersonic, turbulent boundary layer[73] show that u' is greatest and that v' and w' are essentially equal, except near the wall. Thus, the velocity fluctuations are definitely three-dimensional even for the most "two-dimensional" of turbulent boundary layers. The fluctuations of the streamwise velocity component across a turbulent boundary layer, as calculated by Harvey and Bushnell[74] using the data of Wallace,[75] are reproduced in Fig. 7.42. It should be noted that, although the velocity fluctuations may be only several percent of the local streamwise values, they have a decisive effect on the overall motion. The size of these macroscopic lumps determines the scale of turbulence.

For many applications, it is assumed that time-averaged values of a product involving the density fluctuation and another fluctuating parameter times a mean value will be assumed to be significantly smaller than a product of two (other) fluctuating parameters times the mean value of the density. However, significant fluctuations in

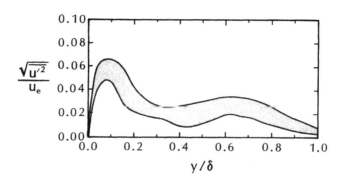

Fig. 7.42 Fluctuations of the streamwise velocity component,
$M_e = 8.25 - 8.82$, $\delta = 5.2 - 5.9$ in., $Re_\delta = 5.37 - 10.75 \times 10^4$, **as taken from Ref. 74.**

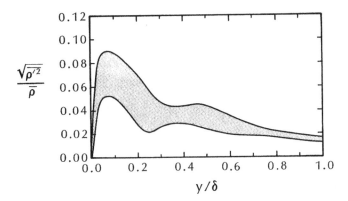

Fig. 7.43   Intensity of the density fluctuations, M_e = 8.25 – 8.82, δ = 5.2 – 5.9 in., Re_δ = 5.37 – 10.75 × 10⁴, h_w = 0.07H_e – 0.16 H_e, as taken from Ref. 74.

the density can occur when the local Mach number is very large. Using the data of Wallace,[75] Harvey and Bushnell[74] calculated the density fluctuations for a turbulent boundary layer where the edge Mach number exceeded 8. As shown in Fig. 7.43, the density fluctuations exceed 8 percent of the average value of the density when y is approximately 0.1δ.

Exercise 7.6:

Using the time-averaging technique, derive the x-component of the momentum equation for a "steady," "two-dimensional" turbulent boundary layer.

Solution:

Using index notation, the instantaneous momentum equation for the i-direction is given by Eq. (2-13):

$$\frac{\partial}{\partial t}[\rho u_i] + \frac{\partial}{\partial x_j}[\rho u_i u_j] = -\frac{\partial p}{\partial x_i}$$

$$+ \frac{\partial}{\partial x_j}\left[\mu\left(\frac{\partial u_i}{\partial x_j} + \frac{\partial u_j}{\partial x_i}\right) - \frac{2}{3}\mu\frac{\partial u_k}{\partial x_k}\delta_{ij}\right]$$

Taking $i = 1$ and summing over j, for a steady two-dimensional flow, Eq. (2-13) becomes:

$$\frac{\partial}{\partial x}(\rho u^2) + \frac{\partial}{\partial y}(\rho uv) = -\frac{\partial p}{\partial x}$$

$$+ \frac{\partial}{\partial x}\left[\mu\left(2\frac{\partial u}{\partial x}\right) - \frac{2}{3}\mu\left(\frac{\partial u}{\partial x} + \frac{\partial v}{\partial y}\right)\right] + \frac{\partial}{\partial y}\left[\mu\left(\frac{\partial u}{\partial y} + \frac{\partial v}{\partial x}\right)\right]$$

Using Eq. (7-26) to describe each variable as the sum of a time-averaged mean value plus an instantaneous fluctuating component,

$$\frac{\partial}{\partial x}\left[(\bar{\rho}+\rho')\,(\bar{u}+u')^2\right] + \frac{\partial}{\partial y}\left[(\bar{\rho}+\rho')\,(\bar{u}+u')\,(\bar{v}+v')\right] =$$

$$= -\frac{\partial(\bar{p}+p')}{\partial x} - \frac{2}{3}\frac{\partial}{\partial x}\left[(\bar{\mu}+\mu')\left(\frac{\partial(\bar{u}+u')}{\partial x} + \frac{\partial(\bar{v}+v')}{\partial y}\right)\right]$$

$$+ 2\frac{\partial}{\partial x}\left[(\bar{\mu}+\mu')\,\frac{\partial(\bar{u}+u')}{\partial x}\right]$$

$$+ \frac{\partial}{\partial y}\left[(\bar{\mu}+\mu')\left(\frac{\partial(\bar{u}+u')}{\partial y} + \frac{\partial(\bar{v}+v')}{\partial x}\right)\right] \qquad (7\text{-}28)$$

The next steps are:

1. Expand the products;

2. Take the time average of each term using the rules:

$$\bar{\bar{f}} = \bar{f}; \quad \overline{f'} = 0; \quad \overline{f'\bar{g}} = 0; \quad \overline{\bar{f}\bar{g}} = \bar{f}\bar{g};$$
$$\overline{f+g} = \bar{f}+\bar{g}; \quad \overline{fg} = \bar{f}\bar{g} + \overline{f'g'}; \quad \text{and}$$

3. Neglect terms involving triple products of fluctuating parameters, i.e., higher-order terms such as $\overline{\rho'u'v'}$.

Thus, Eq. (7-28) becomes:

$$\frac{\partial}{\partial x}\left[\bar{\rho}\bar{u}^2 + \bar{\rho}\overline{u'^2} + 2\bar{u}\overline{\rho'u'}\right] + \frac{\partial}{\partial y}\left[\bar{\rho}\bar{u}\bar{v} + \bar{\rho}\overline{u'v'} + \bar{v}\overline{\rho'u'} + \bar{u}\overline{\rho'v'}\right]$$

$$= -\frac{\partial\bar{p}}{\partial x} - \frac{2}{3}\frac{\partial}{\partial x}\left[\bar{\mu}\frac{\partial\bar{u}}{\partial x} + \overline{\mu'\frac{\partial u'}{\partial x}} + \bar{\mu}\frac{\partial\bar{v}}{\partial y} + \overline{\mu'\frac{\partial v'}{\partial y}}\right]$$

$$+ 2\frac{\partial}{\partial x}\left[\bar{\mu}\frac{\partial\bar{u}}{\partial x} + \overline{\mu'\frac{\partial u'}{\partial x}}\right] + \frac{\partial}{\partial y}\left[\bar{\mu}\frac{\partial\bar{u}}{\partial y} + \overline{\mu'\frac{\partial u'}{\partial y}} + \bar{\mu}\frac{\partial\bar{v}}{\partial x} + \overline{\mu'\frac{\partial v'}{\partial x}}\right] \qquad (7\text{-}29)$$

Not all of the terms in Eq. (7-29) are of equal importance. Many of them are negligible in general applications; others are negligible in specific applications. Thus, at this point, one usually simplifies the equation to be solved by eliminating the terms of lesser importance. Using order-of-magnitude approximations, the x-momentum equation for a compressible, turbulent boundary layer as developed by Bertin et al.[76] is:

$$\bar{\rho}\bar{u}\frac{\partial \bar{u}}{\partial x} + \left(\bar{\rho}\bar{v} + \overline{\rho'v'}\right)\frac{\partial \bar{u}}{\partial y} + \frac{\partial}{\partial y}\left[\overline{\bar{\rho}u'v'}\right]$$

$$= -\frac{\partial \bar{p}}{\partial x} + \frac{\partial}{\partial y}\left[\bar{\mu}\frac{\partial \bar{u}}{\partial y}\right] \qquad (7\text{-}30)$$

The term $(\overline{\bar{\rho}u'v'})$ represents the turbulent transport of momentum and is known as the turbulent inertia tensor. Thus, whereas momentum transport occurs on a microscopic (or molecular) scale in a laminar boundary layer, it occurs on a macroscopic scale in a turbulent boundary layer.

The energy equation derived in a similar manner by Bertin et al.[76] is:

$$\bar{\rho}\bar{u}\frac{\partial \bar{h}}{\partial x} + \left(\bar{\rho}\bar{v} + \overline{\rho'v'}\right)\frac{\partial \bar{h}}{\partial y} + \frac{\partial}{\partial y}\left[\overline{\bar{\rho}v'h'}\right]$$

$$= \frac{\partial}{\partial y}\left[\bar{k}\frac{\partial \bar{T}}{\partial y}\right] + \bar{u}\frac{\partial \bar{p}}{\partial x} + \bar{\mu}\left(\frac{\partial \bar{u}}{\partial y}\right)^2 - \overline{\bar{\rho}u'v'}\frac{\partial \bar{u}}{\partial y} \quad (7\text{-}31)$$

For most applications where the boundary layer is thin, the pressure variation normal to the surface is negligible. Thus, the partial derivative of the pressure in the streamwise direction which appears in Eqs. (7-30) and (7-31) is replaced by an ordinary derivative. A simplified Euler equation can then be used to relate the pressure gradient to the acceleration at the edge of the boundary layer; i.e.,

$$\frac{\partial \bar{p}}{\partial x} = \frac{d\bar{p}}{dx} = \frac{dp_e}{dx} = -\rho_e u_e \frac{du_e}{dx} \qquad (7\text{-}32)$$

This assumption breaks down for turbulent boundary layers at very high Mach numbers. Bushnell, Cary, and Harris[77] cite data for which the wall pressure is significantly greater than the edge value for turbulent boundary layers where the edge Mach number is approximately 20. Shang[78] computed the turbulent boundary layer for $M_e = 9.37$, $T_w = 0.385T_{te}$, and $Re_\theta = 3.68 \times 10^4$. Shang notes, "The pressure decreases very rapidly from the wall and reaches the minimum value near the outer edge of the sublayer, then approaches the freestream value asymptotically. The pressure variation across the turbulent boundary layer is directly proportional to the turbulent shear stress."

7.4.4.1 Modeling concepts.

The determination of the turbulent inertia term $(\bar{\rho}\overline{u'v'})$, which appears in Eqs. (7-30) and (7-31), and the turbulent convection term $(\bar{\rho}\overline{v'h'})$, which appears in Eq. (7-31), is the critical problem in the analysis of turbulent shear flows. However, these "new variables" can be defined only through an understanding of the detailed turbulent structure. These terms are related not only to physical fluid properties but also to local flow conditions. Since there are no further physical laws available to evaluate these terms, empirical correlations are introduced to model them. There is a hierarchy of techniques for "closure" which have been developed from nonrigorous postulates.

Neumann[79] notes, "Turbulence models employed in computational schemes to specify the character of turbulent flows are just that...models, non-physical ways of describing the character of the physical situation of turbulence. These models are the result of generalizing and applying fundamental experimental observations; they are not governed by the physical principles of turbulence and they are not unique."

In accordance with Boussinesq's concept, the local turbulent shear stress can be represented by the product of an effective turbulent viscosity coefficient (or eddy viscosity) and the local mean-velocity gradient. Thus,

$$\tau_t = -\bar{\rho}\,\overline{u'v'} = \mu_t\,\frac{\partial \bar{u}}{\partial y} \qquad (7\text{-}33)$$

Similarly, the turbulent convective flux can be related to the local mean-temperature gradient by introducing an effective turbulent thermal conductivity:

$$\dot{q}_t = -\bar{\rho}\,\overline{v'h'} = k_t\,\frac{\partial \bar{T}}{\partial y} = \frac{\mu_t}{Pr_t}\,\frac{\partial \bar{h}}{\partial y} \qquad (7\text{-}34)$$

The turbulent Prandtl number, Pr_t, incorporates both the effective turbulent viscosity (μ_t) and the turbulent thermal conductivity (k_t).

Based on the Kolmogorov-Prandtl model of turbulence, the turbulent transport is defined in terms of a mixing length which appears in the eddy viscosity model. For the zero-equation models, the mixing length is expressed in terms of the mean flow. Zero-equation models are well adapted to simple attached flows where local turbulence equilibrium exists, i.e., the local production of turbulence is balanced by the local dissipation of turbulence. However, there are many complex flows, such as those involving shock/boundary-layer interactions, where one-equation models or two-equation models are used.

The effect of compressibility for a turbulent boundary layer past an adiabatic, flat plate, as taken from Ref. 80, is reproduced in Fig. 7.44. Computations using a zero-equation model and a two-equation model agree well with the correlation of Van Driest,[81] as represented by the solid line. Additional computations showed that the use of mass-averaging or of time-averaging had no significant effect on the predicted results. To account for the effect of compressibility in the zero-equation models, the local mean density is used in the definition of the eddy viscosity.

Cebeci and Bradshaw[82] state:

> Although the transport equations for Reynolds stress are not soluble as they stand, they have long been used as a guide to turbulence processes. In particular, they tell us that Reynolds stresses are not directly linked to the

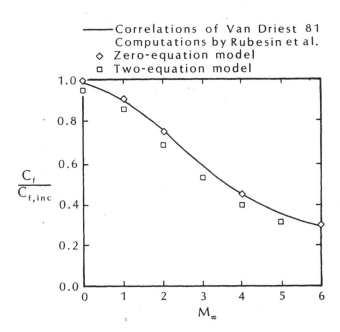

Fig. 7.44   The effect of compressibility on turbulent skin friction on a flat-plate, adiabatic wall, $Re_L = 10^7$, as taken from Ref. 80.

local mean-velocity gradient as it is assumed in the eddy-viscosity and mixing length formulas. The Reynolds stress at a given point depends on the whole history of turbulence passing that point. The Reynolds-stress transport equation is a DE for Reynolds stress, whereas the eddy-viscosity formula is an algebraic equation. However, in self-similar to slowly changing flows, the history of the turbulence is in some sense uniform, so that history can be related to local conditions.

In these cases, the eddy viscosity is simply behaved. For more demanding flows, empirical correlations are sought for the unknown turbulence quantities in the Reynolds-stress transport equations. This process is called *modeling*.

As noted in Eq. (7-34), the turbulent heat flux $(\bar{\rho}\,\overline{v'h'})$ is modeled using a turbulent Prandtl number Pr_t, by which the turbulent thermal conductivity is related to the turbulent viscosity as:

$$Pr_t = \frac{\mu_t c_p}{k_t} \tag{7-35}$$

Numerous investigators assume that the turbulent Prandtl number is constant across the boundary layer, e.g., Hodge and Adams,[83] who use a value of 0.90.

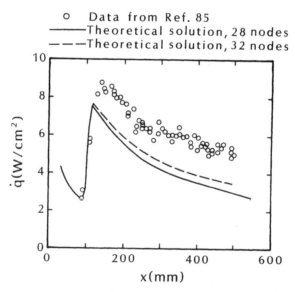

Fig. 7.45   Heat-transfer distribution for hypersonic flow past a flat plate, $M_\infty = 9.26$, as taken from Ref. 86.

7.4.4.2 Computed turbulent boundary layers. Solutions of compressible turbulent boundary layers have been obtained using the computer code described in Refs. 39 and 76. The governing differential equations of motion have been written representing the flow properties as the sum of a time-averaged mean and an instantaneous fluctuation. Then, the equations for the conservation of mass, the conservation of linear momentum, Eq. (7-30), and the conservation of energy, Eq. (7-31), were derived using Reynolds averaging. The closure procedure was similar to that described by Hodge and Adams.[83] The turbulence length scale was calculated by solving the integral form of the kinetic-energy-of-turbulence (IKET) equation. The IKET equation is obtained by multiplying the ith component of the Navier-Stokes momentum equation by the ith component of the velocity, then summing the three averaged equations. The use of the IKET equation allows the "history" of the turbulent state to be considered and the "mixing-length correlation" to reflect the development of the turbulent boundary layer.

As given by Wolfshtein,[84] the turbulent viscosity is given by:

$$\mu_t = C_\mu \, \bar{\rho} \, k^{0.5} \ell_\mu \qquad (7\text{-}36)$$

where C_μ is an empirical constant, ℓ_μ is a length scale, and k is the turbulent kinetic energy:

$$k = \frac{1}{2}\left[\overline{(u')^2} + \overline{(v')^2} + \overline{(w')^2}\right]$$

Thus, the formulation assumes that the local state of turbulence is dependent upon a length scale and the local kinetic energy contained in the turbulent velocity fluctuations. The essential advantage of this model over a more simple mixing length hypothesis is its applicability when local turbulence equilibrium does not exist, since it does not require that the local production of turbulence be balanced by the local dissipation of turbulence.

Theoretical solutions were computed for a hypersonic, turbulent boundary layer in order to compare with measurements presented by Bartlett et al.[85] The data are for hypersonic flow past a flat plate with $M_\infty = 9.26$, $Re_\infty/m = 5.71 \times 10^7/m$, $T_{te} = 1064$ K, and $T_w = 298$ K. Using the experimental heat-transfer distribution, which will be discussed subsequently, the onset of transition was assumed to be at an x of 89.6 mm. Theoretical solutions were computed using either 28 or 32 nodes in the y-direction. There were 900 stations in the x-direction, with step sizes of 3.048 mm in the laminar region, 0.152 mm in the transitional region, and 0.671 mm in the fully turbulent region. The step sizes were selected based on user experience and are not necessarily the largest or the most efficient values.

The theoretical heating rates are compared with the experimental values in Fig. 7.45. The fact that the heating-rate measurements from the thin-film gage located nearest the leading edge agree with the theoretical laminar values substantiates the assumption that $x_{tr,i} = 89.6$ mm. The theoretical heating rates closely follow the rapid streamwise increase exhibited by the heat-transfer measurements in the transitional region. The maximum value of the theoretical heating rate occurs at $x = 118.1$ mm. The relatively short transition length, i.e., $x_{tr,f} = 1.32 x_{tr,i}$, as determined from the computed solution, is supported by the data. The number of nodes had a negligible effect on the computed heating rates for the laminar and for the transitional regions, i.e., $x < 118.1$ mm.

Although the theoretical heating rates for the turbulent region parallel the measured values, they are 10 to 33 percent below the experimental values. The present author cannot comment on the accuracy of the data, but will identify a possible source of error in the computed values. Note that the theoretical heating rates in the turbulent region did depend on the number of nodes. This dependence on the number of nodes apparently reflects the accuracy of the flowfield solution for the laminar sublayer. The values of y^+ at the first nodal point are presented in Fig. 7.46 as a function of x, where

$$y^+ = \frac{y}{\nu_w} \left(\frac{\tau_w}{\rho_w} \right)^{0.5} \tag{7-37}$$

The values of y^+ at the first nodal point exceed 10 for this high Reynolds number flow for $x > 300$ mm. Calculations and data pre-

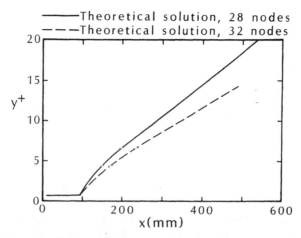

Fig. 7.46 **Streamwise distribution of y^+ at the first nodal point for the $M_\infty = 9.26$ boundary layer, as taken from Ref. 86.**

sented by Stalmach[87] indicate upper bound values for y^+ of approximately 15 for the laminar sublayer for supersonic flows past flat plates. Thus, the first nodal point is near the edge of the laminar sublayer, affecting the resolution of the theoretical flowfield near the wall. Furthermore, the y^+ coordinate of the first nodal point of the 28 node solution is significantly greater than that for the 32 node solution. Thus, it is not surprising that theoretical heating rates for the 32 node solution are closer to the measured values. Note further, that the y^+-coordinate of the first nodal point of the 32 node solution is 2.2 at the end of the transitional region, i.e., $x = 118.1$ mm, and is 5.3 at an x of 200 mm. Marvin and Coakley[88] note, "The procedure of applying no-slip boundary conditions is frequently referred to as the integration-to-the-wall procedure. To be applicable, the numerical mesh spacing normal to the wall must be chosen such that the value of y^+ at the first point off the wall is of the order of unity, placing it well within the viscous sublayer." The condition of Neumann[79] is even more restrictive, "Substantial differences in heating are noted and the trend is apparent. Improving the quality of the solution requires that the y^+ parameter be everywhere less than 1 and, ideally, closer to 0.1."

The analytical model of the boundary-layer transition process should include the precursor effect. This effect is characterized by the existence of appreciable length scales in the outer region of compressible boundary layers well upstream of the location at which transition is indicated by surface measurables, such as skin friction or heat transfer. Thus, the outer portion of the mean-flow profiles may already be distorted by turbulence effects at the conventional transition location. This precursor effect is modeled in the computations, as shown in Fig. 7.47, in which the profiles of the turbulent

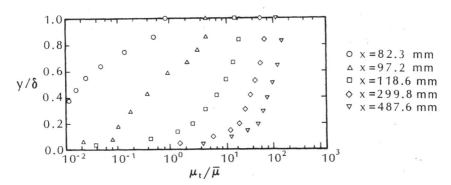

Fig. 7.47 Streamwise development of the turbulent viscosity for the $M_\infty = 9.26$ boundary layer, as taken from Ref. 86.

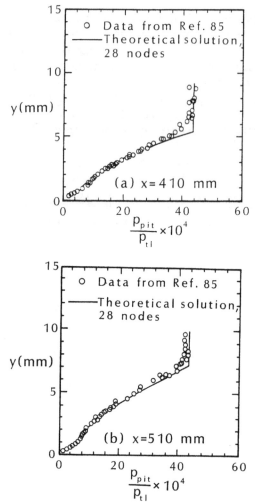

Fig. 7.48 Comparison of the theoretical and of the experimental pitot-pressure profiles for the $M_\infty = 9.26$ boundary layer, as taken from Ref. 86.

viscosity divided by the mean value of the molecular viscosity are presented for several streamwise locations. At the first station, which is upstream of the onset of transition (as determined from the heat-transfer data of Fig. 7.45), a sharp "spike" in the turbulent viscosity occurs near the outer edge of the boundary layer. Thus, the effects of turbulence are evident in the outer region of the boundary layer, whereas flow near the surface is laminar.

Using the theoretical values of the Mach number and the assumption that the static pressure is constant across the boundary layer, the pitot-pressure distributions across the boundary layer were calculated. The theoretical pitot-pressure profiles, thus computed, are compared in Fig. 7.48 with the pitot-pressure measurements reported by Bartlett et al.[85] The correlation between the theoretical values and the experimental values of the pitot-pressure ratio is considered very good from the wall through the middle portion of the boundary layer. The differences are greatest near the edge of the boundary layer. The y-coordinate at which the computed value of u was $0.995u_e$, i.e., δ, was below the experimental value of δ. Furthermore, significant differences exist between the theoretical and the experimental values of the pitot-pressure ratio at the boundary-layer edge at some stations. Bartlett et al.[85] noted that there were run-to-run variations in the data, and the pitot pressures measured at the boundary-layer edge exhibited a scatter for which no explanation could be presented.

7.4.4.3 Concluding remarks.   Eddy viscosity models constitute the simplest but most widely used class of turbulence models. However, since such "models" employ a great deal of empiricism and do not represent the mixture of random and of deterministic elements of the turbulent structure, a model might be successfully applied for one flow but not another. Numerical exercises should be undertaken to calibrate the code over the range of conditions for which the code is to be applied. It is difficult to sort out discrepancies between computation and experiment that arise separately from modeling and computational procedures and from numerical errors. Although turbulence modeling is often cited as the main deficiency in computations, the algorithm, the grid, the convergence criteria, and the experiment itself may be the source of the discrepancies.

7.5   THE EFFECTS OF SURFACE CATALYCITY

As noted in Chap. 2 (refer to Sec. 2.1.2), the flow over the Space Shuttle is in chemical nonequilibrium during that portion of the flight when significant aerodynamic heating occurs. Here, *nonequilibrium* is used synonymously with finite-rate chemical reactions, as opposed to *equilibrium* in which the reaction times are assumed to be very short in comparison with flow times. Calculations have shown that both dissociation nonequilibrium (i.e., there are not sufficient collisions for equilibrium chemistry, as the gas passes through the bow shock wave) and recombination nonequilibrium (i.e., there are not sufficient collisions for the atoms to recombine, achieving equilibrium composition, as the gas expands around the vehicle or approaches the cool surface) exist during the high heating phase of the Shuttle re-entry.

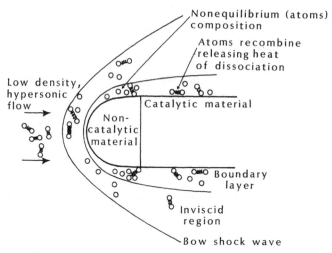

Fig. 7.49 Sketch of nonequilibrium flow around a sphere/ cylinder with a noncatalytic/catalytic surface.

The nonequilibrium character of the flow will have some effect on the pressure distribution (see Section 6.7). However, it will have a much greater impact on the heat transfer. During flight through the atmosphere, the dissociated atoms tend to recombine near the relatively cool surface in a rate-controlled process. If the surface is catalytic, the reaction goes to completion, and the released heat of dissociation increases the heating to the surface (so that it is essentially equal to the equilibrium value). If the surface is noncatalytic, the recombination does not take place, and the heating to the surface is relatively low.

7.5.1 Introductory Remarks

For hypervelocity flight at relatively high altitudes, where the density is low, nonequilibrium effects can be significant. Rakich et al.[89] note that "The high heating portion of the trajectory lies well below the viscous interaction regime, and our numerical calculations indicate that above an altitude of 40 km and at velocities greater than about 4 km/sec some nonequilibrium effects can occur." See Fig. 2.3. Scott[10] states that "Nonequilibrium effects on the heating are not significant below about 65 km even though the flow may not be in equilibrium." Specific differences between the lower-limit altitudes cited in these two quotations may be due in part to the exact wording, e.g., "some nonequilibrium effects" as compared with significant effects on heating even though the flow may not be in equilibrium. Nevertheless, there is no simple altitude criterion below which nonequilibrium effects can be neglected. In fact, the importance of nonequilibrium effects depends on many parameters, including vehicle size, since the size affects the time required for the air particles to flow around the vehicle. The reader should understand, as often has been the case in discussions throughout this volume, that the significance of specific phenomena for a given application often depend on the analyst's experience and the physical models inherent in the tools used by the analyst.

As illustrated in Fig. 7.49, air molecules passing through the bow shock wave dissociate relatively quickly (although not at the rate required for equilibrium) and then slowly recombine as they move downstream through a decreasing pressure field. At the low densities associated with the flow depicted in Fig. 7.49, the recombination of atoms at a catalytic surface can significantly affect the heat transfer. Thus, the heat transfer depends on the catalytic nature of the surface, i.e., k_w, the surface reaction-rate parameter (see Fig. 7.4). However, when a dissociated gas that has been moving past a noncatalytic surface encounters a surface that has a high surface reaction-rate parameter, the heat-transfer rate will jump abruptly, exceeding the heating that would have existed if k_w had been constant for the entire surface. This is depicted in the sketch of Fig. 7.49. Thus, the heat flux to a catalytic surface depends on the character of the upstream surface, as well as on the local catalytic efficiency.

Experimental heat-transfer rates obtained by Sheldahl and Winkler[90] in an arc-heated $M_\infty = 5.6$ stream of nitrogen illustrate the effect of variations in the surface catalytic efficiency. These data are reproduced in Fig. 7.50. For a copper nose followed by a copper afterbody, i.e., a Cu/Cu configuration, the heat-transfer distribution is continuous and corresponds to the equilibrium distribution (since copper is known to be catalytic). For an SiO nose and afterbody, the heat-transfer distribution is continuous, but at a lower level than the Cu/Cu body, since SiO is relatively noncatalytic. When the flow goes from the SiO nose to the Cu afterbody, the heating increases discontinuously to levels exceeding those for the Cu/Cu level. Apparently, the energy of dissociation remained in the flow as it passed over the noncatalytic SiO nose until it reached the Cu afterbody. Then it was released, producing an "overshoot" in heating. The afterbody heating was lowest for the Cu/SiO configuration, being slightly lower than that for the SiO/SiO configuration. Apparently, the catalytic recombination at the surface of the Cu nose depleted the atom concentration of the flow over the SiO afterbody. Thus, it is important to know the history of the surface recombination-rate parameter, since the upstream surface can significantly affect the atom concentration within the boundary layer.

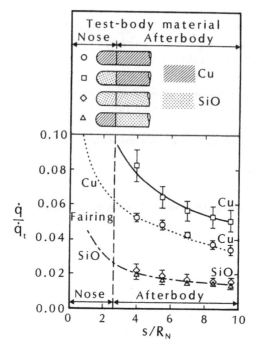

Fig. 7.50 Effect of surface-catalytic discontinuities on measured heating of a hemisphere / cylinder, using data taken from Ref. 90.

7.5.2 Shuttle Catalytic Surface Effects (CSE) Experiment

The tiles used in the thermal-protection system (TPS) for the windward surface of the Space Shuttle Orbiter are made up of a rigidized fibrous ceramic with a reaction-cured glass (RCG) coating, which constitutes a high-temperature reusable insulation (HRSI). Using information from arc-tunnel tests and from flowfield computations, the Orbiter designers expected significant reductions in the heat transfer to the noncatalytic surface exposed to the nonequilibrium flow. However, "because of uncertainties about the composition and the contamination of the arc-heated airstreams,"[89] designers tended to rely on the more conservative, fully catalytic analysis.

Because the thermal conductivity of the Shuttle TPS tiles was so low, the heat flux could be determined directly from the temperature histories provided by a thermocouple located just below the surface coating. Using Eq. (7-6),

$$\dot{q} = 1.06 \, \varepsilon \, \sigma \, T_w^4$$

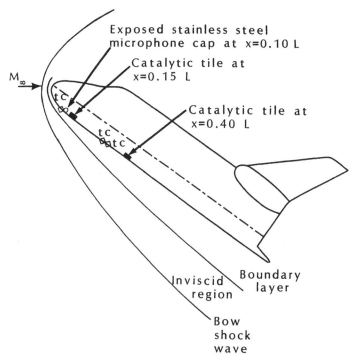

Fig. 7.51 Sketch of Shuttle Orbiter illustrating the locations of selected thermocouple, an exposed microphone cap, and the catalytic-coated tiles for STS-2.

Based on an extensive data base, the emissivity is 0.89 ± 0.02.[89] As noted by Rakich et al.[89]:

> Since a 10–20% variation in the measured surface temperature might be misinterpreted as a variation of ambient density, a three-dimensional effect, or some other uncertainty, an experiment is needed that especially focuses on the catalytic efficiency of the reaction-cured glass (RCG) coating of the Space Shuttle heat protection tiles.

> The catalytic surface effects experiment, which is part of the OEX program, utilizes the base-line flight instrumentation to measure catalytic efficiency of the Shuttle TPS, without affecting flight operations. An overcoat, with high catalytic efficiency, is sprayed onto selected tiles (test tiles) and side-by-side surface temperature measurements are made on the test and baseline tiles. If the temperature of the test tile is higher than the surroundings, then the standard tiles are relatively noncatalytic; if the test tile is

at the same temperature as the surroundings, then they have the same catalytic efficiency. Thus, one can get an immediate and positive indication of the relative catalytic efficiency.

The NASA Orbiter Experiments (OEX) program provided a mechanism for using the Shuttle Orbiter as an entry aerothermodynamic flight-research vehicle, as an adjunct to its normal operational missions. The OEX experiment packages included instrumentation for: (1) the determination of the Orbiter aerodynamic characteristics and (2) the determination of the convective heating rates to the vehicle surface.

For the CSE experiment of the Orbiter Experiments program, selected tiles were coated with a highly catalytic material consisting of iron-cobalt-chromia spinel in a vinyl acetate binder. This coating, designated C742, was developed at the Ames Research Center (NASA). For the STS-2 flight, tiles at $x = 0.15L$ and at $x = 0.40L$ were sprayed with the catalytic coating. See Fig. 7.51. For the STS-3 flight, the catalytic tiles were at $x = 0.30L$ and at $x = 0.40L$. For the STS-5 flight, the catalytic-coated tiles were at $x = 0.10L, 0.15L, 0.20L, 0.30L, 0.40L,$ and $0.60L$. The heat-transfer rates determined using temperature measurements from these flights in Eq. (7-6), as taken from Ref. 91, are reproduced in Fig. 7.52. An effect of the catalytic coating on the heating is obvious.

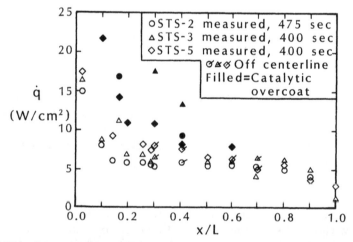

Fig. 7.52 Comparison of measured radiation equilibrium heat fluxes near the windward centerline of Orbiter. $U_\infty = 7.16\,km/s$ and altitude of 74.7 km STS-2, STS-3, and STS-5, as taken from Ref. 91.

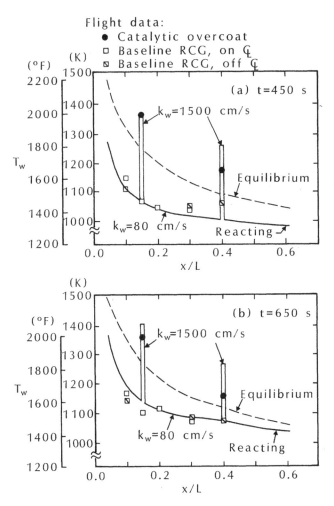

Fig. 7.53 Streamwise surface-temperature distributions for windward pitch-plane for STS-2 flight, as taken from Ref. 89.

However, one must compare the data with computed heat-transfer rates in order to obtain a quantitative measure of the catalytic efficiency. Scott[10] noted, "Nitrogen recombination was found to be significant early in the entry especially in areas dominated by normal shock flow such as near the nose. This makes knowing the nitrogen recombination phenomena important." Rakich et al.[89] noted, "In the flow regime for the Space Shuttle, the gas temperature varies typically between 2000 and 5000K. In this temperature range, negligible ionization occurs; hence ionized species are excluded from the model. Five species are considered: O, N, O_2, N_2, and NO."

Rakich et al.[89] divided the flowfield into an inviscid region and a boundary layer (using the axisymmetric analogue to account for three-dimensional effects). Accounting for nonequilibrium effects and entropy-layer swallowing, heat-transfer distributions along the windward pitch-plane were computed for selected conditions. Surface-temperature distributions were measured 450 s and 650 s after the start-of-entry, where the start-of-entry is defined as passage through an altitude of 122 km (400 kft), as taken from Rakich et al.,[89] and are reproduced in Fig. 7.53. The results for 450 s illustrate the situation when the nonequilibrium effects are greatest near the nose. For 650 s, U_∞ is 6.73 km/s at an altitude of 71.29 km. For both times, the boundary-layer analysis predicts a discontinuous rise in temperature on the test tiles with a catalytic overcoat. Because the energy of dissociation which remained in the flow as it passed over the noncatalytic surface is released as the flow encounters a catalytic tile, the temperature exceeds the equilibrium value.

Thus, these flight data show that the surface temperatures of the baseline RCG tiles are consistently below the equilibrium computations (by 15 percent to 35 percent), while the surface temperatures of the catalytic tiles are above the equilibrium computations. However, these results would change if the number and the extent of the tiles with a catalytic overcoat were changed. As discussed in the next section, contamination would change the temperature distribution, since the heat flux to a catalytic surface depends strongly on the upstream surface as well as its own catalytic efficiency.

Calculations presented by Scott,[91] which are reproduced in Fig. 7.54, show that the overshoot in the heat flux decreases quickly after a discontinuity in the surface recombination rate. Scott[91] notes:

> The distance required to reach the heating value that would exist had the entire surface been highly catalytic is very long. So long that it is impractical to say that it is ever reached. Shuttle experience indicates that the relaxation zone is longer than 1.5 meters. However, the very high part of the overshoot does not persist very far, only a few centimeters from the leading edge of the overcoat. This phenomena can be seen clearly in Fig. 7.54. This implies that if one has a discontinuity in materials such as a ceramic-to-metal joint, then one might experience temperature at the leading edge of the metal to overheat if the overshoot is not properly accounted for in the design.

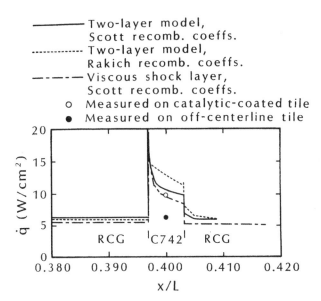

Fig. 7.54   Heat fluxes illustrating the effect of catalytic coating (C742) on tile at $x = 0.40L$, for $t = 475$ s of STS-2 flight, $U_\infty = 7.16$ km/s, alt = 74.4 km, as taken from Ref. 91.

7.5.3 The Effect of Surface Contamination

The temperature histories provided by three thermocouples located near the windward plane-of-symmetry of the Shuttle Orbiter in the region $0.099L \leq x \leq 0.299L$ are reproduced in Fig. 7.55. These data are taken from Ref. 92. All three of these thermocouples were located on baseline tiles. Sudden increases in temperature occurred at three times during the re-entry flight of the STS-2, corresponding to the integers placed in Fig. 7.55.

1. The sudden increase in temperature, which was sensed by only two of the thermocouples (not the one at $x = 0.099L$), was due to contamination.

2. The sudden increase in temperature, sensed by all three thermocouples approximately 800 s after entry, was due to the pullup/pushover maneuver. As discussed in Sec. 3.7, a pullup/pushover maneuver was conducted to investigate the body-flap effectiveness.

3. The increased heating due to the onset of boundary-layer transition is reflected by the sudden increase in the surface temperature evident for times in excess of 1250 s. The thermocouple

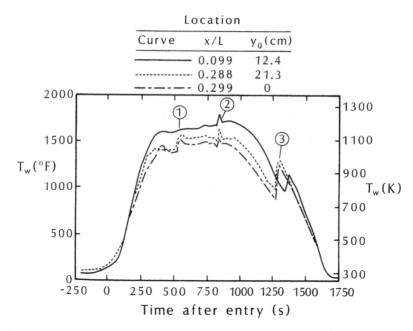

Fig. 7.55 Temperature histories from thermocouples in the region $0.099L \leq x \leq 0.299L$ from STS-2 flight, as taken from Ref. 92.

at $x = 0.099L$ is the last to experience the transition-induced increase in heating, since it is the farthest upstream of the three.

The temperature jump of interest to this section is the one that occurs at approximately 500 s. This jump is sensed by thermocouples that are downstream of an exposed stainless steel microphone cap which was located at $x = 0.10L$ (see Fig. 7.51) and within ± 1.2 deg. of the symmetry plane. Both Scott and Derry[92] and Throckmorton et al.[93] reported that a deposition of oxidation products was visible on tiles downstream of the acoustic sensor for at least 3 m. Since iron oxide and nickel oxide are highly catalytic to oxygen and to nitrogen recombination, the coating caused increased heating on the contaminated tiles.

Throckmorton et al.[93] note that:

On STS-2, the temperature "jump" phenomenon was observed at the centerline measurement locations at $0.194 \leq x/L \leq 0.402$. At the most aft of these locations ($x/L = 0.402$), the STS-2 anomaly response was a temperature *decrease* as opposed to the increase observed at other locations. This tile was catalytically coated as part of the CSE experiment. If surface contamination had caused a sudden increase in the catalytic efficiency of the TPS surface upstream of this location, as is suggested by the available evidence, a sudden depletion in the number of dissociated oxygen atoms reaching the location of the catalytic-coated tile would result. Therefore, with suddenly fewer oxygen atoms available for recombination, the sudden temperature decrease which was observed would be expected—not due to local surface contamination, but rather the residual effect of upstream surface contamination.

The heat-transfer measurements from the first five Shuttle flights indicate that the surface catalycity (i.e., the surface reaction-rate parameter) of the RCG coating increased from flight to flight. After the STS-5 flight, tiles from $x = 0.138L$ and from $x = 0.40L$ were removed and chemically analyzed. Stewart et al.[94] reported:

> ...the presence of aluminum, silicon, sodium, and magnesium on the surface of the two tiles. The aluminum, probably in the form of alumina, is attributed to by-products deposited from burning solid rocket fuel during launch. The other elements are commonly found in sea salt. Sea salt tends to increase the reaction-rate constant because of its effect on ion mobility and viscosity of the borosilicate glass, and alumina decreases the emittance.

Clearly, the designer of a hypersonic vehicle intended for multimission, "airplane-like" operation must consider many factors which may affect the aerothermodynamic performance of the vehicle.

7.6 BASE HEAT TRANSFER IN SEPARATED FLOW

As discussed in Sec. 6.8, the pressure in a separated region is relatively low. Similarly, the heat transfer to the base region is relatively

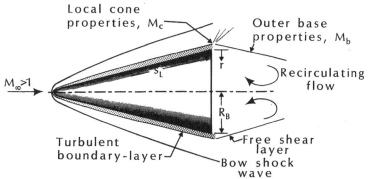

Fig. 7.56 Nomenclature for the base heat-transfer correlation of Eq. (7-38).

low, as compared to the heat transfer to surfaces where the boundary layer is attached. Although the base heating is low, the weight of the thermal-protection system required to protect that surface may be significant when the base area is a significant fraction of the total vehicle surface. Slender, high-performance re-entry vehicles generally experience severe heating environments and, therefore, require ablative thermal-protection systems for the base-cover design. Although the base area for these slender configurations represents only 10 percent to 20 percent of the total surface area for $5° \leq \theta_c \leq 15°$, optimization of the base heat-shield design is needed to reduce the base-cover weight and to ensure an acceptable vehicle static margin.

The separated flows, such as depicted in Fig. 7.56, are difficult to model accurately. Therefore, predictions of the base-region heat transfer are based on empirical correlations, which are developed using wind-tunnel data or flight-test data. Heat-transfer measurements obtained during the flights of two slender, slightly blunted RVs having an approximately flat base geometry, which were reported by Bulmer,[95] are reproduced in Fig. 7.57. Note that there is a sudden jump in each of the heat-transfer histories. The jumps occurred simultaneously with the onset of boundary-layer transition, as determined using accelerometer data and using surface-pressure data.

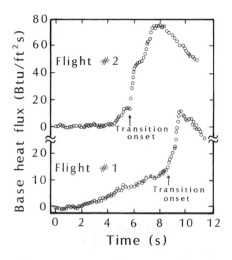

Fig. 7.57 Base heat-transfer rate measurements from flight tests of two slender, slightly blunted RVs, as taken from Ref. 95.

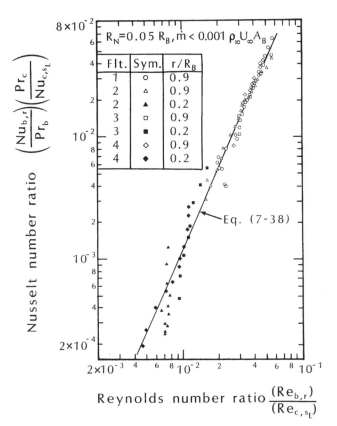

Fig. 7.58 Correlation of flight-test, turbulent, base heat-transfer data, as taken from Ref. 96.

Thus, the base heating is much greater when the boundary layer on the conical surface is turbulent.

Heat-transfer data from the (approximately flat) base of relatively sharp $(R_N = 0.05R_B)$, slender cones were correlated by Bulmer.[96] The base heat-transfer data were obtained during four flights, when the boundary layer was turbulent, the heat-shield ablation rates were very low $(0.0002\,\rho_\infty U_\infty A_B \leq \dot{m} \leq 0.001\,\rho_\infty U_\infty A_B)$, and the angle-of-attack was less than 0.5 deg. As shown in Fig. 7.58, an empirical equation that represents the best visual, linear fit of the data is:

$$\left(\frac{Nu_{b,r}}{Pr_b}\right)\left(\frac{Pr_c}{Nu_{c,s_L}}\right) = 35.5\left(\frac{Re_{b,r}}{Re_{c,s_L}}\right)^{2.2} \qquad (7\text{-}38)$$

The local cone-flow properties, i.e., the boundary-layer-edge properties at s_L, which are designated by the subscript c, assume that the flow is in equilibrium and include the effects of entropy swallowing and of the nose-region pressure overexpansion. The outer, base-region flow properties, which are designated by the subscript b, were calculated by expanding the local cone-flow through a Prandtl-Meyer expansion to the measured base pressure. Although the cone flow and the base flow are axisymmetric, the boundary layers on the full-scale vehicles during re-entry are sufficiently small relative to the base radius, so that a two-dimensional Prandtl-Meyer expansion provides a reasonable method for calculating the outer-base conditions. The outer-base thermal properties were evaluated at the outer-base static temperature (T_b), and the outer-base velocity (U_b) was used in the calculation of the base Reynolds numbers and of the base Nusselt numbers. The characteristic length r, which is illustrated in Fig. 7.56, correlates the heat-transfer measurements across the base, since the heat flux increases with r and the maximum value occurs at the centerline $(r = R_B)$.

7.7   CONCLUDING REMARKS

The determination of the skin friction and of the heat transfer acting on configurations in hypersonic streams represents a real challenge to the analyst. Real-gas effects, including nonequilibrium chemistry and surface catalycity, add to the complexity of the computation algorithm. The problem of defining the aerothermodynamic environment is further complicated by the uncertainties associated with boundary-layer transition and with turbulence modeling. Because of the many, interrelated factors that affect boundary-layer transition, transition criteria used in vehicle designs are largely empirical and often should not be expected to have a general application, e.g., see Stetson's discussion regarding $(Re_\theta/M_e)_t$ in Ref. 3. Thus, when it becomes necessary to predict transition on a new configuration or at new flow conditions, empirical prediction methods may not provide valid criteria. Although turbulence models used in numerical simulations of the flow may represent the gross behavior of the flow, they do not model the details of the flow structure. Furthermore, as noted earlier, the question of modeling is intertwined with the numerical procedures. Although turbulence modeling is often cited as the main deficiency in a computation, the algorithm, the grid, the convergence criteria, and the experimental data themselves can be contributing factors.

Despite the apparently overwhelming nature of the challenge of predicting the aerothermodynamic environment, designers have always dealt with the challenge. Often the thermal-protection system is overdesigned or the flight envelope overrestricted to allow for the uncertainties. The author is familiar with many such examples. After an unmanned flight of the Gemini capsule, small holes were observed in the metal heat-shield panels of the conical surface. Subsequent analysis traced the problem to locally high heating rates that were produced by vortices from the flow over the umbilical fairing. The remedy was to restrict the flight corridor to ensure the safety of the astronauts.

Thus, the penalties for ignorance include vehicle weight (and, therefore, increased launch costs) and restricted operational capabilities (to avoid those flight conditions that would lead to the severe environment). Improvements in our understanding of the flowfield and in our ability to model it will continue to come from the research and from the design communities. Meanwhile, the designers will have at their disposal a variety of experimental techniques, of analytical relations, and of computational codes (of varying rigor).

REFERENCES

[1] Li, C. P., Private discussion, 5 Sep. 1991.

[2] Aupoix, B., Brazier, J. Ph., Cousteix, J., and Monnoyer, F., "Second Order Effects in Hypersonic Boundary Layers," J. J. Bertin, J. Periaux, and J. Ballmann (eds.), *Advances in Hypersonics, Volume 3: Computing Hypersonic Flows*, Birkhäuser Boston, Boston, MA, 1992.

[3] Stetson, K. F., "Hypersonic Boundary-Layer Transition," J. J. Bertin, J. Periaux, and J. Ballmann (eds.), *Advances in Hypersonics, Volume 1: Defining the Hypersonic Environment*, Birkhäuser Boston, Boston, MA, 1992.

[4] McWherter, M., Noack, R. W., and Oberkampf, W. L., "Evaluation of Inviscid/Boundary Layer and Parabolized Navier-Stokes Solutions for Design of Reentry Vehicles," AIAA Paper 84-0486, Reno, NV, Jan. 1984.

[5] Adams, J. C., Jr., Martindale, W. R., Mayne, A. W., Jr., and Marchand, E. O., "Real-gas Scale Effects on Shuttle Orbiter Laminar Boundary-Layer Parameters," *Journal of Spacecraft and Rockets*, Vol. 14, No. 5, May 1977, pp. 273–279.

[6] Gupta, R. N., Lee, K. P., Zoby, E. V., Moss, J. N., and Thompson, R. A., "Hypersonic Viscous Shock-Layer Solutions over Long Slender Bodies – Part I: High Reynolds Number Flows," *Journal of Spacecraft and Rockets*, Vol. 27, No. 2, Mar.–Apr. 1990, pp. 175–184.

[7] Hirschel, E. H., Mundt, C., Monnoyer, F., and Schmatz, M. A., "Reynolds-Number Dependency of Radiation-Adiabatic Wall Temperature," Messerschmitt-Bölkow-Blohm MBB-FE122-AERO-MT-872, June 1990.

[8] Hopkins, E. J., and Inouye, M., "An Evaluation of Theories for Predicting Turbulent Skin Friction and Heat Transfer on Flat Plates at Supersonic and Hypersonic Mach Numbers," *AIAA Journal*, Vol. 9, No. 6, June 1971, pp. 993–1003.

[9] Williams, S. D., and Curry, D. M., "An Analytical and Experimental Study for Surface Heat Flux Determination," *Journal of Spacecraft and Rockets*, Vol. 14, No. 10, Oct. 1977, pp. 632–637.

[10] Scott, C. D., "A Review of Nonequilibrium Effects and Surface Catalysis on Shuttle Heating," *Shuttle Performance: Lessons Learned*, NASA CP-2283, Part 2, Mar. 1983.

[11] Goodrich, W. D., Li, C. P., Houston, C. K., Chiu, P. B., and Olmedo, L., "Numerical Computations of Orbiter Flowfields and Laminar Heating Rates," *Journal of Spacecraft and Rockets*, Vol. 14, No. 5, May 1977, pp. 257–264.

[12] Rakich, J. V., and Lanfranco, M. J., "Numerical Computation of Space Shuttle Laminar Heating and Surface Streamlines," *Journal of Spacecraft and Rockets*, Vol. 14, No. 5, May 1977, pp. 265–272.

[13] Zoby, E. V., Lee, K. P., Gupta, R. N., and Thompson, R. A., "Nonequilibrium Viscous Shock Layer Solutions for Hypersonic Flow Over Slender Bodies," Paper No. 71, Eighth National Aero-Space Plane Technology Symposium, Monterey, CA, Mar. 1990.

[14] Cooke, J. C., "An Axially Symmetric Analogue for General Three-Dimensional Boundary Layers," British Aeronautical Research Council R & M 3200, 1961.

[15] Hildebrand, F. B., *Advanced Calculus for Engineers*, Prentice-Hall, Englewood Cliffs, NJ, 1957.

[16] Brong, E.A., and Leigh, D. C., "Surface Streamlines in Three-Dimensional Hypersonic Flows," *Journal of the Aerospace Sciences*, Vol. 28, No. 7, July 1961, pp. 585–587.

[17] Fivel, H. J., "Numerical Flow Field Program for Aerodynamic Heating Analysis, Vol I – Equations and Results," Air Force Flight Dynamics Laboratory AFFDL-TR-79-3128, Vol. I, Dec. 1979.

[18] Hecht, A. M., and Nestler, D. E., "A Three-Dimensional Boundary-Layer Computer Program for Sphere-Cone Type Re-entry Vehicles, Volume I – Engineering Analysis and Description," Air Force Flight Dynamics Laboratory AFFDL-TR-78-67, June 1978.

[19] Leigh, D. C., and Ross, B. B., "Surface Geometry of Three-Dimensional Inviscid Hypersonic Flows," *AIAA Journal*, Vol. 7, No. 1, Jan. 1969, pp. 123–129.

[20] Brandon, H. J., and DeJarnette, F. R., "Three-Dimensional Turbulent Heating on an Ogive at Angle of Attack Including Effects of Entropy-Layer Swallowing," AIAA Paper 77-754, Albuquerque, NM, June 1977.

[21] Riley, C. J., DeJarnette, F. R., and Zoby, E. V., "Surface Pressure and Streamline Effects on Laminar Heating Calculations," *Journal of Spacecraft and Rockets*, Vol. 27, No. 1, Jan.–Feb. 1990, pp. 9–14.

[22] Eckert, E. R. G., "Engineering Relations for Friction and Heat Transfer to Surface in High Velocity Flow," *Journal of the Aeronautical Sciences,* Aug. 1955, Vol. 22, No. 8, pp. 585–587.

[23] Lee, D. B., and Harthun, M. H., "Aerothermodynamic Entry Environment of the Space Shuttle Orbiter," P. E. Bauer and H. E. Collicott (eds.), *Entry Vehicle Heating and Thermal Protection Systems: Space Shuttle, Solar Starprobe, Jupiter Galileo Probe,* Vol. 85 of Progress in Astronautics and Aeronautics, AIAA, New York,1983, pp. 3–20.

[24] Hayes, W. D., and Probstein, R. F., *Hypersonic Flow Theory,* Academic Press, New York, 1959.

[25] Fay, J. A., and Riddell, F. R., "Theory of Stagnation Point Heat Transfer in Dissociated Air," *Journal of the Aeronautical Sciences,* Vol. 25, No. 2, Feb. 1958, pp. 73–85, 121.

[26] Dorrance, W. H., *Viscous Hypersonic Flow,* McGraw-Hill, New York, 1962.

[27] Cleary, J. W., "Effects of Angle of Attack and Bluntness on Laminar Heating-Rate Distributions of a 15° Cone at a Mach Number of 10.6," NASA TN D-5450, Oct. 1969.

[28] Staff, "Equations, Tables, and Charts for Compressible Flow," NACA Rept. 1135, 1953.

[29] Lees, L., "Laminar Heat Transfer Over Blunt-Nosed Bodies at Hypersonic Flight Speeds," *Jet Propulsion,* Vol. 26, No. 4, Apr. 1956, pp. 259–269.

[30] Kemp, N. H., Rose, P. H., and Detra, R. W., "Laminar Heat Transfer Around Blunt Bodies in Dissociated Air," *Journal of the Aerospace Sciences,* Vol. 26, No. 7, July 1959, pp. 421–430.

[31] Bertin, J. J., "The Effect of Protuberances, Cavities, and Angle of Attack on the Wind Tunnel Pressure and Heat-Transfer Distributions for the Apollo Command Module," NASA TM X-1243, Sep. 1966.

[32] Stoney, W. E., Jr., "Aerodynamic Heating of Blunt-Nose Shapes at Mach Numbers Up to 14," NACA RML 58E05a, 1958.

[33] DeJarnette, F R., "Calculation of Inviscid Surface Streamlines and Heat Transfer on Shuttle-Type Configurations. Part I– Description of Basic Method," NASA CR-111921, North Carolina State University, Aug. 1971.

[34] Cleary, J. W., and Duller, C. E., "Effects of Angle of Attack and Bluntness on the Hypersonic Flow Over a 15° Semiapex Cone in Helium," NASA TN D-5903, Aug. 1970.

[35] Griffith, B. J., Majors, B. M., and Adams, J. C., Jr., "Blunt-Body Turbulent Boundary-Layer Parameters Including Shock Swallowing Effects," T. E. Horton (ed.), *Thermophysics of Atmospheric*

Entry, Vol. 82 of Progress in Astronautics and Aeronautics, AIAA New York, ppf. 90, 1982.

[36] Zoby, E. V., "Approximate Heating Analysis for the Windward-Symmetry Plane of Shuttle-Like Bodies at Large Angle-of-Attack," AIAA Paper 81-1042, Palo Alto, CA, June 1981.

[37] Bertin, J. J., Idar, E. S., and Galanski, S. R., "Effects of Surface Cooling and of Roughness on the Heating (Including Transition) to the Windward Plane-of-Symmetry of the Shuttle Orbiter," The University of Texas at Austin Aerospace Engineering Rept. 77002, Apr. 1977.

[38] Zoby, E. V., Moss, J. N., and Sutton, K., "Approximate Convective-Heating Equations for Hypersonic Flows," *Journal of Spacecraft and Rockets,* Vol. 18, No. 1, Jan.–Feb. 1981, pp. 64–70.

[39] Bertin, J. J., and Cline, D. D., "Variable-Grid-Size Transformation for Solving Nonsimilar Laminar and Turbulent Boundary Layers," M. Gerstein and P. R. Choudhury (eds.) *Proceedings of the 1980 Heat Transfer and Fluid Mechanics Institute,* Stanford University Press, Stanford, CA, 1980.

[40] Bartlett, E. P., and Kendall, R. M., "An Analysis of the Coupled Chemically Reacting Boundary Layer and Charring Ablator, Part III, Nonsimilar Solutions of the Multicomponent Laminar Boundary Layer by an Integral Matrix Method," NASA CR-1062, June 1968.

[41] Miller, C. G., Wilder, S. E., Gnoffo, P. A., and Wright, S. A., "Measured and Predicted Vortex Induced Leeward Heating on a Biconic at Mach 6 and 10," J. N. Moss and C. D. Scott (eds.), *Thermophysical Aspects of Re-entry Flows,* Vol. 103 of Progress in Astronautics and Aeronautics, AIAA, New York, 1985, pp. 310–340.

[42] Miller, C. G., Gnoffo, P. A., and Wilder, S. E., "Measured and Predicted Heating Distributions for Biconics at Mach 10," *Journal of Spacecraft and Rockets,* Vol. 23, No. 3, May–June 1986, pp. 251–258.

[43] Chaussee, D. S., "NASA Ames Research Center's Parabolized Navier-Stokes Code: A Critical Evaluation of Heat-Transfer Predictions," AIAA Paper 87-1474, Honolulu, HI, June 1987.

[44] James, C. S., "Observations of Turbulent-Burst Geometry and Growth in Supersonic Flow," NACA TN-4235, 1958.

[45] Owen, F. K., Horstman, C. C., Stainback, P. C., and Wagner, R. D., "Comparison of Wind Tunnel Transition and Freestream Disturbances Measurements," *AIAA Journal,* Vol. 13, No. 3, Mar. 1975, pp. 266–269.

[46] Pate, S. R., and Schueler, C. J., "Radiated Aerodynamic Noise Effects on Boundary-Layer Transition in Supersonic and Hypersonic Wind Tunnels," *AIAA Journal,* Vol. 7, No. 3, Mar. 1969, pp. 450–457.

[47] Pate, S. R., "Effects of Wind Tunnel Disturbances on Boundary-Layer Transition with Emphasis on Radiated Noise: A Review," AIAA Paper 80-0431, Colorado Springs, CO, Mar. 1980.

[48] Beckwith, I. E., and Bertram, M. H., "A Survey of NASA Langley Studies on High-Speed Transition and the Quiet Tunnel," NASA TM X-2566, July 1972.

[49] Morkovin, M. V., "Critical Evaluation of Transition From Laminar to Turbulent Shear Layers with Emphasis on Hypersonically Traveling Bodies," Air Force Flight Dynamics Laboratory, AFFDL-TR-68-149, Mar. 1969.

[50] Reshotko, E., "A Program for Transition Research," *AIAA Journal*, Vol. 13, No. 3, Mar. 1975, pp. 261–265.

[51] Mack, L. M., "Linear Stability Theory and the Problem of Supersonic Boundary-Layer Transition," *AIAA Journal*, Vol. 13, No. 3, Mar. 1975, pp. 278–289.

[52] Mack, L. M., "Boundary-Layer Stability Analysis for Sharp Cones at Zero Angle-of-Attack," Air Force Wright Aeronautical Laboratories, AFWAL-TR-86-3022, Aug. 1986.

[53] Herbert, T., "Boundary-Layer Transition Analysis and Prediction Revisited," AIAA Paper 91-0737, Reno, NV, Jan. 1991.

[54] Chang, C. L., Malik, M. R., Erlebacher, G., and Hussaini, M. Y., "Compressible Stability of Growing Boundary Layers Using Parabolized Stability Equations," AIAA Paper 91-1636, Honolulu, HI, June 1991.

[55] Bertolotti, F. P., "Compressible Boundary Layer Stability Analyzed with the PSE Equations," AIAA Paper 91-1637, Honolulu, HI, June 1991.

[56] Beckwith, I. E., "Development of a High Reynolds Number Quiet Tunnel for Transition Research," *AIAA Journal*, Vol. 13, No. 3, Mar. 1975, pp. 300–306.

[57] Potter, J. L., "Boundary-Layer Transition on Supersonic Cones in an Aeroballistic Range," *AIAA Journal*, Vol. 13, No. 3, Mar. 1975, pp. 270–277.

[58] Stetson, K. F., "Hypersonic Laminar Boundary Layer Transition, Part I: Nosetip Bluntness Effects on Cone Frustrum Transition, Part II: Mach 6 Experiments of Transition on a Cone at Angle of Attack," Air Force Wright Aeronautical Laboratories, AFWAL-TR-86-3089, Dec. 1986.

[59] Reda, D. C., "Boundary-Layer Transition Experiments on Sharp Slender Cones in Supersonic Freeflight," Naval Surface Weapon Center, NSWC/WOL-77-59, Sep. 1977.

[60] Stetson, K. F., "Mach 6 Experiments on a Cone at Angle of Attack," *Journal of Spacecraft and Rockets*, Vol. 19, No. 5, Sep.–Oct. 1982, pp. 397–403.

[61] Potter, J. L., and Whitfield, J. D., "Effects of Slight Nose Bluntness and Roughness on Boundary Layer Transition in Supersonic Flows," *Journal of Fluid Mechanics,* Vol. 12, Pt. IV, 1962, pp. 501–535.

[62] Sterrett, J. R., Morrisette, E. L., Whitehead, A. H., Jr., and Hicks, R. M., "Transition Fixing for Hypersonic Flow," NASA TND-4129, Oct. 1967.

[63] Nestler, D. E., and McCauley, W. D., "A Study of a Boundary-Layer Trip Concept at Hypersonic Speeds," AIAA Paper 81-1086, Palo Alto, CA, June 1981.

[64] Bertin, J. J., Tedeschi, W. J., Kelly, D. P., Bustamante, A. C., and Reece, E. W., "Analysis of the Expansion-Fan Flowfield for Holes in a Hypersonic Configuration," *AIAA Journal,* Vol. 27, No. 9, Sep. 1989, pp. 1241–1248.

[65] Boudreau, A. H., "Artificially Induced Boundary-Layer Transition on Blunt-Slender Cones at Hypersonic Speeds," *Journal of Spacecraft and Rockets,* Vol. 16, No. 4, July–Aug. 1979, pp. 245–251.

[66] Bertin, J. J., Idar, E. S., III, and Goodrich, W. D., "Effect of Surface Cooling and Roughness on Transition for the Shuttle Orbiter," *Journal of Spacecraft and Rockets,* Vol. 15, No. 2, Mar.– Apr. 1978, pp. 113–119.

[67] Bertin, J. J., Hayden, T. E., and Goodrich, W. D., "Shuttle Boundary-Layer Transition Due to Distributed Roughness and Surface Cooling," *Journal of Spacecraft and Rockets,* Vol. 19, No. 5, Sep.–Oct. 1982, pp. 389–396.

[68] Goodrich, W. D., Derry, S. M., and Bertin, J. J., "Shuttle Orbiter Boundary Layer Transition at Flight and Wind Tunnel Conditions," *Shuttle Performance: Lessons Learned,* NASA CP-2283, Part 2, Mar. 1983.

[69] Braslow, A. L., "A Review of Factors Affecting Boundary-Layer Transition," NASA TND-3384, Apr. 1966.

[70] Stalmach, C. J., Bertin, J. J., Pope, T. C., and McCloskey, M. H., "A Study of Boundary Layer Transition on Outgassing Cones in Hypersonic Flow," NASA CR-1908, Dec. 1971.

[71] Hearne, L. F., Chin, J. H., and Woodruff, L. W., "Study of Aerothermodynamic Phenomena Associated with Reentry of Manned Spacecraft," Lockheed Missile and Space Company Rept. Y-78-66-1, May 1966.

[72] Marvin, J. G., and Akin, C. M., "Combined Effects of Mass Addition and Noise Bluntness on Boundary-Layer Transition," *AIAA Journal,* Vol. 8, No. 5, May 1970, pp. 857–863.

[73] Acharya, M., Kussoy, M. I., and Horstman, C. C., "Reynolds Number and Pressure Gradient Effects on Compressible Turbulent Boundary Layers," AIAA Paper 78-199, Huntsville, AL, Jan. 1978.

[74] Harvey, W. D., and Bushnell, D. M., "Velocity Fluctuation Intensities in a Hypersonic Turbulent Boundary Layer," *AIAA Journal*, Vol. 7, No. 4, Apr. 1969, pp. 760–762.

[75] Wallace, J. E., "Hypersonic Turbulent Boundary-Layer Measurements Using an Electron Beam," *AIAA Journal*, Vol. 7, No. 4, Apr. 1969, pp. 757–759.

[76] Bertin, J. J., Amirkabirian, I., and Cline, D. D., "The Numerical Analysis of Nonsimilar Boundary Layers: Laminar, Transitional, and Turbulent – Volume II, Derivation of Equations," The University of Texas at Austin Aerospace Engineering Rept. 80003, June 1980.

[77] Bushnell, D. M., Cary, A. M., Jr., and Harris, J. E., "Calculation Methods for Compressible Turbulent Boundary Layers–1976," NASA SP-422, 1977.

[78] Shang, J. S., "Computation of Hypersonic Turbulent Boundary Layers with Heat Transfer," *AIAA Journal*, Vol. 12, No. 7, July 1974, pp. 883–884.

[79] Neumann, R. D., "Defining the Aerothermodynamic Environment," J. J. Bertin, R. Glowinski, and J. Periaux (eds.), *Hypersonics, Volume I: Defining the Hypersonic Environment*, Birkhäuser Boston, Boston, MA, 1989.

[80] Marvin, J. G., "Turbulence Modeling for Computational Aerodynamics," *AIAA Journal*, Vol. 21, No. 7, July 1983, pp. 941–955.

[81] Van Driest, E. R., "Turbulent Boundary Layer in Compressible Fluids," *Journal of the Aerospace Sciences*, Vol. 18, No. 3, Mar. 1951, pp. 145–160.

[82] Cebeci, T., and Bradshaw, P., *Momentum Transfer in Boundary Layers*, McGraw-Hill, New York, 1977.

[83] Hodge, B. K., and Adams, J. C., "The Calculation of Compressible Transitional, Turbulent, and Relaminarization Boundary Layers Over Smooth and Rough Surfaces Using an Extended Mixing-Length Hypothesis," Arnold Engineering Development Center, AEDC-TR-77-96, Feb. 1978.

[84] Wolfshtein, M., "The Velocity and Temperature Distribution in One-Dimensional Flow with Turbulence Augmentation and Pressure Gradient," *International Journal of Heat and Mass Transfer*, Vol. 12, No. 3, Mar. 1969, pp. 301–318.

[85] Bartlett, R. P., Edwards, A. J., Harvey, J. K., and Hillier, R., "Pitot Pressure and Total Temperature Profile Measurements in a Hypersonic, Turbulent Boundary Layer at $M = 9$," Imperial College of Science and Technology I. C. Aero Rept. 79-01, Feb. 1979.

[86] Bertin, J. J., and Cline, D. D., "The Numerical Analysis of Nonsimilar Boundary Layers: Laminar, Transitional, and Turbulent – Volume I, General Results," The University of Texas at Austin Aerospace Engineering Rept. 80002, June 1980.

[87] Stalmach, C. J., Jr., "Experimental Investigation of the Surface Impact Probe Method of Measuring Local Skin Friction at Supersonic Speeds," The University of Texas Defense Research Laboratory DRL-410, Jan. 1958.

[88] Marvin, J. G., and Coakley, T. J., "Turbulence Modeling for Hypersonic Flows," J. J. Bertin, J. Periaux, and J. Ballmann (eds.), *Advances in Hypersonics, Volume 2: Modeling Hypersonic Flows,* Birkhäuser Boston, Boston, MA, 1992.

[89] Rakich, J. V., Stewart, D. A., and Lanfranco, M. J., "Catalytic Efficiency of the Space Shuttle Heat Shield," P. E. Bauer and H. E. Collicott (eds.), *Entry Vehicle Heating and Thermal Protection Systems: Space Shuttle, Solar Starprobe, Jupiter Galileo Probe,* Vol. 85 of Progress in Astronautics and Aeronautics, AIAA, New York, pp. 97–122, 1982.

[90] Sheldahl, R. E., and Winkler, E. L., "Effect of Discontinuities in Surface Catalytic Activity on Laminar Heat Transfer in Arc-Heated Nitrogen Streams," NASA TND-3615, 1965.

[91] Scott, C. D., "Wall Catalytic Recombination and Boundary Conditions in Nonequilibrium Flows – with Applications," J. J. Bertin, J. Periaux, and J. Ballmann (eds.), *Advances in Hypersonic Flows, Volume 2: Modeling Hypersonic Flows,* Birkhäuser Boston, Boston, MA, 1992.

[92] Scott, C. D., and Derry, S. M., "Catalytic Recombination/Space Shuttle Heating," P. E. Bauer and H. E. Collicott (eds.), *Entry Vehicle Heating and Thermal Protection Systems: Space Shuttle, Solar Starprobe, Jupiter Galileo Probe,* Vol. 85 of Progress in Astronautics and Aeronautics, AIAA, New York, pp. 123–148, 1982.

[93] Throckmorton, D. A., Zoby, E. V., and Hamilton, H H., II, "Orbiter Catalytic/Noncatalytic Heat Transfer as Evidenced by Heating to Contaminated Surfaces on STS-2 and STS-3," *Shuttle Performance: Lessons Learned,* NASA CP-2283, Part 2, Mar. 1983.

[94] Stewart, D. A., Rakich, J. V., and Lanfranco, M. J., "Catalytic Surface Effects on Space Shuttle Thermal Protection System During Earth Entry of Flights STS-2 through STS-5," *Shuttle Performance: Lessons Learned,* NASA CP-2283, Part 2, Mar. 1983.

[95] Bulmer, B. M., "Detection of Flight Vehicle Transition From Base Measurements," *Journal of Spacecraft and Rockets,* Vol. 10, No. 4, Apr. 1973, pp. 280–281.

[96] Bulmer, B. M., "Flight Test Correlation Technique for Turbulent Base Heat Transfer with Low Ablation," *Journal of Spacecraft and Rockets,* Vol. 10, No. 3, Mar. 1973, pp. 222–224.

PROBLEMS

7.1 Using the temperature histories presented in Fig. 7.55 and Eq. (7-6), calculate the heating rates at all three locations, i.e., $x = 0.099L$, $x = 0.288L$, and $x = 0.299L$, 1000 s after entry.

Problems 7.2 – 7.4

For the following three flow conditions, you are to use Fig. 7.28 to define the location of boundary-layer transition and use Eckert's reference temperature relations and Reynolds' analogy to calculate the total skin-friction drag. Using Reynolds' analogy

$$C_f = \frac{\tau}{\frac{1}{2}\rho_e u_e^2} = \frac{1.150}{(Re_x^*)^{0.5}}$$

for a laminar boundary-layer and

$$C_f = \frac{\tau}{\frac{1}{2}\rho_e u_e^2} = \frac{0.0583}{(Re_x^*)^{0.2}}$$

for a turbulent boundary-layer.

If the transition location is within 15 percent of the apex, use the turbulent expression for the entire length of the model. If the transition location is within 15 percent of the end of the model, use the laminar expression for the entire length of the model. Assume the perfect-gas model.

7.2 The flow conditions are: a sharp cone with $\theta_c = 15$ deg. and the total length of a conical generator of 10.0 ft at an angle-of-attack of 0 deg. The cone is flying at Mach 6 at 50,000 ft. $T_w = 600\,°R$.

7.3 The flow conditions are: a sharp cone with $\theta_c = 15$ deg. and the total length of a conical generator of 1.0 m at zero angle-of-attack. The cone is flying at Mach 6 at 32 km. $T_w = 400$ K.

7.4 A sharp cone ($\theta_c = 15$ deg., $x_L = 6.0$ in., where x_L is the total length of the conical generator) is to be tested in the Langley Research Center's 20-Inch Mach 6 Wind Tunnel. The test conditions are $p_{t_1} = 30$ psia, $T_{t_1} = 845\,°R$, $T_w = 520\,°R$.

8

AERODYNAMIC FORCES
AND MOMENTS

8.1 INTRODUCTION

The pressures and the shear stresses can be integrated over the surface on which they act in order to yield the resultant force (R) which acts at the center-of-pressure (cp) of the vehicle. For convenience, the total force vector is usually resolved into components, as shown in Fig. 8.1. When the application is concerned primarily with the vehicle response, e.g., aerodynamics or structural dynamics, body-oriented force components are used. For the pitch-plane forces depicted in Fig. 8.1, the body-oriented components are the force parallel to the vehicle axis (A) and the force perpendicular to the vehicle axis (N). For applications such as trajectory analysis, the force components are taken relative to the velocity vector, i.e., the flight path. Thus, the resultant force is divided into a component parallel to the velocity vector, the drag (D), and a component perpendicular to the velocity vector, the lift (L).

When the resultant force does not act through the center-of-gravity, a moment is created. Consider the case where the position of the center-of-pressure relative to the center-of-gravity is as shown in Fig. 8.1. If the angle-of-attack increases due to a disturbance, the increase in the pressures acting on the windward surface will produce an increase in the normal force. Because the center-of-pressure is aft of the center-of-gravity, the increased normal force will produce a nose-down (negative) pitching moment, which decreases the angle-of-attack, restoring the vehicle to its original position. Thus, the vehicle is said to be statically stable. Conversely, if the center-of-gravity is aft of the center-of-pressure, the vehicle is statically unstable. That is, a perturbation which produces an increased angle-of-attack causes a nose-up (positive) pitching moment, which further increases the angle-of-attack.

The parameter,

$$S.M. = \frac{x_{cp} - x_{cg}}{L} \qquad (8\text{-}1)$$

is the static margin. The static margin must be positive for uncontrolled vehicles. For high-performance hypersonic vehicles, the static margin is usually 3 to 5 percent. Note, however, as illustrated in Fig. 8.1, both the axial force and the normal force contribute to the pitching moment. Thus, it is possible that the vehicle is statically stable when $x_{cp} = x_{cg}$, if y_{cp} is below y_{cg} (as is the case in the sketch

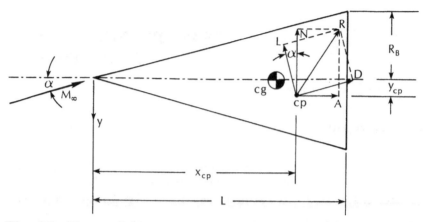

Fig. 8.1 Nomenclature for aerodynamic forces in the pitch plane.

of Fig. 8.1). In this case, the axial force will produce the required restoring moment.

8.1.1 The Value of Lift

Lift can be used to reduce the heating rates during re-entry and to reduce the level of deceleration. Correlations presented by Neumann[1] show that an (L/D) ratio of 1 is sufficient to reduce both the heating rate and the deceleration to a point where they are essentially independent of the level of (L/D) above 1. The effect of (L/D) on the reference (or stagnation-point) heat-transfer-rate histories, as presented by Neumann, is reproduced in Fig. 8.2. Lift allows the configuration to decelerate at higher altitudes. The lower value of the freestream density at these altitudes means that the aerodynamic heating will be lower at a given velocity. Note, however, that the configurations that achieve relatively high values for (L/D) may be slender, complex shapes flying at relatively low angle-of-attack. As a result, the local heating rates for these configurations may be relatively high, as discussed in Chap. 7. As noted by Neumann, "The reader is cautioned that the simple statement that increased lift reduces aerodynamic heating requires verification for geometries of current interest."

A relatively high value for the (L/D) ratio provides the capability of arbitrary recall from orbit, as shown in Fig. 8.3 which is taken from Ref. 1, and the capability for high cross range during the re-entry phase of flight. The lateral (or cross) range performance capabilities of several configurations studied at the Wright Laboratory, Flight Dynamics Directorate, as taken from Ref. 2, are reproduced in Fig. 8.4. Similarly, large values for (L/D) enable efficient orbital

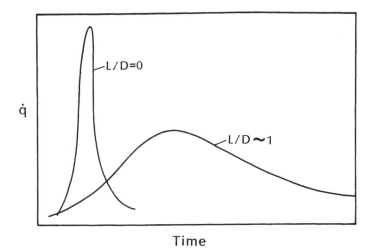

Fig. 8.2   The effect of (*L/D*) on the reference heating-rate history, as taken from Ref. 1.

plane changes through the combined use of a propulsion system and of aerodynamic forces. The use both of aerodynamic forces and of a propulsion system to accomplish plane change measurements is generally associated with the need to achieve larger values of the orbital plane change.

8.1.2 Typical Values of (L/D)

Because hypersonic vehicles have such a broad range of mission/performance/cost criteria and can be subjected to a wide variety of aerothermal environments, there are a wide variety of (L/D) ratios. Typical lift-to-drag ratios (L/D) and ballistic coefficients $(W/C_D A)$ for a range of generic configurations, as modified from the original presentation by Walberg,[3] are reproduced in Fig. 8.5. Walberg noted, "L/D greater than 1.5 requires lifting bodies and blended wing/body configurations that have relatively poor volumetric efficiency and higher ballistic coefficients (based on cross-sectional area). Accordingly, most studies of planetary aerocapture missions have recommended on-axis or bent biconic configurations."

The correlation presented in Fig. 8.5 indicates that for sphere/cone shapes, (L/D) is 1.0, or less. The space capsules of the early U.S. manned spacecraft program, i.e., the Mercury, the Gemini, and the Apollo Command Modules, were sphere/cone shapes. As shown in Fig. 6.11, the capsules were axisymmetric shapes consisting of a truncated spherical cap followed by a conical afterbody. If the center-of-gravity is offset, i.e., $y_{cg} \neq 0$, the axial force would create a nose-up pitching moment at zero angle-of-attack. Thus, the vehi-

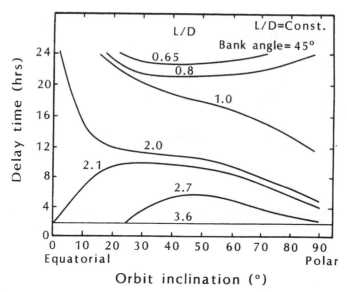

Fig. 8.3 Maximum time for return to Edwards AFB, as taken from Ref. 1.

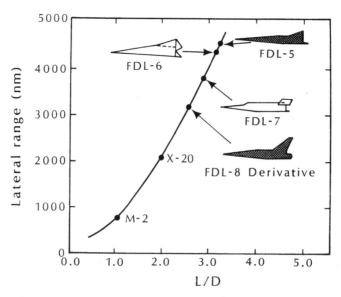

Fig. 8.4 Lateral (or cross) range as a function of (L/D) for typical configuration designs, as taken from Ref. 2.

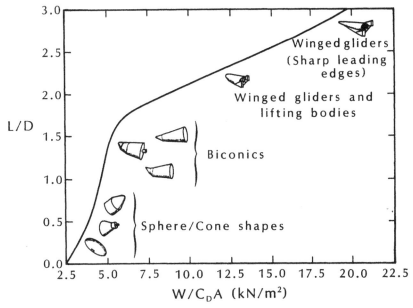

Fig. 8.5 Lift-to-drag ratios (*L/D*) and ballistic coefficients for re-entry configurations, as taken from Ref. 3.

cle would trim at a nonzero angle-of-attack, and generate lift when trimmed. The center-of-gravity was offset both for the Gemini and for the Apollo Command Modules, producing lift during re-entry. See Table 8.1. Note that the reference area for the capsule-like vehicles is the maximum cross-section area. This is consistent with the practice for missiles. For the Space Shuttle Orbiter, however, the reference area is the wing area and is usually represented by the aerodynamicist by S_{ref} or S. The reader is reminded that, just as there are a diverse variety of problems facing the hypersonic aerothermodynamicists, there are a variety of symbols that they have brought with them from their particular backgrounds.

At the Program Requirements Review (PRR), the requirements for the Shuttle Orbiter aerodynamics included minimize heating and provide adequate (L/D) for cross range. The pitch plane aerodynamic forces for the Shuttle Orbiter in a hypersonic stream (neglecting viscous interaction effects), as taken from Ref. 5, are reproduced in Fig. 8.6. As noted by Surber and Olsen,[5] "Values of hypersonic lift coefficient and lift-to-drag ratio have been maintained at levels consistent with aeroheating constraints and trajectory requirements. The pitch trim range is sufficiently wide and well centered with respect to the design center of gravity range."

The side force necessary for generating cross range was obtained for the Shuttle Orbiter (as it was obtained for the simpler capsule

Table 8.1   Aerodynamic parameters for early spacecraft programs (Ref. 4).

Program	L^{*}_{ref} (in.)	$A^{\dagger}_{ref}$ (ft^2)	Entry C_D	Entry (L/D)	$\dfrac{W}{C_D A_{ref}} \left(\dfrac{lbf}{ft^2}\right)$
Mercury	74.5	30.27	1.60	0.00	54.98
Gemini	90.0	44.18	1.60	0.15	67.94
Apollo	154.0	129.35	1.45	0.35	66.63
Space Shuttle Orbiter	1290.3	2690.00	0.84	1.00	88.07

* L_{ref} is the maximum cross-section diameter for the Mercury, for the Gemini, and for the Apollo (see Fig. 6.11), and is the length of the Space Shuttle Orbiter.

† A_{ref} is the maximum cross-section area for the Mercury, for the Gemini, and for the Apollo, and is the wing planform area for the Space Shuttle Orbiter. A_{ref} is designated S_{ref}, or simply S, by the airplane aerodynamicist.

shapes) by rotating the lift vector out of the vertical plane. Thus, although the Shuttle Orbiter was designed to be a reusable vehicle that lands horizontally, for most of the hypersonic phase of re-entry flight, the Orbiter flies like a spacecraft rather than like an airplane.[6] Some fighter concepts incorporate direct side force, e.g., the AFTI F-16.

As indicated in Fig. 8.5, configurations with relatively high lift-to-drag ratios are generally characterized by highly swept winged configurations possessing low-bluntness ratios and high-fineness ratios. The maximum values of the (L/D) ratio for these configurations occur at comparatively low angles-of-attack. Viscous effects, i.e., the skin-friction component of drag, become a significant fraction of the overall drag for slender vehicles with very small leading-edge bluntness operating at low angle-of-attack. As a result, wetted area becomes important. High values of the volumetric efficiency $(V^{0.667}/S_w)$ translate into reduced wetted area, which means reduced skin-friction drag. Wetted area is also a first-order indicator of the acreage which must be thermally protected. Consequently, reduced wetted surface area yields higher values for the volumetric efficiency and reduced thermal-protection system weight.

Carefully designed configurations can achieve values of $(L/D)_{max}$ of approximately 6 at hypersonic speeds. Representative values[7] of $(L/D)_{max}$ are presented as a function of the Mach number in Fig. 8.7.

A vehicle designed for hypersonic flight at Mach 12 with a range of 16,490 km at an altitude of 40 km is shown in Fig. 8.8. The

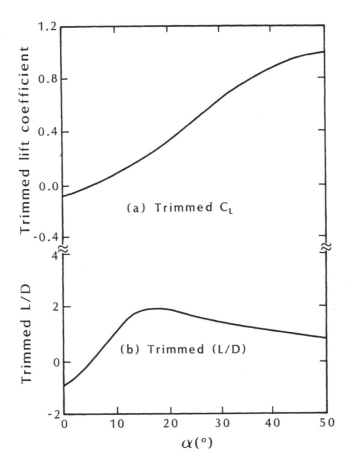

Fig. 8.6 Space Shuttle Orbiter longitudinal aerodynamic characteristics for hypersonic speeds, S_{ref} = 2690 ft^2, x_{cg} = 0.65L, δ_{BF} = − 11.7 deg., δ_{SB} = 25 deg., as taken from Ref. 5.

vehicle has deployable canards for takeoff and for transonic flight. The design also includes "telescoping" wing tips, which increase the wing span from 35 m to 48 m. The vehicle is designed to generate a hypersonic lift-to-drag ratio of approximately 5 at an angle-of-attack of 8 deg.

Gottmann[8] notes that waveriders could be the only relevant configuration, if the corresponding long ranges are to be achieved at hypersonic Mach number. Anderson et al.[9] note:

A waverider is a supersonic or hypersonic vehicle which, at the design point, has an *attached* shock wave all along its leading edge, as sketched in Fig. 8.9a. Because of this, the vehicle appears to be riding on top of its shock wave,

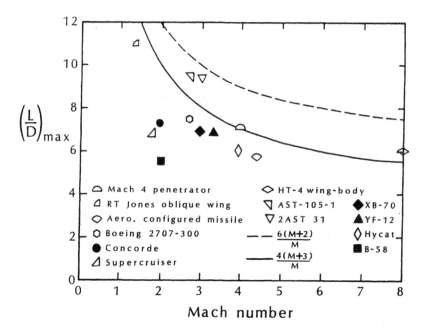

Fig. 8.7   Representative values of (L/D)ₘₐₓ, as taken from Ref. 7.

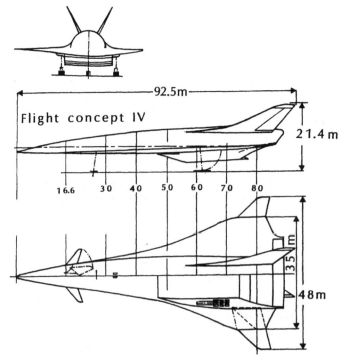

Fig. 8.8   Hypersonic cruise configuration, as taken from Ref. 8.

hence the term "waverider." This is in contrast to a more conventional hypersonic vehicle, where the shock wave is usually detached from the leading edge, as sketched in Fig. 8.9b. The aerodynamic advantage of the waverider in Fig. 8.9a is that the high pressure behind the shock wave under the vehicle does not "leak" around the leading edge to the top surface; the flowfield over the bottom surface is contained, and the high pressure is preserved. In contrast, for the vehicle shown in Fig. 8.9b, there is communication between the flows over the bottom and top surfaces; the pressure tends to "leak" around the leading edge, and the general integrated pressure level on the bottom surface is reduced, resulting in less lift. Because of this, the generic vehicle in Fig. 8.9b must fly at a larger angle of attack, α, to produce the same lift as the waverider in Fig. 8.9a. This is illustrated in Fig. 8.10, where the lift curves (L versus α) are sketched for the two vehicles·in Fig. 8.9. At the same lift, points 1a and 1b in Fig. 8.10 represent the waverider and generic vehicles respectively. Also shown in Fig. 8.10 are typical variations of L/D versus α, which for slender hypersonic vehicles are not too different for shapes in Fig. 8.9a and 8.9b. However, note that because

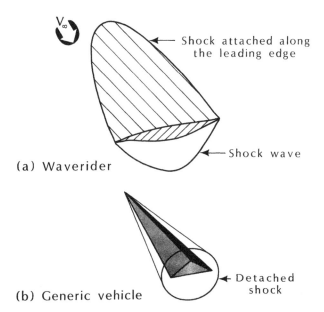

(a) Waverider

Shock attached along the leading edge

Shock wave

(b) Generic vehicle

Detached shock

V_∞

Fig. 8.9   Comparison of a waverider with a generic hypersonic configuration, as taken from Ref. 9.

the waverider generates the same lift at a smaller α (point 1a in Fig. 8.10), the L/D for the waverider is considerably higher (point 1aa) than that for the generic shape (point 1bb).

Anderson et al.[9] note further that "When waveriders are optimized for maximum L/D, the previous studies have demonstrated that the driving parameter that alters the L/D ratio is the skin friction drag."

For waveriders operating at Mach numbers less than the design value or at angles-of-attack greater than the design value, the shock wave separates from the wing leading edges. As a result, the flow around the leading edges separates, rolling up to form spiral vortex sheets. The vortices cause low pressures on the upper surface of the wing, producing a considerable amount of lift, thus, improving the lift-to-drag ratio. At subsonic speeds, low pressures due to leading-edge vortices provide roughly one half the total lift.

Increases in the lift-to-drag ratio provide increased operational flexibility. The flowfields for low (L/D) configurations are relatively simple. The flowfields for high (L/D) configurations are complex, containing viscous/inviscid interactions which can cause locally severe heat fluxes and locally severe, fluctuating aerodynamic loads. However, as complex as the flowfield for a high (L/D) hypersonic ve-

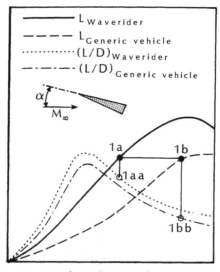

Angle-of-attack, α

Fig. 8.10 Curves of lift and (L/D) versus angle-of-attack: Comparison between a waverider and a generic vehicle, as taken from Ref. 9.

hicle is, one cannot develop the optimum configuration design considering only the aerodynamic efficiency. As discussed by Blankson,[10] "the integration and interaction of the propulsion system, aerodynamics, aerodynamic heating, stability and control, materials and structures, fuel, thermal management, energy availability, missions, trajectories, etc., are unique and very important aspects of hypersonic vehicle design." High (L/D) hypersonic configurations require considerable time and effort to design and to build, the costs escalating rapidly with (L/D).

8.2 NEWTONIAN AERODYNAMIC COEFFICIENTS

The advantages and shortcomings of using the (modified) Newtonian flow model to calculate the pressure distribution were discussed in Chap. 6. A principal advantage is simplicity. The calculation of the local pressure coefficient requires only the knowledge of the angle between the surface normal and the freestream velocity vector and, therefore, is essentially a geometry problem. However, there are many flowfield phenomena that produce significant differences between the actual pressures and those given by the (modified) Newtonian flow model. Recognizing its limitations, the Newtonian flow model finds use as a tool to evaluate quickly the effect of geometric features or to generate preliminary design information. The Newtonian flow model was used in the preliminary design of the Shuttle Orbiter. Surber and Olsen[5] noted that:

> The hypersonic wing lift is primarily a function of exposed wing area, and is not appreciably affected by other wing geometrical parameters ... The wing-alone normal force coefficient was developed from a Newtonian flow model ... The wing center of pressure location was obtained from a comparison of hypersonic wind tunnel data and Newtonian flow predictions ... The hypersonic fuselage center of pressure was determined from a Newtonian flow simulation and was located at 49.8% of the fuselage length at an angle of attack of 35°.

8.2.1 Sharp Cones

Sharp cones, such as depicted in Fig. 8.1, make excellent subjects for code-calibration and for code-validation exercises. Because the flowfield is relatively simple, those who develop CFD models of physical phenomena and those who develop instrumentation to obtain the verification measurements should have *fewer* problems. The word *fewer* is italicized because substantial problems exist even for relatively simple applications, e.g., a turbulent boundary layer on a sharp cone in a perfect-gas, Mach 9 wind-tunnel flow. Because of the uncertainties associated with grid generation and with turbulence modeling, four computations[11] yielded four different solutions for the heat transfer.

Severe heat-transfer rates at the apex of a sharp cone in hyper-velocity flight could cause the nose to degrade if the flight lasted for more than a few seconds. Nevertheless, because sharp conical shapes (of various cross sections) have high ballistic coefficients, they are used as hypervelocity penetrators, where the flight times are very short.

The relations for hypersonic flow past a sharp cone, where $-\theta_c \leq \alpha \leq +\theta_c$, are derived in Ref. 12 using the pressure distributions of the modified Newtonian flow model. The pressure distribution is given by:

$$C_p = C_{p,t2} \left(\cos \alpha \sin \theta_c + \sin \alpha \cos \theta_c \cos \beta\right)^2 \qquad (8\text{-}2)$$

The axial force coefficient is found by integrating the pressure force over the entire (closed) surface of the cone:

$$C_A = \frac{1}{\left(\frac{1}{2}\rho_\infty U_\infty^2\right)\left(\pi R_B^2\right)} \oiint_S (p - p_\infty)(\hat{n}dS) \cdot \hat{\imath} \qquad (8\text{-}3)$$

$$C_A = C_{p,t2} \left[\sin^2 \theta_c + 0.5 \sin^2 \alpha \left(1 - 3\sin^2 \theta_c\right)\right] \qquad (8\text{-}4)$$

Similarly, the normal force coefficient due to pressures acting over the closed surface of a cone is:

$$C_N = \frac{1}{\left(\frac{1}{2}\rho_\infty U_\infty^2\right)\left(\pi R_B^2\right)} \oiint_S (p - p_\infty)(\hat{n}dS) \cdot (-\hat{\jmath}) \qquad (8\text{-}5)$$

The $(-\hat{\jmath})$ is used, since the normal force is positive in the negative y-direction. Refer to Fig. 8.1. The resultant normal force coefficient is:

$$C_N = \frac{C_{p,t2}}{2} \sin 2\alpha \cos^2 \theta_c \qquad (8\text{-}6)$$

The lift coefficient is:

$$C_L = C_N \cos \alpha - C_A \sin \alpha \qquad (8\text{-}7)$$

so that

$$C_L = C_{p,t2} \sin \alpha \left[\cos^2 \alpha \cos^2 \theta_c - \sin^2 \theta_c - 0.5 \sin^2 \alpha \left(1 - 3\sin^2 \theta_c\right)\right] \qquad (8\text{-}8)$$

Similarly, the drag coefficient is:

$$C_D = C_N \sin \alpha + C_A \cos \alpha \qquad (8\text{-}9)$$

so that:

$$C_D = C_{p,t2} \cos \alpha \left[\sin^2 \alpha \cos^2 \theta_c + \sin^2 \theta_c + 0.5 \sin^2 \alpha \left(1 - 3 \sin^2 \theta_c \right) \right]$$
$$(8\text{-}10)$$

To calculate the moments due to the incremental pressure force acting at a radial distance r from the origin,

$$d\boldsymbol{M} = \boldsymbol{r} \times p \, \hat{\boldsymbol{n}} \, dS \tag{8-11}$$

If we want to take moments about the apex of the sharp cone, the moment arm to a point on the surface is:

$$\boldsymbol{r} = x \, \hat{\boldsymbol{i}} + x \tan \theta_c \cos \beta \, \hat{\boldsymbol{j}} + x \tan \theta_c \sin \beta \, \hat{\boldsymbol{k}} \tag{8-12}$$

and the unit inward normal is:

$$\hat{\boldsymbol{n}} = \sin \theta_c \, \hat{\boldsymbol{i}} - \cos \theta_c \cos \beta \, \hat{\boldsymbol{j}} - \cos \theta_c \sin \beta \, \hat{\boldsymbol{k}} \tag{8-13}$$

Note that the incremental moment $d\boldsymbol{M}$ is a vector, which can be written in terms of its components:

$$d\boldsymbol{M} = dL \, \hat{\boldsymbol{i}} + dN \, \hat{\boldsymbol{j}} + dM \, \hat{\boldsymbol{k}} \tag{8-14}$$

In Eq. (8-14), L is the rolling moment (which is positive when causing the right wing to move down), N is the yawing moment (which is positive when causing the nose to move to the right), and M is the pitching moment (which is positive when causing a nose-up motion). The rolling moment is zero in the absence of viscous forces, since the pressure forces act normal to the surface and, therefore, act through the axis-of-symmetry for a body-of-revolution. For motions in the pitch plane, only a pitching moment will exist.

Although the resultant pitching moment is produced by the integration of the distributed pressures acting over the vehicle surface, it can be represented as composed of two components, one effectively due to the normal-force component and the other due to the axial-force component. The pitching moment about the apex can be written as the sum of the two components:

$$C_{M_0} = -\frac{2}{\pi R_B^2 R_B} \int_0^L \left(\int_0^\pi C_p x \tan \theta_c \, d\beta \frac{dx}{\cos \theta_c} x \cos \theta_c \cos \beta \right)$$

$$-\frac{2}{\pi R_B^2 R_B} \int_0^L \left(\int_0^\pi C_p x \tan \theta_c \, d\beta \frac{dx}{\cos \theta_c} x \sin \theta_c \tan \theta_c \cos \beta \right) \tag{8-15}$$

The negative sign results because a pressure force acting in the first and in the fourth quadrants ($\pi/2 \geq \beta \geq -\pi/2$) produces a nose-down (negative) pitching moment. Integrating the two terms in Eq. (8-15) yields:

$$C_{M_0} = -\frac{C_{p,t2} \sin 2\alpha \cos^2 \theta_c}{3} \frac{L}{R_B} - \frac{C_{p,t2} \sin 2\alpha \sin^2 \theta_c}{3} \frac{L}{R_B} \quad (8\text{-}16)$$

Continuing the division of the pitching moment into two terms, one due to the normal force and one due to the axial force, and adding the concept that these forces act through the center-of-pressure (see Fig. 8.1):

$$C_{M_0} = \frac{M_0}{q_\infty A_{\text{Base}} R_B} = -C_N \frac{x_{cp}}{R_B} - C_A \frac{y_{cp}}{R_B} \quad (8\text{-}17)$$

Since the first term in Eq. (8-16) is due to the normal-force component, we can use the expression for the normal-force coefficient, i.e., Eq. (8-6), together with Eq. (8-17), to determine the x-coordinate of the center-of-pressure.

$$x_{cp} = \frac{2}{3}L \quad (8\text{-}18)$$

Similarly, the second term in Eq. (8-16) is due to the axial-force component. Thus, substituting the expression for the axial-force coefficient, i.e., Eq. (8-4), into Eq. (8-17) yields the expression for the y-coordinate of the center-of-pressure

$$y_{cp} = \frac{\sin 2\alpha \sin^2 \theta_c L}{3 \left[\sin^2 \theta_c + 0.5 \sin^2 \alpha \left(1 - 3 \sin^2 \theta_c\right)\right]} \quad (8\text{-}19)$$

Note that, since
$$\cos^2 \theta_c + \sin^2 \theta_c = 1$$

and
$$R_B = L \tan \theta_c$$

the two terms of Eq. (8-16) can be easily combined to yield:

$$C_{M_0} = -\frac{C_{p,t2} \sin 2\alpha}{3 \tan \theta_c} \quad (8\text{-}20)$$

Although Eq. (8-20) is more compact, the two contributions to the pitching moment, i.e., that due to the normal force and that due to the axial force, can be more easily identified in Eq. (8-16).

Exercise 8.1:

Force and moment coefficients were obtained[13] for a sharp cone ($\theta_c = 9$ deg.) located in a Mach 6.77 stream over an angle-of-attack range from 0 deg. (sharp end into the freestream) to 180 deg. (flat conical base facing the oncoming flow). The freestream unit Reynolds number was 1.35×10^5/in. In a compilation of longitudinal aerodynamic characteristics, Foster[14] reproduced these data and compared them with the corresponding coefficients computed using the Newtonian flow model. The data and the Newtonian-based computations for C_A, C_N, C_L, C_D, (L/D), and C_M, as taken from Ref. 14, are reproduced in Fig. 8.11. The reference point for the pitching moment is a point on the axis-of-symmetry located $0.6355L$ from the apex. The reference area for the coefficients is the cone base area (πR_B^2), as was the case for Eqs. (8-3) through (8-16). However, the reference length for the pitching moment was D_B, not R_B as was used for Eqs. (8-15) and (8-16).

Using the Newtonian flow model and the modified Newtonian flow model, calculate values for these six parameters for an angle-of-attack of 60 deg.

Solution:

Since the angle-of-attack exceeds the cone half angle, at least part of the leeward side lies in the shadow region where $C_p = 0$ (see Fig. 6.4). Nevertheless, the oncoming flow "sees" at least part of the cone to the lee of the yaw plane, which corresponds to $\beta = (\pi/2)$ and $\beta = -(\pi/2)$. This can be seen in the sketch of Fig. 8.12, which depicts the configuration of interest. The Newtonian flow pressure coefficients remain positive in the region between the yaw plane (dashed line of the front view) and the outer projection of the cone. To determine the circumferential angle β where the Newtonian pressure coefficient first goes to zero, set C_p equal to zero in Eq. (8-2) and solve for β. This value of β will serve as the upper limit in the integration of the pressures to determine the aerodynamic coefficients and is designated β_u. Thus,

$$\beta_u = \cos^{-1}\left[-\frac{\tan\theta_c}{\tan\alpha}\right] = 95.2466 \text{ deg.} = 1.6624 \text{ rad}$$

The value of β_u is depicted in Fig. 8.13.

To calculate the axial-force coefficient, refer to Eq. (8-3). Referring to Fig. 8.13, the surface area dS is:

$$dS = r\, d\beta\, ds = x\tan\theta_c\, d\beta\frac{dx}{\cos\theta_c}$$

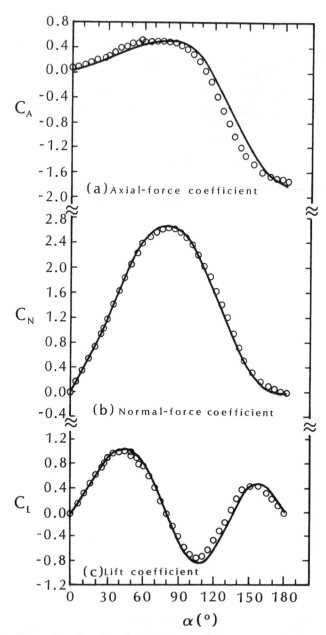

Fig. 8.11 Longitudinal aerodynamic characteristics for a sharp cone as a function of angle-of-attack, $M_1 = 6.77$, Re/in. $= 1.35 \times 10^5$, $\theta_c = 9$ deg., as taken from Refs. 13 and 14.

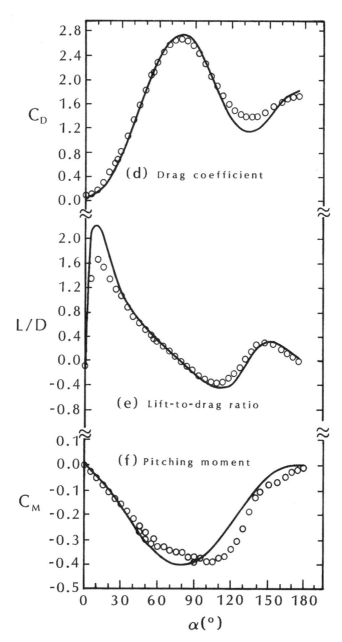

Fig. 8.11 (continued) Longitudinal aerodynamic characteristics for a sharp cone as a function of angle-of-attack, $M_1 = 6.77$, Re/in. $= 1.35 \times 10^5$, $\theta_c = 9$ deg., as taken from Refs. 13 and 14.

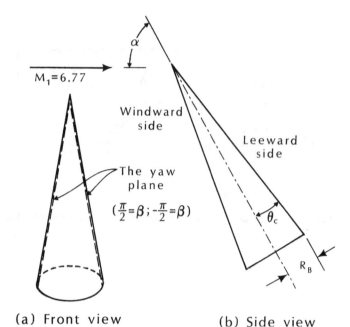

(a) Front view   (b) Side view

Fig. 8.12 Sketch of sharp cone (θ_c = 9 deg.) at a high angle-of-attack (α = 60 deg.) for Exercise 8.1.

where ds is the wetted length along the surface. Also,

$$\hat{n} \cdot \hat{i} = \sin \theta_c$$

For $\beta \geq \beta_u$ and on the flat base of the cone, the pressure coefficient is zero, i.e., $p = p_\infty$. Thus,

$$C_A = \frac{2}{\left(\frac{1}{2}\rho_\infty U_\infty^2\right)\left(\pi R_B^2\right)} \int_0^L \int_0^{\beta_u} (p - p_\infty)\, x \tan \theta_c \, d\beta \frac{dx}{\cos \theta_c} \sin \theta_c$$

$$C_A = \frac{2C_{p,t2}}{\pi R_B^2} \int_0^L \left[\int_0^{\beta_u} \left(\cos^2 \alpha \sin^2 \theta_c + 0.5 \sin 2\alpha \sin 2\theta_c \cos \beta \right. \right.$$

$$\left. \left. + \sin^2 \alpha \cos^2 \theta_c \cos^2 \beta \right) \tan^2 \theta_c \, d\beta \right] x \, dx$$

$$C_A = \frac{2C_{p,t2}}{\pi R_B^2} \tan^2 \theta_c \frac{L^2}{2} \left[\cos^2 \alpha \sin^2 \theta_c \beta_u \right.$$

$$\left. + 0.5 \sin 2\alpha \sin 2\theta_c \sin \beta_u + \sin^2 \alpha \cos^2 \theta_c \left(\frac{\beta_u}{2} + \frac{\sin 2\beta_u}{4} \right) \right]$$

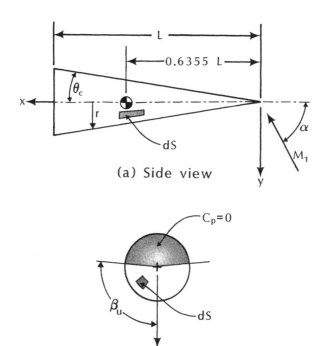

(a) Side view

(b) End view

Fig. 8.13 Nomenclature for aerodynamic calculations of Exercise 8.1.

Thus,

$$C_A = 0.2286 C_{p,t2}$$

To calculate the normal-force coefficient,

$$\hat{n} \cdot (-\hat{\jmath}) = \cos \theta_c \cos \beta$$

so that:

$$C_N = \frac{2}{\left(\frac{1}{2}\rho_\infty U_\infty^2\right)\left(\pi R_B^2\right)} \int_0^L \int_0^{\beta_u} (p - p_\infty)\, x \tan \theta_c \; d\beta \; dx \cos \beta$$

$$C_N = \frac{2 C_{p,t2}}{\pi R_B^2} \int_0^L \int_0^{\beta_u} \left(\cos^2 \alpha \sin^2 \theta_c + 0.5 \sin 2\alpha \sin 2\theta_c \cos \beta \right.$$

$$\left. + \sin^2 \alpha \cos^2 \theta_c \cos^2 \beta \right) \cos \beta \; d\beta \; (\tan \theta_c x \; dx)$$

$$
C_N = \frac{2C_{p,t2}}{\pi R_B^2} \tan \theta_c \frac{L^2}{2} \Bigg[\cos^2 \alpha \sin^2 \theta_c \sin \beta_u
$$

$$
+ \, 0.5 \sin 2\alpha \sin 2\theta_c \left(\frac{\beta_u}{2} + \frac{\sin 2\beta_u}{4} \right)
$$

$$
+ \sin^2 \alpha \cos^2 \theta_c \left(\frac{1}{3} \sin \beta_u \right) \left(\cos^2 \beta_u + 2 \right) \Bigg]
$$

Thus,

$$
C_N = 1.2038 \, C_{p,t2}
$$

$$
C_L = C_N \cos \alpha - C_A \sin \alpha = 0.4039 \, C_{p,t2}
$$

$$
C_D = C_N \sin \alpha + C_A \cos \alpha = 1.1568 \, C_{p,t2}
$$

$$
\frac{L}{D} = \frac{C_L}{C_D} = 0.3492
$$

To calculate the pitching moment about a point on the axis-of-symmetry, which is $0.6355L$ from the apex:

$$
C_{Mref} = -\frac{2}{\pi R_B^2 D_B} \int_0^L \int_0^{\beta_u} C_p \, x \tan \theta_c \, d\beta \frac{dx}{\cos \theta_c}
$$

$$
\times \left[(x - x_{ref}) \cos \theta_c \cos \beta + x \sin \theta_c \tan \theta_c \cos \beta \right]
$$

where $x_{ref} = 0.6355L$, the location of the moment reference point.

$$
C_M = -\frac{2C_{p,t2}}{\pi R_B^2 D_B} \Bigg[\int_0^L (x - x_{ref}) x \tan \theta_c \, dx \int_0^{\beta_u} \Big(\cos^2 \alpha \sin^2 \theta_c \cos \beta
$$

$$
+ \, 0.5 \sin 2\alpha \sin 2\theta_c \cos^2 \beta + \sin^2 \alpha \cos^2 \theta_c \cos^3 \beta \Big) \, d\beta
$$

$$
+ \int_0^L x^2 \tan^3 \theta_c \, dx \int_0^{\beta_u} \Big(\cos^2 \alpha \sin^2 \theta_c \cos \beta
$$

$$
+ \, 0.5 \sin 2\alpha \sin 2\theta_c \cos^2 \beta + \sin^2 \alpha \cos^2 \theta_c \cos^3 \beta \Big) \, d\beta \Bigg]
$$

So that:

$$C_M = -\frac{2C_{p,t2}}{\pi R_B^2 D_B}\left[\frac{L^3}{3}\tan\theta_c - \frac{L^2(0.6355L)}{2}\tan\theta_c + \frac{L^3}{3}\tan^3\theta_c\right]$$

$$\times\left[\cos^2\alpha\sin^2\theta_c\sin\beta_u + 0.5\sin 2\alpha\sin 2\theta_c\left(\frac{\beta_u}{2} + \frac{\sin 2\beta_u}{4}\right)\right.$$

$$\left.+ \sin^2\alpha\cos^2\theta_c\left(\frac{1}{3}\sin\beta_u\right)\left(\cos^2\beta_u + 2\right)\right]$$

$$C_M = -0.1906\,C_{p,t2}$$

For the Newtonian flow model, $C_{p,t2} = 2$. For the modified Newtonian flow model, when the freestream Mach number is 6.77,

$$C_{p,t2} = \left(\frac{p_{t2}}{p_1} - 1\right)\frac{2}{\gamma M_1^2} = 1.8230$$

Thus,

Parameter	General	Newtonian flow model	Modified Newtonian flow model
C_A	$0.2286C_{p,t2}$	0.4572	0.4168
C_N	$1.2038C_{p,t2}$	2.4075	2.1945
C_L	$0.4039C_{p,t2}$	0.8078	0.7363
C_D	$1.1568C_{p,t2}$	2.3136	2.1088
L/D	0.3492	0.3492	0.3492
C_M	$-0.1906C_{p,t2}$	-0.3812	-0.3474

The aerodynamic coefficients calculated using the Newtonian flow model are in better agreement with the data than those calculated using the modified Newtonian pressures. Such was the case for the pressure distributions for a sharp cone that were presented in Fig. 6.9. This relatively simple flow model provides reasonable estimates of the magnitude and of the angle-of-attack dependence for the aerodynamic parameters except for the lift-to-drag ratio at low angles-of-attack and for the pitching moment coefficient at angles-of-attack of 60 deg. and greater. Neal[13] said that the unusual behavior evident in the variation of the pitching moment coefficients may be due to the flow between the shock wave and the body changing from supersonic to subsonic in this angle-of-attack range. When this occurs, cone surface-pressure "bleed-off" near the base can become appreciable, yielding the experimental trends evident in these data.

However, one should not assume that the good agreement that exists between the measured coefficients and the calculated values based on this relatively simple flow model would always be obtained. The coefficients calculated in Exercise 8.1 did not take into account the base pressure or the skin friction. Glover and Hagan[15] note, "At $M = 10$ approximately 60% of C_A results from pressure drag (C_{A_p}), 30% from skin friction (C_{A_F}), and about 10% from base drag (C_{A_B}). The percentages vary somewhat with M, Re, and type of boundary layer, but the given values are typical of the relative importance of the three sources of axial force for low-drag reentry bodies. For blunted bodies, the pressure drag may account for nearly 100% of the total drag." Of course, a sharp cone at 60 deg. angle-of-attack should be classified as a blunt body. Nevertheless, one should always carefully evaluate the validity of the assumptions inherent in the flow models used in the computations.

Exercise 8.2:

Adams[16] computed the aerodynamic parameters for a sharp cone ($\theta_c = 7$ deg.) in a hypersonic stream of perfect air ($M_\infty = 16$ and $Re_{\infty,L} = 1 \times 10^6$). The boundary layer was assumed to be laminar at these conditions. The three-dimensional flowfield solutions were obtained using a Hypersonic Viscous Shock Layer (HVSL) code employing an implicit finite-difference numerical solution and marching in the streamwise direction. The equations model the entire shock layer (both the inviscid region and the viscous, laminar region) including a possible circumferential separation zone which develops on the leeward side at the higher angles-of-attack. The model would not be valid if streamwise separation occurs, i.e., at very high angles-of-attack. The aerodynamic parameters computed using the HVSL code are compared with those calculated using the Newtonian flow model.

For the pitching moment about the apex, Adams[16] uses the notation that:

$$C_{M_0} = \frac{M_0}{q_\infty A_{\text{Base}} L} \tag{8-21}$$

and that:

$$M_0 = -x_{cp} N \tag{8-22}$$

Using the Adams notation and Eq. (8-6) for the normal force and Eq. (8-20) for the pitching moment, develops expressions for C_{M_0} and for x_{cp}.

Solution:

Solving Eq. (8-6) for the normal forces:

$$N = \frac{C_{p,t2}}{2} \sin 2\alpha \cos^2 \theta_c \, q_\infty A_{\text{Base}}$$

Solving Eq. (8-20) for the pitching moment:

$$M_0 = -\frac{C_{p,t2} \sin 2\alpha}{3 \tan \theta_c} q_\infty A_{\text{Base}} R_B$$

Substituting this expression for the pitching moment about the apex into Eq. (8-21), the pitching moment coefficient is:

$$C_{M_0} = -\frac{C_{p,t2} \sin 2\alpha}{3} \tag{8-23}$$

since $R_B = L \tan \theta_c$. Note that, using L as the characteristic length instead of R_B (or D_B), yields an expression for the pitching moment coefficient that is independent of θ_c.

Combining the expressions for M_0 and for N into Eq. (8-22) and solving for x_{cp} yields:

$$x_{cp} = -\frac{\left(-\dfrac{C_{p,t2} \sin 2\alpha}{3 \tan \theta_c}\right) \left(q_\infty A_{\text{Base}} R_B\right)}{\left(\dfrac{C_{p,t2} \sin 2\alpha \cos^2 \theta_c}{2}\right) \left(q_\infty A_{\text{Base}}\right)}$$

Thus,

$$\frac{x_{cp}}{L} = \frac{2}{3 \cos^2 \theta_c} \tag{8-24}$$

The expression for x_{cp} given by Eq. (8-24) differs from that given by Eq. (8-18), because Eq. (8-22) from which it is derived neglects the contribution of the axial force to the pitching moment. However, the simplified model of Eq. (8-22) is commonly used to define the center-of-pressure location.

The aerodynamic parameters computed by Adams[16] using the HVSL code are compared in Fig. 8.14 with the corresponding values calculated using the Newtonian flow model over a range of α from 0 deg. to 25 deg. The two techniques yield values for the normal-force coefficients and for the pitching moment coefficients that are in

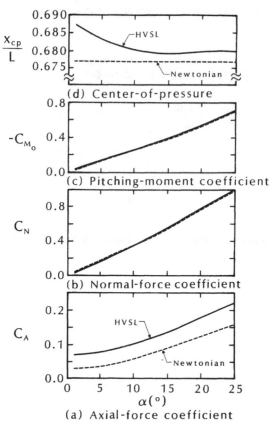

(d) Center-of-pressure

(c) Pitching-moment coefficient

(b) Normal-force coefficient

(a) Axial-force coefficient

Fig. 8.14 Comparison of HVSL and Newtonian aerodynamic parameters for a sharp cone (θ_c = 7 deg.) in a hypersonic stream (M_∞ = 16), as taken from Ref. 16.

good agreement. The differences between the axial-force coefficients reflect the importance of skin friction for this slender configuration. Based on the HVSL computations, the skin-friction force is approximately 47 percent of the total axial force for an angle-of-attack of 1 deg. and is approximately 25 percent of the total axial force for an angle-of-attack of 25 deg. Note that the center-of-pressure locations computed using the HVSL code are always aft the Newtonian locations. The increasing values of (x_{cp}/L) as the angle-of-attack decreases are consistent with data obtained in Tunnel F at AEDC at these conditions.[17]

8.2.2 Spherically Blunted Cones

In Chap. 5, we learned that the convective heating to the stagnation point of a sphere is inversely proportional to the square root of the nose radius. In Chap. 6, we learned that blunting the nose of a cone caused significant changes in the pressure distribution over a

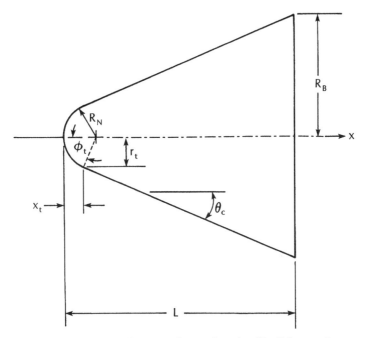

Fig. 8.15 Nomenclature for spherically blunted cones.

sphere/cone and that blunting could affect the properties at the edge of the boundary layer for more than 100 nose radii downstream. The effect of blunting on boundary-layer transition was discussed in Chap. 7. In this chapter, the effect of nose blunting on the aerodynamic parameters of conic configurations will be discussed. The characteristic geometry parameters for these spherically blunted cones are defined in Fig. 8.15.

As noted by Zoby and Thompson,[18] nose bluntness can be used to minimize drag and convective heat transfer. The optimum nose bluntness can be based on parameters such as the volume, stability derivatives, and flowfield effects such as pressure losses and mass-flow capture, in addition to the drag or to the heating rate. The optimum shape may also depend on the freestream conditions, such as the Reynolds number. Zoby and Thompson[18] presented the results of detailed computations of the flow over a blunted cone ($\theta_c = 5$ deg.). The flowfield solutions were obtained for a freestream Mach number of 15 at an altitude of 150,000 ft using a three-dimensional viscous shock layer (VSL3D) code. Computations were made for two different nose radii (R_N of 0.125 ft and of 0.750 ft) for fixed cone half angle and base radius. Although the increased blunting results in a physically shorter body, i.e., 135 ft as compared with 140 ft, with a smaller volume, these changes are a small fraction of the total

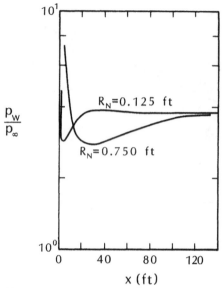

Fig. 8.16   The effect of nose radius on the surface-pressure distribution for a slender cone, θ_c = 5 deg., M_∞ = 15, altitude = 150,000 ft., γ = 1.4, α = 0 deg., as taken from Ref. 18.

vehicle size. Flowfield solutions were obtained for two assumptions for the boundary layer: for one, the boundary layer was assumed to be entirely laminar and, for the second, boundary-layer transition occurred using a transition correlation based on Re-entry F flight data.

   The effect of the nose radius on the predicted surface pressure distribution over the 5 deg. cone with laminar flow is illustrated in Fig. 8.16. The pressure over the smaller nose-radius body is dominated by the nose radius for the first 25 feet only. Downstream, the pressure is constant and equal to the sharp cone pressure. For the larger nose radius, bluntness affects the pressure over the entire configuration, as shown in Fig. 8.16, with the pressure being relatively low over most of the conical surface. The pressure-drag coefficients which were computed by integrating the pressure distributions are given in Table 8.2. Although the pressure-drag coefficient increased as the nose radius was increased, as expected, the increase was relatively small. However, the skin-friction-drag coefficient was significantly lower for the blunter configuration. Thus, when the skin-friction drag and the pressure-drag components are added together, the total drag is lower for the blunter cone.

   Coupled with the benefit of reduced drag coefficient, increasing the nose radius also reduces the stagnation-point heating rate and the laminar heating to the conical frustrum. The heat-transfer distri-

Table 8.2   Predicted drag coefficient and heating on a 5 deg. blunted cone, M_∞ = 15, altitude = 150,000 ft, γ = 1.4, α = 0 deg., as taken from Ref. 18.

R_N (ft)	C_{D_P}	C_{D_F}	C_D	$\dot{Q}$ (Btu/s)
(a) Boundary layer is wholly laminar				
0.125	0.01771	0.00472	0.02243	9800
0.750	0.01855	0.00358	0.02213	8390
(a) Transition occurs				
0.125	0.01885	0.01687	0.03572	35070
0.750	0.01940	0.01078	0.03018	23990

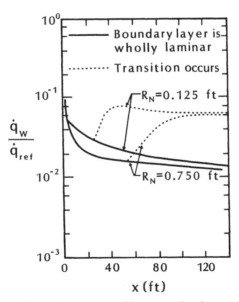

Fig. 8.17   The effect of nose radius on the heat-transfer distribution for a slender cone, θ_c = 5 deg., M_∞ = 15, altitude = 150,000 ft., γ = 1.4, α = 0 deg., as taken from Ref. 18.

butions computed by Zoby and Thompson[18] are reproduced in Fig. 8.17. Comparing the total heating rates to the vehicle surface (integrated over the surface) given in Table 8.2 indicates a significant heating decrease for the larger nose radius for both boundary-layer assumptions. Note that both the convective heat transfer and the skin friction exhibit similar dependence on the nose radius, as one would expect from Reynolds analogy.

8.2.2.1 The equations for the Newtonian flow model.

In this section, we will develop the force coefficients for a spherically blunted cone over an angle-of-attack range of $-\theta_c \leq \alpha \leq \theta_c$ using the (modified) Newtonian flow model to define the pressure distribution. For simplicity, we will develop expressions for the forces in a body-fixed coordinate system, i.e., the axial force and the normal force. The resultant computed force coefficients will be compared with data from the Trisonic Wind Tunnel (TWT) at the U.S. Air Force Academy.

The axial-force coefficient is given by Eq. (8-3). The wetted area dS is given by:

$$dS = r \, d\beta \, ds$$

On the spherical cap (refer to Fig. 8.15),

$$ds = R_N d\phi \quad \text{and} \quad r = R_N \sin \phi$$

where $\phi = 90 \text{ deg.} - \theta$. On the conical surface,

$$dS = r \, d\beta \frac{dx}{\cos \theta_c} \quad \text{and} \quad r = r_t + (x - x_t) \tan \theta_c$$

Since the angle-of-attack range is $\pm \theta_c$, $\beta_u = \pi$. Thus,

$$
\begin{aligned}
C_A = \frac{2C_{p,t2}}{\pi R_B^2} \Bigg[& \int_0^{\phi_t} \int_0^\pi \left(\cos^2 \alpha \cos^2 \phi + 2 \sin \alpha \cos \alpha \cos \phi \sin \phi \cos \beta \right. \\
& \left. + \sin^2 \alpha \sin^2 \phi \cos^2 \beta \right) R_N \sin \phi \, d\beta \, R_N \, d\phi \cos \phi \\
& + \int_{x_t}^L \int_0^\pi \left(\cos^2 \alpha \sin^2 \theta_c + 2 \sin \alpha \cos \alpha \sin \theta_c \cos \theta_c \cos \beta \right. \\
& \left. + \sin^2 \alpha \cos^2 \theta_c \cos^2 \beta \right) (r_t + (x - x_t) \tan \theta_c) \, d\beta \frac{dx}{\cos \theta_c} \sin \theta_c \Bigg]
\end{aligned}
$$

$$(8\text{-}25)$$

where the first term of Eq. (8-25) is for the spherical cap and the second term is for the conical surface. Integrating, first with respect to β and then with respect to ϕ or to x for the nose or for the conical frustrum, respectively, yields:

$$C_A = \frac{2C_{p,t2}}{R_B^2} \left[R_N^2 \left(0.25 \cos^2 \alpha (1 - \cos^4 \phi_t) \right. \right.$$
$$+ \left. 0.125 \sin^2 \alpha \sin^4 \phi_t \right) + \tan \theta_c \left(\cos^2 \alpha \sin^2 \theta_c \right.$$
$$+ \left. \left. 0.50 \sin^2 \alpha \cos^2 \theta_c \right) \left((L - x_t) r_t + 0.5 (L - x_t)^2 \tan \theta_c \right) \right]$$

This equation can be written in terms of three configuration parameters: θ_c, R_N, and R_B, by noting that:

$$\cos \phi_t = \sin \theta_c; \quad \sin \phi_t = \cos \theta_c; \quad r_t = R_N \cos \theta_c;$$

$$L - x_t = \frac{R_B - r_t}{\tan \theta_c} = \frac{R_B - R_N \cos \theta_c}{\tan \theta_c}$$

Thus,

$$C_A = 2C_{p,t2} \left(\frac{R_N^2}{R_B^2} \right) \left[\left(0.25 \cos^2 \alpha (1 - \sin^4 \theta_c) \right. \right.$$

$$+ \left. 0.125 \sin^2 \alpha \cos^4 \theta_c \right) + \tan \theta_c \left(\cos^2 \alpha \sin^2 \theta_c + 0.50 \sin^2 \alpha \cos^2 \theta_c \right)$$

$$\times \left. \left(\frac{(R_B/R_N) - \cos \theta_c}{\tan \theta_c} \cos \theta_c + \frac{((R_B/R_N) - \cos \theta_c)^2}{2 \tan \theta_c} \right) \right] \qquad (8\text{-}26)$$

The normal-force coefficient is given by Eq. (8-5).

$$C_N = \frac{2C_{p,t2}}{\pi R_B^2} \left[\int_0^{\phi_t} \int_0^{\pi} \left(\cos^2 \alpha \cos^2 \phi + 2 \sin \alpha \cos \alpha \cos \phi \sin \phi \cos \beta \right. \right.$$
$$+ \left. \sin^2 \alpha \sin^2 \phi \cos^2 \beta \right) R_N \sin \phi \, d\beta R_N \, d\phi \sin \phi \cos \beta$$
$$+ \int_{x_t}^{L} \int_0^{\pi} \left(\cos^2 \alpha \sin^2 \theta_c + 2 \sin \alpha \cos \alpha \sin \theta_c \cos \theta_c \cos \beta \right.$$
$$+ \left. \sin^2 \alpha \cos^2 \theta_c \cos^2 \beta \right) (r_t + (x - x_t) \tan \theta_c) \, d\beta \frac{dx}{\cos \theta_c} \cos \theta_c \cos \beta \right]$$

$$(8\text{-}27)$$

Again, the first term represents the contribution of the spherical cap and the second term that of the conical section. Integrating yields:

$$C_N = \frac{C_{p,t2}}{2} \left(\frac{R_N}{R_B} \right)^2 \sin \alpha \cos \alpha \sin^4 \phi_t$$
$$+ \frac{2C_{p,t2}}{R_B^2} \sin \alpha \cos \alpha \sin \theta_c \cos \theta_c \left[(L - x_t) r_t + 0.5 (L - x_t)^2 \tan \theta_c \right]$$

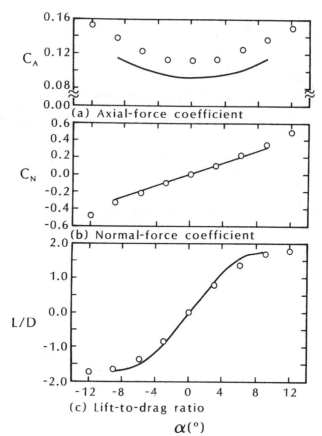

(a) Axial-force coefficient

(b) Normal-force coefficient

(c) Lift-to-drag ratio

$\alpha(°)$

Fig. 8.18 Aerodynamic coefficients for blunted sphere/cone in the USAFA TWT, $M_1 = 4.3$, $R_N = 0.2152\,R_B$.

Writing this equation in terms of the three configuration parameters: θ_c, R_N, and R_B, we obtain:

$$C_N = 2C_{p,t2} \left(\frac{R_N}{R_B}\right)^2 \left[0.25 \sin\alpha \cos\alpha \cos^4\theta_c + \sin\alpha \cos\alpha \sin\theta_c \cos\theta_c \right.$$

$$\left. \times \left(\frac{(R_B/R_N) - \cos\theta_c}{\tan\theta_c} \cos\theta_c + \frac{((R_B/R_N) - \cos\theta_c)^2}{2\tan\theta_c} \right) \right] \quad (8\text{-}28)$$

Exercise 8.3:

Aerodynamic forces were measured on a blunted sphere/cone in the Mach 4.3 stream of the Trisonic Wind Tunnel (TWT) at the U.S.

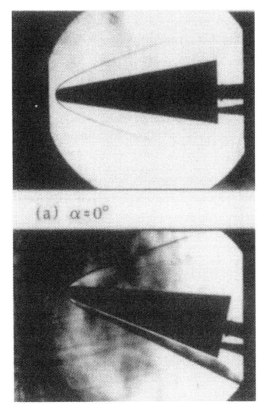

(a) $\alpha = 0°$

(b) $\alpha = 12°$

Fig. 8.19   Blunted sphere/cone in the USAFA TWT, $M_1 = 4.3$, $R_N = 0.2152\,R_B$.

Air Force Academy. For the cone: $\theta_c = 9$ deg., $R_N = 0.355$ in., and $R_B = 1.65$ in.. The experimental values were as follows.

$\alpha(°)$	-12	-9	-6	-3	0
C_A	$+0.154$	$+0.138$	$+0.123$	$+0.113$	$+0.113$
C_N	-0.476	-0.336	-0.213	-0.104	-0.002
L/D	-1.735	-1.642	-1.371	-0.826	-0.015

$\alpha(°)$	$+3$	$+6$	$+9$	$+12$
C_A	$+0.114$	$+0.125$	$+0.136$	$+0.150$
C_N	$+0.103$	$+0.221$	$+0.349$	$+0.489$
L/D	$+0.813$	$+1.397$	$+1.715$	$+1.798$

Using the Newtonian flow model to define the pressure distribution, calculate values for C_A, C_N, and (L/D) over the angle-of-attack range: $-9° \leq \alpha \leq +9°$.

Solution:

$$R_N = 0.2152R_B \quad \text{and} \quad C_{p,t2} = 2$$

Use Eq. (8-26) to calculate C_A and Eq. (8-28) to calculate C_N.

$$\frac{L}{D} = \frac{C_N \cos \alpha - C_A \sin \alpha}{C_N \sin \alpha + C_A \cos \alpha} \tag{8-29}$$

The calculated values are:

$\alpha(°)$	-9	-6	-3	0	$+3$	$+6$	$+9$
C_A	$+0.114$	$+0.102$	$+0.095$	$+0.093$	$+0.095$	$+0.102$	$+0.114$
C_N	-0.295	-0.198	-0.100	0.000	$+0.100$	$+0.198$	$+0.295$
L/D	-1.721	-1.521	-0.941	0.000	$+0.941$	$+1.521$	$+1.721$

The calculated values are compared with the experimental values in Fig. 8.18. Because viscous effects are not included in the computed axial-force coefficients, they lie below the measured values.

Schlieren photographs for $\alpha = 0$ deg. and for $\alpha = 12$ deg. are presented in Fig. 8.19.

Aerodynamic forces and moments illustrating the effects of nose bluntness were obtained by Neal.[13] The geometry of the three conic configurations ($\theta_c = 9$ deg.) is illustrated in Fig. 8.20. The reference point for the pitching moment, which is $0.098D_B$ ahead of the centroid of the planform area, is indicated in the sketches. The aerodynamic coefficients for a Mach 6.77 stream, as taken from Ref. 14, are reproduced in Fig. 8.21. The axial-force coefficient increases with nose bluntness at the lower angles-of-attack, as would be expected. At higher angles-of-attack, the axial-force coefficient is essentially the same for all three configurations. When the angle-of-attack is such that the flat base is forward, they become identical.

The normal force coefficient is very dependent on bluntness for the angle-of-attack range from 50 deg. $\leq \alpha \leq 130$ deg. Note that, for α of 90 deg., the ratio of the normal-force coefficients (see Fig. 8.21b) is roughly the same as the ratio of the projected planform area (see Fig. 8.20). Note that the reference area for the normal-force coefficients is the same for all three configurations, i.e., the base area. However, the normal force is due to the pressure acting primarily on the conical surface, which has the same semi-vertex an-

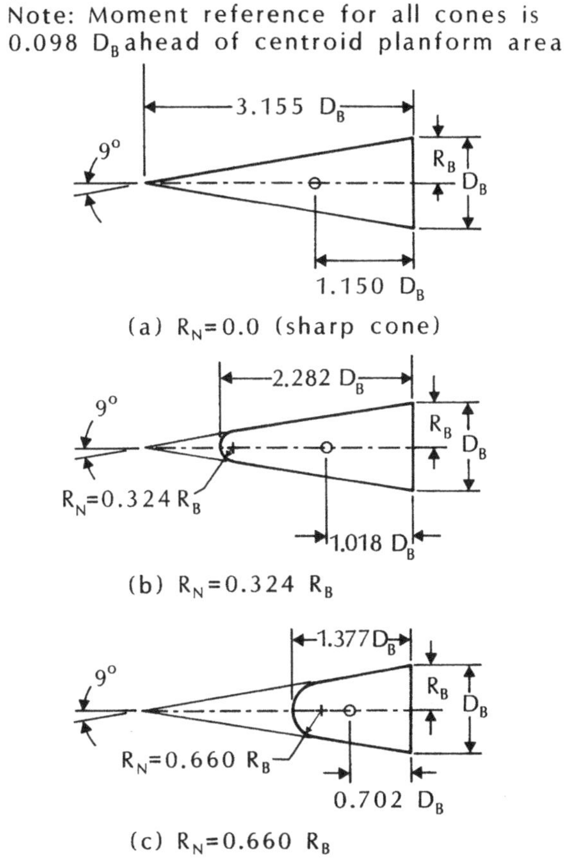

Note: Moment reference for all cones is
0.098 D_B ahead of centroid planform area

(a) $R_N = 0.0$ (sharp cone)

(b) $R_N = 0.324 \; R_B$

(c) $R_N = 0.660 \; R_B$

Fig. 8.20   Sketch of 9 deg. cones for which data are presented in Fig. 8.21.

gle ($\theta_c = 9$ deg.) for all three configurations, projected normal to the axis-of-symmetry. Thus, it should not be surprising that the normal force and the normal-force coefficient are (to a first approximation) proportional to the projected planform area.

As evident in Fig. 8.21e, bluntness has a significant effect on the (L/D) ratio. For $R_N = 0.660 \; R_B$, $(L/D)_{max}$ is approximately 0.5 and occurs at an α of 30 deg. For $R_N = 0.324 R_B$, $(L/D)_{max}$ is approximately 1.1 and occurs at an α of 20 deg.

8.2.2.2 Multi-conic configurations.   In order to satisfy constraints relating to the drag and to the heat transfer, to volumetric packaging, and to the static margin, the designer may use multi-conic configurations. For some multi-conic configurations, the flow expands from one conical surface to the next. Such is the case for the blunted

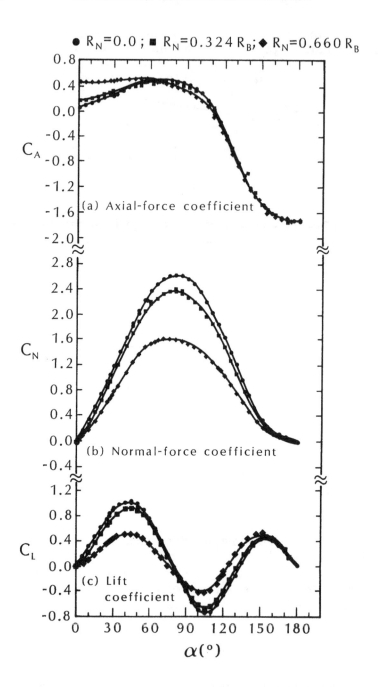

Fig. 8.21 The effect of nose bluntness on the aerodynamic coefficients of a cone (θ_c = 9 deg.) M_1 = 6.77, as taken from Refs. 13 and 14.

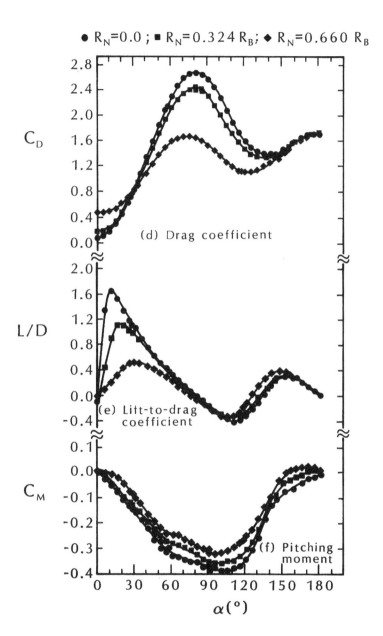

$\bullet R_N = 0.0$; $\blacksquare R_N = 0.324\, R_B$; $\blacklozenge R_N = 0.660\, R_B$

(d) Drag coefficient

C_D

(e) Lift-to-drag coefficient

L/D

(f) Pitching moment

C_M

$\alpha(°)$

Fig. 8.21   (continued) The effect of nose bluntness on the aerodynamic coefficients of a cone (θ_c = 9 deg.) M_1 = 6.77, as taken from Refs. 13 and 14.

biconic configurations illustrated in Fig. 7.19, where the semi-vertex angle for the fore cone is 12.84 deg. that for the aft cone is 7 deg. Heat-transfer data for this blunted biconic were presented in Figs. 7.20 and 7.21. For other configurations, the flow undergoes a compression as it encounters the last conic segment; i.e., the last conic segment is a flare. Because the pressure acting on the flare is relatively high, an afterbody flare causes the center-of-pressure to be relatively far aft. Thus, an afterbody flare serves to increase the static margin, i.e., to increase the static stability, of the vehicle.

Exercise 8.4:

Using the Newtonian flow model for the pressure distribution, develop expressions for the x- and y-coordinates of the center of pressure for the sharp biconic configuration illustrated in Fig. 8.22. The dimensions of the biconic are such that it could be placed within the volume occupied by a sharp, 10 deg. half-angle cone (as indicated by the broken line in the sketch of Fig. 8.22). (For certain applications, the designer is faced with the constraint of designing a vehicle that fits within the heat-shield shroud of an existing payload.) Limit the calculation to the angle-of-attack range: $-\theta_{cf} \leq \alpha \leq +\theta_{cf}$.

Solution:

In order to obtain expressions both for x_{cp} and for y_{cp}, the contributions of the normal force and of the axial force to the pitching moment will be kept separate. For simplicity, the pitching moments will be referenced to a point on the axis-of-symmetry at the base, i.e., $x = L$ and $y = 0$, as indicated in the sketch of Fig. 8.23.

Expressions for the forces and for the moments acting on the fore cone will be written in a coordinate system originating at its

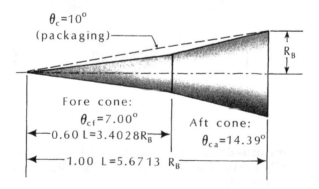

Fig. 8.22   Sketch of the sharp biconic used in Exercise 8.4.

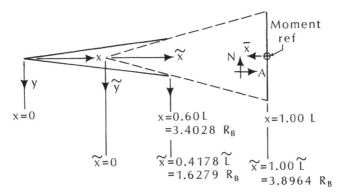

Fig. 8.23 Coordinate system for the calculations of Exercise 8.4.

apex (x,y in Fig. 8.23). Expressions for the forces and the moments acting on the aft cone will be written in the $\tilde{x},\tilde{y}$ coordinate system, originating at its apex. Thus,

$$
M_L = 2 \int_0^{0.6L} \int_0^{\pi} p\, x \tan\theta_{cf}\, d\beta \frac{dx}{\cos\theta_{cf}}(L-x)\cos\theta_{cf}\cos\beta
$$

$$
- 2 \int_0^{0.6L} \int_0^{\pi} p\, x \tan\theta_{cf}\, d\beta \frac{dx}{\cos\theta_{cf}} x \tan\theta_{cf} \sin\theta_{cf} \cos\beta
$$

$$
+ 2 \int_{0.4178\tilde{L}}^{\tilde{L}} \int_0^{\pi} p\, \tilde{x} \tan\theta_{ca}\, d\beta \frac{d\tilde{x}}{\cos\theta_{ca}}(\tilde{L}-\tilde{x})\cos\theta_{ca}\cos\beta
$$

$$
- 2 \int_{0.4178\tilde{L}}^{\tilde{L}} \int_0^{\pi} p\, \tilde{x} \tan\theta_{ca}\, d\beta \frac{d\tilde{x}}{\cos\theta_{ca}} \tilde{x} \tan\theta_{ca} \sin\theta_{ca} \cos\beta
$$

$$
(8\text{-}30)
$$

The four terms represent (in order) the contributions of the normal force acting on the fore cone, of the axial force acting on the fore cone, of the normal force acting on the aft cone, and of the axial force acting on the aft cone. Note that the normal forces contribute to a nose-up (positive) pitching moment. A positive axial force contributes to a nose-down (negative) pitching moment.

Introducing the definitions for the Newtonian pressure coefficient, noting that $C_p = 0$ on the base, and for the pitching moment coefficient:

$$
C_{M_L} = M_L/(q_\infty \pi R_B^2 R_B)
$$

and rearranging the terms:

$$
C_{M_L} = \frac{2C_{p,t2}}{\pi R_B^2 R_B} \left[\int_0^{0.6L} \int_0^\pi \left(\cos^2 \alpha \sin^2 \theta_{cf} \right. \right.
$$

$$
+ 2 \sin \alpha \cos \alpha \sin \theta_{cf} \cos \theta_{cf} \cos \beta
$$

$$
\left. + \sin^2 \alpha \cos^2 \theta_{cf} \cos^2 \beta \right) \tan \theta_{cf} \cos \beta d\beta (L - x) x dx \Big]
$$

$$
+ \frac{2C_{p,t2}}{\pi R_B^2 R_B} \left[\int_{0.4178\tilde{L}}^{\tilde{L}} \int_0^\pi \left(\cos^2 \alpha \sin^2 \theta_{ca} \right. \right.
$$

$$
+ 2 \sin \alpha \cos \alpha \sin \theta_{ca} \cos \theta_{ca} \cos \beta
$$

$$
\left. + \sin^2 \alpha \cos^2 \theta_{ca} \cos^2 \beta \right) \tan \theta_{ca} \cos \beta d\beta (\tilde{L} - \tilde{x}) \tilde{x} d\tilde{x} \Big]
$$

$$
- \frac{2C_{p,t2}}{\pi R_B^2 R_B} \left[\int_0^{0.6L} \int_0^\pi \left(\cos^2 \alpha \sin^2 \theta_{cf} \right. \right.
$$

$$
+ 2 \sin \alpha \cos \alpha \sin \theta_{cf} \cos \theta_{cf} \cos \beta
$$

$$
\left. + \sin^2 \alpha \cos^2 \theta_{cf} \cos^2 \beta \right) \tan^3 \theta_{cf} \cos \beta d\beta x^2 dx \Big]
$$

$$
- \frac{2C_{p,t2}}{\pi R_B^2 R_B} \left[\int_{0.4178\tilde{L}}^{\tilde{L}} \int_0^\pi \left(\cos^2 \alpha \sin^2 \theta_{ca} \right. \right.
$$

$$
+ 2 \sin \alpha \cos \alpha \sin \theta_{ca} \cos \theta_{ca} \cos \beta
$$

$$
\left. + \sin^2 \alpha \cos^2 \theta_{ca} \cos^2 \beta \right) \tan^3 \theta_{ca} \cos \beta d\beta \tilde{x}^2 d\tilde{x} \Big] \quad (8\text{-}31)
$$

In Eq. (8-31), the first two terms represent the contributions of the normal force and the last two terms represent the contributions of the axial force. Noting that:

$$
L = 5.6713 R_B; \quad 0.4178\tilde{L} = 1.6279 R_B; \text{ and } \tilde{L} = 3.8964 R_B
$$

We can evaluate the four terms in Eq. (8-31):

$$
C_{M_L} = 0.2926 \sin 2\alpha C_{p,t2} + 0.3789 \sin 2\alpha C_{p,t2}
$$

$$
- 0.00294 \sin 2\alpha C_{p,t2} - 0.07433 \sin 2\alpha C_{p,t2}
$$

$$
C_{M_L} = [0.6715 \sin 2\alpha] C_{p,t2} - [0.0773 \sin 2\alpha] C_{p,t2}
$$

$$
= C_N \frac{\bar{x}_{cp}}{R_B} - C_A \frac{\bar{y}_{cp}}{R_B} \quad (8\text{-}32)
$$

where $\bar{x}_{cp}$ is measured forward from the moment reference, as shown in Fig. 8.23.

Expressions are needed for C_A and C_N.

$$C_A = \frac{2C_{p,t2}}{\pi R_B^2} \left[\int_0^{0.6L} \int_0^\pi \left(\cos^2 \alpha \sin^2 \theta_{cf} \right. \right.$$

$$+ 2 \sin \alpha \cos \alpha \sin \theta_{cf} \cos \theta_{cf} \cos \beta$$

$$\left. + \sin^2 \alpha \cos^2 \theta_{cf} \cos^2 \beta \right) x \tan \theta_{cf} d\beta \frac{dx}{\cos \theta_{cf}} \sin \theta_{cf} \right]$$

$$+ \frac{2C_{p,t2}}{\pi R_B^2} \left[\int_{0.4178\tilde{L}}^{\tilde{L}} \int_0^\pi \left(\cos^2 \alpha \sin^2 \theta_{ca} \right. \right.$$

$$+ 2 \sin \alpha \cos \alpha \cos \theta_{ca} \sin \theta_{ca} \cos \beta$$

$$\left. + \sin^2 \alpha \cos^2 \theta_{ca} \cos^2 \beta \right) \tilde{x} \tan \theta_{ca} d\beta \frac{d\tilde{x}}{\cos \theta_{ca}} \sin \theta_{ca} \right]$$

Evaluating the integrals:

$$C_A = C_{p,t2} \left[0.05355 \cos^2 \alpha + 0.47295 \sin^2 \alpha \right] \qquad (8\text{-}33)$$

$$C_N = \frac{2C_{p,t2}}{\pi R_B^2} \left[\int_0^{0.6L} \int_0^\pi \left(\cos^2 \alpha \sin^2 \theta_{cf} \right. \right.$$

$$+ 2 \sin \alpha \cos \alpha \sin \theta_{cf} \cos \theta_{cf} \cos \beta$$

$$\left. + \sin^2 \alpha \sin^2 \theta_{cf} \cos^2 \beta \right) x \tan \theta_{cf} d\beta \frac{dx}{\cos \theta_{cf}} \cos \theta_{cf} \cos \beta \right]$$

$$+ \frac{2C_{p,t2}}{\pi R_B^2} \left[\int_{0.4178\tilde{L}}^{\tilde{L}} \int_0^\pi \left(\cos^2 \alpha \sin^2 \theta_{ca} \right. \right.$$

$$+ 2 \sin \alpha \cos \alpha \sin \theta_{ca} \cos \theta_{ca} \cos \beta$$

$$\left. + \sin^2 \alpha \sin^2 \theta_{ca} \cos^2 \beta \right) \tilde{x} \tan \theta_{ca} d\beta \frac{d\tilde{x}}{\cos \theta_{ca}} \cos \theta_{ca} \cos \beta \right]$$

Evaluating the integrals:

$$C_N = C_{p,t2} \left[0.47300 \sin 2\alpha \right] \qquad (8\text{-}34)$$

Using Eq. (8-32),

$$C_N \frac{\bar{x}_{cp}}{R_B} = [0.6715 \sin 2\alpha]\, C_{p,t2}$$

Using Eq. (8-34) to define C_N,

$$\frac{\bar{x}_{cp}}{R_B} = 1.41966$$

Since $L = 5.6713 R_B$,

$$\bar{x}_{cp} = 0.2503 L$$

Since $\bar{x}_{cp}$ is measured from the base, one can calculate the distance from the apex (x_{cp}) to be:

$$x_{cp} = L - \bar{x}_{cp} = 0.7497 L \qquad (8\text{-}35)$$

Similarly, using Eq. (8-32),

$$-C_A \frac{\bar{y}_{cp}}{R_B} = -[0.0773 \sin 2\alpha]\, C_{p,t2}$$

Using Eq. (8-33) to define C_A and noting that $y_{cp} = \bar{y}_{cp}$, the y-coordinate of the center-of-pressure is:

$$\frac{y_{cp}}{R_B} = \frac{0.0773 \sin 2\alpha}{0.05355 \cos^2 \alpha + 0.47295 \sin^2 \alpha} \qquad (8\text{-}36)$$

    The (modified) Newtonian flow model is most valid when the bow shock wave follows the body contour relatively closely. However, when the bow shock wave does not follow the body contour, large errors can result. Such was the case in Sec. 6.3.3, where a flap deflected into a locally supersonic flow formed a compression corner, generating a secondary (embedded) shock wave. It was noted that, if a hypersonic flow is turned 20 deg. in increments of 10 deg. (such as would be the case for a flap deflected 10° at the trailing edge of a "wing" at 10 deg. angle-of-attack), the pressure would be higher than if that same hypersonic flow was turned 20 deg. in a single turn (i.e., the "wing" itself was at 20 deg. angle-of-attack). The (modified) Newtonian flow model does not account for the fact that the flow properties upstream of the "flap-induced" shock wave differ from the freestream properties because the flow has passed through the bow shock wave. Nor does the (modified) Newtonian flow model account for a separation induced at the compression corner by a shock wave/boundary-layer interaction. As discussed in Sec. 6.4, the extent and the effect of the interaction depend on the following parameters of the local flow: the Reynolds number (and whether the approach boundary layer is laminar or turbulent), the Mach number, the chemical state of the gas, and the turning angle.

Using data for a blunt-nosed, flare-stabilized body that were obtained in a counter-flow ballistic range, Kirk and Chapman[19] reported a decrease in stability with increasing Mach number and with decreasing Reynolds number. Seiff[20] noted, "...on flare stabilizers on blunt-nosed bodies of revolution, pressures are lower than Newtonian and diminish with increasing flight speed in the hypersonic speed range. The calculated pressures vary over the flare surface as a result of the nonuniformity of the incident stream, and depend on the axial location of the flare." The nonuniformity of the incident stream is caused, to a large degree, by the curved bow-shock wave, as indicated in the sketch of Fig. 8.24. As a result, gradients in the entropy and in the total pressure exist normal to the streamlines, and vary in the radial direction and in the axial direction. Accounting for the variations in the flowfield parameters in the inviscid flow upstream of the embedded shock wave and assuming that the embedded shock layer is thin, Seiff[20] calculated the pressures on the flare using an impact model. To calculate the pressure, Seiff used the equation:

$$C_{p_2} = C_{p_1} + \frac{q_1}{q_\infty} (C_{p_2})_{\text{Newt}} \qquad (8\text{-}37)$$

where $(C_{p_2})_{\text{Newt}}$ is the pressure on the flare using the usual Newtonian impact theory. For the applications to which Seiff applied this equation, the pressure tended to rise with distance from the cor-

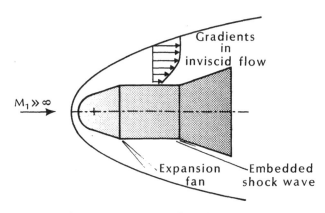

Fig. 8.24 Blunted multi-conic.

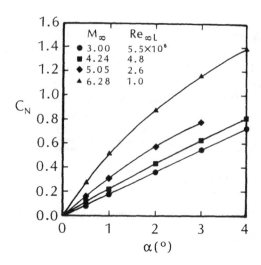

Fig. 8.25 The effect of boundary-layer separation on the normal force coefficients for a cone/cylinder/flare configuration, as taken from Ref. 22.

ner, since q_1 was a minimum in the low-energy air which had passed through the normal-shock wave near the axis.

Ericsson[21] noted that, for the very blunt-nosed shapes used in the first generation of re-entry bodies, the loss of mean dynamic pressure over the flare was the main effect. And for these very blunt shapes, "the entropy gradient over the radial extent of the flare was very minute. However, when a higher performance re-entry body is used, exemplified by the spherical and elliptical nose shapes, the entropy gradients become large."[21] Furthermore, for a hemispherical nose shape, the bow shock shape is not affected by the flare or by the aft body location. As a result, translation of the hemispherical nose results in the translation of the entropy wake across the flare that is not affected by the presence of the cylindrical segment. Thus, one can compute the contribution caused by the translation of the flare in the inviscid shear flow.[21] When the nose bluntness is not of the hemispherical type, the bow shock shape and the associated entropy wake depend on the angle-of-attack.

Thus, the pressure acting on the flare may be either greater than or less than the Newtonian value. Whether the pressure acting on the flare is above or below the Newtonian level depends on many parameters including the nose bluntness.

The effect of the shock wave/boundary-layer interaction on the pressure distribution on a compression ramp was discussed in Sec. 6.4. As noted in that discussion, the shock-induced pressure rise feeds forward through the subsonic portion of the approach boundary layer. The upstream influence may be relatively small, e.g., a slight thickening of the approach boundary layer, or it may be dramatic, e.g., a large separation bubble at the corner with a complex, lambda-shaped shock wave. Wind-tunnel measurements reported by Dennis and Syvertson[22] of the low angle-of-attack normal-force coefficients for a cone/cylinder/flare are reproduced in Fig. 8.25. The measured normal-force coefficients increase by a factor of two, as the Reynolds number based on the model length decreases from 5.5×10^6 to 1.0×10^6. (Although the Mach number varies simultaneously, the normal-force coefficient should, if anything, decrease by a few percent over this range of Mach number.) The Reynolds number-dependent change in the experimental normal-force coefficients is due to increased separation at the cylinder/flare junction as the Reynolds number decreases.

The corresponding change in the center-of-pressure, as presented by Dennis and Syvertson,[22] is reproduced in Fig. 8.26. Note that the Reynolds-number-dependent aft movement of the center-of-pressure is as much as 0.25L.

Based on discussion in this section, the reader should realize that, although simple flow models may be useful in conceptual configuration studies, more rigorous flow models are needed to generate realistic values for the aerodynamic parameters.

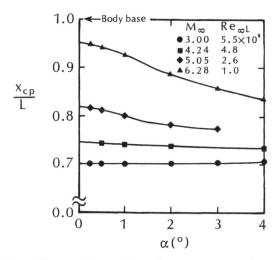

Fig. 8.26 The effect of boundary-layer separation on the center of pressure for a cone/cylinder/flare configuration, as taken from Ref. 22.

8.3   RE-ENTRY CAPSULE AERODYNAMICS

The term *capsule* can be used to describe a variety of spacecraft, including: those of the early U.S. Manned Spacecraft Program, e.g., the Mercury, the Gemini, and the Apollo; planetary probes, e.g., the Viking; and research vehicles, e.g., the Aeroassist Flight Experiment (AFE) configuration.

8.3.1 Achieving a Trimmed Angle-of-Attack

For axisymmetric configurations, lift is generated when the vehicle flies at a nonzero angle-of-attack. Although the (L/D) of capsule configurations is relatively small, it is sufficient to provide the vehicle with atmospheric maneuvering capability. As noted by Crowder and Moote,[23] the atmospheric maneuvering capability for the Apollo Command Module, i.e., the re-entry capsule, was required "to 1) prevent skip-out in the event of a shallow entry flight path angle, 2) prevent excessive atmospheric decelerations in the event of a steep entry flight path angle, and 3) provide ranging capability for any combination of entry interface flight conditions within the tolerances of the lunar return trajectory."

As indicated in the sketch of Fig. 8.27, with the center-of-gravity located off the centerline (axis) and relatively near the blunt, spherical heat shield, the Apollo Command Module will trim at a nonzero angle-of-attack. Since the velocity vector is not aligned with the configuration's axis-of-symmetry, lift is produced. A reaction-control system was used to rotate the Apollo CM and, thereby, point the lift vector in order to maneuver the spacecraft.

Pitching moment data for the Apollo CM reported by Moseley et al.[24] are reproduced in Fig. 8.28. Note that, for these aerodynamic data, the angle-of-attack is measured relative to the apex forward.

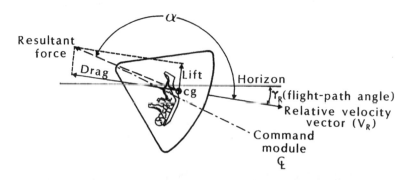

Fig. 8.27   Apollo Command Module (CM) shape and trim attitude, aerodynamic-force relationships, as taken from Ref. 23.

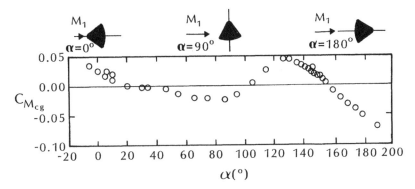

Fig. 8.28 Pitching moment for the Apollo Command Module.
$x_{cg} = +0.263D_b$, $z_{cg} = +0.035D_b$, $M_1 = 3.40$, as taken from Ref. 24.

This differs from the convention used for the pressure data (Chap. 6) and for the heat-transfer measurements (Chap. 7). For the data of Chaps. 6 and 7, zero angle-of-attack corresponded to the blunt, spherical heat shield forward with the axis-of-symmetry aligned with the freestream velocity. Thus, to the aerodynamicist, the preflight estimate of the trim angle-of-attack for the Apollo 2 flight would have been 159 deg. (see Ref. 25), while to the aerothermodynamicist, it would have been 21 deg. Note that, for the center-of-gravity location for the data of Fig. 8.28, there are two statically stable trim conditions. That is, there are two angles-of-attack for which $C_M = 0$ and $C_{M_\alpha} < 0$. One occurs at an angle-of-attack of 23 deg.; the other at 157 deg. The trim condition with the apex forward, i.e., $\alpha = 23$ deg., is undesirable, since it would result in heating rates to the conical surface exceeding the heat-shield design. Furthermore, entry under these conditions could result in excessive g-loads on the spacecraft occupants.

Using data obtained in the Ames Hypervelocity Free-Flight Aerodynamic Facility, De Rose[26] studied the effect of the displacement of the center-of-gravity on the trim angle and on the lift-to-drag ratio. Shadowgraphs, such as those presented in Fig. 8.29, of the 0.95-cm model of the Apollo CM provided the history of the attitude and of the position from which the aerodynamic forces could be determined, e.g., Ref. 27. De Rose[26] noted that the cg positions were measured from the centerline of the blunt-face heat shield rather than from the structural centerline. This is significant when comparing wind-tunnel measurements to flight data, since the varying thickness of the ablative heat shield causes the exterior surface of the full-scale spacecraft to be different from the substructure and from the wind-tunnel models. The differences are indicated in the sketch of Fig. 8.30, which is taken from Ref. 23. As noted by De Rose,[26] "The heat-shield

(a) $\alpha \approx 147°$

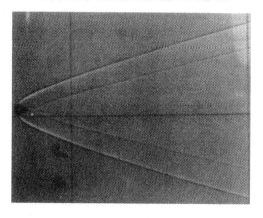

(b) $\alpha \approx 180°$

Fig. 8.29   Shadowgraphs of the Apollo Command Module in the Ames Hypervelocity Free-Flight Aerodynamic Facility, provided by G. T. Chapman of NASA.

centerline was used to reference the cg because, when flying heat-shield forward, this surface supports significant aerodynamic forces ... using the heat-shield centerline to locate the cg provides a good correlation of the flight data [of Ref. 28] with the present free-flight data and earlier wind-tunnel data." The trim angle measurements presented in Fig. 8.31 are independent of the Mach number. However, data from Apollo flights AS-202, Apollo 4, and Apollo 6 indicated that the trim angle-of-attack was consistently greater than the predicted values at flight speeds in excess of 18,000 fps.[29] Since the trim angle is a function of the center-of-gravity offset, the designer can avoid the undesired trim condition. Moseley et al.[24] conclude, "It has been shown that center-of-gravity placement can provide, within limits, the lift-drag ratio necessary for entry while still maintaining stability through the Mach number range tested."

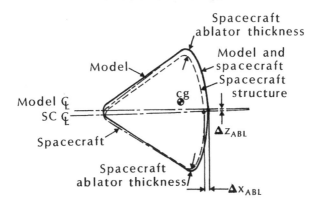

Fig. 8.30 Comparison of reference surfaces of wind-tunnel models and of full-scale spacecraft, as taken from Ref. 23.

Since aeroassisted orbit transfer vehicles (AOTVs) generate modest lift-to-drag ratios, e.g., 0.2 to 0.4, their aerodynamic maneuverability can be used to reduce the amount of propellant required to make orbital plane changes. These vehicles will typically operate in the upper reaches of the atmosphere at relatively high velocities. Since real-gas effects will have a significant effect on the flowfield, it is important to understand how to model the flow in ground-based tests and in CFD formulations. Weilmuenster and Hamilton[30] compared wind-tunnel measurements and CFD computations of the aerodynamic characteristics of the Aeroassist Flight Experiment (AFE) vehicle (see Fig. 6.6). The wind-tunnel measurements were obtained in a Mach 10 airstream ($\gamma = 1.4$), in a Mach 6.29 stream of tetrafluoromethane, CF_4, and in a Mach 21 stream of helium ($\gamma = 1.667$). The flowfield computations were made using the Newtonian pressures and an inviscid flowfield code, High Alpha Inviscid Solution (HALIS). Weilmuenster and Hamilton[30] concluded that the coefficients computed using Newtonian pressures were not useful for predicting wind-tunnel data for this vehicle, whose flowfield was dominated by subsonic flow in the shock layer.

Weilmuenster and Hamilton[30] generated flowfield solutions at the point of maximum dynamic pressure, which corresponds to the minimum altitude, so that the use of equilibrium chemistry in a continuum flow model would be a reasonable simulation of the flow. The pitching moments computed for this flight condition are compared with those generated using the modified Newtonian flow model and with those computed for the three wind-tunnel conditions. These computed pitching moments are reproduced in Fig. 8.32. The HALIS computations indicate that the vehicle trims at zero angle-of-attack at these flight conditions.

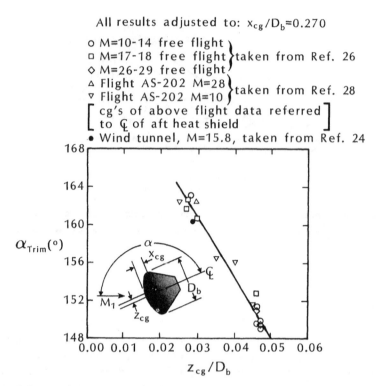

All results adjusted to: $x_{cg}/D_b=0.270$

o M=10-14 free flight ⎫
□ M=17-18 free flight ⎬ taken from Ref. 26
◇ M=26-29 free flight ⎭
△ Flight AS-202 M=28 ⎫ taken from Ref. 28
▽ Flight AS-202 M=10 ⎭
[cg's of above flight data referred]
[to C_L of aft heat shield]
• Wind tunnel, M=15.8, taken from Ref. 24

Fig. 8.31 Trim angle as a function of the center-of-gravity displacement, as taken from Ref. 26.

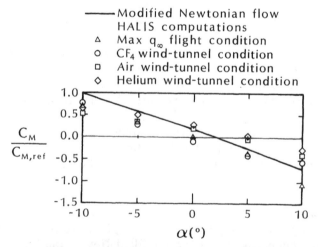

——— Modified Newtonian flow
HALIS computations
△   Max q_∞ flight condition
o CF_4 wind-tunnel condition
□   Air wind-tunnel condition
◇   Helium wind-tunnel condition

Fig. 8.32 Comparison of HALIS pitching moment results for tunnel conditions and for the maximum dynamic pressure flight point with the Newtonian pitching moment for the AFE configuration, as taken from Ref. 30.

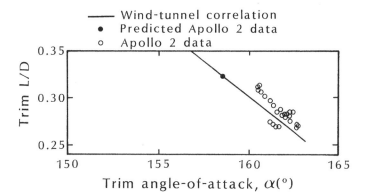

Fig. 8.33 L/D **as a function of the trim angle-of-attack for the Apollo CM, as taken from Ref. 23.**

Surprisingly, from a stability standpoint, the $C_{M\alpha}$ values at zero degree angle-of-attack are similar for the Newtonian, flight, and CF_4 curves even though in magnitude, the Newtonian C_M is seriously in error, while the C_M in CF_4 is considerably closer to the flight value. From the standpoint of magnitude, the CF_4 results are the best approximation to the flight values of C_M for $\alpha \leq 5$ deg. However, over the same values of α, there are considerable differences in C_{M_α}.

The statement of Weilmuenster and Hamilton,[30] "The ability to predict the gross characteristics of a flowfield are not sufficient to validate a computer code." is worth repeating. To validate a computer code, one needs measurements that can be used to evaluate the numerical models of the flow physics and of the flow chemistry.

8.3.2 Aerodynamic Coefficients

The inertial measurement units (IMUs) of the Apollo spacecraft provide the three acceleration components, which are resolved into two components: one parallel to the relative velocity, the other perpendicular to it. Since these accelerations are proportional to the aerodynamic forces, their ratio is equal to L/D. The (L/D) ratio, thus determined for the Apollo 2 (AS 202), as taken from Ref. 23, is reproduced in Fig. 8.33. As indicated in Fig. 8.33, the cg-location was such that the pre-flight correlation indicated that the Apollo CM should have statically trimmed at an angle-of-attack of 159 deg., generating an (L/D) of 0.32. However, the trim angle-of-attack measured during the flight was consistently larger and the (L/D) ratio was smaller. Nevertheless, the flight values of (L/D) followed the α-dependence

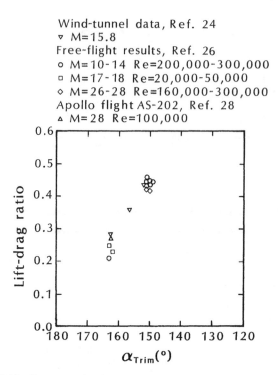

Fig. 8.34 Lift-drag ratio for the Apollo CM as a function of the trim angle-of-attack, as taken from Ref. 26.

exhibited by the wind-tunnel data. It was, therefore, concluded that the C_M dependence on α and the resulting trim attitude were affected by some unknown flow phenomena during the flight.[23]

The (L/D) ratios determined using free-flight data[26] obtained in the Ames counter-flow ballistic range are presented in Fig. 8.34. The free-flight results of De Rose[26] agree with the values obtained in the wind tunnel[24] and during the Apollo AS-202 flight.[28] The free-flight data show no dependence either on the Mach number or on the Reynolds number.

The drag coefficients of the Viking Entry Vehicle measured during ballistic range tests[31] are presented in Fig. 8.35. The blunt Viking Entry Vehicle, which is shown in Fig. 8.36, was at an angle-of-attack of 11.2 deg. for these data. A significantly smaller drag coefficient was measured for Reynolds numbers between 4×10^3 and 4×10^4. A detailed analysis indicated that the increase in drag was due to a transition from nonequilibrium flow to equilibrium flow in the shock layer, because the bow shock wave was on the borderline between attached conical-type flow and detached curved bow shock blunt-body flow. Shadowgraphs from the ballistic range tests indicated that the bow shock wave shape was very sensitive to the ratio of specific heats in the shock layer, which depended on the gas chemistry. The dashed line at very low Reynolds numbers indicates the expected behavior as slip flow and then free-molecular flow occurs.

Measurements of the drag of spheres in the hypersonic low-density wind tunnels of the Deutsche Forschungs- und Versuchsanstalt für Luft- und Raumfahrt (DFVLR) are reproduced in Fig. 8.37. Koppenwallner and Legge[32] presented the drag-coefficient data as a function of the Reynolds number downstream of a normal-shock wave:

$$Re_2 = \frac{\rho_2 U_2 D}{\mu_2} \qquad (8\text{-}38)$$

At the high end of the Reynolds number scale, i.e., for continuum flow (cont), the drag is assumed to be due to the pressure only and given by the modified Newtonian correlation[32]:

$$[C_D]_{\text{cont}} = C_{DP} = \frac{1}{2}\frac{\gamma+3}{\gamma+1} \qquad (8\text{-}39)$$

Deviation from the modified Newtonian pressure drag starts due to the skin-friction contribution at Reynolds numbers (Re_2) near 500. The drag approaches the free-molecular (FM) values when Re_2 approaches 1. The drag coefficient for free-molecular flow, i.e., $Re_2 \to 0$ or $Kn \gg 10$ is given by:

$$[C_D]_{FM} = C_{DP} + C_{DF}$$

$$[C_D]_{FM} = \left[2 - \sigma_N + \frac{2}{3}\sigma_N\left(\pi\frac{\gamma-1}{\gamma}\frac{T_w}{T_t}\right)^{0.5}\right]_P + [\sigma_T]_F \qquad (8\text{-}40)$$

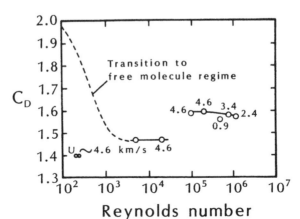

Fig. 8.35 Effect of Reynolds number on Viking drag coefficient, as taken from Ref. 31.

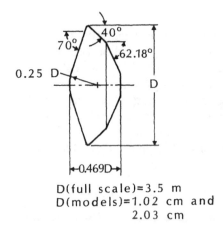

D(full scale)=3.5 m
D(models)=1.02 cm and
2.03 cm

Fig. 8.36   Sketch of Viking entry vehicle, as taken from Ref. 31.

where σ_N and σ_T represents the accommodation coefficients for the normal and tangential components of momentum, respectively.

The three-dimensional hypersonic rarefied flow for the Aeroassist Flight Experiment (AFE) vehicle at altitudes of 100 km and greater were computed by Celenligil et al.[33] using the Direct Simulation Monte Carlo method (DSMC). The resultant lift-to-drag ratio for zero angle-of-attack is presented as a function of altitude in Fig. 8.38. Clearly, the DSMC results approach the free-molecular limit (denoted by FM) very slowly but, even at 150 km, the flow has not

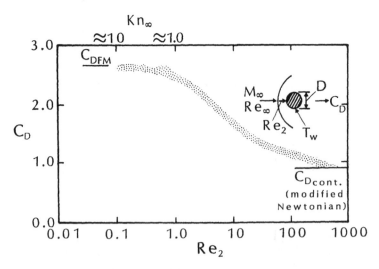

Fig. 8.37   Wind-tunnel values of the sphere drag coefficient (air) for $8.6 < M_1 < 13.0$, $300\ K < T_t < 600\ K$, as taken from Ref. 32.

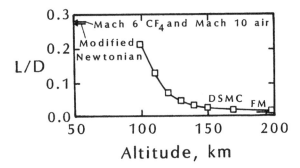

Fig. 8.38 DSMC Computations of the (L/D) ratio for the AFE vehicle at zero angle-of-attack, as taken from Ref. 33.

reached the free-molecular limit. The DSMC computations of Celenligil et al.[33] indicate that the drag increases and the lift decreases with altitude for this blunt vehicle.

The Newtonian/viscous blunt-body theory is compared in Fig. 8.39 with flight data for the Apollo Command Module, as taken from Ref. 34. The data are in reasonable agreement in the range $\overline{V'_{\infty}}$ from 0.005 to 0.01. Wilhite et al.[34] note that the reduction in L/D in this range is caused by a 2 deg. reduction in the angle-of-attack. In the viscous regime ($\overline{V'_{\infty}} > 0.08$), the flight data remain constant, but the theoretical values decrease because of the viscous effects. The degradation of the theoretical (L/D) ratio occurs because the blunt-body viscous effects were modeled using data from the Gemini capsule for which the flow separates at the sharp edge between the spherical heat shield and the conical afterbody. However, the flow remains attached for the Apollo Command Module, and the viscous shear is recovered on the afterbody at these low Reynolds numbers. Thus, for blunt bodies with a faired afterbody similar to the Apollo Command Module, the viscous effects on lift are negligible at these flow conditions.[34]

8.4 SHUTTLE ORBITER AERODYNAMICS

The physical characteristics of the Space Shuttle Orbiter are illustrated in Fig. 8.40. As discussed by Romere and Whitnah,[35] the body flap is the predominant longitudinal trim device, while the wing-mounted elevons are used for longitudinal stability and as ailerons for lateral trim and control. For velocities in excess of 12,000 fps, the body flap is used for pitch trim, while maintaining elevon deflections within thermal constraint limits, and protects the main propulsion system. From 1000 fps to 12,000 fps, the body flap is used for trim control, while maintaining elevon deflections within limits for

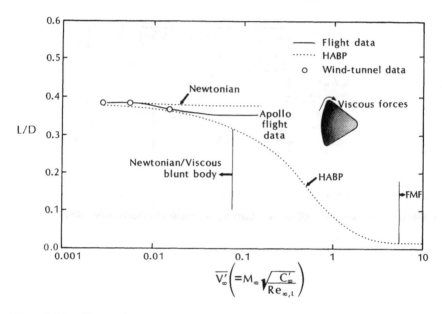

Fig. 8.39　Comparison of HAPB estimates with Apollo flight and wind-tunnel data, as taken from Ref. 34.

aileron effectiveness and elevon hinge moment control. The vertical tail consists of the fin and a combination rudder/speedbrake, with the speedbrake providing lift-to-drag ratio modulation during the terminal area energy management and the approach and landing phases of the flight. Aft-mounted side-firing reaction-control jets are used to supplement yaw stability from entry down to Mach 1.0. For speeds in excess of 10,000 fps, the speedbrake is kept closed because of thermal seal problems. From 4,000 fps to 10,000 fps, the speedbrake is kept full open to provide pitch trim. From 1,000 fps to 4,000 fps, the speedbrake is partially deflected and used as a rudder. At speeds below 1,000 fps, it is used for speed control.

The Space Shuttle Orbiter is designed to perform an unpowered, gliding entry from orbit at an angle-of-attack of 40 deg., which is modulated depending upon crossrange requirements. A gradual pitchdown is initiated at Mach 14 and is completed at Mach 2. From Mach 2 to touchdown, more conventional angles-of-attack, i.e., 3 to 10 deg., are flown. At the beginning of entry, downrange modulation is achieved by periodically performing roll reversals across the prescribed ground track.

Aerodynamically, during the major portions of the flight from entry to touchdown, the vehicle is longitudinally and laterally stable. In certain flight regimes where the vehicle is statically unstable, the stability is artificially provided by the flight control system. The de-

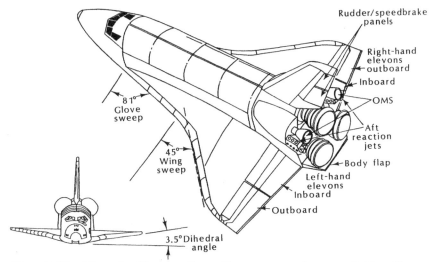

Fig. 8.40 Sketch illustrating the aerosurfaces of the Space Shuttle Orbiter.

sign concept of using a stability augmented flight control system has increased the need to accurately define the aerodynamic characteristics beyond those for a conventional aircraft development program.

8.4.1 Pre-flight Predictions of the Orbiter Aerodynamics

Although the pre-flight predictions of the Orbiter aerodynamics were based on an extensive wind-tunnel data base as discussed in Sec. 4.2, even the most valid set of wind-tunnel measurements must be adjusted for unsimulated parameters. Romere and Whitnah[35] noted, "The major adjustments applied to the Space Shuttle wind tunnel data base involved corrections for nonsimulation of structural deformation, flowfield parameters, and the profile drag due to thermal protection system roughness and minor protuberances."

The traditional freestream Reynolds number was selected for the flowfield scaling parameter below Mach 15, while a viscous interaction parameter $(\overline{V'_\infty})$ was utilized at higher Mach numbers. The parameter $(\overline{V'_\infty})$ was used for those conditions where the boundary-layer thickness was significant with respect to the shock stand-off distance. The selection was further based upon the assumption that simulation of the shock/boundary-layer interaction with the flow on the windward side of the vehicle also provides adequate flight-to-wind-tunnel correlation for the leeside flowfield.

Romere and Whitnah[35] continue:

In general, no attempt was made to obtain a wind tunnel simulation of the effects of structural deformation on the longitudinal aerodynamics through testing of an aeroelastic or deformed model. Since, at higher dynamics pressures these effects are significant, some adjustment to the wind tunnel data to account for structural deformation must be made to provide adequate estimates of the flight aerodynamics. The approach used to evaluate the aeroelastic effects is unique. These effects were derived using a sensitivity analysis performed with the aid of a structural/aerodynamic analysis program. The program was used to stiffen systematically various portions of the vehicle structure and evaluate analytically the effect of the stiffness changes on the aerodynamics. The results indicated that the major longitudinal aeroelastic effects were produced by deformation of the elevon about its hinge line as a result of the aerodynamic hinge moments. The effect was modeled by combining a rotary spring constant, as determined from vehicle loading tests, with wind tunnel derived aerodynamic hinge moment characteristics to produce an elastically deformed elevon deflection angle. The elastic elevon angle is subsequently used with the rigid aerodynamic characteristics in determining the vehicle longitudinal aeroelastic characteristics.

8.4.2 Flight Measurements of the Orbiter Aerodynamics

Gamble and Young[36] noted:

The decision for Shuttle to perform an orbital, manned mission on the first launch raised the general question of how to maximize the mission safety without the benefit of either a graduated flight test program (as used by the aircraft industry) or an initial unmanned flight concept (as used in the early space program). The consequence of this decision was to adopt a philosophy of providing a reasonable estimate of maximum possible errors in the preflight predicted aerodynamics, and certify the flight control system (FCS) using the errors prior to STS-1. However, the estimated errors must not be so great as to completely invalidate the FCS design. Thus, a set of "worst case" aerodynamic uncertainties, defined as variations, was developed. As part of the first flight certification, variations, combined with other system uncertainties, were used to "stress" the FCS through a multitude of simulations. As a consequence, the initial entry was flown at a center of gravity and with FCS gains which maximized the aerodynamic margins thereby maximizing mission safety for these systems.

Kirsten et al.[37] note that:

Three types of test maneuvers specifically designed for obtaining aerodynamic data were performed during orbiter re-entries: (1) pushover-pullup maneuvers to obtain longitudinal performance data as a function of angle-of-attack, (2) body-flap sweeps to obtain body-flap effectiveness, and (3) longitudinal and lateral-directional control pulses to obtain stability and control derivatives. All three maneuvers were designed on ground based simulators and practiced by flight crews prior to flight.

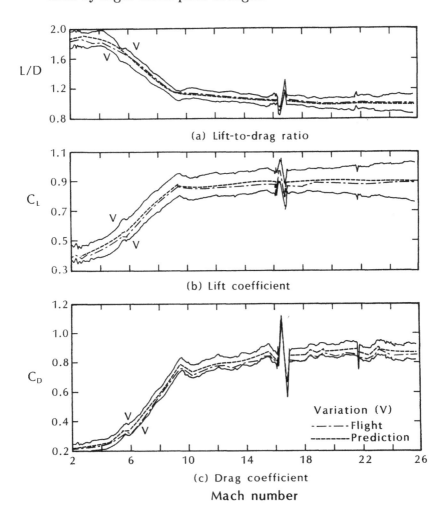

(a) Lift-to-drag ratio

(b) Lift coefficient

Variation (V)
—·—·—Flight
————Prediction

(c) Drag coefficient

Mach number

Fig. 8.41 The lift coefficient and the drag coefficient as a function of Mach number for STS-5 flight, as taken from Ref. 38.

The lift-to-drag ratio, the lift coefficient, and the drag coefficient for STS-5, as reported by Compton et al.[38] are presented as a function of Mach number in Fig. 8.41. The effect of the pushover-pullup (POPU) maneuver at Mach 16 is clearly evident in the aerodynamic coefficients. The flight values for the (L/D) ratio are slightly greater than the predictions for Mach numbers in excess of 6. For Mach numbers below 6, the flight values are slightly smaller. However, the differences are typically less than 2 percent. The (L/D) ratio is independent of the freestream density and of the dynamic pressure. However, there are some significant differences between the flight values and the predictions for the lift coefficients and for the drag coefficients at the higher Mach numbers. Compton et al.[38] note, "These differences are probably due to errors in the measured density but may be due to a flow-field phenomenon as yet unexplained."

As evident in the data presented in Fig. 1.7, viscous effects can have a pronounced effect on the drag at high altitudes. Actually, Maslen[39] has shown that the maximum L/D degrades with increasing altitude. The results of applying the theories in the Hypersonic Arbitrary Body Program[40] to the methods in the Shuttle aerodynamic data book were compared with flight results for the Shuttle Orbiter

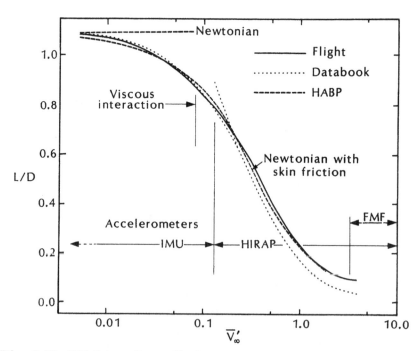

Fig. 8.42 Flight and prediction comparisons of (L/D) for the Space Shuttle Orbiter at α = 40 deg., as taken from Ref. 34.

by Wilhite et al.[34] Their results are reproduced in Fig. 8.42. A substantial reduction in L/D occurs in the viscous-interaction regime $(0.005 < \overline{V'_\infty} < 0.08)$ and a substantial reduction from the continuum regime through the noncontinuum regime. The L/D error using the inviscid Newtonian analysis is approximately 20 percent at $\overline{V'_\infty} = 0.08$.

For each of the first five flights of the Space Transportation System (STS), i.e., Space Shuttle Orbiter, significant differences between the predicted trim with the known control-surface deflections and that derived from flight at hypersonic speeds were noted. These differences were discussed at length in Sec. 3.7. As noted by Cooke,[41] "Anomalies in the actual flight data have extended the test requirements as originally conceived. These anomalies have in some cases accentuated the need for certain data already planned for. Others have pointed to a need for more concentrated investigation of certain flight regimes. A summary of flight anomalies is shown [in Fig. 8.43, items 1 through 7]."

An important concept for the reader to take from Fig. 8.43 is that there are both *known unknowns* and *unknown unknowns*. Certain critical unknowns are identified during the design process, e.g., the uncertain effect of misaligned tiles on boundary-layer transition (see Chap. 7) and questions relating to the control authority for lateral/directional trim (see Fig. 8.43). Although tests have been conducted and flowfields have been computed, questions remain about the validity of the models when applied to flight conditions. Thus, the designers compensate for the known unknowns (or pre-flight concerns) by restricting the flight envelope, by adding to the thermal-protection system, etc. Maneuvers are designed and data are obtained during the early flights to answer questions. In reality, of equal or of greater concern to the designer are the unknown unknowns, i.e., those phenomena that were not identified during the design process. The list of STS-1 anomalies given in Fig. 8.43 includes the fact that the pitching moment was outside the expected variations, causing body-flap trim problems. These were discussed in Sec. 3.7.

8.5 X-15 AERODYNAMICS

The power-off lift and drag characteristics of the X-15 Research Aircraft were presented in Ref. 42 for Mach numbers from 0.65 to 6.0 and for freestream Reynolds numbers from 0.2×10^6 to 2.8×10^6 per foot. The aerodynamic coefficients presented in this section were obtained during a flight-test program that used two variants of the X-15, which is shown in Fig. 8.44. For its earliest flights, the X-15 was powered by two four-chamber rocket engines (LR11) with a combined

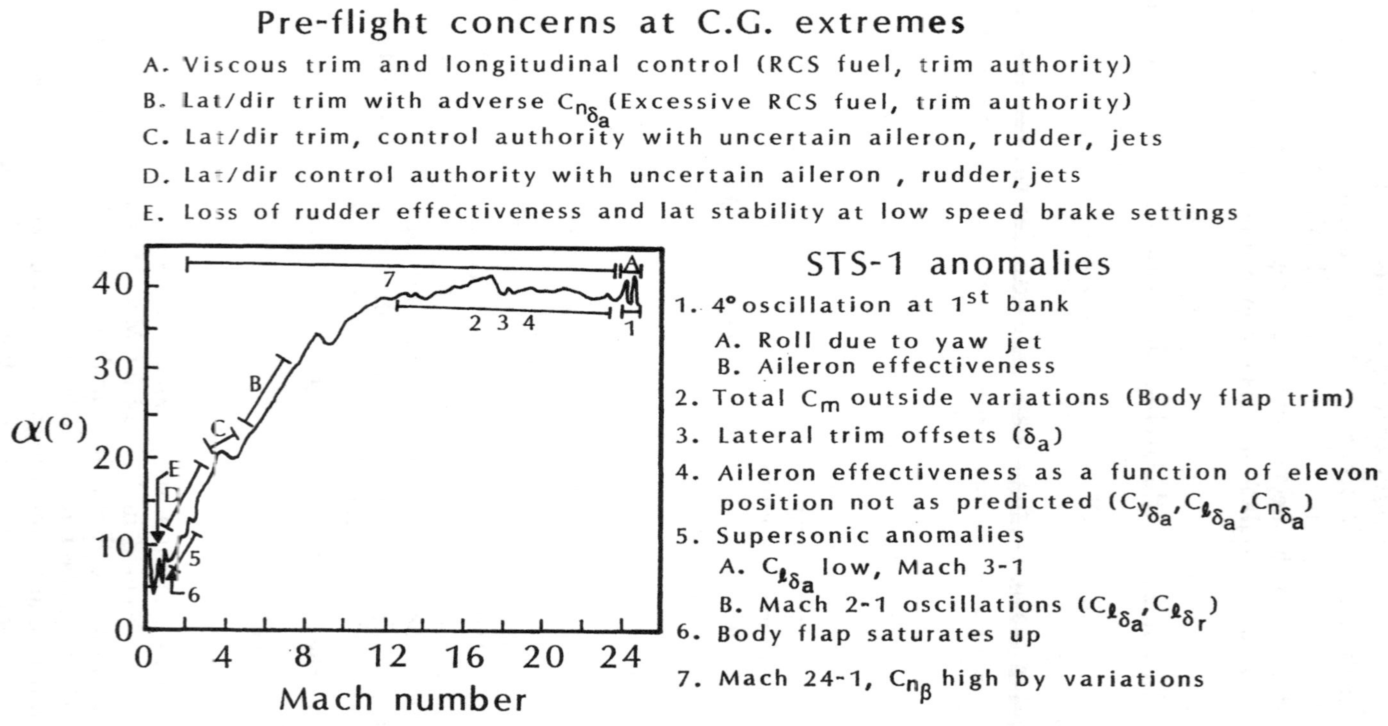
Pre-flight concerns at C.G. extremes
A. Viscous trim and longitudinal control (RCS fuel, trim authority)
B. Lat/dir trim with adverse $C_{n_{\delta_a}}$ (Excessive RCS fuel, trim authority)
C. Lat/dir trim, control authority with uncertain aileron, rudder, jets
D. Lat/dir control authority with uncertain aileron , rudder, jets
E. Loss of rudder effectiveness and lat stability at low speed brake settings
STS-1 anomalies
1. 4° oscillation at 1st bank
A. Roll due to yaw jet
B. Aileron effectiveness
2. Total C_m outside variations (Body flap trim)
3. Lateral trim offsets (δ_a)
4. Aileron effectiveness as a function of elevon position not as predicted ($C_{y_{\delta_a}}, C_{\ell_{\delta_a}}, C_{n_{\delta_a}}$)
5. Supersonic anomalies
A. $C_{\ell_{\delta_a}}$ low, Mach 3-1
B. Mach 2-1 oscillations ($C_{\ell_{\delta_a}}, C_{\ell_{\delta_r}}$)
6. Body flap saturates up
7. Mach 24-1, C_{n_β} high by variations
α(°)
0
10
20
30
40
0
4
8
12
16
20
24
Mach number
A
B
C
D
E
1
2 3 4
5
6
7

Fig. 8.43 Pre-flight concerns and anomalies observed during the first Shuttle Orbiter re-entry (STS-1), as taken from Ref. 41.

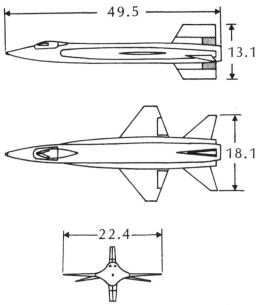

Fig. 8.44 Three-view drawing of the basic X-15 airplane. Shaded areas denote speed brakes. All dimensions in feet, as taken from Ref. 42.

thrust of approximately 16,000 pounds. This configuration, known as the *interim configuration*, was capable of about 4 minutes of powered flight, providing a maximum Mach number somewhat greater than 3. In 1960, a large, single-chamber rocket engine (LR99) was installed that provided approximately 58,000 pounds of thrust. At this thrust level, the *basic configuration* could achieve a maximum Mach number of 6 after 85 seconds of powered flight. The fuselage base area of the basic configuration with its LR99 engine was approximately 10 percent greater than that of the interim configuration.

Values for lift-curve slope, C_{L_α}, are presented as a function of the Mach number in Fig. 8.45. The flight measurements of C_{L_α} represent derivatives of the data from the lowest values obtained for the lift coefficient to those near the maximum lift-to-drag ratio. Wind-tunnel values are in good agreement with the flight data at supersonic Mach numbers.

The drag-due-to-lift increases significantly with Mach number. However, the total zero-lift drag decreases significantly with Mach number. With large, blunt base areas, the base drag for the un-powered portion of the flight represents a significant fraction of the zero-lift drag and exhibits the same Mach number dependence as the total zero-lift drag measurements. The net result is an $(L/D)_{max}$ curve that is depressed by the base drag at the lower supersonic

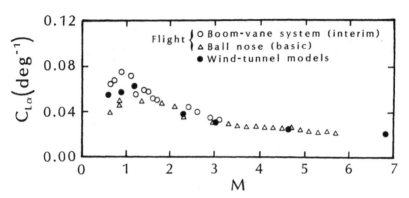

Fig. 8.45   Lift-curve-slope variation with Mach number for the trimmed X-15, as taken from Ref. 42.

Mach numbers and by the high drag-due-to-lift at the higher Mach numbers. As shown in Fig. 8.46, the values for the maximum lift-to-drag ratio are relatively insensitive to Mach number throughout the supersonic speed range, i.e., Mach numbers in excess of 2. The physical differences between the basic configuration and the interim configuration had only a slight effect on $(L/D)_{max}$ at Mach 3.

The lift-to-drag ratio is presented in Fig. 8.47 as a function of the angle-of-attack for two Mach numbers. The ticks on the curves indicate the range of angle-of-attack within which 95 percent of the maximum lift-to-drag can be obtained. The lift-to-drag ratio is relatively flat and independent of Mach number over this angle-of-attack range.   Thus, for the X-15, a near-optimum post-burnout gliding range can be approximated by flying at an angle-of-attack between about 7 deg. to 12 deg. throughout the supersonic speed range.

After having been extensively damaged during an emergency landing on its thirty-first flight, the X-15-2 aircraft was rebuilt and

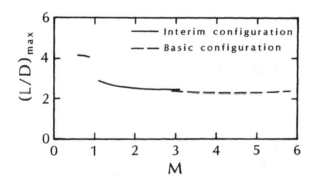

Fig. 8.46   Variation of $(L/D)_{max}$ with Mach number for trimmed X-15 flight, as taken from Ref. 42.

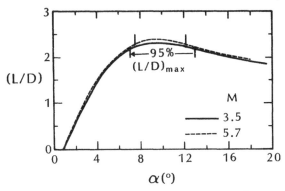

Fig. 8.47 The variation of lift-drag ratio with angle-of-attack of the X-15, as taken from Ref. 42.

modifications were incorporated to increase the vehicle performance capability to allow flight testing of a hypersonic ramjet engine. The flight-test program of the modified aircraft, the X-15A-2, is described by Armstrong.[43] On the last flight of this aircraft, the vehicle achieved a maximum Mach number of 6.7. However, locally severe heating rates due to an unexpected shock/shock interaction caused extensive damage to the dummy ramjet and to the lower ventral fin (to which the engine was attached). Shock/shock interactions will be discussed in Chap. 9. This phenomenon, which was unexpected at the time, i.e., an unknown unknown, serves as a warning of the potential problems when operating in the hypersonic aerothermodynamic environment.

8.6 HYPERSONIC AERODYNAMICS OF RESEARCH AIRPLANE CONCEPTS

Penland et al.[44] evaluated the aerodynamic performance of several conceptual hypersonic research airplanes, designed to be launched from a B-52 airplane and to cruise at Mach 6. Sketches of three vehicle concepts are presented in Fig. 8.48. They are: (A) an

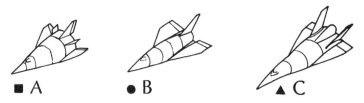

Fig. 8.48 Sketches of hypersonic research airplane concepts, as taken from Ref. 44.

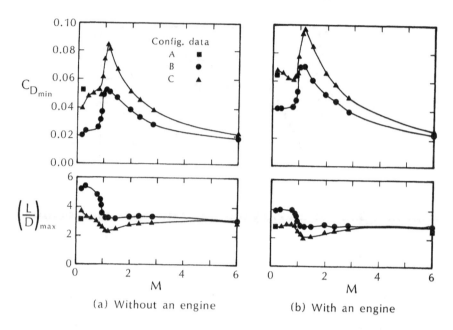

(a) Without an engine   (b) With an engine

Fig. 8.49   Variation of $C_{D_{min}}$ and of $(L/D)_{max}$ with Mach number, as taken from Ref. 44.

early lifting body concept, (B) a wing/body concept, and (C) a preliminary design concept. The variation of the minimum drag and the maximum lift-to-drag ratio with Mach number is reproduced in Fig. 8.49. Penland et al.[44] note, "Configuration C has excessive drag compared to the other concepts, partly because it had the largest toed-in vertical tails, the largest wetted area, and the greatest nozzle expansion angle."

The variation of the longitudinal aerodynamic center (the static margin) with Mach number is reproduced in Fig. 8.50 for the untrimmed configurations A, B, and C, both with and without a scramjet engine, and both at $C_L = 0$ and at $C_L = 0.2$. Penland et al.[44] note, "All three concepts exhibit a high level of static longitudinal stability at speeds up to $M = 3$ and satisfactory stability at $M = 6$. Since the mean aerodynamic chord of the present models is approximately one-half of the fuselage length, a static margin of 2% fuselage length would translate into 4% mean aerodynamic chord. The large static margins at off-design speeds were a consequence of the design, i.e., to maintain positive longitudinal stability up to $M = 6$. This excessive stability results in large elevon control forces to trim and leads to excessive trim drag."

The safe flight of an airplane depends on the static directional stability (the weather vane effect) and on the dihedral effect (roll

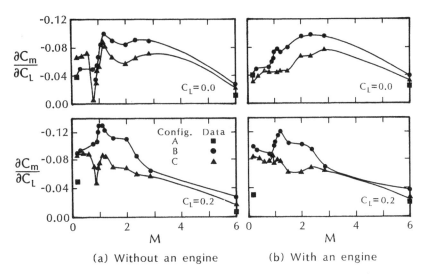

(a) Without an engine   (b) With an engine

Fig. 8.50 Variation of the longitudinal aerodynamic center with Mach number for three research airplane concepts, as taken from Ref. 44.

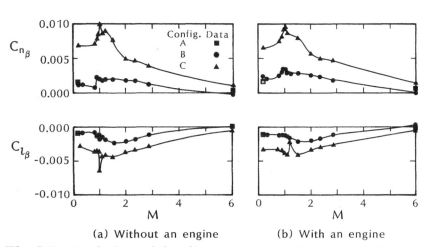

(a) Without an engine   (b) With an engine

Fig. 8.51 Variation of the directional stability and of dihedral effect with Mach number for three research airplane concepts, as taken from Ref. 44.

due to yaw). For directional stability, $C_{n_\beta} > 0$. For dihedral effect, $C_{\ell_\beta} < 0$. These parameters are presented at $C_L = 0$ as a function of Mach number in Fig. 8.51 for the three configurations both with and without a scramjet engine. All three configurations have satisfactory characteristics through subsonic and low supersonic speeds. However, only configuration C has satisfactory characteristics at $M = 6$. The satisfactory degree of stability is obtained at the expense of the drag coefficient. Recall that configuration C had the largest drag of the three design concepts due in part to the large toed-in vertical tails.

8.7 DYNAMIC STABILITY CONSIDERATIONS

The concept of dynamic stability is concerned with the motion of a vehicle perturbed from a steady-state condition. The vehicle is said to be dynamically stable if it returns to its original orientation following a disturbance (or a perturbation).

8.7.1 Stability Analysis of Planar Motion

Consider a planar motion in which the vehicle can pitch about its center of gravity, i.e., pure rotation, and/or undergo a translational acceleration. Three types of oscillatory motions for a sharp cone are presented in Fig. 8.52. For the motion depicted in Fig. 8.52a, the

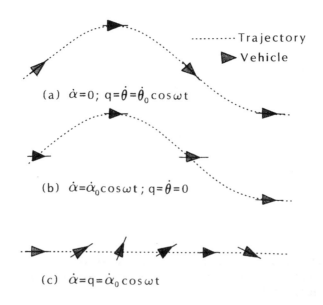

(a) $\dot{\alpha}=0;\ q=\dot{\theta}=\dot{\theta}_0\cos\omega t$

(b) $\dot{\alpha}=\dot{\alpha}_0\cos\omega t\,;\ q=\dot{\theta}=0$

(c) $\dot{\alpha}=q=\dot{\alpha}_0\cos\omega t$

Fig. 8.52   Three types of oscillating planar motions.

cone pitches about its center-of-gravity while it follows the oscillating flight path so that the angle-of-attack remains constant, i.e., $\alpha = 0$, $\dot{\alpha} = 0$. For the motion of Fig. 8.52b, the axis of the cone remains parallel to the horizon so that $q = \dot{\theta} = 0$. The angular velocity about the pitch axis, or rate of pitch, is represented by the symbol q. However, because the vehicle undergoes transverse accelerations, the angle-of-attack is a function of time. For the motion of Fig. 8.52c, the cone rotates about its center-of-gravity, which moves in a straight line at constant velocity. This fixed-axis oscillation is characteristic of wind-tunnel tests where dynamic stability data are obtained using the free-to-tumble technique. In the free-to-tumble technique, the model is mounted on a transverse rod which passed through the center-of-gravity of the model. For the motion of Fig. 8.52c, the rate-of-change of angle-of-attack ($\dot{\alpha}$) and the rate of pitch (q) are equal.

The angular motion of the vehicle is given by:

$$\dot{q}(t) = \frac{\rho_\infty U_\infty^2 S L}{2 I_{yy}} C_M(t) \tag{8-41}$$

where I_{yy} is the pitch moment of inertia. In order to solve for $\dot{q}$ and, therefore, to determine the stability of the vehicle, one must determine the pitching moment coefficient, $C_M(t)$. Using the linearized approximation:

$$C_M(t) = C_{M_0} + C_{M_\alpha} \alpha(t)$$

$$+ C_{Mq} \frac{L q(t)}{2 U_\infty} + C_{M_{\dot{\alpha}}} \frac{L \dot{\alpha}(t)}{2 U_\infty} \tag{8-42}$$

where $C_{M_\alpha} = (\partial C_M)/(\partial \alpha)$ is the slope of the static pitching moment curve,

$$C_{Mq} = \frac{\partial C_M}{\partial \left(\dfrac{q L}{2 U_\infty} \right)}$$

is the rotary pitching derivative defined as the variation of the pitching moment coefficient with respect to the pitch rate parameter ($q L / 2 U_\infty$), and

$$C_{M_{\dot{\alpha}}} = \frac{\partial C_M}{\partial \left(\dfrac{\dot{\alpha} L}{2 U_\infty} \right)}$$

is the dynamic pitching moment coefficient derivative with respect to the rate of change of the angle-of-attack. For small perturbations from the steady-state configuration, where the stability derivatives C_{M_α}, $C_{M_{\dot\alpha}}$, and C_{M_q} are constants, Eq. (8-42) provides a reasonable aerodynamic model.

As noted by Bustamante,[45] the damping moment coefficient is the sum of C_{M_q} and of $C_{M_{\dot\alpha}}$, i.e.,

$$C_{M_d} = C_{M_q} + C_{M_{\dot\alpha}} \qquad (8\text{-}43)$$

The vehicle is dynamically stable when the damping moment coefficient is negative

In a general longitudinal motion, the damping in pitch is determined by the combined effects of C_{M_q} and $C_{M_{\dot\alpha}}$, since in general both the angle of pitch and the angle-of-attack are changing. For fixed-axis oscillation experiments in a wind tunnel, such as depicted in Fig. 8.52c, the resulting oscillatory motion contains both a pitching rate q and a rate of change of angle-of-attack $\dot\alpha$, which are equal. Therefore, their individual effects on the resulting model motion are superimposed and inseparable. It is not possible to determine C_{M_q} and $C_{M_{\dot\alpha}}$ separately using the free-to-tumble technique.

8.7.2 Stability Data for Conic Configurations

The damping-in-pitch derivatives, i.e., $(C_{M_q} + C_{M_{\dot\alpha}})$, for a slightly-blunted cone with a semi-vertex angle of 10 deg. were obtained in Range G of the von Karman Gas Dynamics Facility (VKF) at

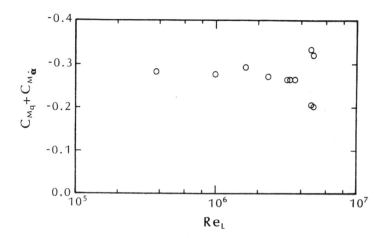

Fig. 8.53 **Damping-in-pitch derivatives as a function of the Reynolds number for a slightly blunted cone, $\theta_c = 10$ deg., $M_\infty \approx 6.5$, as taken from Ref. 46.**

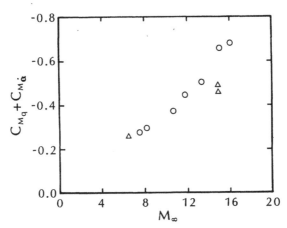

Fig. 8.54 Damping-in-pitch derivatives as a function of the Mach number for a slightly blunted cone, θ_c = 10 deg., Re_L = 0.4 × 10⁶, as taken from Ref. 46.

AEDC. The nose-to-base radius ratio (R_N/R_B) was 0.032 and the center-of-gravity was located at 0.65L. The experimentally determined damping-in-pitch derivatives, as reported by Welsh et al.,[46] are presented as a function of the Reynolds number in Fig. 8.53 and as a function of the Mach number in Fig. 8.54. The experimental values of $(C_{Mq} + C_{M_{\dot\alpha}})$ that are presented in Fig. 8.53 represent data obtained for Mach numbers from 5.7 to 7.5 and mean effective angles-of-attack from 1 deg. to 6 deg. As noted by Welsh et al.[46]:

> The level and trend of the damping values for a laminar boundary layer $(Re_L < 4.5 \times 10^6)$ are well defined. For higher Reynolds number shots $(Re_L > 4.5 \times 10^6)$, the location of transition is in the region of the model base. The increased spread in the measurements at the highest Reynolds numbers is believed to be related to fluctuations in the location of transition and is indicative of the increased difficulty in obtaining consistent damping measurements for this test condition.

The experimental values of $(C_{Mq} + C_{M_{\dot\alpha}})$ that are presented in Fig. 8.54 represent data for Reynolds numbers from 0.38×10⁶ to 0.45×10⁶ and mean effective angles-of-attack from 2.5 deg. to 12 deg. The circular symbols represent shots for which the models experienced nonplanar motion patterns, but the transverse component of the velocity was small at the maximum amplitude. For the three shots represented by the triangular symbols, the model had combinations of a rolling velocity and a wider elliptic motion pattern such that the model would tend to have a larger transverse velocity component at its maximum amplitude. The measurements indicate that the damping-in-pitch derivatives for the cone increase appreciably with increasing Mach number between 8 and 16 at a Reynolds number (based on the freestream conditions and the model length) of about 0.4×10^6 (for which the boundary layer was laminar).

Welsh et al.[46] also noted, "C_{M_α} for the cone decreases significantly as the nose-radius to base-radius ratio of the cone is increased up to 0.1 for amplitudes greater than about 5°."

The effect of nose blunting on the damping derivative for slightly blunted cones was correlated by East and his co-workers in terms of the bluntness scaling parameter,

$$\chi - 1 = \frac{2(C_{DN})^{0.5}}{\tan \theta_c} \left(\frac{R_N}{R_B} \right) \qquad (8\text{-}44)$$

Using the approximation for spherically-blunted cones that:

$$2(C_{DN})^{0.5} \approx 1.9$$

Khalid and East[47] presented damping-derivative data for slender blunted cones over the range of parameters:

$8 \leq M_\infty \leq 14.2$; $5.6° \leq \theta_c \leq 20°$; $0.56L \leq x_{cg} \leq 0.67L$; and

$0.0 \leq R_N \leq 0.4R_B$.

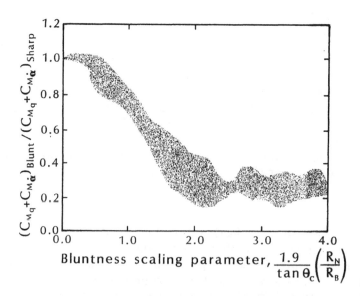

Fig. 8.55   The effect of nose blunting on the damping, as taken from Ref. 47.

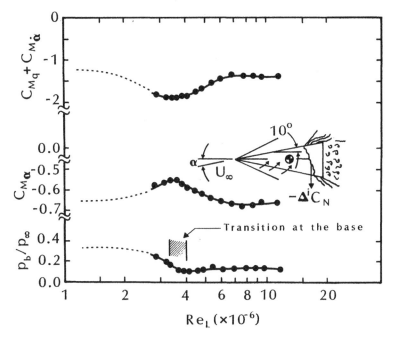

Fig. 8.56 The aerodynamic characteristics of a sharp cone (θ_c = 10 deg.) as a function of the Reynolds number, as taken from Ref. 49.

The band of data are reproduced in Fig. 8.55. By presenting the data as the ratio of the blunted-cone damping derivative to the sharp-cone value, viscous effects that are common to the flowfields for blunted and for sharp cones do not affect the correlation.

As noted by Ericsson[21]:

> When the nose bluntness is not of the hemispherical type, the bow shock shape and associated entropy wake are no longer insensitive to angle of attack. The drag for different circumferential segments of the nose are not the same at $\alpha \neq 0$, and the entropy wake is reshaped accordingly. Of particular interest is the effect of a change from hemispherical to conical nose bluntness, such as is taking place due to ablative nosetip recession during missile re-entry into the earth's atmosphere.

The large changes in the damping-derivative data presented in Fig. 8.53 was attributed to the occurrence of transition in the region of the model base. Ericsson[48] states:

> Among the many problems the test engineer faces when trying to simulate full scale vehicle dynamics in a wind

tunnel test is the fact that the test usually will be performed at Reynolds numbers far below those existing on
the full scale vehicle. It is found that even in the case
of attached flow a severe scaling problem may exist. The
strong coupling existing between boundary layer transition and vehicle motion can cause the wind tunnel results
to be very misleading ...For example, the subscale test
can fail to show a dynamic stability problem existing in
full scale flight or, conversely, show one that does not exist
on the full scale vehicle. When flow separation occurs together with boundary layer transition, the scaling problem
becomes more complicated, and the potential for danger-
ously misleading subscale test results increases.

The effect of boundary-layer transition on the aerodynamic characteristics of a sharp cone with a semi vertex angle of 10 deg. is
shown in Fig. 8.56, using the data of Ward.[49] Significant changes
are evident in the moment coefficients in the same Reynolds number
range that the base pressures indicate for the onset of transition at
the model base. As indicated in the sketch, the transition front is
asymmetric as it moves on to the aft end of the cone. An incremental negative normal force ($\Delta^i C_N$) is generated, causing a decrease in
the static stability. Note that C_{M_α} becomes less negative. Because
there is a convective flow time lag, the damping moment coefficient
becomes more negative. That is, there is an increase in the dynamic
stability. However, as noted by Ericsson,[21] "the effect changes sign,
becoming statically stabilizing and dynamically destabilizing, when
the Reynolds number is increased enough to cause transition to occur
forward of the oscillation center or body C.G."

Note that, since the effect changes sign, becoming statically stabilizing and dynamically destabilizing, the designer should determine
if this phenomena actually has a significant effect on the flight of the
vehicle. In many applications, a researcher will identify phenomena
which produce significant changes in the flowfield but which have
only a second-order effect on the vehicle design/performance.

8.7.3 Additional Considerations

Ericsson and Reding[50] note:

The aerodynamic forces generated by ablative mass addition may be classified as follows: local forces generated
by local ablative mass addition (C_N), downstream forces
due to cross flow effects on a boundary layer that has
been thickened by upstream ablation (transport effects,
$\Delta^i_{tr} C_N$), and downstream forces caused by cross flow ef-

fects on downstream ablation (shielding effects, $\Delta^i_{sh} C_N$). Not included in this list are the loads generated by changes in the body shape due to ablation, the largest of which on a slender conical body is caused by increased nose bluntness, which increases drag and influences the body pitching moment due to a time-dependent nose asymmetry.

Hodapp and Clark[51] note:

A slender ballistic re-entry vehicle that has a lateral center-of-gravity offset (mass asymmetry) may exhibit anomalous roll behavior as a result of the roll torque due to lateral aerodynamic force and the mass asymmetry. The aerodynamic force is caused by the total angle of attack, which consists of transient and trim components.

8.8 CONCLUDING REMARKS

Computational fluid dynamics (CFD) of varying degrees of rigor and experimental programs can be used to obtain aerodynamic force and moment coefficients. Relatively simple flow models, e.g., Newtonian flow model and Euler equations, provide reasonable estimates of the normal-force coefficient and of the pitching moment coefficient for many applications. However, in many cases, viscous effects and/or real-gas effects can have a dramatic effect on the aerodynamic coefficients. Seemingly second-order differences in the pressure distribution for the Space Shuttle Orbiter had a significant effect on the pitching moments in the high-velocity/high altitude regime. Simulation of the Reynolds number can be important, since boundary-layer transition can affect the forces and moments either directly or through its influence on viscous/inviscid interactions.

REFERENCES

[1] Neumann, R. D., "Defining the Aerothermodynamic Methodology," J. J. Bertin, R. Glowinski, and J. Periaux (eds.), *Hypersonics, Volume I: Defining the Hypersonic Environment,* Birkäuser Boston, Boston, MA, 1989.

[2] Draper, A. C., and Buck, M. L., "Lifting Bodies – An Attractive Aerodynamic Configuration Choice for Hypervelocity Vehicles," Paper 30 in AGARD Conference Proceedings No. 428, *Aerodynamics of Hypersonic Lifting Vehicles,* Nov. 1987.

[3] Walberg, G. D., "A Survey of Aeroassociated Orbit Transfer," *Journal of Spacecraft and Rockets,* Vol. 22, No. 1, Jan.–Feb. 1985, pp. 3–18.

[4] Romere, P. O., "Private Transmittal," Feb. 1992.

[5] Surber, T. E., and Olsen, D. C., "Space Shuttle Orbiter Aerodynamic Development," *Journal of Spacecraft and Rockets,* Vol. 15, No. 1, Jan.–Feb. 1978, pp. 40–47.

[6] Silveira, M. A., "Private Transmittal," Feb. 1992.

[7] Hunt, J. L., Presentation to the U.S.A.F. Scientific Advisory Board, June 1991.

[8] Gottmann, T., "Aspekte des Hyperschallfluges und Beschreibung des Leitkonzeptes 4 der BMFT – Studie Überschallflugzeug," Messerschmitt-Bölkow-Blohm MBB/LKE127/HYPAC/R/5, Mar. 1987.

[9] Anderson, J. D., Lewis, M. J., and Corda, S., "Several Families of Viscous Optimized Waveriders – A Review of Waverider Research at the University of Maryland," *Proceedings of the First International Hypersonic Waverider Symposium,* University of Maryland, Oct. 1990.

[10] Blankson, I., "Prospects for Airbreathing Hypersonic Waveriders," AIAA Paper 92-0303, Reno, NV, Jan. 1992.

[11] Desideri, J. A., Glowinski, R., and Periaux, J., (eds.), *Hypersonic Flow for Reentry Problems,* Springer-Verlag, Berlin, 1992.

[12] Bertin, J. J., and Smith, M. L., *Aerodynamics for Engineers,* 2nd Ed., Prentice-Hall, Englewood Cliffs, NJ, 1989.

[13] Neal, L., Jr., "Aerodynamic Characteristics at a Mach Number of 6.77 of a 9° Cone Configuration, with and without Spherical Afterbodies, at Angles of Attack up to 180° with Various Degrees of Nose Blunting," NASA TN D-1606, Mar. 1961.

[14] Foster, A. D., "A Compilation of Longitudinal Aerodynamic Characteristics Including Pressure Information for Sharp- and Blunt-Nose Cones Having Flat and Modified Bases," Sandia Corporation SC-R-64-1311, Jan. 1965.

[15] Glover, L. S., and Hagan, J. C., "The Motion of Ballistic Missiles," Johns Hopkins University, Applied Physics Laboratory TG 1164, July 1971.

[16] Adams, J. C., Jr., "Calculated Hypersonic Viscous Aerodynamics of a Sharp 7-Deg Cone at Incidence," AIAA Paper 76-361, San Diego, CA, July 1976.

[17] Adams, J. C., Jr., and Griffith, B. J., "Hypersonic Viscous Static Stability of a Sharp 5-deg Cone at Incidence," *AIAA Journal,* Vol. 14, No. 8, Aug. 1976, pp. 1062–1068.

[18] Zoby, E. V., and Thompson, R. A., "Flowfield and Vehicle Parameter Influence on Hypersonic Heat Transfer and Drag," *Journal of Spacecraft and Rockets,* Vol. 27, No. 4, July–Aug. 1990, pp. 361–368.

[19] Kirk, D. B., and Chapman, G. T., "Free-Flight Tests of a Blunt-Nosed Flare-Stabilized Body at Speeds to 8.2 km/sec," *Journal of Spacecraft and Rockets,* Vol. 3, No. 3, Mar. 1966, pp. 374–377.

[20] Seiff, A., "Secondary Flow Fields Embedded in Hypersonic Shock Layers," NASA TN D-1304, May 1962.

[21] Ericsson, L. E., "Rapid Computation of Missile Dynamics," AIAA Paper 84-0391, Reno, NV, Jan. 1984.

[22] Dennis, D. H., and Syvertson, C A., "Effects of Boundary Layer Separation on Normal Force and Center of Pressure of a Cone-Cylinder Model with a Large Base Flare at Mach Numbers from 3.00 to 6.28," NACA RM A55H09, 1955.

[23] Crowder, R. S., and Moote, J. D., "Apollo Entry Aerodynamics," *Journal of Spacecraft and Rockets,* Vol. 6, No. 3, Mar. 1969, pp. 302–307.

[24] Moseley, W. C., Jr., Moore, R. H., Jr., and Hughes, J. E., "Stability Characteristics of the Apollo Command Module," NASA TN D-3890, Mar. 1967.

[25] Park, C., *Nonequilibrium Hypersonic Aerothermodynamics,* John Wiley, New York, 1990.

[26] De Rose, C. E., "Trim Attitude, Lift and Drag of the Apollo Command Module with Offset Center-of-Gravity Positions at Mach Numbers to 29," NASA TN D-5276, June 1969.

[27] Malcolm, G. N., and Chapman, G. T., "A Computer Program for Systematically Analyzing Free-Flight Data to Determine the Aerodynamics of Axisymmetric Bodies," NASA TN D-4766, May 1968.

[28] Hillje, E. R., "Entry Flight Aerodynamics from Apollo Mission AS-202," NASA TN D-4185, Oct. 1967.

[29] Hillje, E. R., and Savage, R., "Status of Aerodynamic Characteristics of the Apollo Entry Configuration," AIAA Paper 68-1143, 1968.

[30] Weilmuenster, K. J., and Hamilton, H. H., II, "A Comparison of Computed and Measured Aerodynamic Characteristics of a Proposed Aeroassist Flight Experiment Configuration," AIAA Paper 86-1366, Boston, MA, June 1986.

[31] Kirk, D. B., Intrieri, P. F., and Seiff, A., "Aerodynamic Behavior of the Viking Entry Vehicle: Ground Test and Flight Results," *Journal of Spacecraft and Rockets,* Vol. 15, No. 4, July–Aug. 1978, pp. 208–212.

[32] Koppenwallner, G., and Legge, H., "Drag of Bodies in Rarefied Hypersonic Flow," J. N. Moss and C. D. Scott (eds.), *Thermophysical Aspects of Reentry Flows,* Vol. 103 of Progress in Astronautics and Aeronautics, AIAA, New York, 1985, pp. 44–59.

[33] Celenligil, M. C., Moss, J. N., and Blanchard, R. C., "Three-Dimensional Flow Simulation About the AFE Vehicle in the Transitional Regime," AIAA Paper 89-0245, Reno, NV, Jan. 1989.

[34] Wilhite, A. W., Arrington, J. P., and McCandless, R. S., "Performance Aerodynamics of Aeroassisted Orbital Transfer Vehicles," H. F. Nelson (ed.), *Thermal Design of Aeroassisted Orbital Transfer Vehicles*, Vol. 96 of Progress in Astronautics and Aeronautics, AIAA, New York, 1985, pp. 165–185.

[35] Romere, P. O., and Whitnah, A. M., "Space Shuttle Entry Longitudinal Aerodynamic Comparisons of Flights 1–4 with Preflight Predictions," in *Shuttle Performance: Lessons Learned*, NASA CP 2238, Part 1, Mar. 1983.

[36] Gamble, J. D., and Young, J. C., "The Development and Application of Aerodynamic Uncertainties in the Design of the Entry Trajectory and Flight Control System of the Space Shuttle Orbiter," AIAA Paper 82-1335, San Diego, CA, Aug. 1982.

[37] Kirsten, P. W., Richardson, D. F., and Wilson, C. M., "Predicted and Flight Test Results of the Performance, Stability and Control of the Space Shuttle from Reentry to Landing," in *Shuttle Performance: Lessons Learned*, NASA CP 2283, Part 1, Mar. 1983.

[38] Compton, H. R., Schiess, J. R., Suit, W. T., Scallion, W. I., and Hudgins, J. W., "Stability and Control over the Supersonic and Hypersonic Speed Range," in *Shuttle Performance: Lessons Learned*, NASA CP 2283, Part 1, Mar. 1983.

[39] Maslen, S. H., "Synergetic Turns with Variable Aerodynamics," *Journal of Spacecraft and Rockets*, Vol. 4, No. 11, Nov. 1967, pp. 1475–1482.

[40] Gentry, A. E., Smyth, D. N., and Oliver, W. R., "The Mark IV Supersonic-Hypersonic Arbitrary-Body Program, Volume II – Program Formulation," Air Force Flight Dynamics Laboratory, AFFDL-TR-73-159 Vol II, Nov. 1973.

[41] Cooke, D. R., "Minimum Testing of the Space Shuttle Orbiter for Stability and Control Derivatives," in *Shuttle Performance: Lessons Learned*, NASA CP 2283, Part 1, Mar. 1983.

[42] Saltzman, E. J., and Garringer, D. J., "Summary of Full-Scale Lift and Drag Characteristics of the X-15 Airplane," NASA TN D-3343, Mar. 1966.

[43] Armstrong, J. G., "Flight Planning and Conduct of the X-15A-2 Envelop Expansion Program," Air Force Flight Test Center FTC-TD-69-4, July 1969.

[44] Penland, J. A., Dillon, J. A., and Pittman, J. L., "An Aerodynamic Analysis of Several Hypersonic Research Airplane Concepts from $M = 0.2$ to 6.0," *Journal of Aircraft*, Vol. 15, No. 11, Nov. 1978, pp. 716–723.

[45] Bustamante, A. C., "Aerodynamic Analysis and Characteristics of the SNAP 10-A Re-Entry Test Vehicle for the RFD-1 Re-Entry Flight Test," Sandia Corporation SC-4927, Oct. 1963.

[46] Welsh, C. J., Winchenbach, G. L., and Madagan, A. N., "Free-Flight Investigation of the Aerodynamic Characteristics of a Cone at High Mach Numbers," *AIAA Journal,* Vol. 8, No. 2, Feb. 1970, pp. 294–300.

[47] Khalid, M. and East, R. A., "Stability Derivatives of Blunt Slender Cones at High Mach Numbers," *British Aeronautical Quarterly,* Vol. 31, Nov. 1979, pp. 559–589.

[48] Ericsson, L. E., "Review of Transition Effects on the Problem of Dynamic Simulation," AIAA Paper 80-2004, Aug. 1980.

[49] Ward, L. K., "Influence of Boundary Layer Transition on Dynamic Stability at Hypersonic Speeds," *Transactions of the Second Technical Workshop on Dynamic Stability Testing,* Paper 9, Vol. II, Arnold Engineering Development Center (AEDC), Apr. 1965.

[50] Ericsson, L. E., and Reding, J. P., "Ablation Effects on Vehicle Dynamics," *Journal of Spacecraft and Rockets,* Vol. 3, No. 10, Oct. 1966, pp. 1476–1483.

[51] Hodapp, A. E., Jr., and Clark, E. L., Jr., "Effects of Products of Inertia on Re-Entry Vehicle Roll Behavior," *Journal of Spacecraft and Rockets,* Vol. 8, No. 2, Feb. 1971, pp. 155–161.

PROBLEMS

8.1 For purposes of estimating the aerodynamic forces, let us represent the Shuttle by a flat plate delta wing whose leading edge is swept 50 deg., as shown in Fig. 8.57

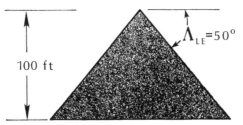

Fig. 8.57   Sketch for Problem 8.1.

Using the modified Newtonian flow assumption, calculate the lift coefficient as a function of the angle-of-attack. At what angle-of-attack is the lift coefficient a maximum? What is the maximum value of the lift coefficient?

8.2

(a) Using the modified Newtonian flow model, develop an expression for the pressure drag coefficient for a sphere in a hypersonic stream.

(b) Using the modified Newtonian flow model, develop an expression for the pressure drag coefficient (per unit span) for a cylinder in a hypersonic stream.

8.3 You are to design a payload shroud to protect a payload during launch. The shroud is to fit in a space that is of length L and base diameter D_B. As indicated in the sketches of Fig. 8.58, the two options are: a sharp cone or a parabolic shape. The parabolic shape is described by the equation

$$x = 4y^2$$

where $y = R_B = 0.5L$, when $x = L$. The parabolic shape has a lower maximum heating rate and a greater volume. Determine expressions for the following aerodynamic parameters for these two configurations: C_D, C_L, and x_{cp}/L for $0° \leq \alpha \leq 10°$.

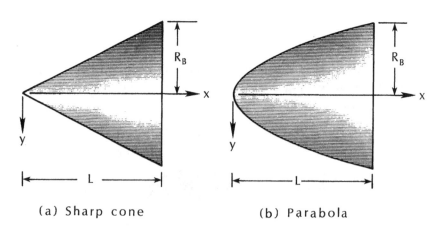

(a) Sharp cone   (b) Parabola

Fig. 8.58 Sketch of configurations for Problem 8.3.

8.4 Consider the sharp biconic shown in the sketch of Fig. 8.59. Using the Newtonian flow model, calculate the pressure coefficient on the two conical surfaces. What is the pressure drag coefficient,

$$C_{D_P} = \frac{D_P}{q_\infty \pi R_B^2} \quad ?$$

Assume that the base pressure is $p_b = p_\infty$.

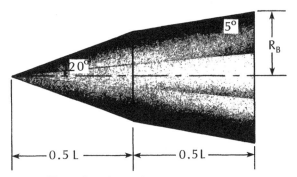

Fig. 8.59   Sketch of configuration for Problem 8.4.

8.5   Consider the sharp biconic shown in the sketch of Fig. 8.60.

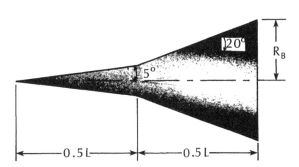

Fig. 8.60   Sketch of configuration for Problem 8.5.

Using the Newtonian flow model, calculate the pressure coefficient on the two conical surfaces. What is the pressure drag coefficient,

$$C_{D_P} = \frac{D_p}{q_\infty \pi R_B^2} \quad ?$$

Assume that the base pressure is $p_b = p_\infty$. How does the (actual) flow for this configuration differ from that of Problem 8.4, if one includes the effects of viscosity (i.e., there is a boundary layer)?

8.6   Consider the sharp cone in a hypersonic stream, which was discussed in this chapter. If $M_\infty = 10$ and $\theta_c = 10$ deg., prepare a graph of C_D as a function of α for $-10° \leq \alpha \leq 10°$.

8.7 Consider the sharp cone in a hypersonic stream, which was discussed in this chapter. If $M_\infty = 10$ and $\theta_c = 10$ deg., prepare a graph of C_L as a function of α for $-10° \leq \alpha \leq 10°$.

8.8 Consider the sharp cone in a hypersonic stream, which was discussed in this chapter. If $M_\infty = 10$ and $\theta_c = 10$ deg., prepare a graph of L/D as a function of α for $-10° \leq \alpha \leq 10°$.

8.9 Consider the sharp cone in a hypersonic stream, which was discussed in this chapter. If $M_\infty = 10$ and $\theta_c = 10$ deg., prepare a graph of C_M as a function of α for $-10° \leq \alpha \leq 10°$. Is the configuration statically stable if $x_{cg} = 0.6L$?

8.10 A configuration that would generate lift at low angles of attack is a half-cone model, such as shown in Fig. 8.61. Neglecting the effects of skin friction and assuming the pressure is that of the Newtonian flow model, develop an expression for the lift-to-drag ratio as a function of θ_c for $0 \leq \alpha \leq \theta_c$.

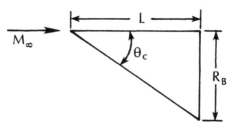

Fig. 8.61 Sketch of configuration for Problems 8.10 and 8.11.

8.11 Develop expressions for y_{cp} and x_{cp} for the flow described in Problem 8.10.

8.12 Consider the blunted sphere cone, shown in the sketch of Fig. 8.62.

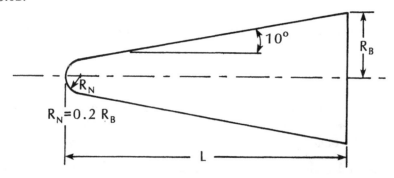

Fig. 8.62 Sketch of configuration for Problems 8.12 and 8.13.

If $M_\infty = 8$, use the modified Newtonian flow model to prepare a graph of C_D as a function of α for $-10° \leq \alpha \leq 10°$.

8.13 Consider the blunted sphere cone shown in the sketch of Fig. 8.62. If $M_\infty = 8$, prepare a graph of C_L as a function of α for $-10° \leq \alpha \leq 10°$.

VISCOUS
INTERACTIONS

9.1 INTRODUCTION

Slender vehicles that cruise at hypersonic speeds and gliding re-entry vehicles designed to generate significant lift have complex three-dimensional flowfields which contain a variety of viscous interactions and shock/shock interactions. The sources of typical interactions are illustrated in the sketch of Fig. 9.1. The interactions that would result when the vehicle shown in Fig. 9.1 is located in a hypersonic stream include the following.

a. Shock waves that are generated when there is a compressive turning of the local flow create shock/boundary-layer interactions. As indicated in Fig. 9.1, the canopy and deflected control surfaces can produce the required flow deflections. The shock/boundary-layer interactions which result are described in Sec. 9.2.

b. Shock/shock interactions occur when two shock waves intersect. Two examples for the vehicle of Fig. 9.1 are (1) the "shock on lip," which occurs when the bow shock wave intersects the shock wave generated at the inlet cowl leading edge, and (2) the wing leading-edge interaction, which occurs when the bow shock wave intersects the wing leading-edge shock wave. As will be discussed in Sec. 9.3, the six types of shock/shock interactions which can occur produce a wide variety of flowfield perturbations.

c. The flowfield perturbations produced by a swept fin mounted on a flat plate contain features of the interactions described in paragraphs a and b. These flowfields will be discussed in Sec. 9.4.

d. Axial corners formed by compression surfaces, such as occur in airbreathing engine inlets and at wing/body or at fin/wing junctions can produce complex flow patterns which contain vortices and embedded shock waves. Corner flows will be discussed in Sec. 9.5.

The viscous interactions of paragraphs a through d produce complex flowfields, which may contain separation bubbles, vortices, impinging jets, etc. As a result, the viscous interactions described above can cause boundary layers to separate, can produce regions of locally high pressure, and can produce regions of locally high heat transfer. Flow separation can result in a loss of control effectiveness or flow degradation in an engine inlet. The heating rates and the pressures in the interaction regions can be locally severe, being orders of magnitude greater than the stagnation-point values. Large gradients exist in the heat transfer and in the pressure distributions, with locally severe values affecting extremely small areas. Unless the correct instrumentation is located at the proper location, the experimentalist may fail to "measure" the peak heat-transfer rate or the peak pressure. For instance, a pressure orifice which is too large will sense a space-averaged and/or time-averaged value, and not the maximum value, even if it is correctly positioned. A properly sized orifice will not sense the maximum value if not correctly positioned. Failure to properly model the flow physics or to develop a computational grid with the proper resolution may produce similar shortcomings in the CFD solutions.

The designers of vehicles that are to fly at hypersonic speeds have long recognized that the locally severe heating rates produced by viscous interactions and by shock/shock interactions can cause catastrophic failures. One of the first in-flight confirmations of the severity of shock-impingement heating occurred in October 1967, when the

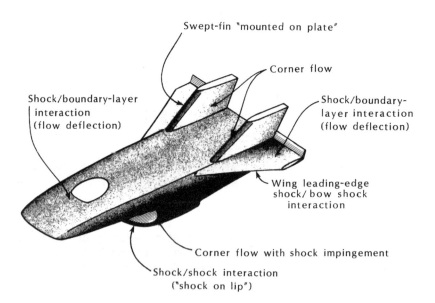

Fig. 9.1 Sources of viscous interactions and of shock/shock interactions for a hypersonic vehicle.

X-15A-2 suffered severe damage to its ventral fin (pylon) during a high-altitude flight at Mach 6.7. The ventral fin (or pylon) supported a dummy model of the Hypersonic Research Engine (HRE), as shown in the photographs of Fig. 9.2, for the last three flights of the X-15 flight-test program. On 5 May 1967, the X-15A-2 with no ablative coating (see Fig. 9.2a) was flown to a maximum velocity of 4,750 fps (1,448 m/s). The main purpose of this flight was to evaluate the handling qualities of the aircraft with the dummy ramjet engine installed (Ref. 1). Two flights were made with an ablative heat shield protecting the X-15A-2 and with a dummy ramjet attached. On the

(a) X-15A-2 (no ablative coating)

(b) X-15A-2 (full ablative coating)

Fig. 9.2 Dummy Hypersonic Research Engine (HRE) mounted on pylon on X-15 (photographs supplied by NASA, one by W. J. D. Escher).

second flight in October 1967, the aircraft achieved 6,630 fps (2,021 m/s). Although the ablative heat shield protected the aircraft structure, a shock/shock interaction produced considerable damage on the dummy ramjet and on the ramjet pylon. Gaping holes were burned in the pylon, and four probes were lost.

Heat transfer and surface-pressure measurements were obtained to determine the effects of the cavities and of the protuberances, such as the shear pads (cylinders of dense ablator material designed to transmit shear loads between the spacecraft and the service module during the launch phase of the mission) on the Apollo Command Module. Interference heating factors, i.e., the ratio of $\dot{q}_{pert}/\dot{q}_{smooth}$, as large as 9 were measured in the vicinity of protuberances on the windward (spherical) heat shield.[2] Bertin[2] noted further , "The heat transfer to the leeward afterbody was apparently unaffected by the cavities and protuberances." The flow was separated over the leeward afterbody.

Potentially severe heating rates to the wing leading edge where the bow shock wave intersects the wing leading-edge shock wave concerned the designers of the Space Shuttle Orbiter. The sketch of Fig. 9.3 illustrates a shock/shock interaction.

As noted by Gaitonde and Shang,[3] the forebodies of proposed hypersonic aircraft powered by airbreathing propulsion systems form

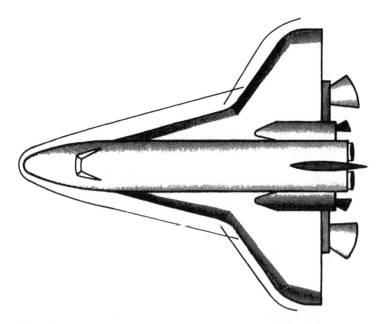

Fig. 9.3   Interactions between the bow shock wave and the wing leading-edge shock wave for the Space Shuttle Orbiter.

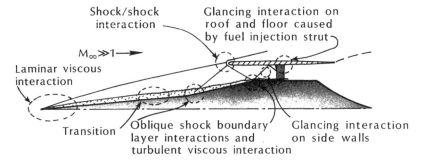

(a) Sketch of overall flowfield, as taken from Ref. 4.

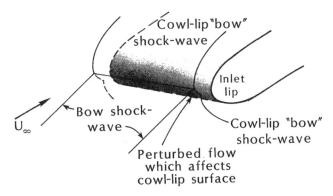

(b) Sketch of local flow, based on the presence of only two shock waves.

Fig. 9.4 Sketch of flowfield showing typical viscous interactions associated with airbreathing propulsion systems.

ramp-like structures designed to compress the incoming air with oblique shocks. They, therefore, act as the compressor system for the inlet. "For optimum mass flow through the inlet, it is desirable that these compression system shocks, which may form a relatively strong oblique shock in conjunction with the vehicle bow shock, be positioned to converge on the inlet cowl leading edge where they interact with the bow shock produced by the cowl lip. Viscous hypersonic shock-on-shock interactions (often denoted 'interfering' flows) can significantly affect the performance of the inlet through the creation of anomalous pressure and heat-transfer peaks on the cowl leading edge." The hypothetical hypersonic intake flowfield depicted in the sketch of Stollery,[4] which is reproduced in Fig. 9.4a, indicates that three shock waves interact at the cowl lip: (1) the "bow" shock wave produced by the cowl lip, (2) the bow shock wave originating at the vehicle leading edge, and (3) an oblique shock produced by the compressive turning of the flow at the inlet ramp. Most of the computations of the flowfield for the cowl lip are based on the interaction of two shock waves: (1) the shock wave produced by the cowl lip and (2) the bow shock wave originating at the vehicle leading edge. As indicated in the sketch of Fig. 9.4b, the interaction of these two shock waves produces a flowfield perturbation which affects the cowl-lip surface.

The viscous interactions and the shock/shock interactions described in the previous paragraphs and depicted in Figs. 9.3 and 9.4 can produce catastrophic failures. The designers of hypersonic vehicles often use conservative estimates of the effect of these phenomena in order to allow for the uncertainties in our understanding of them. However, too much conservatism may cause unacceptable weight penalties in the thermal-protection system (TPS) or in restrictions on the allowable flight corridor. It is the intent of the presentations in this chapter to identify the characteristics of the viscous-interaction phenomena represented in Figs. 9.1 through 9.4.

When the Mach number is relatively high and the Reynolds number is relatively low, the viscous boundary layer is a relatively large fraction of the shock layer, as shown in the sketch of Fig. 9.5. The displacement effects caused by the boundary layer produce flowfield perturbations, such as the pressure changes that were described in Sec. 3.2 and in Sec. 6.6. The flowfield perturbations were correlated in terms of the hypersonic viscous interaction parameter (V or $\bar{V}_\infty$). The discussion will not be repeated in this chapter.

9.2 COMPRESSION-RAMP FLOWS

The pressure distribution in the region of the shock-wave/boundary-layer interaction that occurs when a supersonic flow

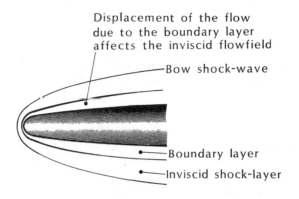

Displacement of the flow
due to the boundary layer
affects the inviscid flowfield

Bow shock-wave

Boundary layer

Inviscid shock-layer

Fig. 9.5 Hypersonic viscous interaction.

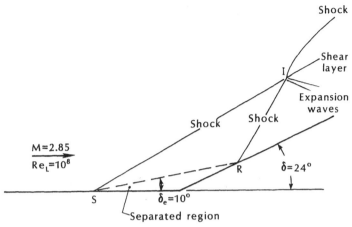

Fig. 9.6   Sketch of simplified model of separated flow at a compression ramp, based on the data of Ref. 6.

encounters a compression ramp was discussed in Sec. 6.4. It was noted that the strength of the interaction depended on the size of the subsonic portion of the approach boundary layer and the strength of the shock wave produced by the turning of the flow. Therefore, the shock/boundary-layer interaction depends on the Mach number, on the Reynolds number (and whether the approach boundary layer is laminar or turbulent), on the deflection angle, and on the chemical state of the gas.

At relatively large deflection angles, i.e., when appreciable separation of the boundary layer occurs, the pressure distributions exhibit three inflection points. The three inflection points are: (1) that associated with the separation of the boundary layer, (2) that associated with the onset of reattachment, and (3) that associated with the reattachment compression. As noted by Delery,[5] at very large deflection angles, the pressure distribution exhibits a decrease following the rise corresponding to reattachment. Then it tends to the constant level of the inviscid solution, with the tendency that the pressure overshoot increases as the deflection angle increases. A sketch of a simplified model of the separated flow at a compression ramp, which is based on the data of Settles et al.,[6] is presented in Fig. 9.6. The relatively large turning angle of the ramp ($\delta = 24$ deg.) causes the turbulent boundary layer to separate at S, upstream of the compression corner. It is as if the oncoming $M = 2.85$ flow encountered a ramp whose effective deflection angle (δ_e) was 10 deg. When the flow along the effective ramp encounters the actual ramp at R, the reattachment point, it turns another 14 deg. The shock waves produced by the two incremental compressive turns intersect at point

I, as shown in Fig. 9.6. Downstream of I, a single, curved shock wave eventually turns the oncoming flow so that it is parallel to the compression surface. Expansion waves originating at I impinge on the ramp producing the pressure distribution described by Delery.[5] The flowfield depicted in the sketch of Fig. 9.6 is essentially that of a Type VI shock/shock interaction that will be described in Sec. 9.3.

In 1968, Markarian[7] reviewed the literature relating to the pressure distributions and to the heat-transfer distributions in the vicinity of shock wave/boundary-layer interactions. The interactions considered included: (1) an externally-generated planar shock wave impinging on and reflecting from a flat plate or cylinder and (2) a planar shock wave extending outward from the boundary layer when the flow encounters a compression ramp. The peak measured heating rate usually occurred in the vicinity of reattachment where the boundary layer is the thinnest or, if separation does not occur, immediately downstream of the shock interaction. Markarian[7] found that the ratio of the peak perturbed value of the heat-transfer coefficient to the undisturbed, flat-plate value, i.e., (h_{pk}/h_{fp}), was approximately proportional to the ratio of the pressure rise across the shock interaction to the power n, i.e., $(p_f/p_{fp})^n$. The final pressure, p_f, is that downstream of the interaction and, for the majority of data reviewed by Markarian, was approximately equal to the peak pressure. When the approach boundary layer is laminar:

$$\left(\frac{h_{pk}}{h_{fp}}\right)_{\text{lam}} = \left(\frac{p_f}{p_{fp}}\right)^{1.3} \tag{9-1}$$

When the approach boundary layer is turbulent:

$$\left(\frac{h_{pk}}{h_{fp}}\right)_{\text{turb}} = \left(\frac{p_f}{p_{fp}}\right)^{0.85} \tag{9-2}$$

Although the peak heat-transfer coefficients calculated using Eqs. (9-1) and (9-2) are very approximate, these equations provide a rapid means of estimating the heat transfer in the shock-interaction region.

Simeonides and Wendt[8] note that for a relation of the form:

$$\frac{h_{pk}}{h_{\text{ref}}} = \left(\frac{p_3}{p_1}\right)^n$$

to be strictly valid, h_{ref} must be determined using the same virtual origin as the reattaching boundary layer. Simeonides and Wendt note that not using the same virtual origin "may partly explain the wide scatter of data exhibited in such correlations and the wide variety of exponents found in the literature."

The wide scatter using correlations such as Eqs. (9-1) and (9-2) may also occur when the interaction causes the premature onset of boundary-layer transition. In cases where the interaction causes transition, increases of over 30 times the undisturbed value have been measured. Furthermore, simple correlations break down when the interaction promotes the onset of boundary-layer transition.

Rizzetta and Mach[9] conducted a comparative numerical study of hypersonic flow ($M_1 = 14.1$) past compression ramps. Using four different numerical algorithms with a variety of grid spacings, solutions were obtained for the steady, compressible, two-dimensional, laminar, thin-layer Navier-Stokes equations. The compression ramps produced flowfields which varied in nature from having nearly incipient separation (for $\delta = 15$ deg.) to having a large recirculating region (for $\delta = 24$ deg.). It was found that all four methods developed essentially identical wall-pressure distributions. However, the computed heat-transfer distributions depended both on the method and on the grid resolution. The computed heat-transfer distributions were more sensitive to the grid resolution normal to the wall than that in the streamwise direction.

Rizzetta and Mach[9] concluded that "The algorithms considered here have resulted in numerical solutions which varied widely. Great sensitivity was observed with respect to both ramp angle and to grid point distribution. In the limit of grid independence, none of the methods compared extremely well with the experiment, which clearly evidences a need for the improvement of numerical techniques as applied to separated hypersonic flows."

The grid resolution is also of considerable importance for ramp-compression flowfields where the approach boundary layer is turbulent. As noted by Shang et al.,[10] "A major difficulty encountered in the numerical analysis is the determination of the resolution required for engineering accuracy. The basic requirement is to adequately resolve all significant features of the flow. The commonly accepted criteria of a mesh Reynolds number of the order of two is impractical for the present analysis, but also seems to be unnecessary. For turbulent flow, the high-velocity gradients near the wall dictate an extremely fine mesh spacing to achieve adequate numerical resolution."

Moss et al.[11] compared the direct simulation Monte Carlo (DSMC) solutions for rarefied, hypersonic flows about a compression corner with data from experiments conducted in the low-density hypersonic wind tunnels at the DLR in Göttingen. The flow conditions and model size were such that

$$V' = M_\infty \sqrt{\frac{C'}{Re_{\infty,w_c}}} \qquad (9\text{-}3)$$

ranged from 0.09 to 0.22. Note that Re_{∞,x_c} is the freestream Reynolds number based on the distance from the leading edge to the ramp corner (x_c). Furthermore,

$$C' = \frac{\rho'\mu'}{\rho_\infty\mu_\infty} = \left(\frac{T_\infty}{T'}\right)^{0.25} \qquad (9\text{-}4)$$

where ρ' and μ' are calculated at the reference temperature,[12]

$$\frac{T'}{T_\infty} = 0.468 + 0.532\frac{T_w}{T_\infty} + 0.195\left(\frac{\gamma - 1}{2}\right)M_\infty^2 \qquad (9\text{-}5)$$

The calculations and the measurements were consistent in that separation was not observed for any of the ramp angles, i.e., δ of 15 deg., 25 deg., and 35 deg., for those flow conditions for which V' was 0.22 at the ramp corner. The computed heat transfer for this rarefied flow increased (with reference to the flat-plate value) as the ramp angle increased. In contrast, measurements of the heat transfer for laminar continuum flows at a deflection angle of 35 deg. exhibit a reduction in comparison to the corresponding flat-plate values due to separated flow in the corner region.

Another interesting observation of Moss et al.[11] is that "the comparisons can be viewed as qualitative at best, since the experimental measurements show the flow to be three-dimensional while the computations assume the flow to be two-dimensional." In many instances, careful review of the data has indicated that the flow was three dimensional even when the model was "two-dimensional," e.g., a finite-width flat plate with a ramp.

9.3 SHOCK/SHOCK INTERACTIONS

9.3.1 Introductory Information

A simplified model to describe the flowfield that occurs when a supersonic flow encounters a compression ramp whose deflection angle is sufficient to produce a separation bubble at the corner was presented in Fig. 9.6. Two left-running shock waves occur: one originating at S, the other originating at R. They intersect at I. According to the sketch, the flowfield downstream of the shock/shock intersection contains a curved shock wave, a shear layer, and an expansion fan impinging on the surface of the ramp.

Exercise 9.1:

A Mach 2.85 flow encounters a two-segment wedge, as shown in Fig. 9.7. The flow is assumed to be an inviscid flow of perfect air. The

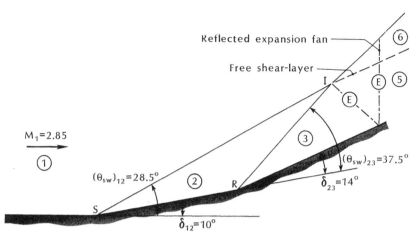

Fig. 9.7   Nomenclature for Exercise 9.1.

flow is deflected by 10 deg. as it proceeds from region 1 to region 2, i.e., $\delta_{12} = 10$ deg. The flow is deflected by 14 deg. as it proceeds from region 2 to region 3, i.e., $\delta_{23} = 14$ deg. The left-running shock wave originating at the intersection I is curved due to its interaction with the reflected expansion waves. However, since the flow is planar, this shock wave eventually reaches the strength (or inclination) required to turn the flow parallel to the second wedge, i.e., $\delta_{16} = 24$ deg. In fact, since the flow is two dimensional, the streamlines are parallel to the second wedge downstream of the reflected expansion fan, i.e., from immediately downstream of the shock wave (region 6) to the wall (region 5). Although the static pressure is constant across the interface between regions 5 and 6, i.e., $p_5 = p_6$, other parameters, such as the temperature, the velocity, etc. are not.

Assume that conditions in region 1 include a static pressure of 10 psia and a static temperature of 180 °R. Calculate the static pressure, the velocity, and the Mach number in regions 1, 2, 3, 5, and 6.

Solution:

For region 1:

$$U_1 = M_1 a_1 = 2.85(49.02\sqrt{180}) = 1874.4 \text{ fps}$$

Neglecting the viscous effects and assuming that the air behaves as a perfect gas, the total temperature will be constant throughout the flow. Thus,

$$T_{t1} = T_{t2} = T_{t3} = T_{t5} = T_{t6} = T_t$$

$$T_{t1} = \left(1 + \frac{\gamma - 1}{2}M_1^2\right)T_1 = 472 \text{ °R} = T_t$$

To calculate the flow in region 2, use the charts of Ref. 13 with the input parameters: $M_1 = 2.85$ and $\delta_{12} = 10$ deg. The shock wave angle $(\theta_{sw})_{12}$ is 28.5 deg., $C_{p2} = 0.172$, and $M_2 = 2.38$. The shock wave angle is not needed to calculate the requested parameters. However, it is needed to develop the sketch of Fig. 9.7, which is a "shock-fitting" scheme, i.e., one in which the shock wave location and inclination can be determined readily (as they are in this exercise). To calculate the remaining parameters:

$$\frac{p_2}{p_1} = 1 + \frac{\gamma}{2} M_1^2 C_{p2} = 1.978 \; ; \quad p_2 = 19.78 \, \text{psi}$$

$$T_2 = T_t \left(1 + \frac{\gamma - 1}{2} M_2^2\right)^{-1} = 221.3 \, °\text{R}$$

so that

$$U_2 = M_2(49.02\sqrt{T_2}) = 1735.6 \text{ fps}$$

Note that the percentage change in the Mach number from region 1 to region 2 is much greater than the percentage change in velocity. The velocity in region 2 is only 7.4 percent less than the velocity upstream of the ramp. These results will also occur on the compression ramp of an airbreathing propulsion system, where the Mach number at the inlet is much lower than the freestream Mach number, whereas the inlet velocity is only slightly lower than the freestream velocity.

To calculate the flow in region 3, again use the charts of Ref. 13. Note that region 2 represents the upstream conditions (region 1 in the nomenclature of Ref. 13) and region 3 represents the downstream conditions (region 2 in the nomenclature of Ref. 13). For upstream conditions where $M_2 = 2.38$ and $\delta_{23} = 14$ deg., $(\theta_{sw})_{23} = 37.5$ deg., $M_3 = 1.82$, and $C_{p3} = 0.326$.

$$\frac{p_3}{p_2} = 1 + \frac{\gamma}{2} M_2^2 C_{p3} = 2.293 \; ; \quad p_3 = 45.36 \, \text{psi}$$

$$T_3 = T_t \left(1 + \frac{\gamma - 1}{2} M_3^2\right)^{-1} = 283.9 \, °\text{R}$$

so that

$$U_3 = M_3(49.02\sqrt{T_3}) = 1503.2 \text{ fps}$$

The flow in region 6 constitutes a "downstream boundary condition," since the shock wave inclination angle must be that required to turn the two-dimensional flow parallel to the second wedge, i.e., $\delta_{16} = 24$ deg. Thus, once again, the charts of Ref. 13 are used to calculate the flow across an oblique shock wave. For $M_1 = 2.85$ and $\delta_{16} = 24$ deg., $(\theta_{sw})_{16} = 44$ deg., $C_{p6} = 0.60$, and $M_6 = 1.69$.

$$\frac{p_6}{p_1} = 1 + \frac{\gamma}{2} M_1^2 C_{p6} = 4.411 \quad ; \quad p_6 = 44.11 \, \text{psi}$$

$$T_6 = T_t \left(1 + \frac{\gamma - 1}{2} M_6^2 \right)^{-1} = 300.4 \, ^\circ\text{R}$$

so that
$$U_6 = M_6 (49.02\sqrt{T_6}) = 1435.9 \, \text{fps}$$

The flow in region 6 has turned 24 deg. from the freestream direction by passing through a single shock wave. The flow in region 3 has turned 24 deg. while passing through two weak shock waves. The entropy in region 6 is greater than that in region 3. As a result an entropy layer, or a free shear layer, divides the flow that has passed through two shock waves from the flow that has passed through a single shock wave. Thus,

$$s_3 = s_5 < s_6$$

Note that $p_6 < p_3$. Since the pressure is constant across the entropy layer, $p_5 = p_6$. Thus,
$$p_3 > p_5 = p_6$$

Although the pressure decreases as the flow goes from region 3 to region 5, the entropy remains constant. For this isentropic acceleration:
$$p_{t3} = p_{t5}$$

Therefore, since
$$\frac{p_5}{p_3} = \frac{(p_5/p_{t5})}{(p_3/p_{t3})} = \frac{44.11}{45.36} = 0.9724$$

for $M_3 = 1.82$:
$$\frac{p_3}{p_{t3}} = 0.1688$$

Thus,
$$\frac{p_5}{p_{t5}} = 0.1641$$

so that the Mach number in region 5 is 1.84.

$$T_5 = \left(1 + \frac{\gamma - 1}{2} M_5^2 \right)^{-1} 472 = 281.4 \, ^\circ\text{R}$$

and
$$U_5 = M_5 (49.02\sqrt{T_5}) = 1513.0 \, \text{fps}$$

Note that $U_5 \neq U_6$. Thus, as stated, a free shear layer develops along the interface between region 5 and region 6.

In summary,

Region	p (psi)	U (ft/s)	M
1	10.00	1874.4	2.85
2	19.78	1735.6	2.38
3	45.36	1503.2	1.82
5	44.11	1513.0	1.84
6	44.11	1435.9	1.69

Exercise 9.2:

If the deflection of the second wedge is increased, there will be two distinct changes in the shock/shock interaction. The first occurs at the deflection angle for which the turning of the freestream flow cannot be accomplished by a single, weak shock wave. Thus, the shock wave outboard of the interaction is strong and the downstream flow is mixed subsonic/supersonic.

If the deflection angle is increased further still, a value is reached when the flow in region 2 can no longer be turned by a weak, oblique shock wave. Thus, a strong, curved shock wave will divide regions 2 and 3.

Using the information presented in Fig. 9.8, which is taken from Ref. 13, determine the approximate values of the deflection angles at which these two changes occur.

Solution:

Note that the maximum deflection angle for which an $M = 2.85$ can be turned by a weak shock wave is approximately 33 deg. Thus, if the second wedge is inclined by more than 33 deg. to the original freestream direction, the outboard shock wave becomes a strong, curved shock wave. With $\delta_{16} = 33$ deg,. $\delta_{23} = 23$ deg. Note that it is possible to turn the $M_2 = 2.38$ flow through a weak, oblique shock wave. When $M_2 = 2.38$ and $\delta_{23} = 23$ deg., $(\theta_{sw})_{23} = 49.2$ deg. and $M_3 > 1$.

The second dramatic change occurs when the flow in region 2 can no longer be turned parallel to the second wedge by a weak, oblique shock wave. Referring to Fig. 9.8, it can be seen that since $M_2 = 2.38$, the maximum deflection angle for which a weak shock wave exists is just over 28 deg. Thus, if the second wedge is inclined 39 deg. ($\delta_{12} = 10$ deg. and $\delta_{23} = 29$ deg.) relative to the freestream, a weak shock wave for $(\theta_{sw})_{23}$ is no longer possible.

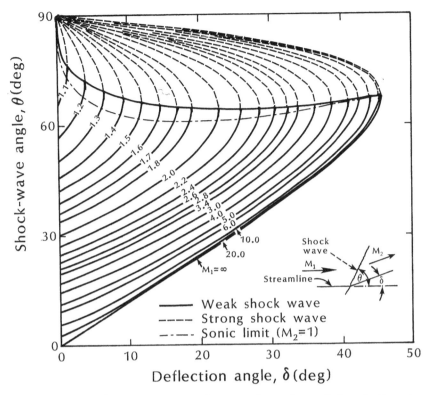

Fig. 9.8 The shock-wave angle as a function of the deflection angle and upstream Mach number for a perfect gas with γ =1.4, as taken from Ref. 13.

As evident in the results of Exercise 9.2, the character of the shock/shock interaction is sensitive to the configuration geometry. Whereas the flow was everywhere supersonic with an impinging expansion fan for the flowfield of Exercise 9.1, the flowfields of Exercise 9.2 would contain impinging jets and shear layers. As a result, the heat transfer and the pressures in the interaction region of the flowfields of Exercise 9.2 exceed by orders of magnitude the corresponding values for the flowfield of Exercise 9.1. In addition, the character of the shock/shock interaction and its effect on the neighboring surface are sensitive to the Mach number, to the specific-heat ratio (or real-gas effects), and to the strength of the impinging shock waves.

9.3.2 The Six Interference Patterns of Edney

The definitive treatise on shock/shock interaction patterns was done by Edney. The complete presentation is given in the FFA report, Ref.

14. A summary of the work is more generally available as Ref. 15. Surface pressures, heat-transfer rates, oil-flow patterns on the model surface, and schlieren photographs were obtained as symmetric models were injected into a hypersonic stream through a slot in a variable-incidence flat plate which generated the impinging shock wave. The interaction between the impinging shock wave and the bow shock wave was a function of the angle between the impinging shock wave and the bow shock wave. Edney[14,15] found that there were six different interference patterns. A sketch illustrating the approximate relation between the location where the impinging shock wave intersects the bow shock wave of a hemisphere and the shock-interference patterns, as taken from Ref. 16, is presented in Fig. 9.9. Researchers have found that the impingement of a shear layer (Type III) or of a supersonic jet (Type IV) can produce locally severe heating rates. Heating rates in the impingement region can be more than an order of magnitude greater than the unperturbed values, i.e., those which would exist in the absence of a shock/shock interaction.

Note that a Type VI pattern occurs when the impinging shock wave and the bow shock wave are of the same family, i.e., either both are left-running (as depicted in Fig. 9.9) or both are right-running, with a relatively small angle between the two shock waves. Note also that the shock/shock interaction depicted in Fig. 9.9 is generated when an impinging shock wave encounters the bow shock

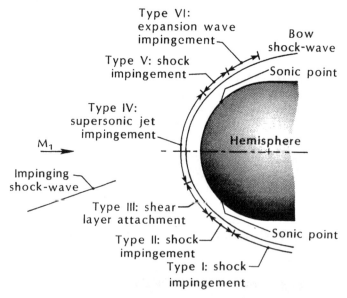

Fig. 9.9 Approximate relation of the shock/shock geometry and the type of interference pattern, as taken from Ref. 16.

wave ahead of a hemisphere, that the interaction depicted in Fig. 9.7 was generated by inviscid flow over a double wedge, and that the interaction depicted in Fig. 9.6 was generated by viscous flow over a ramp. Nevertheless, all three shock/shock interactions exhibit the characteristics of a Type VI interaction.

Over the years, the design of specific vehicles provided the focus for research which advanced the understanding of shock/shock interaction phenomena. There was concern about the potential for locally high heating rates for the wing leading edge of the Space Shuttle Orbiter. As noted by Keyes and Hains,[16] "Heating due to interfering shocks may also appear on the leading edge of wings and control surfaces depending on the amount of sweep." Bertin et al.[17] examined surface-pressure and heat-transfer rate data for a variety of Shuttle Orbiter configurations over an angle-of-attack range from 0 deg. to 60 deg. The correlations for these three-dimensional flows indicated that the type of shock-interaction pattern was dominated by the effective sweep angle of the leading edge. For the relatively low sweep angles of the straight-wing Orbiters (which were an early design option), the interaction between the bow-generated shock wave and the wing-generated shock wave exhibited the characteristics of a Type V shock/shock-interaction pattern. For delta-wing Orbiters, the shock/shock interaction pattern exhibited the characteristics of a Type VI pattern for all angles-of-attack. The effect of gas properties, i.e., the perfect-air model or equilibrium-air properties, were examined using numerical codes based on the shock-fitting scheme[18,19] as discussed for Fig. 9.7 of Exercise 9.1. Using the equilibrium-air model to determine the real-gas properties, it was found that the minimum sweep angle for which a Type VI pattern existed decreased as the freestream velocity increased.

The most intense heating rates for an airbreathing hypersonic vehicle are expected to occur on the blunted lip of the engine-inlet cowl. Thus, the effect of viscous shock-on-shock interactions attracted considerable attention during the design of the National Aerospace Plane (NASP). Klopfer and Yee[20] examined the six types of shock/shock interaction patterns which occur when an externally generated shock wave intersects the cowl-lip bow shock wave. Klopfer and Yee note that:

> Due to the complex flow patterns that occur for shock-on-shock flows three types of finite difference grids were studied. These ranged from simple body conforming orthogonal meshes to semi-adapted meshes and full solution adaptive meshes... While the numerical results obtained with semi-adapted meshes were much improved over those obtained with the simple meshes, the details in the inter-

action regions were still not fully resolved. This was especially true for the Type III and Type IV flows where
a small and highly localized supersonic jet grazes or impinges on the blunt body. In this localized region the
jet consists of two parallel shear surfaces which are difficult to capture very accurately without sufficient grid
resolution...The main difficulty is that in both the experiment and the computation, the pressure and heat-transfer
peaks are very sensitive to small changes in the impinging
shock location.

In a study focusing on Type III and Type IV interactions,
Gaitonde and Shang[3] made similar observations.

The computation of shock-on-shock interactions is a challenging task requiring the resolution of intense shock
waves, shear layers, and slip streams and their interactions with each other and the boundary layer...The introduction of viscous terms (typically through central differencing) introduces contradictory requirements on the
numerical scheme, i.e., shock-capturing requires a finite
amount of numerical dissipation which must not however,
overwhelm the physical dissipation in the boundary layer.

The sketch of the viscous interaction associated with an airbreathing propulsion system, which is taken from Stollery[4] and presented in Fig. 9.4a, depicts two distinct shock waves impinging on
the cowl-lip bow shock wave. Wieting[21] presented the results of an
experimental study of multiple shock wave interference patterns at
a cylindrical leading edge. The tests were conducted in the 48-Inch
Hypersonic Shock Tunnel at the Calspan/University of Buffalo Research Center. A double-wedge configuration was used to generate
the two impinging shock waves. The test apparatus was such that
the instrumented cylinder could be positioned so that the impinging shock waves either coalesced just prior to intersecting the bow
shock wave or remained distinct. The test conditions were primarily
for a Type IV pattern, since "it represents the most severe pressure
and heat transfer rate condition." The heat-transfer and surface-
pressure data presented in Fig. 9.10 were taken from Ref. 21. These
data represent a condition where the two impinging oblique shock
waves coalesced just prior to intersecting the cylindrical bow shock
wave. This configuration produced the highest local heating rate for
the measurements presented in Ref. 21.

9.3.2.1 A Type I shock/shock interaction. A typical Type I shock/shock
interaction pattern is illustrated in Fig. 9.11. This pattern occurs

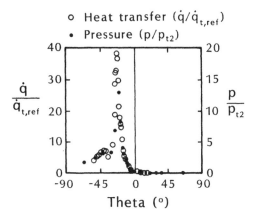

Fig. 9.10 Heat-transfer rate and surface-pressure distributions on a cylinder subjected to dual incident shock waves, as taken from Ref. 21.

when the strength of the two intersecting shock waves is such that the flow remains supersonic in regions 4 and 5. A Type I interaction may occur when the two weak intersecting shock waves are generated at the inlet of an engine or when they are generated during staging, when two elements of a mated configuration separate.

Exercise 9.3:

Use Eq. (9-2) to calculate the heat-transfer ratio across the impinging shock wave on the body surface, as shown in the sketch of Fig. 9.11. By using Eq. (9-2), we are assuming that the boundary layer on the body surface is turbulent. For this exercise, $M_1 = 6.0$, $p_1 = 10^{-3}$ atm, $\delta_{12} = 15$ deg., and $\delta_{13} = 5$ deg. Assume perfect air and neglect viscous effects, except for the use of Eq. (9-2) to account for the interaction between the wall boundary layer and the impinging shock wave.

Solution:

To determine the heat-transfer ratio, it will be necessary to solve for the strengths of the shock waves in Fig. 9.11 and, thus, determine the ratio of p_6/p_2. As a first step, let us calculate the flow in regions 2 and 3. Using the charts of Ref. 13 to calculate the flow in region 2,

$$M_2 = 4.00$$

$$C_{p2} = \frac{p_2 - p_1}{0.5\gamma p_1 M_1^2} = 0.200$$

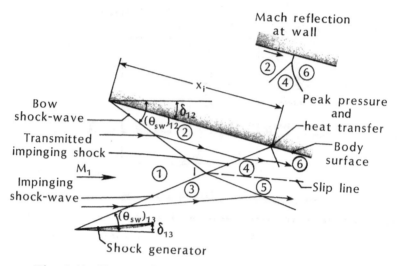

Fig. 9.11   Type I shock/shock interaction pattern.

Thus,

$$\frac{p_2}{p_1} = 1 + \frac{\gamma}{2} M_1^2 C_{p_2} = 6.04 \quad ; \quad p_2 = 6.04 \times 10^{-3}\,\text{atm}$$

Similarly, for the flow in region 3

$$M_3 = 5.32$$

$$C_{p_3} = 0.043 \quad ; \quad p_3 = 2.08 \times 10^{-3}\,\text{atm}$$

Note that the pressures and the flow directions in regions 4 and 5 are equal. These two requirements together with the flow conditions in regions 2 and 3 uniquely determine the strength of the shock wave dividing regions 2 and 4 and the strength of the shock wave dividing regions 3 and 5. Note also that the flows in regions 4 and 5 have passed through crossing shock waves of the opposite family. Since these shock waves are of differing strength, properties, such as the entropy, the velocity, the temperature, etc., differ from region 4 to region 5. As a result, the streamline that passes through I (the intersection of the shock waves) is a surface of discontinuity, or a slip line. Neglecting viscous effects, the slip line is a very thin region of concentrated vorticity, which is sometimes called a vortex sheet. If one accounts for viscous effects, a free-shear layer will develop along the interface between regions 4 and 5.

Since the flow properties in regions 2 and 3 are known, an iterative process can be used to determine the flow in regions 4 and 5.

In this exercise, θ_f will be used to designate the flow direction, i.e., the subscript f will distinguish this symbol from that designating the shock wave. The iterative process would be (1) assume a value for the flow directions in regions 4 and 5 (i.e., assume $\theta_{f,4}$ and $\theta_{f,5}$), (2) calculate δ_{24} and δ_{35}, (3) use the charts of Ref. 13 to determine the flow in regions 4 and 5, and (4) iterate on θ_f until $p_4 = p_5$ within a specified tolerance. To automate the process, a single reference system for the flow directions should be used, e.g., $\theta_f = 0$ when aligned with the freestream and is positive when turned counterclockwise from the reference. Thus,

$$M_2 = 4.00 \quad, \quad p_2 = 6.04 \times 10^{-3}\,\text{atm} \quad, \quad \theta_{f,2} = -15 \text{ deg.}$$

$$M_3 = 5.32 \quad, \quad p_3 = 2.08 \times 10^{-3}\,\text{atm} \quad, \quad \theta_{f,3} = +5 \text{ deg.}$$

Iterating on θ_f to the nearest tenth of a degree,

$$\theta_{f,4} = \theta_{f,5} = -9.7 \text{ deg.}$$

Thus,

$$\delta_{24} = 5.3 \text{ deg.}, \quad \delta_{35} = 14.7 \text{ deg.}, \quad \text{and} \quad p_4 = p_5 = 0.0103\,\text{atm}$$

Note that the flow direction in region 4 is -9.7 deg., whereas the wedge is inclined -15 deg. to the freestream. Thus, the shock wave dividing regions 4 and 6 must turn the flow 5.3 deg., i.e.,

$$|\delta_{46}| = |\theta_{f,6} - \theta_{f,4}| = |-15 \text{ deg.} - (-9.7 \text{ deg.})| = 5.3 \text{ deg.}$$

Using the charts of Ref. 13 for:

$$M_4 = 3.60 \quad, \quad p_4 = 0.0103\,\text{atm} \quad, \quad \text{and} \quad \delta_{46} = 5.3 \text{ deg.}$$

$$C_{p6} = \frac{p_6 - p_4}{0.5\gamma p_4 M_4^2} = 0.062$$

Thus,

$$\frac{p_6}{p_4} = 1 + \frac{\gamma}{2} M_4^2 C_{p6} = 1.56$$

The pressure ratio across the impinging oblique shock wave that reflects off the wedge is:

$$\frac{p_6}{p_2} = \frac{p_6}{p_4}\frac{p_4}{p_2} = (1.56)(1.71) = 2.67$$

The heat-transfer ratio due to shock/shock interaction is:

$$\left(\frac{h_{pk}}{h_{fp}}\right)_{turb} = \left(\frac{p_6}{p_2}\right)^{0.85} = 2.30$$

Note that, when the conditions are such that a regular, oblique shock wave reflection is no longer possible at the surface of the wedge, i.e., the values of M_4 and δ_{46} do not allow a weak solution for the reflected shock wave, a Mach reflection will occur (as shown in the insert of Fig. 9.11).

9.3.2.2 A Type II shock/shock interaction.   As discussed in the previous section, when two shock waves of opposite families intersect such that the flow in regions 4 and 5 is supersonic, a Type I pattern occurs. However, if the strengths of the intersecting shock waves are increased or if the upstream Mach number (M_1) is reduced such that the flow downstream at the intersection is subsonic, a Type II shock/shock interaction occurs. The features of a Type II pattern are

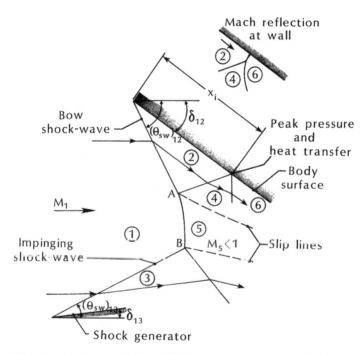

Fig. 9.12   Type II shock/shock interaction pattern.

illustrated in the sketch of Fig. 9.12. The intersecting shock waves are weak (as was the case for the Type I pattern), but are of such a strength that, in order to turn the flow, a Mach reflection must exist in the center of the flowfield with an embedded subsonic region located between the intersection points, A and B of Fig. 9.12, and the accompanying slip lines.

Note that, for the situation depicted in Fig. 9.9, a Type II pattern occurs when the impinging shock wave intersects the bow shock wave just below the sonic line.

9.3.2.3 Type III and Type IV shock/shock interactions.
Type III and Type IV shock/shock interactions occur when the impinging shock wave intersects the subsonic portion of the cowl-lip bow shock wave. Sketches of the Type III pattern and of the Type IV pattern are presented in Figs. 9.13 and 9.14, respectively. The conditions are such that the flow in region 2 is subsonic and the flow in region 4 is supersonic for both Type III and for Type IV. Thus, these two regions are divided by a slip line that originates at the point where the impinging shock wave intersects the bow shock wave, i.e., point A in Figs. 9.13 and 9.14. Of course, when one accounts for the viscous effects, a free-shear layer develops along the slip line.

For a Type III interaction, the slip line (or free-shear layer) attaches to the surface at point C. The actual flowfield near the attachment point C is much more complicated than depicted in the sketch of Fig. 9.13. The flowfield for the Type IV interaction is the same as that for the Type III interaction for regions 1 through

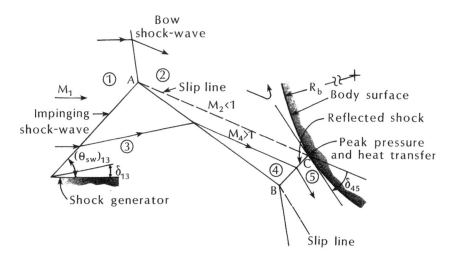

Fig. 9.13 Type III shock/shock interaction pattern.

4. However, for a Type IV pattern, the inclination of the model surface relative to the oncoming flow in region 4 exceeds the angle for which a supersonic flow can be turned through an oblique shock wave. Thus, a dramatic change occurs in the flowfield. As depicted in Fig. 9.14, a supersonic jet is formed, which separates two subsonic regions, i.e., regions 2 and 5. The jet, which terminates in a "jet" bow shock wave, produces locally severe pressures and heating rates in the impingement region.

Surface-pressure distributions and heat-transfer rate distributions computed by Klopfer and Yee[20] for a Type III interaction and for a Type IV interaction are reproduced in Figs. 9.15 and 9.16, respectively. The data, as taken from Klopfer and Yee, were originally presented by Wieting and Holden.[22] The peak heat-transfer rate occurs where the jet impinges on the surface and depends not only on the peak pressure generated by the jet, but also on the width of the jet, on the angle at which the jet impinges on the surface, and on whether the jet shear layers are laminar or turbulent.

For these shock/shock interactions, the impinging shock wave dramatically alters the bow shock wave generated by the cowl lip. This is illustrated in the shock shapes presented by Tannehill et al.[23] for a Type IV interaction and reproduced in Fig. 9.17. The numerical flowfield was computed using a time-dependent finite-difference method to solve the complete set of the Navier-Stokes equations. The experimental shock shape was obtained in the Langley 20-Inch Hypersonic Tunnel at $M_1 = 5.94$. The impinging shock wave caused the stagnation point to move a considerable distance from its original location.

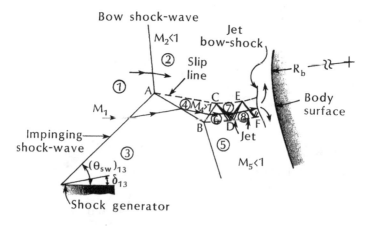

Fig. 9.14 Type IV shock/shock interaction pattern.

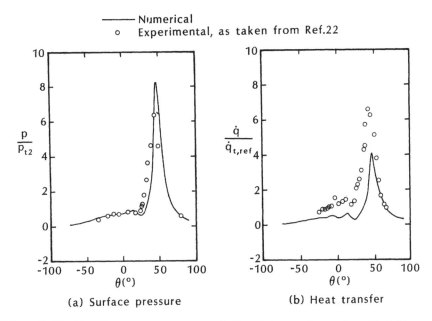

Fig. 9.15 Surface-pressure and heat-transfer-rate distributions for a Type III shock/shock interaction, M_1 = 8.03, Re/ft =1.55 × 10^6, as taken from Ref. 20.

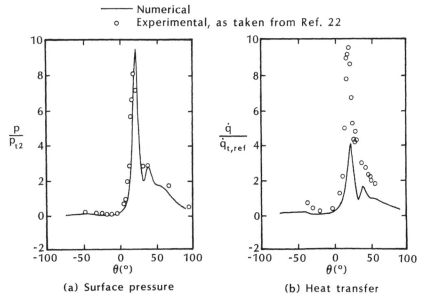

Fig. 9.16 Surface-pressure and heat-transfer-rate distributions for a Type IV shock/shock interaction, M_1 = 8.03, Re/ft =1.55 × 10^6, as taken from Ref. 20.

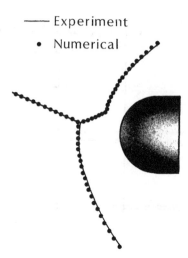

Fig. 9.17 A comparison of the computed and of the experimental shock shapes for a Type IV interaction, $M_1 = 5.94$, as taken from Ref. 23.

9.3.2.4 A Type V shock/shock interaction. A Type V pattern occurs when two oblique shock waves of the same family interact such that the impinging shock wave intersects the bow shock wave just above the sonic point, as depicted in the sketch of Fig. 9.9. Because the impinging shock wave intersects the bow shock wave not far from the sonic point, the Type V pattern is similar to the Type II pattern. A sketch of a Type V shock/shock interaction is presented in Fig. 9.18. Because the two intersecting shock waves are of the same family for the Type V pattern, a jet originates at point A. However, the jet is very thin and, for all practical purposes, it is indistinguishable from a free-shear layer. The jet and the free-shear layer (that develops along the slip line) converge as the subsonic flow in region 4 accelerates to sonic speed. Both the jet and the free-shear layer diffuse rapidly, meeting the body (only if it is sufficiently long) far outboard of the impingement point. Thus, although the jet and the free-shear layer have much less influence on the heat transfer, their influence should not be ignored.

9.3.2.5 A Type VI shock/shock interaction. A Type VI pattern occurs when two weak shock waves of the same family intersect such that a single, weak shock wave occurs outboard of the intersection. As indicated in the sketch presented in Fig. 9.19, a slip line (or free-shear layer) originates at the intersection, dividing the flow that has passed through two shock waves from that passing through a single shock wave. Furthermore, as indicated in the calculations of Exercise 9.1, an impinging expansion fan causes a streamwise decrease in surface pressure. Similarly, the heat-transfer rate decreases across the interaction.

A numerical model of the Type VI shock/shock interaction flowfield was developed for a wedge/cylinder configuration, such as the one shown in Fig. 9.20. The simulated bow shock wave was generated when the hypersonic freestream encountered the wedge. The cylinder, i.e., the simulated wing leading edge, was inclined relative to the wedge such that the flow downstream of the wing-root shock wave remained supersonic, i.e., the flow in region 3 of Fig. 9.19 was supersonic. Cross flow away from the plane-of-symmetry causes the shock-layer flow in region 3 to be three dimensional. The cross flow velocity was calculated assuming that the modified Newtonian flow model could be used to define the pressure distribution perpendicular to the leading-edge plane-of-symmetry. Agnone et al.[24] found that "the surface limiting streamline patterns appear to leave the leading edge nearly perpendicular to it."

A sketch of the streamtube patterns that were used to calculate[19] the inviscid flowfield in the plane-of-symmetry inboard of the shock/shock interaction is presented in Fig. 9.21. Note that the shock wave and the boundary layer for the wing leading edge begin at the juncture between the wedge, i.e., the simulated fuselage, and

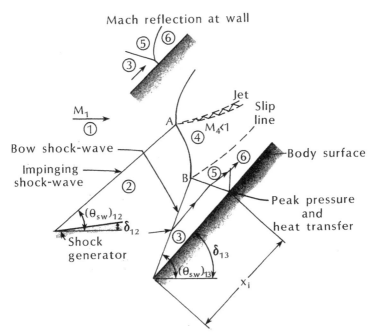

Fig. 9.18 Type V shock/shock interaction pattern.

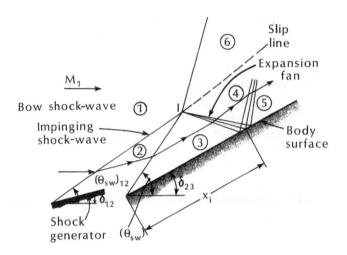

Fig. 9.19   Type VI shock/shock interaction pattern.

the cylinder, i.e., the simulated wing. Thus, the effect of the fuselage boundary layer and the possibility that the viscous/inviscid interaction caused separation at the wing root were not included in the flow model. This omission could be of importance at very low Reynolds numbers (where the fuselage boundary layer would be sensitive to shock-induced perturbations) or for very high sweep angles.

A single wave was used to model the expansion process in the present model for flow past the swept cylinder that represents the wing leading edge. This flow-model simplification allowed the mass balance (including cross flow) for the streamtube elements to be more easily calculated while maintaining an accuracy consistent with this technique. An iterative procedure was employed in the calculation of the streamline curvature, the computation of the pressure gradient across the shock layer, and the application of the mass-flow-balance requirement to solve for the inviscid shock-layer flow. The iterative procedure was repeated until successive values of the surface pressure agree within 0.1 percent.

The reflected expansion wave was assumed to be of equal strength to the impinging expansion wave but was a left-running wave. Once the reflected wave encountered the wing leading-edge shock wave, it was canceled and there were no further waves crossing the shock layer. Thus, the procedure used to compute the flow downstream of this station in region 5 was identical to the procedure used to calculate the flow in region 3.

The shock shape and the pressure distribution computed for region 3 are compared with experimental values in Figs. 9.22 and 9.23,

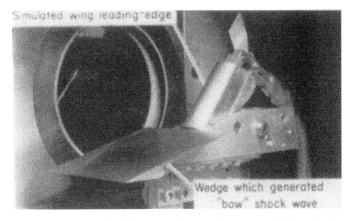

Fig. 9.20 Wedge/cylinder configuration used to study shock/ shock interactions for a wing leading edge, as taken from Ref. 19.

respectively, as taken from Ref. 19. The experimental values were obtained in the University of Texas Supersonic Wind Tunnel using the model shown in Fig. 9.20. The wedge was inclined 15 deg. to the freestream, i.e., $\delta_{12} = 15$ deg. The sweep angle of the cylinder (Λ) was 45 deg. During the tests, a constant gap of 0.178 cm (0.070 in.) was maintained between the wedge and the upstream end of the cylinder to allow for boundary-layer bleed-off and, thus, to minimize the possibility of separation in the corner interaction. The freestream Mach number for the tests was 4.97 ± 0.02. The stagnation pressure was 2.01×10^6 N/m^2 with a maximum fluctuation

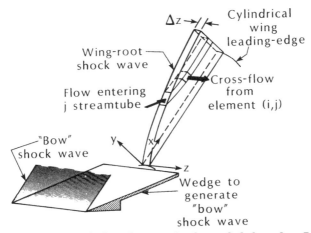

Fig. 9.21 Sketch of the theoretical model for the flowfield inboard of the shock/shock interaction, i.e., region 3, as taken from Ref. 19.

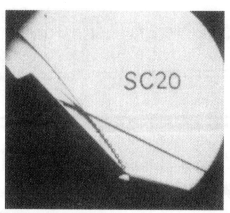

Fig. 9.22   Schlieren photograph of the Type VI shock/shock interaction pattern for M_1 = 4.97, δ_{12} = 15 deg., Λ = 45 deg. Theoretical locations of the wing-root shock wave are indicated by the symbol "O," as taken from Ref. 19.

during a run of $\pm 1.38 \times 10^4$ N/m^2. The stagnation temperature was 325 K. As a result, the nominal freestream Reynolds number was 0.517×10^6/cm.

As evident in Fig. 9.22, the agreement between the experimental and the theoretical shock waves in the wing-root region, i.e., region 3, was very good. Furthermore, weak (Mach) waves which occurred when the flow at the surface of the swept cylinder was perturbed by the static-pressure orifices were evident in many of the schlieren

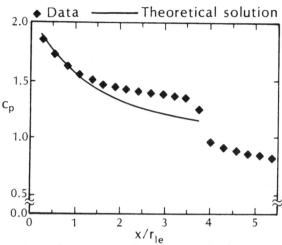

Fig. 9.23   Static-pressure distribution in the plane-of-symmetry for the cylindrical leading edge, M_1 = 4.97, δ_{12} = 15 deg., Λ = 45 deg., as taken from Ref. 19.

photographs taken during the program. Experimental values for the local Mach number were calculated using the measured wave angles. The experimentally determined values for the local Mach number were in good agreement with theory over the range of sweep angles tested. Note that, because of the three-dimensional effects due to cross flow, the free-shear layer moves toward the cylinder leading edge. The author is familiar with situations where the test model was sufficiently long that the free-shear layer impinged on the wing leading edge. Locally severe heating rates were found when the impinging shear layer caused the boundary layer on the wing leading edge to transition from laminar to turbulent.

The static pressure measurements from the plane-of-symmetry of the swept cylinder are presented in Fig. 9.23. Also included is the theoretical distribution, as calculated using the numerical code described above. For the orifice nearest the root, the experimental pressure was within 1.4 percent of the theoretical value. For the orifice at the downstream end of region 3, the experimental pressure was 18 percent higher than the theoretical value. The correlation between the measured and the computed pressures in Fig. 9.23 is consistent with the shock wave comparison in Fig. 9.22, , since the measured shock-layer thickness was slightly greater than the theoretical value at the downstream end of region 3. The effect of the boundary-layer displacement thickness would contribute to the difference between experiment and theory. Thus, the actual shock inclination is greater than the theoretical inclination, resulting in higher pressures downstream of the shock wave.

The heat-transfer rates computed using the numerical code described in Refs. 19 and 25 were compared with those measured on a 0.025-scale model of the Space Shuttle Orbiter in the Calspan Hypersonic Shock Tunnel.[26] As shown in the sketch of Fig. 9.24, a wedge/cylinder configuration was used to represent the relevant geometry of the Shuttle Orbiter. A schlieren photograph presented in Ref. 26 was used to define the initial coordinates and the approximate inclination angle of the fuselage-generated shock wave. As shown in Fig. 9.24, the wedge was inclined 9 deg. to the freestream flow, thus matching the planform inclination angle of the fuselage-generated shock wave. The cylinder was swept 45 deg., matching the wing leading-edge sweep angle. Note that the wedge/cylinder geometry does not simulate the gradual turning of the flow in the wing-root region produced by the glove on the Orbiter (the planform trace of which is indicated by the broken line of Fig. 9.24).

The computed heat-transfer distribution for the wedge/cylinder configuration described previously is reproduced in Fig. 9.25. The computed heating rates[25] are compared with those measured[26] in the Calspan Hypersonic Shock Tunnel using a 0.025-scale Orbiter at zero

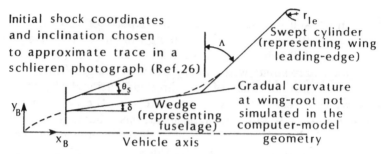

Fig. 9.24 Sketch of the nomenclature for the coordinate systems used in the numerical model, as taken from Ref. 25.

angle-of-attack. The heat-transfer rates for the fuselage and for the wing leading edge are presented as a function of x_B/L. Thus, a sketch of the model planform is included in Fig. 9.25 to provide a physical reference for the heating-rate distribution. The freestream Reynolds number based on a wing leading-edge diameter of 1.27 cm (0.50 in.), which is the approximate leading-edge diameter at $2y_B = 0.55b$ and which serves as the "reference value" for the present discussion of the wind-tunnel data, is 0.66×10^5. Based on Bushnell and Huffman's

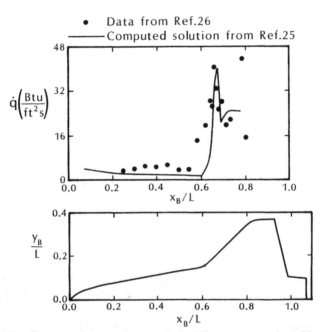

Fig. 9.25 Comparison of computed heat transfer distribution from Ref. 25 with the measurements of Ref. 26 for $\alpha = 0$ deg., $M_1 = 9.857$, $Re/\mathrm{ft} = 1.584 \times 10^6$.

transition criteria for swept cylinders or for wing leading edges,[27] the boundary layer should be laminar for this case. The "reasonable" agreement between the theoretical laminar values and the experimental values substantiate this assumption.

9.4 FLOWFIELD PERTURBATIONS AROUND SWEPT FINS

Many objects protrude from the vehicle surface through the boundary layer. The protuberance might be a tail surface, such as shown in Fig. 9.1, a strut within the intake of an airbreathing engine, or a brace used in mating multiple vehicles. The flowfields around these protuberances contain shock-wave/boundary-layer interactions, complex vortical flows, and shock/shock interactions. Locally severe heating rates can occur on the vehicle surface or on the protuberance itself. The protuberance also affects the pressure distribution on the vehicle surface, as well as adding to the wave drag. Westkaemper[28] noted, "A cylindrical protuberance mounted on a flat plate will cause the plate boundary layer to separate if the cylinder is long compared to the boundary-layer height." The shock wave pattern for the stagnation plane of the cylinder that was proposed by Westkaemper [28] is reproduced in Fig. 9.26. A bow shock wave (stand-off distance Δ in Fig. 9.26) is produced when the freestream flow encounters the cylinder. The pressure rise produced by the bow shock wave feeds upstream through the subsonic portion of the plate's boundary layer, causing it to thicken and to separate. The oncoming flow is deflected by the thickened boundary layer, producing an oblique shock wave (which makes an angle θ_{sw} with the plate). When the oblique shock intersects the cylinder's bow shock wave (at the triple point A), a shock/shock interaction occurs (a Type IV interaction would occur for the unswept cylinder shown in Fig. 9.26). Because the flow downstream of the oblique shock wave is supersonic, another shock wave is generated when the flow encounters the cylinder below the triple point A. Thus, a lambda-shaped shock wave is produced in the plane-of-symmetry.

A simplified model for the three-dimensional flowfield near an unswept, blunt-nosed strut, as presented by Stollery,[4] is reproduced in Fig. 9.27. Note that, in the plane-of-symmetry, the lambda shock shape presented by Stollery[4] in Fig. 9.27 is essentially the same as that presented by Westkaemper[28] in Fig. 9.26. McMaster and Shang[29] note, "The primary mechanism associated with diverting fluid past the blunt fin is a horseshoe vortex which forms upstream of the fin." Although only two vortices are shown in the sketch of Fig. 9.27, Stollery[4] notes that "in practice as many as six have been indicated by surface oil flow patterns. So far as the wall is concerned

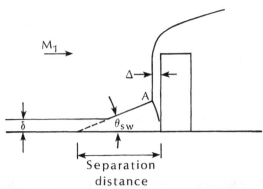

Fig. 9.26　Shock-wave pattern in the plane-of-symmetry for an unswept cylinder mounted on a plate, as taken from Ref. 28.

the maximum pressures and heat-transfer rates are recorded near the attachment lines where the flow passing over the vortices returns to the surface." Neumann and Hayes[30] developed a correlation for the peak heat-transfer coefficient on the surface ahead of the protuberance:

$$h_{pk} = 4.369 \times 10^{-4} M_L (p_{t2})^{1.176} \qquad (9\text{-}6)$$

where: h_{pk} is the peak value for the heat-transfer coefficient in Btu/(ft^2· s · °R); M_L is the local Mach number; and p_{t2} is the stagnation pressure behind the normal shock wave above the triple point (in psia). Neumann and Hayes [30] caution that deviations from the semi-infinite cylinder model are small until the triple point occurs above the cylinder. "In all cases the data assuming a semi-infinite cylinder are conservative."

The upstream and spanwise influence of the fin decreases significantly with increasing sweepback. Analyzing data obtained with heat-flux sensors that have a high spatial resolution and fast response, Aso et al.[31] noted, "As the sweep angle of the blunt fin is increased, the significant decrease of the extent of the interaction region is observed. A secondary separation region is observed for the sweep angles of 0 and 15 degrees and only a primary separation region is observed for 30 and 45 degrees... Two heat-flux peaks are observed along the freestream for sweep angle of 0 and 15 degrees and a single heat-flux peak for much larger sweep angle."

Local heat-transfer rates and pressures were measured with a cylinder at sweep angles of 45 deg. and 60 deg. with respect to the freestream mounted on a 12 deg. half-angle wedge. Bushnell[32] noted, "The extent of the flow separation in the cylinder-wedge juncture was small for the present test conditions." That is, there was no vortical pattern as depicted for the unswept strut in Fig. 9.27. With-

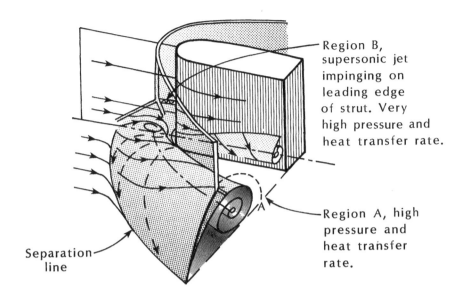

Fig. 9.27 Model for three-dimensional flowfield near an unswept, blunt-nosed strut, as taken from Ref. 4.

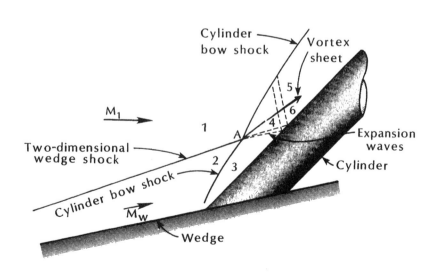

Fig. 9.28 Schematic sketch of typical schlieren photograph showing inviscid flowfield phenomena associated with intersection of two shocks, as taken from Ref. 32.

out the complex flow at the base of the cylinder, the sketch that Bushnell[32] presented for the leading-edge flowfield for these highly swept cylinders, which is reproduced in Fig. 9.28, is that of a Type VI shock/shock interaction.

9.5 CORNER FLOWS

A complex, three-dimensional flowfield containing a variety of viscous interactions is generated in the corner formed by the inter-section of two surfaces, such as occur where a tail or a wing joins the fuselage (see Fig. 9.1), where a control surface such as the rudder is located on a vertical surface at the tip of the wing (as on Hermes), at the inlet of an airbreathing engine, and for certain wind-tunnel supports. As noted by Charwat and Redekopp,[33] "The region is characterized by cross flows induced in the boundary layers on each of the intersecting surfaces by their interaction at the corner. This cross flow is not simple; theoretical considerations and experiments (in turbulent flow) indicate the existence of a pair of secondary vor-tices on each side of the bisector such that there is outflow along the surface from the corner opposed by inflow (along the surface) from the outer boundary layer...Pressure gradients and shock dis-continuities in the corner zone would be particularly important in hypersonic low-density flows in compressive corners."

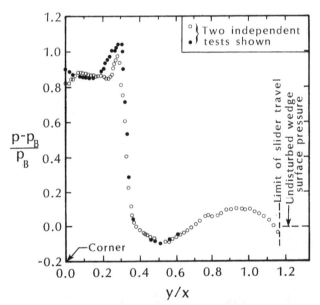

Fig. 9.29 Surface-pressure distribution on the base wedge, $\delta_B = \delta_I = 12.2$ **deg.,** $M_1 = 3.17$, **as taken from Ref. 33.**

Using measurements of the surface pressure and of the impact pressure and flow-visualization photographs, Charwat and Redekopp[33] developed a model for interference flow in the corner formed by two intersecting wedges. The surface-pressure distribution for the base wedge is presented in Fig. 9.29. The pressure measurements are presented in dimensionless form as $(p - p_B)/p_B$, where p_B is the average undisturbed pressure on the base wedge. The flow pattern postulated by Charwat and Redekopp is reproduced in Fig. 9.30. The flow is divided into four zones, all supersonic. The flow in Zone I (the central sector) appears to be nearly conical. Note that Zone I is not bounded by either wall and, therefore, does not affect the pressure measurements in Fig. 9.29. The adjustment between the conical flow and the two-dimensional flow far from the corner takes place in two distinct zones. The inner region, Zone II, is separated from the conical flow of Zone I by slip lines, and from the flow in Zone III by a relatively strong, curved, "inner" shock wave.

Referring to Fig. 9.29, the pressure is greatest in Zone II, i.e., $y < 0.3x$, which is downstream of the shock wave that divides Zone II from Zone III. The fact that the pressure distribution exhibits significant overshoots, both upstream and downstream of the shock wave, indicates that it is not a simple shock wave.

As indicated in the sketch of Fig. 9.30, the intersection of the "transmitted" central-sector shock wave and the surface is the outer boundary of Zone III. As a result, the locally high pressure at $y \simeq 0.9x$ (in Fig. 9.29) corresponds to the outer edge of Zone III. Thus, Zone III, which is the outer perturbation region, corresponds to $0.3x < y < 0.9x$. An oil-accumulation line was clearly evident in the flow-visualization photographs along the edge of the outer perturbation region. The oil-accumulation line did not exhibit either stagnant oil or vortices, which are characteristic of the two types of separation. It does define the outer boundary of the region of strong cross flow. Beyond Zone III, the flow is two-dimensional.

The viscous interaction in a corner flow is a function of the Mach number, the Reynolds number (being very sensitive to whether the boundary layer is laminar or turbulent), and the configuration geometry.

For the inlet of an airbreathing hypersonic vehicle, additional shock waves generated by the forebody and by the cowl could change the flow dramatically. Venkateswaran et al.[34] studied the strong, viscous interaction between shock waves and the boundary layer in the axial compression corner regions characteristic of an engine inlet. Crossing, oblique shock waves were generated by the forebody and by the cowl lip. Locally severe heat-transfer rates were observed. As noted by Venkateswaran et al.,[34] "In addition to the intensity of the heating, the complex nature of the flow makes it difficult to predict the peak heating location and the attendant gradients."

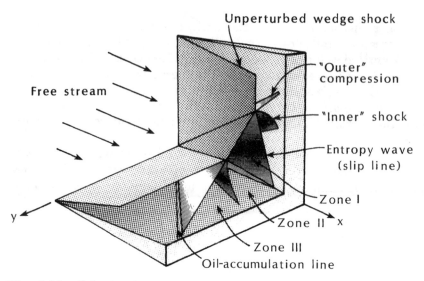

Fig. 9.30 Schematic of the characteristic wave structure, as taken from Ref. 33.

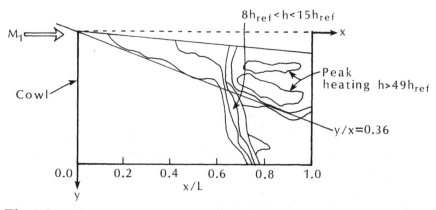

Fig. 9.31 Heat transfer contours on the cowl for the sharp strut model, $M_1 = 6.0$, $Re/\text{ft} = 3.35 \times 10^6$, as taken from Ref. 34.

Representative heat-transfer contours on the cowl from these tests are reproduced in Fig. 9.31. For orientation, the reader views the heat-transfer contours on the inner cowl surface from a location in the plane of the forebody (with the shock-generating surface to the viewer's left). The wedge-like surface centered about the x-axis at the top of the figure represents the sharp strut. The heat-transfer

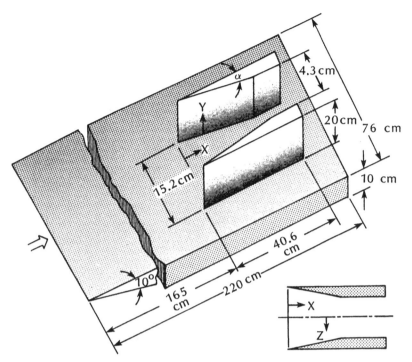

Fig. 9.32 Test-body configuration and coordinate system, as taken from Ref. 35.

contours near $x \simeq 0.7L$, reflect the impingement of the forebody-generated shock wave. The local heat-transfer coefficient (h) in this region is 8 to 15 times the reference value (h_{ref}, the value calculated using the Eckert's reference temperature method). Downstream of the shock-impingement region, two regions of locally severe heating occur for $0.75L < x < 0.90L$. The heat-transfer coefficients in these two regions are in excess of $49h_{ref}$. These two peaks are associated with strong corner vortices. Venkateswaran et al.[34] note, "The present data downstream of the shock impingement region show similarity with the results from corner flow without impingement."

Kussoy and Horstman[35] present data for the three-dimensional flow created by two intersecting shock waves interacting with a turbulent boundary layer. As shown in Fig. 9.32, the test bodies were composed of two sharp fins fastened to a flat-plate test bed. This experiment is one of the forty-one building-block experiments designated by Marvin[36] as elements of a code-validation program for airbreathing-vehicle design codes. Apparently, Kussoy and Horstman[35] concur, because they state, "The data obtained during this test program (undisturbed flowfield surveys, surface pressure and heat transfer distributions, and extensive flowfield surveys for two inlet configura-

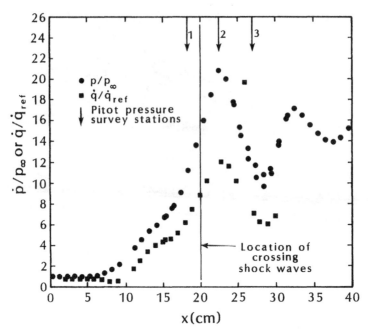

Fig. 9.33 Streamwise variation of pressure and heat transfer on flat-plate surface (y, z = 0 cm), 15 deg. double-fin configuration, as taken from Ref. 35.

tions) can be used as a data base against which existing computer codes should be verified." In addition, they have included a 3.5-inch diskette with Ref. 35 to enhance the reader's ability to use the data "to validate existing or future computational models of these hypersonic flows." Reference 35 is a clear example of a test program whose principal objective is to generate experimental data that can be used to validate CFD codes.

The experimentally determined pressure distributions and heat-transfer distributions from the $z = 0$ plane are reproduced in Fig. 9.33.

9.6 EXAMPLES OF VISCOUS INTERACTIONS FOR HYPERSONIC VEHICLES

As indicated in the sketch of Fig. 9.1, there are a variety of sources for the viscous interactions described in Secs. 9.2 through 9.5 that can occur in the flowfields of lifting bodies. Examples of such viscous interactions for the X-15, for the Space Shuttle Orbiter, and for an airbreathing aircraft, such as the National Aerospace Plane (NASP) will be discussed in this section.

9.6.1 The X-15

Two views of an X-15 model in free flight in a ballistic range are presented in Fig. 9.34. The photographs illustrate the large number

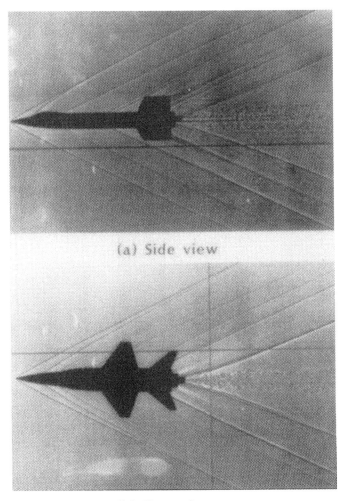

(a) Side view

(b) Top view

Fig. 9.34 X-15 model in free flight in a ballistic range, as provided by NASA.

of shock waves generated during high-speed flight of this vehicle.

During the late 1960s, NASA was engaged in the development of a Hypersonic Research Engine (HRE). An initial objective of the program was to conduct ground-based tests and flight tests on a hydrogen-burning ramjet engine over the Mach number range 3 to 8. In preparation for these tests, the NASA Flight Research Center conducted a flight program on the X-15-2 airplane with a dummy ramjet attached. Two views of the X-15-2 with the dummy ramjet installed are presented in Fig. 9.35. The ventral fin of earlier X-15

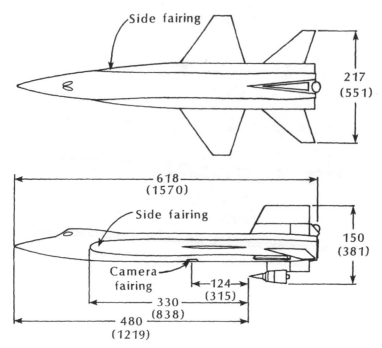

Fig. 9.35 Two-view sketch of the X-15-2 with the dummy ramjet installed. All dimensions in inches (centimeters), as taken from Ref. 37.

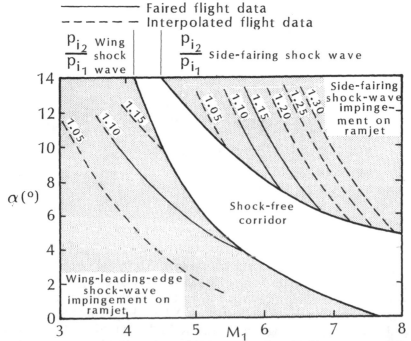

Fig. 9.36 The effect of M_1 and α on the flight-test corridor free of shock-wave impingement on the dummy ramjet for the X-15-2, as taken from Ref. 37.

configurations has been replaced with an instrumented pylon which supports the dummy ramjet.

Of concern to the designers of this experiment is the possibility that some of the shock waves evident in the photographs of Fig. 9.34 would impinge on the dummy ramjet. In addition, the dummy ramjet mounted on the pylon would create a complex shock/shock-interaction flowfield of its own. Burcham and Nugent[37] discussed the local flowfield around a pylon-mounted dummy ramjet. They determined the combinations of the freestream Mach number and of the angle-of-attack for which either the shock waves generated by the wing leading edge or by the fuselage side-fairing impinged on the pylon/ramjet region. The correlation showing the M_1/α combinations for which either the wing leading-edge shock wave or the side-fairing shock wave impinges on the ramjet is reproduced in Fig. 9.36. For the correlations of Fig. 9.36, p_{i_2}/p_{i_1} is the ratio of impact pressures. Burcham and Nugent[37] noted that, "The side-fairing shock wave was detected only at Mach number of 6.2 and greater." Furthermore, "A shock-free corridor exists above the wing shock-wave-impingement region and below the side-fairing shock-wave-impingement region. For Mach numbers below the corridor at the lower angles of attack, the wing shock wave is weak and probably would not affect the HRE inlet. The side-fairing shock wave is stronger than that of the wing and should be avoided." The wing leading-edge shock wave can be seen impinging on the pylon in the schlieren photograph presented by Burcham and Nugent[37] and reproduced in Fig. 9.37.

It was noted earlier that the pylon of the X-15-2 was instrumented. A typical profile for the measured impact pressures for the pitot probes located on the pylon leading edge is presented in Fig. 9.38. The pressures sensed by the pitot probes closest to the fuselage were essentially equal to the static pressure sensed at an orifice located on the fuselage. This suggests that the flow was separated in this region. The impact pressure measurements from pitot probes further from the fuselage showed abrupt pressure changes, indicating that an oblique shock wave crossed the plane of the pitot probes. The flow model presented in Fig. 9.38 was developed using these impact pressures and other data from the wind-tunnel tests and from the flight tests. Note the flow model, which includes a separation region and a lambda-shock structure, is similar to that proposed by Westkaemper[28] (Fig. 9.26) and Stollery[4] (Fig. 9.27).

The shock/shock interactions and viscous/inviscid interactions produced locally severe heating that became critical when the vehicle reached Mach 6.7, instead of 4.9 on previous flights where no damage occurred.

Fig. 9.37   Wind-tunnel schlieren photograph of the X-15-2 with the dummy ramjet installed, $M_1 = 6.7$, $\alpha = 8$ deg., as taken from Ref. 37.

9.6.2 The Space Shuttle Orbiter

A surface oil-flow pattern for the Space Shuttle Orbiter at an angle-of-attack of 35 deg. in the Mach 8 airstream of Tunnel B (AEDC) is presented in Fig. 9.39. The effect of the interaction between the fuselage-generated bow shock wave and the wing leading-edge shock wave is evident in the oil-flow pattern. There are other phenomena evident in the oil-flow pattern. A free-vortex-layer type of separation (such as that depicted in the sketch of Fig. 3.25) occurred near the nose. Thus, the circumferential component of the flow, which was

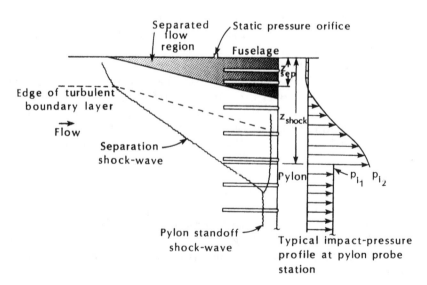

Fig. 9.38   Flow model for the viscous/inviscid interaction at the pylon/fuselage juncture, as taken from Ref. 37.

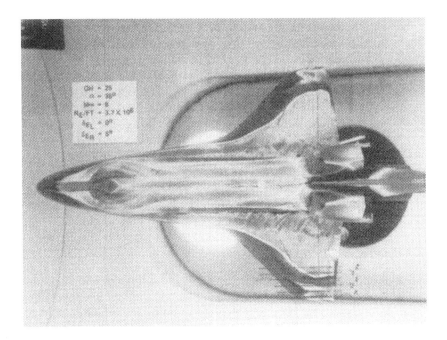

Fig. 9.39   The oil-flow pattern for a Space Shuttle Orbiter at an angle-of-attack of 35 deg. in a Mach 8 stream.

initially directed toward the leeward plane-of-symmetry, reversed direction. At the separation line, oil accumulated and proceeded to travel down the separation line, indicating that a strong axial component of the flow persisted. Hence, the term *free-vortex-layer type of separation.* The fact that there is little oil near the leeward plane-of-symmetry upstream of the canopy indicates that the longitudinal component of skin friction and, by Reynolds' analogy, the heat tranfer in this region were high.

A free-vortex-layer type of separation also occurred downstream of the canopy. The vortex pair reattaches near the leeward plane-of-symmetry. Since a strong axial component of the leeward flow persists for a free-vortex-layer separation, the heat transfer near the leeward plane-of-symmetry can be significant. Thus, boundary-layer transition could be an important consideration for these flows. Developing a transition criteria for this region presents a stiff challenge to the designer. The effect of the primary and the secondary vortices, such as depicted in Fig. 3.7, are evident in the surface oil-flow patterns near the leeward plane-of-symmetry.

The fact that the oil has been scrubbed from the Orbital-Maneuvering-System (OMS) pods indicates that high heating and high shear occurred due to a viscous/inviscid interaction. However, as evident in the data presented by Neumann[38] and reproduced in Fig. 9.40, the heating to the OMS pod is a function of the angle-of-attack. The correlation between the local heating and the angle-of-attack is important, since the Space Shuttle Orbiter employs *ramping* during entry. That is, the angle-of-attack of the Orbiter during entry is initially high, i.e., approximately constant at 40 deg. until Mach 12 is reached, but then it is ramped down, reaching approximately 20 deg. when the flight Mach number is 4. Note also that there are significant differences between the heat-transfer/angle-of-attack correlation based on wind-tunnel data and that based on flight data. These differences probably can be traced, at least in part, to real-gas effects, to Reynolds number related effects, and/or to low-density effects.

9.6.3 Hypersonic Airbreathing Aircraft

As noted by Thomas et al.,[39] the flow around a hypersonic airplane is predominantly three dimensional and is dominated by viscous effects. A sketch designating the critical design issues for a hypersonic air-breathing aircraft is reproduced in Fig. 9.41. Note that every type of viscous/inviscid interaction and of shock/shock interaction that has been described in this chapter (and more) has been identified by Thomas et al.[39] as a design issue.

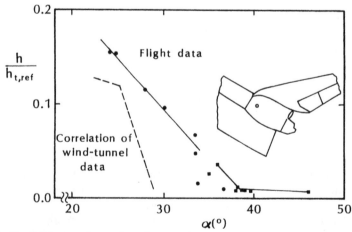

Fig. 9.40 Effect of angle-of-attack on the heating to the OMS pod, as taken from Ref. 38.

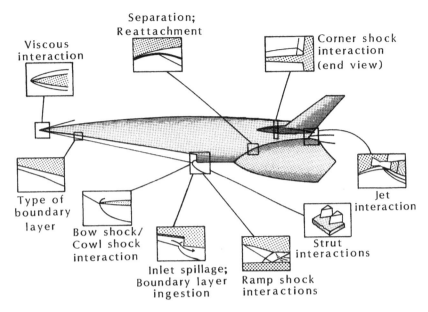

Fig. 9.41 Critical design issues for a hypersonic airbreathing aircraft, as taken from Ref. 39.

9.7 CONCLUDING REMARKS

The shock-wave/boundary-layer interaction for two-dimensional compression ramps was discussed in Sec. 6.4 and in Sec. 9.2. Although some features of the flowfield are captured by the relatively simple flow model depicted in Fig. 9.6, rigorous definition of the flowfield presents a severe challenge both to the experimentalist and to the analyst. Flow over a two-dimensional ramp was one of eight problems challenging the computational community that participated in a workshop held in Antibes, France, in January 1990. See Refs. 40 and 41.

The computations were compared with data obtained by Delery and Coet[42] in wind tunnels R2Ch and R3Ch of ONERA. A sketch of the model is presented in Fig. 9.42. In order to reproduce the ratio of the wall temperature to the recovery temperature typical of hypersonic flight, the model could be cooled by circulation of liquid nitrogen. The data used to define the flowfield include surface-pressure measurements, heat-transfer rate measurements, schlieren photographs, surface flow visualization, and thermosensitive paints.

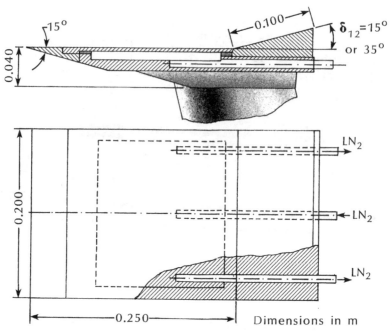

Fig. 9.42 Sketch of the two-dimensional ramp model used to generate data for the Antibes workshop, as taken from Ref. 42.

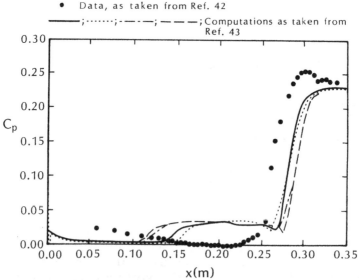

Fig. 9.43 A comparison of the measured and computed surface-pressure distributions for a two-dimensional ramp, $M_1 = 5$, $Re/m = 6 \times 10^6$, $T_w = 290$ K, $\delta_{12} = 15$ deg., as taken from Ref. 43.

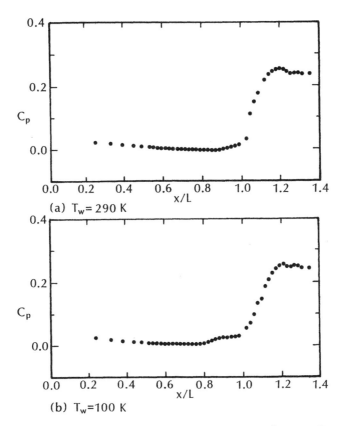

Fig. 9.44 The effect of wall temperature on the surface pressure distributions for a two-dimensional ramp, $M_1 = 5$, $Re/m = 6 \times 10^6$, $\delta_{12} = 15$ deg., as taken from Ref. 42.

The discussion in this section will focus on a nonreacting flow, where:

$$M_1 = 5$$
$$Re/m = 6 \times 10^6$$
$$T_w = 290 \text{ K}$$
$$\delta_{12} = 15 \text{ deg.}$$

Delery and Coet[42] concluded that, for these test conditions, an extended separated region formed, "which is an indication of a laminar boundary layer at separation ... This interaction is certainly laminar over its major part, but transition has a good chance to occur probably just upstream of the reattachment process."

The pressure distributions computed by four of the workshop participants are compared with the data of Delery and Coet[42] in

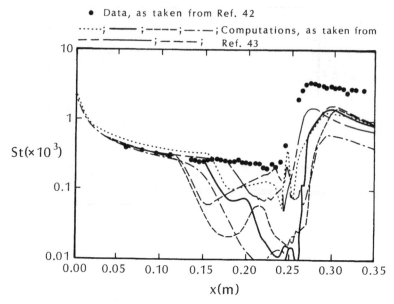

Fig. 9.45   A comparison of the measured and computed heat transfer distributions for a two-dimensional ramp, $M_1 = 5$, $Re/m = 6 \times 10^6$, $T_w = 290$ K, $\delta_{12} = 15$ deg., as taken from Ref. 43.

Fig. 9.43. As noted by Wendt et al.,[43] "All calculations exhibit a plateau pressure not seen in the experiment and differences in the location of separation are evident. On the contrary, the reattachment region shows good agreement between the computational methods."

Note that the pressure measurements of Fig. 9.43, which did not exhibit the pressure plateau, were obtained for $T_w = 290$ K. However, the wall temperature had a significant effect on the pressure distributions presented by Delery and Coet.[42] As indicated in the experimental pressure distributions presented in Fig. 9.44, when the wall is cooled, a well-defined plateau typical of separation forms well ahead of the ramp origin. At the same time, compression on the ramp is more spread out. Delery and Coet[42] state:

> The above tendencies are—at first sight—paradoxical since it is well known that wall cooling tends to contract the interaction domain. In fact, the behavior here observed can be attributed to the fact that transition occurs in the interaction domain itself. Indeed, wall cooling tending to delay transition, the boundary layer which develops on the cooled model is *more laminar* than that on the uncooled model. Hence, the cooled boundary layer offering

a smaller resistance, the separated zone is more extended when $T_w = 100$ K.

The computed Stanton-number distributions, which are presented in Fig. 9.45, show considerable variation among the solution methods regarding the beginning of separation and the detailed structure of the separated region. For Fig. 9.45,

$$St = \frac{\dot{q}}{\rho_\infty U_\infty C_p (T_{t\infty} - T_w)} \tag{9-7}$$

Although the computations are in reasonable agreement regarding the location of the peak heating, the computed values of peak heating are roughly one half of the experimental values. This should be expected, if the flow was indeed transitional at reattachment rather than laminar.

Thomas et al.[39] suggested that (in some cases) the differences between computed solutions and experimental data could be due to the fact that the experimental data were obtained before steady flow had been established during the run time associated with shock-tunnel flow. The flow in the experiment reached steady-state conditions in approximately 4 ms. The total run time was approximately 10 ms. To assess the possibility that transients affected the correlation between the data and the computation, Thomas et al.[39] made time-accurate calculations of the two-dimensional ramp flow. Presented in Fig. 9.46 are computations obtained at five intermediate times between 1 and 5 ms. The separated-flow region is predicted reasonably

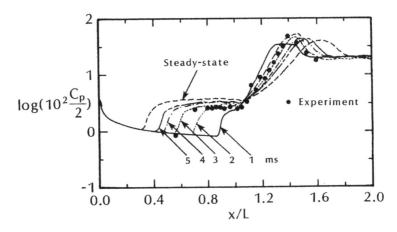

Fig. 9.46 The time evolution of the pressure distribution along a two-dimensional ramp, $M_1 = 14.1$, $Re_{\infty,r} = 1 \times 10^5$, $\delta_{12} = 24$ deg., as taken from Ref. 39.

well at a point in time between 2 and 3 ms, but the size of the region continues to increase as the solution is further advanced in time. It took more than 12 ms to establish steady flow in the computations. Additional computations for other wedge angles indicated that it required significantly more time than 4 ms for the computed flow field to reach its steady-state even for a separated-flow region with a size comparable to that found in the experiments.

Since the experimental data were obtained on a plate with a spanwise width of two feet but with no side plates to constrain the flow, Thomas et al.[39] considered the possibility that three-dimensional effects were significant for this *two-dimensional model.* The pressure contours in the downstream plane on the ramp indicated an expansion of the flow in the spanwise direction near the edge of the plate. The three-dimensional effects produced a smaller separated-flow region in the centerplane than that predicted in the two-dimensional calculations. The time variation of the computed three-dimensional flowfield indicated that steady-state flow is achieved in approximately 4 ms, which is in agreement with the experiment. Thus, the extent of and the time to establish the separation for this flow are strongly influenced by three-dimensional effects.

The points to be made in these "Concluding Remarks" are not related to who was *right* or who was *wrong* and what computational procedure is *correct.* The purpose of these remarks is to emphasize the difficulty of obtaining good experimental data or valid computational solutions. A change in the wall temperature had a significant impact on the experimental pressure distribution. The possibility exists that, in some cases, differences between experiment and computation may be due to transients in the flow or to three-dimensional effects. Finally, as noted by Rizzetta and Mach,[9] great sensitivity was observed with respect to both ramp angle and grid-point distribution. In the limit of grid-point independence, there remained differences between experiment and computation, indicating the continuing need for improved numerical models of the flow physics.

A variety of viscous/inviscid interactions occur in the complex flowfields associated with the hypersonic flight of lifting bodies. The possibility of locally severe aerothermodynamic environments and the difficulty in accurately defining the magnitude and the extent of flow perturbations will continue to challenge the experimental community and the computational-fluid-dynamics community.

REFERENCES

[1] Armstrong, J. G., "Flight Planning and Conduct of the X-15A-2 Envelope Expansion Program," Air Force Flight Test Center FTC-TD-69-4, July 1969.

[2] Bertin, J. J., "The Effect of Protuberances, Cavities, and Angle of Attack on the Wind-Tunnel Pressure and Heat-Transfer Distribution for the Apollo Command Module," NASA TMX-1243, Oct. 1966.

[3] Gaitonde, D., and Shang, J. S., "A Numerical Study of Shock-on-Shock Viscous Hypersonic Flow Past Blunt Bodies," AIAA Paper 90-1491, Seattle, WA, June 1990.

[4] Stollery, J. L., "Some Aspects of Shock-Wave Boundary-Layer Interaction Relevant to Intake Flows," Paper 17 in AGARD Conference Proceedings, No. 428, *Aerodynamics of Hypersonic Lifting Vehicles,* Nov. 1987.

[5] Delery, J. M., "Shock Interference Phenomena in Hypersonic Flows," notes from the *Third Joint Europe/U.S. Short Course in Hypersonics,* Aachen, Germany, Oct. 1990.

[6] Settles, G. S., Vas, I. E., and Bogdonoff, S. M., "Details of a Shock-Separated Turbulent Boundary Layer at a Compression Corner," *AIAA Journal,* Vol. 14, No. 12, Dec. 1976, pp. 1709–1715.

[7] Markarian, C. F., "Heat Transfer in Shock Wave-Boundary Layer Interaction Regions," Naval Weapons Center NWC TP4485, Nov. 1968.

[8] Simeonides, G., and Wendt, J. F., "Compression Corner Shock Wave Boundary Layer Interactions at Mach 14," Preprint 1990-25/AR, *17th Congress International Council of the Aeronautical Science,* Stockholm, Sweden, Sep. 1990.

[9] Rizzetta, D., and Mach, K., "Comparative Numerical Study of Hypersonic Compression Ramp Flows," AIAA Paper 89-1877, Buffalo, NY, June 1989.

[10] Shang, J. S., Hankey, W. L., Jr., and Law, C. H., "Numerical Simulation of Shock Wave – Turbulent Boundary-Layer Interaction," *AIAA Journal,* Vol. 14, No. 10, Oct. 1976, pp. 1451–1457.

[11] Moss, J. N., Price, J. M., and Chun, Ch.-H., "Hypersonic Rarefied Flow About a Compression Corner – DSMC Simulation and Experiment," AIAA Paper 91-1313, Honolulu, HI, June 1991.

[12] Anderson, J. D., Jr., *Hypersonic and High Temperature Gas Dynamics,* McGraw-Hill, New York, 1989.

[13] Ames Research Staff, "Equations, Tables, and Charts for Compressible Flow," NACA Rept. 1135, 1953.

[14] Edney, B. E., "Anomalous Heat Transfer and Pressure Distributions on Blunt Bodies at Hypersonic Speeds in the Presence of an Impinging Shock," Flygtekniska Försöksanstalten (FFA) Rept. 115, 1968.

[15] Edney, B. E., "Effects of Shock Impingement on the Heat Transfer around Blunt Bodies," *AIAA Journal,* Vol. 6, No. 1, Jan. 1968, pp. 15–21.

[16] Keyes, J. W., and Hains, F. D., "Analytical and Experimental Studies of Shock Interference Heating in Hypersonic Flows," NASA TND-7139, May 1973.

[17] Bertin, J. J., Graumann, B. W., and Goodrich, W. D., "Aerothermodynamic Aspects of Shock-Interference Patterns for Shuttle Configurations during Entry," *Journal of Spacecraft and Rockets,* Vol. 10, No. 9, Sep. 1973, pp. 545–546.

[18] Bertin, J. J., Graumann, B. W., and Goodrich, W. D., "High Velocity and Real-Gas Effects on Weak Two-Dimensional Shock-Interaction Patterns," *Journal of Spacecraft and Rockets,* Vol. 12, No. 3, Mar. 1975, pp. 155–161.

[19] Bertin, J. J., Mosso, S. J., Barnette, D. W., and Goodrich, W. D., "Engineering Flowfields and Heating Rates for Highly Swept Wing Leading Edges," *Journal of Spacecraft and Rockets,* Vol. 13, No. 9, Sep. 1976, pp. 540–546.

[20] Klopfer, G. H., and Yee, H. C., "Viscous Hypersonic Shock-on-Shock Interaction on Blunt Cowl Lips," AIAA Paper 88-0233, Reno, NV, Jan. 1988.

[21] Wieting, A. R., "Multiple Shock-Shock Interference on a Cylindrical Leading Edge," AIAA Paper 91-1800, Honolulu, HI, June 1991.

[22] Wieting, A. R., and Holden, M. S., "Experimental Study of Shock Wave Interference Heating on a Cylindrical Leading Edge at Mach 6 and 8," AIAA Paper 87-1511, Honolulu, HI, June 1987.

[23] Tannehill, J. C., Holst, T. L., Rakich, J. V., and Keyes, J. W., "Comparison of Two-Dimensional Shock Impingement Computation with Experiment," *AIAA Journal,* Vol. 14, No. 4, Apr. 1976, pp. 539–541.

[24] Agnone, A. M., Zakkay, V., and Weinacht, P., "Hypersonic Flow Over a Six Finned Configuration," AIAA Paper 85-0453, Reno, NV, Jan. 1985.

[25] Amirkabirian, I., Bertin, J. J., and Mezines, S. A., "The Aerothermodynamic Environment for Hypersonic Flow Past a Simulated Wing Leading-Edge," AIAA Paper 86-0389, Reno, NV, Jan. 1986.

[26] Wittliff, C. E., and Berthold, C. E., "Results of Heat Transfer Testing of an 0.025-Scale Model (66-0) of the Space Shuttle Orbiter Configuration 140B in the Calspan Hypersonic Shock Tunnel (OH66)," Data Management Services DMS-DR-22359 (NASA CR-151,405), Jan. 1978.

[27] Bushnell, D. M., and Huffman, J. K., "Investigation of Heat Transfer to Leading Edge of a 76° Swept Fin With and Without Chordwise Slots and Correlations of Swept-Leading-Edge Transition Data for Mach 2 to 8," NASA TMX-1475, Aug. 1967.

[28] Westkaemper, J. C., "Turbulent Boundary-Layer Separation Ahead of Cylinders," *AIAA Journal,* Vol. 6, No. 7, July 1968, pp. 1352–1355.

[29] McMaster, D. L., and Shang, J. S., "A Numerical Study of Three-Dimensional Separated Flows Around a Sweptback Blunt Fin," AIAA Paper 88-0125, Reno, NV, Jan. 1988.

[30] Neumann, R. D., and Hayes, J. R., "Protuberance Heating at High Mach Numbers—A Critical Review and Extension of the Data Base," AIAA Paper 81-0420, St. Louis, MO, Jan. 1981.

[31] Aso, S., Kuranaga, S., Nakao, S., and Hayashi, M., "Aerodynamic Heating Phenomena in Three-Dimensional Shock Wave/Turbulent Boundary Layer Interactions Induced by Sweptback Blunt Fins," AIAA Paper 90-0381, Reno, NV, Jan. 1990.

[32] Bushnell, D. M., "Interference Heating on a Swept Cylinder in Region of Intersection with Wedge at Mach Number of 8," NASA TN D-3094, Dec. 1965.

[33] Charwat, A. F., and Redekopp, L. G., "Supersonic Interference Flow Along the Corner of Intersecting Wedges," *AIAA Journal*, Vol. 5, No. 3, Mar. 1967, pp. 480–488.

[34] Venkateswaran, S., Witte, D. W., and Hunt, L. R., "Aerothermal Study in an Axial Compression Corner with Shock Impingement at Mach 6," AIAA Paper 91-0527, Reno, NV, Jan. 1991.

[35] Kussoy, M. I., and Horstman, K. C., "Intersecting Shock-Wave/Turbulent Boundary-Layer Interactions at Mach 8.3," NASA TM 103909, Feb. 1992.

[36] Marvin, J. G., "CFD Validation Experiments for Hypersonic Flows," AIAA Paper 92-4024, Nashville, TN, July 1992.

[37] Burcham, F. W., Jr., and Nugent, J., "Local Flow Field Around a Pylon-Mounted Dummy Ramjet Engine on the X-15-2 Airplane for Mach Numbers from 2.0 to 6.7," NASA TN D-5638, Feb. 1970.

[38] Neumann, R. D., "Defining the Aerothermodynamic Methodology," J. J. Bertin, R. Glowinski, and J. Periaux (eds.), *Hypersonics, Volume I: Defining the Hypersonic Environment,* Birkhäuser Boston, Boston,1989.

[39] Thomas, J. L., Dwoyer, D. L., and Kumar, A., "Computational Fluid Dynamics for Hypersonic Airbreathing Aircraft," J. A. Desideri, R. Glowinski, and J. Periaux (eds.), *Hypersonic Flows for Reentry Problems,* Vol. I, Springer Verlag, Berlin, Germany, 1991.

[40] Desideri, J. A., Glowinski, R., and Periaux, J. (eds.), *Hypersonic Flows for Reentry Problems, Volume I,* Springer Verlag, Berlin, Germany, 1991.

[41] Desideri, J. A., Glowinski, R., and Periaux, J. (eds.), *Hypersonic Flows for Reentry Problems, Volume II,* Springer-Verlag, Berlin, Germany, 1991.

[42] Delery, J., and Coet, M. C., "Experiments on Shock-Wave Boundary Layer Interactions Produced by Two-Dimensional Ramps and Three-Dimensional Obstacles," J. A. Desideri, R. Glowinski, and J. Periaux (eds.), *Hypersonic Flows for Reentry Problems, Volume II,* Springer-Verlag, Berlin, Germany, 1991.

[43] Wendt, J. F., Mallet, M., and Oskam, B., "A Synthesis of Results on the Calculation of Flow Over a 2D Ramp and a 3D Obstacle: Antibes Test Cases 3 and 4," J. A. Desideri, R. Glowinski, and J.

Periaux (eds.), *Hypersonic Flows for Reentry Problems, Volume II,*
Springer-Verlag, Berlin, Germany, 1991.

PROBLEMS

9.1   The effect of an impinging shock wave on the heat transfer to
a flat plate is to be studied in Tunnel C at AEDC. As shown in the
sketch of Fig. 9.47, the freestream Mach number is 10 and the total
temperature is 1900 °R. The impinging shock wave is generated by
a plate inclined 15 deg. to the oncoming flow.

(a) If the (total) pressure in the stilling chamber of the tunnel is
    125 psi, i.e., $p_{t1} = 125$ psi, the boundary layer on the flat plate
    is laminar. Using Eq. (9-1) and the charts and tables of Ref. 13,
    what is the ratio of (h_{pk}/h_{fp}) for this flow?

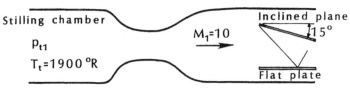

Fig. 9.47   Sketch for Problem 9.1.

(b) You also want to determine the ratio of (h_{pk}/h_{fp}) for a naturally
    turbulent boundary layer. Should p_{t1} be increased or decreased?
    Assuming you can achieve the desired value of p_{t1}, what is the
    value of (h_{pk}/h_{fp})?

Assume the air behaves as a perfect gas.

9.2   In order to obtain the desired flow conditions at a two-
dimensional engine inlet, you are to design a compression surface
that turns the flow 12 deg. from the freestream direction. See Fig.
9.48. The tests are to be conducted in a hypersonic wind tunnel
where

$$p_{t1} = 7 \text{ atm}$$
$$T_t = 450 \text{ K}$$
$$M_1 = 6$$

Determine the velocity at the inlet, the Mach number at the inlet,
and the unit Reynolds number at the inlet for the following assumptions.

(a) The 12 deg. change in the flow direction is accomplished by a single turn, i.e., $\delta_{1f} = 12$ deg.

(b) The 12 deg. change in the flow direction is accomplished by two equal turns, i.e., $\delta_{12} = \delta_{2f} = 6$ deg.

(c) The 12 deg. change in the flow direction is accomplished by four equal turns, i.e., $\delta_{12} = \delta_{23} = \delta_{34} = \delta_{4f} = 3$ deg.

(d) The 12 deg. change in the flow direction is accomplished incrementally so that the flow is isentropic.

Assume inviscid flow of perfect air.

$$\delta_{1f} = 12°$$

$M_1 = 6$?

Fig. 9.48   Sketch for Problem 9.2.

9.3   The Saenger is to stage at a flight Mach number of 7 where the freestream pressure is 100 N/m². At one point in the separation process, the Orbiter is inclined to the freestream at 20 deg. and the first-stage vehicle is inclined at 5 deg. Assuming the vehicles can be represented as flat plates (see Fig. 9.49), what are the pressures and the Mach numbers in regions 4 and 5?

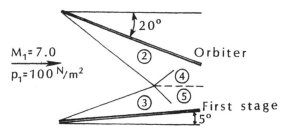

$20°$

$M_1 = 7.0$

$p_1 = 100$ N/m²

②   Orbiter

④

③   ⑤

First stage

$5°$

Fig. 9.49   Sketch for Problem 9.3.

9.4   A double-wedge configuration is to be tested in Tunnel B at AEDC, where

$$p_{t1} = 1.00 \times 10^6 \text{ N/m}^2$$

$$T_t = 725 \text{ K}$$

$$M_1 = 8.0$$

As shown in the sketch of Fig. 9.50, the first wedge deflects the freestream flow 10 deg., i.e., $\delta_{12} = 10$ deg. Using the logic of Exercise 9.2, what is the range of sweep angle (i.e., Λ) for which a Type V shock/shock interaction occurs?

Fig. 9.50 Sketch for Problem 9.4.

AEROTHERMODYNAMICS
AND DESIGN CONSIDERATIONS

10.1 INTRODUCTION

There are a wide variety of vehicles that fly at hypersonic speeds, including vehicles to launch objects into space, vehicles that are designed to cruise through the atmosphere, and vehicles designed to return objects from space. The attributes of a system include performance, safety, cost, operability, reliability, and compatibility with the environment. The attribute, or attributes, which drive a design depend on the application. For instance, a military system may place emphasis on performance; a civil transportation system for humans on safety and on reliability; and a system for delivering commercial payloads to space on operability, on reliability, and on cost. Compatibility with the environment becomes increasingly important to all systems.

To perform its mission, a vehicle must not only exhibit the desired attributes, but its design must satisfy the aerothermodynamic constraints. Aerodynamic heating restricts the envelope in which the vehicle can operate. It also affects the design of the thermal-protection system (TPS) and, hence, the vehicle weight. The body-flap deflection of the Shuttle Orbiter was constrained by limits on the thermal seals in the gap. Because lift allows a configuration to decelerate at higher altitudes for a given velocity, it enables reduced heating.

As noted in Sec. 8.1.1, the lift-to-drag ratio not only affects the convective heating, it also affects the cross-range capability of the vehicle and the ability to reduce the time to recall a vehicle from orbit. The ability to control the vehicle and to meet the mission requirements is also dependent on the aerodynamic characteristics of the vehicle.

10.1.1 Re-entry Vehicles

Allen and Eggers[1] noted that if a *missile* is so light that it will be decelerated to relatively low speeds, even if acted upon by low drag forces, i.e., a low-beta configuration, then the convective heating is minimized by employing shapes with high-pressure drag. Such shapes maximize the amount of heat delivered to the atmosphere and minimize the amount of heat delivered to the body in the deceleration process. The early manned entry vehicles (e.g., the Mercury, the Gemini, and the Apollo Command Module) and winged vehicles that enter at high angles-of-attack (e.g., the Space Shuttle Orbiter and the Hermes) are examples of such vehicles.

On the other hand, if the *missile* is so heavy or has such a relatively low drag that it is only slightly retarded by aerodynamic drag, irrespective of the magnitude of the drag force, i.e., a high-beta configuration, then the convective heating is minimized by minimizing the total shear force acting on the vehicle. Allen and Eggers[1] define this as "the small cone angle case." Indeed, the small half-angle cones typical of high-beta ballistic missiles are examples of these configurations.

Recall, from the presentation in Chap. 8, that the lift-to-drag ratio also has a significant effect on the decelerations and on the lateral maneuverability in flight. A vehicle with a lift-to-drag ratio of 1 may be characterized by adequately low decelerations and adequately high lateral maneuverability. Eggers[2] states, "We are forcefully reminded that L/D should be no higher than that required by considerations of decelerations and manoeuverability." To first-order considerations, the cost of the vehicle and the complexity of the flowfield increase with L/D.

With the lift-to-drag requirements satisfied, Eggers[2] discussed ways to reduce the heating. For a ballistic missile, the high-drag shapes have relatively low heating rates. For airplane-like vehicles, high-lift shapes have reduced heating rates. Thus, Eggers concludes, "We are attracted, therefore, to high-lift, high-drag configurations for sub-satellite applications." Eggers suggests flat-top, blunted cones, such as depicted in Fig. 10.1, as satisfying the requirement for a high-lift, high-drag configuration. As indicated by the vehicles depicted in Fig. 10.2, a variety of blunt, lifting bodies were under development in the 1960s and 1970s. One of the candidate designs[3] to satisfy the requirements for an Assured Crew Return Capability (ACRC) from the Space Station Freedom is shown in Fig. 10.3. The concept, developed in part at the Langley Research Center (NASA) from work on the HL-10 and on the X-24 programs of the late sixties, develops a maximum lift-to-drag ratio of roughly 1.4 for altitudes from 25,000

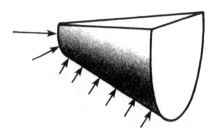

Fig. 10.1 Sketch of a high-lift, high-drag configuration.

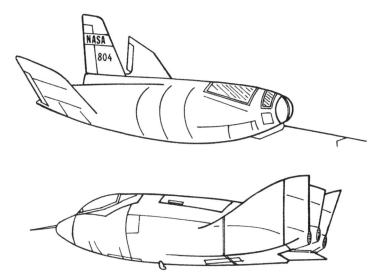

Fig. 10.2   Examples of blunt, lifting bodies under development in 1960s and 1970s.

ft to 250,000 ft. With this lift-to-drag ratio, the vehicle is capable of low-g loadings during entry, which is important for medical emergency recovery, and has sufficient cross-range capability to permit increased landing opportunities to specific sites as compared with ballistic shapes.

The Space Shuttle Orbiters will probably be retired early in the twenty-first century. Concepts for replacement vehicles include the HL-20, a 29-ft orbiter, similar in design to the ACRV depicted in Fig. 10.3. The HL-20 would be launched by a Titan 4, or similar expendable launch vehicle (ELV), and transport up to eight people or a small amount of cargo to and from space. However, since an HL-20-sized vehicle would have significantly less capacity than the current Shuttle, a complementary system would be needed to provide sufficient launch capability.

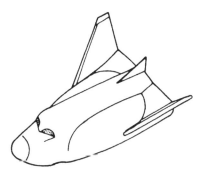

Fig. 10.3 Assured Crew Return Vehicle (ACRV) concept, as taken from Ref. 3.

10.1.2 Design Philosophy

Neumann[4] wrote:

> Hypersonic systems within the Air Force and NASA tend
> to be different because the underlying design criteria stress
> different features. In broad terms, Air Force systems fly
> in regions of the atmosphere where it is difficult to oper-
> ate and not where operating efficiencies are the highest.
> This is true for combat aircraft where the goal often is to
> maneuver decisively at transonic conditions, it is true for
> ballistic missile systems where the goal is to re-enter the
> atmosphere with an extremely low drag body which will
> decelerate at very low altitudes and it is true of hyper-
> sonic lifting entry systems which, for operational reasons,
> may stress aerodynamic efficiency during the entry pro-
> cess. The Air Force is driven by a different set of design
> criteria. In many cases this changes the design process
> somewhat. What may be a design goal in a NASA sys-
> tem design (perhaps minimizing the aerodynamic heating)
> could become only a design constraint in an Air Force sys-
> tem (maximizing aerodynamic performance of a system
> design within the limits of available materials).

10.2 DESIGN CONSIDERATIONS

In the subsequent sections, we will discuss the aerothermodynamic environment of two different types of vehicles. For one type,
the vehicle is placed in orbit by a rocket propulsion system. The vehicle returns to earth as an unpowered glider. Examples of this type
include the Apollo Command Module, the Space Shuttle, and the
Hermes. The second type is powered by an airbreathing propulsion
system, such as the National Aero-Space Plane (NASP). Airbreathing hypersonic vehicles must fly at altitudes which are low enough
that there is sufficient oxygen, i.e., relatively high density, for the
propulsion system to operate effectively. However, the convective
heat transfer and the drag increase as the density increases, i.e., as
the altitude decreases. Thus, the trajectory of an airbreathing hypersonic vehicle must represent a compromise between the propulsion
requirements and the heat-transfer/drag requirements.

Since the environments in which these two types of vehicles operate are dramatically different, the designers must evaluate the importance (or lack of importance) of parameters such as boundary-layer
transition, chemistry, viscous/inviscid interactions, etc. Parameters
which are critical to the design of one type may be relatively unimportant to the other.

10.2.1 Design Considerations for Rocket-Launched / Glide-Re-entry Vehicles

The decision to design and to build the Hermes presented the European aerospace community with the opportunity and with the challenge to apply the tools available at the outset of the 1990s to design a new space transportation vehicle. A sketch of a Hermes configuration is presented in Fig. 10.4. Trella[5] notes:

> A comparison with the Shuttle Orbiter characteristics reveals some major differences and, in particular, the more severe thermal environment which the lower scale of the Hermes will induce. The similar values for the ballistic coefficient and for the hypersonic efficiency imply that the re-entry trajectory will not be substantially different; all other things being equal, the transition to turbulent flow will occur somewhat later on corresponding points of the vehicle, while the heat transfer will be higher, due to the smaller Reynolds number and the smaller radius of curvature of the surfaces. It is necessary, therefore, to round as much as possible both the nose and the leading edges of the wing with corresponding loss on the slenderness of the vehicle. Maximum temperatures are expected to be ~ 100 °C higher for Hermes.

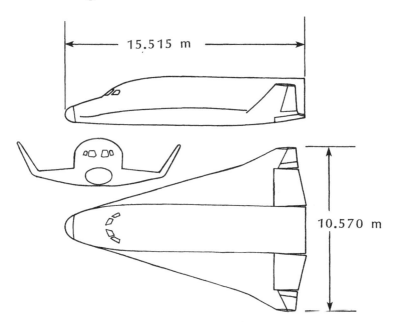

Fig. 10.4 Hermes configuration, as taken from Ref. 5.

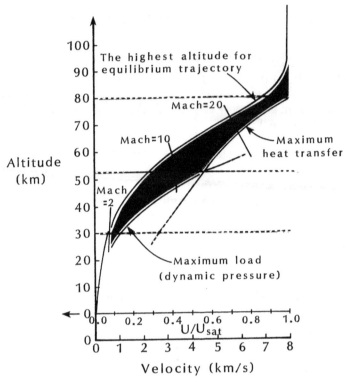

Fig. 10.5 Re-entry corridor for the Hermes, as taken from Ref. 5.

Hermes glides through the earth's atmosphere within the re-entry corridor depicted in Fig. 10.5. The corridor is bounded by lines representing the maximum load, i.e., the maximum dynamic pressure, the maximum heat-transfer rate, and the maximum altitude at which equilibrium can be maintained between the weight and the aerodynamic forces. During the initial phase of re-entry, the Hermes vehicle is at a relatively high angle-of-attack, e.g., 40 deg., so that it acts as a very blunt body ($L/D \sim 0.8$) with most of the energy heating the air in the shock layer and not the vehicle. For this high-alpha configuration, the flow is dominated by a strong bow shock and a large subsonic region covering the entire windward side of the vehicle followed by a largely separated flow for the leeward side. The angle-of-attack is reduced, e.g., 19 deg., for the latter part of the trajectory, so that Hermes acts as a relatively high L/D glider ($L/D \sim$ 1.5 to 2 at hypersonic speeds and 2 to 5 at supersonic speeds). For this lower alpha, the flow around the vehicle is mostly supersonic with the exception of the nose cap.

The fluid mechanic and thermodynamic phenomena encountered by the Hermes during its flight through the earth's atmosphere are indicated in Fig. 10.6. Rocket-launched space vehicles such as the Hermes experience relatively little aerodynamic heating during launch. The large thrust available during ascent results in passage through the atmosphere in a matter of minutes. Burnout velocity is reached at very high altitudes, where the aerodynamic heating is low. Tauber et al.[6] note:

> For instance, the time-integrated heating (total heat load) at the Shuttle orbiter nose stagnation point is almost two orders of magnitude less during launch than during atmospheric entry. In contrast, a vehicle using an airbreathing propulsion system must fly a long time in the denser portion of the atmosphere to develop sufficient thrust to accelerate to orbital speed. This extended period of hypervelocity atmospheric flight exposes the vehicle to high local heating rates and large total heating loads.

From the notations in Fig. 10.6, it is evident that the Hermes experiences a variety of dramatically different flow phenomena during re-entry. At altitudes in excess of 100 km, one must understand rarefied noncontinuum flows. At altitudes in excess of 60 km to 70 km, the aerothermodynamicist must account for nonequilibrium chemistry and surface catalycity. Boundary-layer transition occurs for

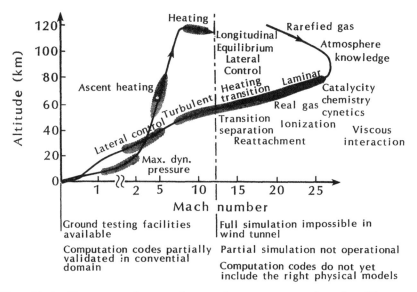

Fig. 10.6 The aerothermodynamic concerns for the Hermes launch and re-entry trajectories, as taken from Ref. 5.

flight Mach numbers on the order of 10, when the altitude is approximately 50 km. As discussed in Chap. 7, the determination of the conditions for boundary-layer transition challenges the most experienced designer. Real-gas effects are not a factor for the subsequent flight at lower Mach numbers/lower altitudes. Nevertheless, determining the solution of the flowfield at these conditions remains a considerable challenge. Problems, such as modeling turbulent boundary layers and viscous/inviscid interactions for the *winglets* (e.g., corner flows and shock/shock interactions), require considerable understanding on the part of the designer.

The comments relating to ground testing and to computational codes, which are presented along the abcissa of Fig. 10.6, are taken from Trella.[5] These comments indicate that ground-testing facilities are available up to Mach 12.5. Above this Mach number, full simulation is not possible in the wind tunnel. Similarly, the comments imply that computation codes do not yet include the right physical models. These comments represent that author's assessment of technology when the paper was first presented in 1987 and are probably configuration/application dependent. Furthermore, it is unlikely that there is a sharp demarcation in our ability to simulate and not to simulate, e.g., that an abrupt change in our ability to fully simulate a flow in a wind tunnel occurs at Mach 12.5. For instance, it is difficult to obtain a naturally turbulent boundary layer on a scale Hermes model in a wind tunnel at Mach 10. Thus, the comments presented in Fig. 10.6 should be considered qualitative, subject to changes in technology and in application. It might be noted that changes are not necessarily advances. Since the number of operational hypersonic wind tunnels decreases continuously, it is possible that a ground-testing capability which was available when this book was written will not be available to the reader.

10.2.2 Design Considerations for Airbreathing Vehicles

In designing a vehicle that is powered by an airbreathing propulsion system, one must simultaneously consider the aerothermodynamic environment, the propulsion system, the structure (including structural dynamics), and the flight control system in the design of the vehicle. As noted by Blankson,[7] "The technical problems are multidisciplinary to first order. Proper resolution of these problems requires the ability to integrate highly-coupled and interacting elements in a fundamental and optimal fashion to achieve the desired system performance."

10.2.2.1 Range considerations.
The Breguet equation indicates the parameters which affect the range for the powered flight of a hyper-

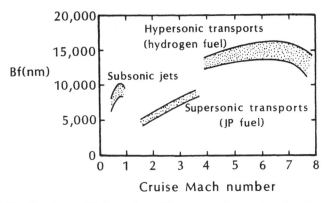

Fig. 10.7   Cruise efficiencies (Breguet factor) of subsonic, supersonic, and hypersonic transports, as taken from Ref. 8.

sonic vehicle:

$$R = I_{sp} U \left(1 - \frac{U^2}{U_{\text{sat}}^2} \right) \frac{L}{D} \, ln \, \frac{W_i}{W_f} \qquad (10\text{-}1)$$

where W_i and W_f are the initial and the final values of the vehicle weight, U is the flight velocity, U_{sat} is the satellite velocity, and I_{sp} is the specific impulse. Gregory et al.[8] conducted a study of the overall mission capabilities for transports powered by hydrogen-fueled subsonic burning ramjets operating in the range $4 \le M_\infty \le 8$.

Gregory et al.[8] calculated the Breguet factor, which was termed the *cruise efficiency*,

$$\text{Bf} = I_{sp} U \left(1 - \frac{U^2}{U_{\text{sat}}^2} \right) \frac{L}{D} \qquad (10\text{-}2)$$

for (existing) subsonic jet aircraft, for (conceptual) JP-fueled, supersonic transports, and for (conceptual) hydrogen-fueled hypersonic transports. Their results are reproduced in Fig. 10.7. It was noted[8] that "With the subsonic burning ramjets considered here, it appears that cooling requirements may be excessive at Mach numbers above 8." However, in the range of Mach number between 4 and 8, "the cruise efficiencies indicated for the hypersonic transports are quite attractive."

10.2.2.2 Propulsion considerations.   As noted in the previous paragraph, the high temperatures that occur in the combustion chambers of subsonic burning ramjets limit the Mach number at which they can be used. Curran[9] presented the approximate specific impulse for various engine cycles using hydrogen as the fuel. The specific impulse for turbojets, subsonic combustion ramjets, supersonic combustion ramjets (i.e., scramjets), and rockets are presented as a function of the Mach number in Fig. 10.8. It is apparent that, as flight speed increases, the turboaccelerator class of engine is supplanted first by the subsonic combustion ramjet and then by the supersonic combustion ramjet. Rocket propulsion may also be needed at the higher flight speeds. Consequently, for a hypersonic flight vehicle operating at a maximum speed above about Mach 5.0, a multimode propulsion system will be required. Bearing in mind the limitations of materials, such an engine system might operate as a turboaccelerator to speeds of the order of Mach 4.0, then transition to a subsonic ramjet operation up to speeds of about Mach 6.0, and then be operating totally as a supersonic combustion engine for speeds above about Mach 7.0.

Curran[9] wrote:

> It is clear that for maximum flight speeds of about Mach 6 the use of a conventional ramjet cycle is imperative and for higher maximum flight speeds the scramjet is the appropriate high speed engine. The lower speed engine cycle must be integrated with the appropriate ramjet element(s) to form a truly combined cycle engine. In the following discussion the term low speed engine applies typically to turboaccelerator or rocket-based systems used to accelerate the vehicle up to speeds of roughly Mach 4.0.
>
> For purposes of discussion, let us consider a hypersonic vehicle with maximum speed in excess of say Mach 7, and consequently using a supersonic combustion ramjet as its dominant propulsion system. A plausible overall propulsion installation for such a vehicle is one which uses a low speed engine of the turboaccelerator class for take-off (and landing), and for acceleration to speeds of about Mach 4.0, followed by an initial transition to ramjet operation. A subsequent transition to scramjet operation takes place at about Mach 6.0. The vehicle then accelerates to a terminal Mach number on scramjet power.
>
> It is appropriate to point out that because of the complexities of mode transition, there is a substantial payoff to eliminating such transitions. For example, the turboaccelerator speed capability may be "stretched" to permit direct transition to a scramjet mode. Alternatively, for say a Mach 7 system, the ramjet performance may be "stretched" to avoid transition to a true scramjet mode. Simultaneously the scramjet take-over Mach number may be reduced or "stretched" down to eliminate the ramjet mode. Similarly in relation to fuels it is desirable where

possible to stretch the performance of hydrocarbon class fuels to higher Mach numbers to defer the logistical problems of operating with cryogenic fuels. Consequently the propulsion engineer must place emphasis on "stretched" capability wherever appropriate, and where mission requirements will not be compromised.

Note, in Fig. 10.7, that the fuel type changed from conventional JP fuels at Mach numbers up to 4 to liquid hydrogen at the higher Mach numbers. As presented in Table 10.1, hydrogen has a high heat of combustion and a high specific heat. Thus, hydrogen is attractive because it can be used to cool the internal (engine) and external flow surfaces, or to cool the air and thus impact the engine's thermodynamic cycle. However, because of the low density of liquid hydrogen, hydrogen-fueled vehicles require large tankage volumes. The large tankage volumes may result in an excessive cross section (with the attendant pressure drag) or excessive surface area (with the attendant skin-friction drag, surface heating, and vehicle weight). Another disadvantage is the boiling point of liquid hydrogen, which is 36 °R. Therefore, the fuel must be carried in cryogenic tanks, creating logistics problems.

For certain applications, endothermic hydrocarbon fuels can be considered as an alternative fuel. As indicated in Table 10.1, hydrocarbon fuels have higher boiling points, so they are storable. However, since they have a significantly lower heat capacity, they may not provide sufficient cooling for Mach numbers above 7 or 8. The specific impulse using endothermic hydrocarbon as the fuel for turbojets, subsonic-combustion ramjets, and scramjets, as provided by Blankson,[10] are included in Fig. 10.8. Note that the specific impulse for the systems using endothermic hydrocarbon fuels is roughly one-third of that for the corresponding hydrogen-fueled system. However, because the density is larger, hydrocarbon fuels require less

Table 10.1   Comparison of fuel characteristics, as taken from Ref. 8.

Property	Hydrocarbon	Hydrogen
Heat of combustion (Btu/lbm)	19,000	51,000
Specific heat (Btu/lbm/°F)	0.46	2.7–3.7
Liquid density (lbm/ft^3)	51	4.5
Boiling temperature at 1 atm (°R)	820–915	36

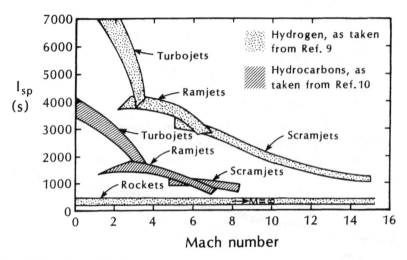

Fig. 10.8 Specific impulse for various propulsion devices fueled by hydrogen or by hydrocarbons.

storage volume. Thus, vehicles using hydrocarbon fuels have smaller cross sections and less surface area. As a result, configurations with lower drag can be developed. Referring to the Breguet equation, the reduced specific impulse and the reduced drag are compensating parameters when determining the range.

By now it should be clear to the reader that vehicles that are to operate efficiently over a wide range of Mach numbers require more than one mode of propulsion. Petley and Jones[11] developed a thermal management concept for a Mach 5 cruise aircraft, using noncryogenic fuel. The modes of operation for their concept are presented in Fig. 10.9. Subsonic propulsion is provided by the turbojet with afterburner, while the ramjet is cold flowing. The ramjet is ignited transonically, and the system operates in dual mode until the turbojet begins to spool down near Mach 2.5. The ramjet has been shown to be of assistance transonically, primarily for drag reduction, even though its low speed efficiency is rather poor. As the aircraft accelerates from the speed of sound to Mach 3, the ramjet becomes more efficient and provides increasing thrust. At the same time, aerodynamic heating becomes an increasing problem for the turbojet. The turbojet remains in full afterburner until it is finally shut down at Mach 3. The ramjet then provides all of the thrust up to Mach 5. The concept proposed by Petley and Jones,[11] which is depicted in Fig. 10.9, reflects the need to carefully integrate the vehicle and the propulsion system. The system consists of an over/under turbo/ramjet design. Foster et al.[12] define a nomenclature convention for "combined cycle engines" as compared to "combined cycle vehicles." "'Combined Cycle Engines' functionally and physically integrate more than one propulsion engine cycle into a single engine assembly. They should not be confused with 'combined cycle vehicles,' 'combination propulsion systems,' 'multi-cycle' propulsion or 'multi-mode vehicles' having more than one physically separate propulsion cycle in a single vehicle."

The correlations presented in Fig. 10.8 afford a relatively simple comparison of various engine cycles based on fuel specific impulse (I_{sp}), which is relevant to cruise applications where the Breguet equation is used to compare ranges. However, for missions where acceleration is the dominant consideration, e.g., earth-to-orbit (ETO) missions, the engine weight becomes an even more important variable. For these missions, the performance of the engine may be evaluated both on the basis of its specific impulse and of its corresponding thrust-to-weight ratio. Indeed, for a truly efficient accelerator design, the installed performance of the propulsion system, such as measured by the installed value of thrust minus drag ($T - D$), should be determined. As noted by Hunt[13]:

> There are differences between configurations dedicated to cruise and those that accelerate to orbit. The accelerator must have a much bigger inlet area relative to body cross section than the cruiser in order to facilitate sufficient thrust margin, and thus sufficient acceleration, to reach orbital speed. Acceleration time must be minimized so that the integrated drag loss in the airbreathing corridor is kept within manageable bounds. On the other hand, the cruiser requires no thrust margin at the design cruise

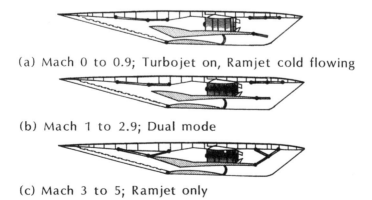

(a) Mach 0 to 0.9; Turbojet on, Ramjet cold flowing

(b) Mach 1 to 2.9; Dual mode

(c) Mach 3 to 5; Ramjet only

Fig. 10.9 Operating modes for turbojet/ramjet engine for cruise up to Mach 5, as taken from Ref. 11.

speed. For the accelerator, the primary aerodynamic issue is minimizing configuration drag near zero angle of attack, while for the cruiser, the task is to maximize configuration lift-to-drag ratio at the design point.

At this point, it should be clear that the designer of a hypersonic, airbreathing vehicle has numerous options regarding the fuel, the propulsion cycle, and the conditions over which a particular propulsion cycle is employed to meet the mission constraints. For detailed information regarding the airbreathing propulsion systems for hypersonic vehicles, the reader is referred to the companion volume in this series by Heiser and Pratt.[14]

10.2.2.3 Aerothermodynamic considerations.
Constraints placed on the vehicle design by trajectory considerations are addressed early in the design process. Some of the considerations that affect the ascent trajectory of an airbreathing, hypersonic transport that were presented in Ref. 8 are reproduced in Fig. 10.10. Gregory et al.[8] note:

> The desired cruise altitudes are indicated by the dashed curve in Fig. 10.10, and the problem is to select a climb trajectory to the cruise altitude that maximizes the payload weight for the specific mission. Since many components of the vehicle are sized during climb, and since up to 40% of the fuel is consumed in this phase of flight, selection

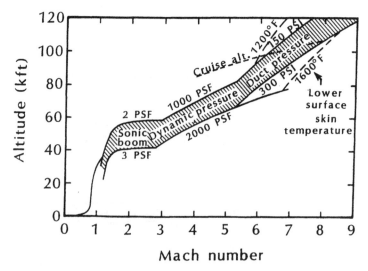

Fig. 10.10 Constraints on climb trajectories for an airbreathing hypersonic transport, as taken from Ref. 8.

of the best available climb trajectory is important. Climb trajectories are affected by a variety of constraints. The trajectories used in this study followed the lines defining the various constraints, since trajectories at higher altitudes resulted in increased fuel consumption and correspondingly lower payloads. The first constraint indicated in Fig. 10.10 is that due to sonic-boom limitations. Typical curves for over-pressures of 2 and 3 psf are shown. The exact location of those curves is affected by many factors; for example, if the vehicle is shaped to minimize sonic boom, the lower curve can serve as an approximation of a 2-psf boundary. As the airplane accelerates to high supersonic speeds along a line of constant overpressure, the dynamic pressure increases until structural considerations dictate some limiting value. The trajectory then follows this maximum allowable dynamic-pressure line until another consideration becomes important. At flight Mach numbers of approximately 5, the internal pressures in the propulsion system increase rapidly causing a corresponding increase in the structural weight of the turboramjet engines. As a result, the trajectories usually are restricted by lines of constant internal pressure. Duct pressures are also dependent on inlet pressure recovery and on the wing angle of attack which affects both the Mach number and the pressure field in which the inlet is located. Thus, the constant duct-pressure lines shown in Fig. 10.10 should be considered only as examples. The next constraint is encountered at higher Mach numbers where aerodynamic heating becomes important. Typical curves for equilibrium temperature of 1200 and 1600 °F are shown. These temperatures are, of course, dependent on such things as angle of attack and distance from the leading edge.

As early as 1970, Becker[15] noted, "The strong interactions between the aerodynamic, structural, and propulsive systems of hypersonic air breathers offer important opportunities for achieving improved vehicles." He continued:

Aircraft configurations based on the dictates of single discipline such as aerodynamics are likely to be naive and impractical. In the early days of hypersonic research, the aerodynamics discipline was pursued very vigorously without benefit of equal effort in the propulsive and structure areas. As a consequence, the attainment of very high lift/drag ratios became an exaggerated goal attended by aerodynamic configurations having very large wings. As more realistic structures, propulsion, and weight inputs became available, the wings began to shrink and the bodies became relatively larger and elliptical or flattened in cross section until at this point in time we see the Mach 8 to 12 cruise vehicles of the future as lifting bodies with only rudimentary wings.

A basic question addressed by Becker[15] was whether the external heat-transfer loads of a cruise aircraft were of such a magnitude that they could be absorbed by the fuel heat sink, while still having a configuration of sufficiently low drag. A correlation of the average heat-transfer/drag ratio as a function of the shape and of the maximum lift-to-drag ratio, which is presented in Fig. 10.11, was used by Becker[15] to help define the characteristics of the configuration. The configuration thus determined by Becker in 1970 is reproduced in Fig. 10.12. The configuration proposed by Becker in 1970 is similar to an early NASP configuration, which is reproduced in Fig. 10.13. This is not surprising since the required technology was not vigorously pursued prior to the start of the NASP program. Furthermore, the general characteristics of a vehicle, e.g., slender or blunt, lift-to-drag ratio, angle-of-attack, etc., depend to first order on the vehicle application, e.g., airbreathing cruiser. The reader should note that the design of an optimum hypersonic cruiser using 1990s technology differs from that depicted in Fig. 10.12.

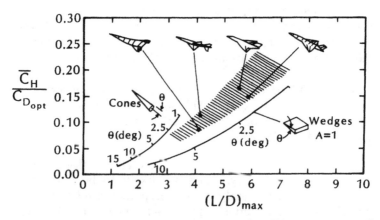

Fig. 10.11 The effect of shape and $(L/D)_{max}$ on heat-transfer/drag ratio evaluations. $M_1 = 6$; $Re_L = 140 \times 10^6$; turbulent boundary layer, as taken from Ref. 15.

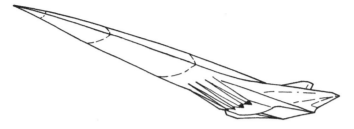

Fig. 10.12   Cruise vehicle concept of 1970, as taken from Ref. 15.

The severity of the aerothermodynamic environment during the ascent phase of the National Aero-Space Plane (NASP) is indicated in Fig. 10.14, which was provided by Blankson.[10] The heating to the cowl lip is orders of magnitude greater than the heating rates experienced by the Space Shuttle Orbiter on entry. The severity of the heat-transfer rate, i.e., 55,000 Btu/ft^{2}s, is due to a combination of factors, including: the relatively high density of the air, the small radius of the cowl lip, and the (assumed) existence of a Type IV shock/shock interaction. Even for the larger surface area of the engine interior, the heat-transfer rate of 1500 Btu/ft^{2}s exceeds the Shuttle Orbiter heat-transfer rates by an order of magnitude.

Schmidt et al.[16] note:

> Most of the research in hypersonics thus far has been concerned with the ascent phase. However, the critical nature of descent may require modification of some ascent

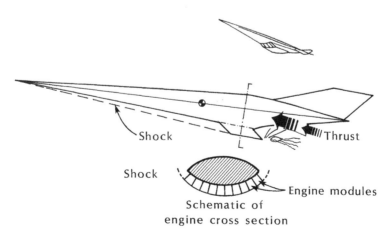

Fig. 10.13   Scramjet vehicle integration as proposed in 1989, as taken from Ref. 13.

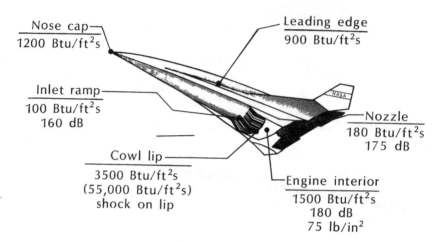

Fig. 10.14 Typical aerothermodynamic constraints for the NASP during ascent, as taken from Ref. 10.

stability characteristics. The NASP will re-enter the atmosphere at high angles of attack similar to the shuttle. This creates strong shock waves to dissipate heat and increase drag. Unlike the shuttle, however, the NASP has a high slenderness ratio and thus will more than likely have highly nonlinear stability characteristics.

Thus, the design of a hypersonic vehicle powered by an airbreathing propulsion system, whether it be a hypersonic cruiser or a transportation system to orbit, presents significant challenges to the designers.

10.2.3 Design Considerations for Combined Rocket/Airbreathing Powered Vehicles

There are a number of hypersonic concepts that employ a combination of rocket propulsion and airbreathing propulsion systems. They include the German Sänger Space Transportation System[17] and the Beta concept.[18] Gord et al.[18] developed a fully reusable two-stage vehicle capable of horizontal take-off and landing. The first stage (booster) employs a multicycle high-Mach-number airbreathing propulsion system complemented by conventional rocket propulsion, as required. The second stage (orbiter) is a high lift-to-drag ratio design with a dedicated high-volume payload bay, powered by a conventional rocket engine.

The orbiter staging Mach number was selected based on several interrelated criteria, which included the theoretical velocity potential (ΔV) capable of being built into the orbiter, the requirements for realistic, near-term airbreathing and rocket propulsion systems for the booster and orbiter components, respectively, and the thrust-to-weight ratio of the orbiter at staging. The second-stage weight is presented as a function of the theoretical velocity potential (ΔV) in Fig. 10.15, which is taken from Ref. 18. Marked on the curve for reference are the orbiter velocity inputs required to reach a 100 n.mi. polar orbit for staging Mach numbers of 0.8, 6.0, and 8.0. Note that the orbiter weight increases rapidly as the staging Mach number drops below 6 to 8.

For staging at Mach 8, the orbiter would weigh approximately 550,000 lb. However, to stage at this Mach number requires that the airbreathing propulsion system includes a scramjet, since, as noted in Sec. 10.2.2.2, a transition to scramjet takes place at about Mach 6.0. Use of a turboaccelerator, i.e., a turbojet propulsion sysem, simplifies the design requirements on the airbreathing propulsion system, but limits the maximum Mach number to 4. Referring to Fig. 10.15, a staging Mach number of 4 results in an orbiter weighing approximately 1 million lb.

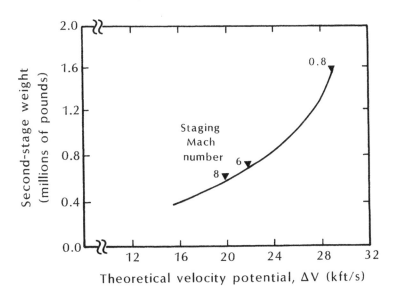

Fig. 10.15 The second-stage (orbiter) weight as a function of ΔV for a 50,000 lb. payload, as taken from Ref. 18.

The thrust-to-weight ratio is an important parameter to achieve orbit. Since the rocket propulsion system of the orbiter was a derivative of the Space Shuttle Main Engine (SSME), the assumed specific impulse (I_{sp}) was 464 s, and the vacuum thrust rating was 516,000 lb. Thus, the thrust-to-weight ratio for Mach 8 staging is adequate to achieve orbit with a single SSME. However, for staging at Mach 6, the orbiter is too heavy for a single SSME, resulting in a more complicated orbiter propulsion system.

10.3 DESIGN OF A NEW VEHICLE

When undertaking a new project, the designer is faced with *known unknowns,* i.e., phenomena of which she/he is aware but cannot quantify, and *unknown unknowns,* i.e., phenomena which have not yet been identified as important to the design. Determining a boundary-layer transition criterion (including the effect of misaligned tiles) was a known unknown for the designers of the Space Shuttle Orbiter. It was of concern from the outset and received considerable attention throughout the design process. The problem related to the pitching moment of the Orbiter was an example of an unknown unknown. It was not discovered until the first flight, STS-1.

Trella[5] notes that the designers of the Hermes must be able to predict, "the fluid mechanical properties: forces and moments (pressure and shear distribution over the body), dynamic equilibrium conditions and stability derivatives (C.G. location, longitudinal and lateral margins), adequacy and efficiency of the aerodynamic control surfaces within maximum deflection; the thermal behaviour: temperature distribution, and their dependence upon surface properties (roughness, catalycity, conductivity, thickness, emissivity, and absorptivity)."

10.3.1 Guidelines

How then does one approach the design of a new vehicle? Perrier[19] suggests:

The simplest answer is:

– to use existing complex technology (for example rocket propulsion) each time it is not mandatory to develop new ones.

– to use existing design or extrapolation of existing design (conventional design preferred to optimized design).

– to select large margins covering uncertainties given by each complex physical phenomenon not well understood and not having blind confidence on scientific advancement.

- to use the first vehicles as test bench of the later ones.
- to reduce as far as possible the flight envelope for remaining inside a small risk but nevertheless efficient domain.

The author has often encountered these guidelines. During the development of the Apollo Command Module, the then existing state-of-the-art for CFD was not capable of generating solutions for the leeward flowfield. Thus, the determination of the leeside heat transfer for the Apollo CM at angle-of-attack was based on wind-tunnel data. However, there was considerable scatter of the heat-transfer measurements from the leeward plane-of-symmetry. Measured values of the heat-transfer rate ranged from 0.01 $\dot{q}_{t,\alpha=0°}$ to 0.06 $\dot{q}_{t,\alpha=0°}$. Although most of the data were in the range $0.02\dot{q}_{t,\alpha=0°}$ to $0.03\dot{q}_{t,\alpha=0°}$, the effect of the model support (which was on the leeward side of the model) on the flowfield and the possibility of transition of the free-shear layer were major contributors to the data scatter. Furthermore, at the time the Apollo Command Module was designed, computational fluid dynamics was relatively crude and could not be used to reduce the uncertainties. Thus, to cover the uncertainties due to poorly understood, complex phenomena and to ensure the safety of the crew, the leeward thermal-protection system was designed to survive heat-transfer rates equal to 0.07 $\dot{q}_{t,\alpha=0°}$. As we improve the tools used in the design process, i.e., improve ground-based test procedures (and our understanding of the relation of ground-test measurements to the flight environment) and improve CFD codes (through code-validation exercises and through code-calibration exercises), we can reduce such conservatism.

During the first (unmanned) flight test of the Gemini capsule, unexpectedly high heating rates occurred on the conical surface near an umbilical fairing. The locally severe heating, which was due to a viscous/inviscid interaction caused by the umbilical fairing, produced a series of small holes in the metal heat shield of the conical surface. The cause of the locally severe heating was verified in a post-flight wind-tunnel test program. Since the problem was not so severe as to warrant redesign at that point in the program, restrictions were placed on the flight envelope.

10.3.2 Tools

However, if one is to achieve significant improvements in performance or to dramatically reduce launch operations costs, one may need designs that differ significantly from previous designs. Thus, one must depart from the simple guidelines of the previous section. To provide the required information, the designer can make use of theoretical analysis, computational fluid dynamics, ground-based tests, and flight tests. Only the flight of a full-scale vehicle at the actual conditions provides a *no compromise* flowfield. However, the flight test of a full-scale vehicle is extremely expensive and requires considerable time to develop. Thus, the design process must carefully integrate the use of the tools to obtain the optimum configuration through a reasonably efficient process.

At this point, the reader should be aware that both ground-based testing and computational fluid dynamics have strengths and weaknesses.

10.4 CONCLUDING REMARKS

Thus, the designer of a hypersonic vehicle has a variety of tools which can be used to define the aerothermodynamic environment. She/he must recognize the strengths and weaknesses of the available tools. Nevertheless, by making intelligent use of analysis, CFD, ground-based tests, and flight tests (both research and development flight tests as well as flights of the prototype), one can determine the aerothermodynamic environment for hypersonic vehicles subjected to diverse requirements.

REFERENCES

[1] Allen, H. J., and Eggers, A. J., Jr., "A Study of the Motion and Aerodynamic Heating of Missiles Entering the Earth's Atmosphere at High Supersonic Speeds," NACA TN-4047, Oct. 1957.

[2] Eggers, A. J., Jr., "Some Considerations of Aircraft Configurations Suitable for Long-Range Hypersonic Flight," A. R. Collar and J. Tinkler (eds.), *Hypersonic Flow, Proceedings of the Eleventh Symposium of the Colston Research Society, April 1959*, Butterworth's Scientific Publications, London, 1960.

[3] Freeman, D. C., Jr., "Assured Crew Return Capability (ACRC) Lifting Body Concept," presentation to the Senior Hypersonics Class at U.S. Air Force Academy, Jan. 1989.

[4] Neumann, R. D., "Defining the Aerothermodynamic Methodology," J. J. Bertin, R. Glowinski, J. Periaux (eds.), *Hypersonics, Volume I: Defining the Hypersonic Environment*, Birkhäuser Boston, Boston, MA, 1989.

[5] Trella, M., "Introduction to the Hypersonic Phenomena of Hermes," J. J. Bertin, R. Glowinski, and J. Periaux (eds.), *Hypersonics, Volume I: Defining the Hypersonic Environment*, Birkhäuser Boston, Boston, MA, 1989.

[6] Tauber, M. E., Menees, G. P., and Adelman, H. G., "Aerothermodynamics of Transatmospheric Vehicles," AIAA Paper 86-1257, Boston, MA, June 1986.

[7] Blankson, I. M., "Air-Breathing Hypersonic Waveriders: A Survey of Research Needs," J. D. Anderson, Jr., M. J. Lewis, S. Corda, and I. M. Blankson (eds.), *Proceedings of the First International Waverider Symposium*, University of Maryland, College Park, MD, Oct. 1990.

[8] Gregory, T. J., Petersen, R. H., and Wyss, J. A., "Performance Tradeoffs and Research Problems for Hypersonic Transports," *Journal of Aircraft*, Vol. 2, No. 4, July–Aug. 1965, pp. 266–271.

[9] Curran, E. T., "The Potential and Practicality of High Speed Combined Cycle Engines," reprinted in Conference Proceedings No. 479, Advisory Group for Aerospace Research and Development (AGARD), NATO, 1990.

[10] Blankson, I. M., Presentation to the U.S.A.F. Scientific Advisory Board, June 1991.

[11] Petley, D. H., and Jones, S. C., "Thermal Management for a Mach 5 Cruise Aircraft Using Endothermic Fuel," AIAA Paper 90-3284, Dayton, OH, Sep. 1990.

[12] Foster, R. W., Escher, W. J. D., and Robinson, J. W., "Studies of an Extensively Axisymmetric Rocket Based Combined Cycle (RBCC) Engine Powered SSTO Vehicle," AIAA Paper 89-2294, Monterey, CA, July 1989.

[13] Hunt, J. L., "Hypersonic Airbreathing Vehicle Design (Focus on Aero-Space Plane)," J. J. Bertin, R. Glowinksi, and J. Periaux (eds.), *Hypersonics, Volume I: Defining the Hypersonic Environment*, Birkhäuser Boston, Boston, MA, 1989.

[14] Heiser, W. H., and Pratt, D., *Hypersonic Airbreathing Propulsion*, AIAA, Washington, DC, 1993.

[15] Becker, J. V., "New Approaches to Hypersonic Aircraft," presented at the Seventh Congress of the International Council of the Aeronautical Sciences, Rome, Italy, Sep. 1970.

[16] Schmidt, D. K., Mamich, H., and Chavez, F., "Dynamics and Control of Hypersonic Vehicles—The Integration Challenge for the 1990's," AIAA Paper 91-5057, Orlando, FL, Dec. 1991.

[17] Högenauer, E., and Loelle, D., "SÄNGER, The German Aerospace Program," AIAA Paper 89-5007, Dayton, OH, July 1989.

[18] Gord, P. R., Langan, K. J., and Stringer, M. E., "Advanced Launch Vehicle Configurations and Performance Trades," Paper from AGARD Conference Proceedings No. 489, *Space Vehicle Flight Mechanics*.

[19] Perrier, P., "Industrial Methodologies for the Design of Hypersonic Vehicles," J. J. Bertin, R. Glowinski, and J. Periaux (eds.), *Hypersonics, Volume I: Defining the Hypersonic Environment*, Birkhäuser Boston, Boston, MA, 1989.

INDEX

CPSIA information can be obtained
at www.ICGtesting.com
Printed in the USA
BVHW032251260719
554489BV00002B/4/P